Genotoxic Effects of Airborne Agents

ENVIRONMENTAL SCIENCE RESEARCH

Recent Volumes in this Series

Genotoxic Effects of Airborne Agents

Edited by
RAYMOND R. TICE
DANIEL L. COSTA
and
KAREN M. SCHAICH
Brookhaven National Laboratory, Upton, New York

PLENUM PRESS • NEW YORK AND LONDON

Library of Congress Cataloging in Publication Data

Associated Universities, Inc./Brookhaven National Laboratory Medical
Department Symposium on Genotoxic Effects of Airborne Agents (1980)
Genotoxic effects of airborne agents.

(Environmental science research; v. 25)
Bibliography: p.
Includes index.
1. Mutagenicity testing—Congresses. 2. Carcinogenicity testing—
Congresses. I. Tice, Raymond R. II. Costa, Daniel L. III. Schaich, Karen M.
IV. Brookhaven National Laboratory. Medical Dept. V. Title. VI. Series.
QH465.A1A87 1980   615.9'02   81-23497
ISBN-13: 978-1-4613-3457-6   e-ISBN-13: 978-1-4613-3455-2
DOI: 10.1007/978-1-4613-3455-2   AACR2

Proceedings of an Associated Universities, Inc./Brookhaven
National Laboratory Medical Department Symposium on Genotoxic
Effects of Airborne Agents, held February 9-11, 1980, at
Brookhaven National Laboratory, Upton, New York

© 1982 Plenum Press, New York
Softcover reprint of the hardcover 1st edition 1982

A Division of Plenum Publishing Corporation
233 Spring Street, New York, N.Y. 10013

FOREWORD

For at least 40 years there has been a great interest in the problems created by infectious airborne agents and other toxic substances transported through the air. During the Second World War, this problem grew out of the very high incidence of upper respiratory infections appearing in new military recruits who were brought together in very large, open quarters. As a result, very interesting methods were developed to measure these airborne agents, especially bacteria, and some important methods were refined for their control. These methods primarily concentrated on ultraviolet radiation, propylene glycol and other means to reduce the dust in an environment. Because of the specialized circumstances at that time the whole consideration of airborne particles became prominent.

Now, with the new strides in the recognition of mutagenic and carcinogenic effects attributed to exposure to airborne chemicals from today's technology, the problem has again become quite prominent. The development of experimental chambers has made it possible to conduct studies under carefully controlled conditions.

For this reason, the organization of this symposium on "Genotoxic Effects of Airborne Agents" is an important milestone in the study of these agents. This symposium brought out many of the new developments and served as an example of the recent, explosive research of environmental mutagenesis and genetic toxicology, of which exposure to airborne chemicals is a very important part in this quickly developing field. The organizers of the symposium and the editors of the symposium proceedings are to be congratulated for having arranged for this important symposium, the proceedings of which serve as a great boost for the study of problems connected with airborne mutagens.

The importance of this field cannot be overemphasized. As a matter of fact, in the area in which I am especially interested, we are considering the quantity and variety of the "mutagen burden" of unavoidable mutagens to which we are regularly exposed. Airborne

chemicals are, of course, a major part of such problems and could
have a considerable influence on our health.

Alexander Hollaender

Associated Universities, Inc.

PREFACE

It is generally accepted, but not proven, that exposure to environmental agents is integrally related to the incidence of most cancers. While these agents may also be absorbed or ingested, inhalation is the most common route of exposure in the workplace and in the ambient environment. The recognition of the potentially important role of airborne agents in carcinogenesis has led to the recent development of a number of techniques for specifically assessing the genotoxicity of these agents. Federal legislation, such as the Clean Air Act of 1970 (PL91-604) which specified that inhalants posing a significant carcinogenic risk to man were to be regulated as air pollutants, provided the initial impetus to this research.

The timeliness of a symposium on the genotoxic effects of airborne agents developed from a series of discusions between Dr. Alexander Hollaender and members of the scientific staff of the Medical Department, Brookhaven National Laboratory. This Symposium would emphasize the state-of-the-art techniques for assessing the genotoxicity of airborne agents in general and the current carcinogenic and mutational knowledge of a selected few. Additionally included as appropriate topics for concern were: (i) the monitoring human populations at risk, (ii) the extrapolation of experimental data to human health effects, and (iii) the regulatory aspects of exposure to airborne toxicants.

A program committee consisting of Drs. Robert Drew (Medical Department, Brookhaven National Laboratory), Michael Waters (U.S. Environmental Protection Agency), Frederick de Serres (National Institutes of Environmental Health Sciences), Alexander Hollaender (Associated Universities, Inc.) and chaired by Dr. Raymond Tice (Medical Department, Brookhaven National Laboratory) was formed to evaluate discussion topics and select speakers. The symposium was divided into three sections. The first day would focus on techniques for assessing genotoxicity, techniques which would include the full range of in vitro and in vivo systems. The second day would review and discuss new data on selected inhalants, such as anesthetics, formaldehyde, styrene, radon, nitropyrenes, organic halides, and benzene. Special attention would be given to benzene because it is one inhalant which had been examined by a number of scientists in different fields and thus could serve as a model airborne agent. The third day would present methods for monitoring human populations. methods for extrapolating experimental data to the human situation and approaches for the regulation of airborne agents.

Any successful symposium and symposium proceedings requires the attention and assistance of many individuals working for a common goal. Special thanks are to be given to Dr. Alexander Hollaender for his guiding touch, to Drs. M. Waters and F. de Serres for their assistance in planning this symposium and for the funds obtained from their respective agencies to support this meeting, to the Session Moderators for thei assistance in obtaining the speakers and in running the program, to the staff of the Medical Department (J. Cutt, K. De Pierro, and T. Smith) and of Brookhaven National Laboratory (D. Schroeder, G. T. Walczyk, H. Boyd, P. Glynn, and others in the housing and travel offices) who did their best to insure a successful meeting. We, the editors of the symposium proceedings, would also like to gratefully acknowledge the prompt attention given to the manuscripts by the contributors. The desire to have a common format for referencing, abbreviations, spelling and presentation necessitated a great many stylistic alterations in some circumstances. We are grateful for the patience of the authors while making these alterations. We are also extremely grateful to the proceedings' secretary, Mrs. Rene Tiernan, for her efforts in completing this book.

Raymond R. Tice
Daniel L. Costa
Karen Schaich

Brookhaven National Laboratory

CONTENTS

SECTION A: ASSAY AND EXPOSURE TECHNOLOGY

SESSION I: ASSAY AND EXPOSURE TECHNOLOGY OF IN VITRO MICROBIAL

ASSAY SYSTEMS APPLIED TO AIRBORNE AGENTS

Joellen Lewtas, Moderator

U.S. Environmental Protection Agency
Research Triangle Park, NC 27711 (U.S.A.)

     Scientists involved with bioassaying air samples tend to view
air as containing two parts: gases and particles. The papers in
this section will include a discussion of the current microbial
mutagenesis bioassay methods applicable to these two phases of air.

     Atmospheric chemists, however, continue to remind us that the
air we breathe is not simply made up of a mixture of gases and
particles. They refer to air as a complex, multiphasic system
consisting of gases, vapors, and a spectrum of particles ranging
in size from 0.01 μm to 100 μm in diameter. Both chemical and
physical reactions are known to occur in this dynamic complex
aerosol mixture. Future research may open the door to applying
microbial bioassays to this complex mixture in a more dynamic exposure system in situ. Current methods, however, limit our discussion to the assay and exposure technology available in the
laboratory setting.

     The first collection and bioassay of air samples in the 1940's
was not performed with microbes but with mice (1). Later, in the
1970's, extracts of urban air were shown to transform rodent cells
in culture (2,3). Microbial mutagenesis studies were not reported
until a year after Ames described the Salmonella typhimurium plate
incorporation assay (4). Three separate groups (5-7) reported
the application of this assay to organic extracts of air particles
collected by the high volume sampler. The intervening three years
have produced a number of studies in this area.

     R. H. Jungers' and J. Lewtas's paper describes recent
research in methods for collecting and extracting the organics
from airborne particles. L. Claxton's paper discusses the biologi-

1

cal and chemical characterization techniques which can be applied to such samples.

The technical problems associated with collecting sufficient quantitites of artifact-free gas samples from ambient air have prevented significant advances in bioassaying such samples. Advances in chemical identification of many gaseous components of air, however, have stimulated the development of laboratory technology to bioassay individual gases. Early studies (4) employed either desiccators or plastic bags for such exposures. E. Barber's paper reports one of the first efforts to develop an exposure system for the quantitative measurement of the microbial mutagenicity of volatile liquids.

REFERENCES

1. J. Leiter, M. B. Shimkin, and M. J. Shear, Production of subcutaneous sarcomas in mice with tars extracted from atmospheric dusts, J. Natl. Cancer Inst. 3:155-165 (1942).

2. A. E. Freeman, P. J. Price, R. J. Bryan, R. J. Gordon, R. V. Gilden, G. J. Kelloff, and R. J. Huebner, Transformation of rat and hamster embryo cells by extracts of city smog, Proc. Natl. Acad. Sci. (U.S.A.) 68:445-449 (1971).

3. R. J. Gordon, R. J. Bryan, J. S. Rhim, C. Demoise, R. G. Wolford, A. E. Freeman, and R. J. Huebner, Transformation of rat and mouse embryo cells by a new class of carcinogenic compounds isolated from city air, Int. J. Cancer 12:223-227 (1973).

4. B. N. Ames, J. McCann, and E. Yamasaki, Methods for detecting carcinogens and mutagens with the Salmonella/mammalian microsome mutagenicity test, Mutat. Res. 31:347-364 (1975).

5. H. Tokiwa, H. Takeyoshi, K. Morita, K. Takahashi, N. Soruta, and Y. Ohnishi, Detection of mutagenic activity in urban air pollutants, Mutat. Res. 38:351-359 (1976).

6. J. N. Pitts, D. Grosjean, J. M. Mischke, V. F. Simmon, and D. Poole, Mutagenic activity of airborne particulate organic pollutants, Toxicol. Lett. 1:65-70 (1977).

7. R. Talcott and E. Wei, Airborne mutagens bioassayed in Salmonella typhimurium, J. Natl. Cancer Inst. 58:449-451 (1977).

AN EXPOSURE SYSTEM FOR QUANTITATIVE MEASUREMENTS OF THE

MICROBIAL MUTAGENICITY OF VOLATILE LIQUIDS

Eugene D. Barber and William H. Donish

Eastman Kodak Co., Rochester, NY 14650 (U.S.A.)

It has been estimated that 5×10^{10} grams per day of vapor-phase chemicals and 10^9 grams per day of particulate matter are released into the atmosphere in the United States (1). Since both of these fractions may contribute to the toxic potential of the atmosphere, it seems important to be able to evaluate them in a precise manner.

The standard Salmonella/microsome mutagenicity assay, or Ames test, is a screening test useful in the evaluation of chemicals and mixtures (2). The test has been applied to a variety of environmental samples including food (3-5), beverages (6,7), air (8-10), water (11-13), tobacco (14) and chemicals such as dyes, drugs and pesticides (15-18). The test measures the ability of a chemical or mixture to cause point mutations (reversions) in the test organisms and the results show a reasonably high degree of correlation with known animal carcinogenicity data (15,19-21). In the standard plate incorporation assay, the test plates are prepared at 45 °C and this warm mixture is spread in a very thin layer over the surface of an agar plate (2). The plates are then incubated at 37 °C for two days. This procedure is almost perfectly designed for driving off volatile chemicals and, indeed, this shortcoming has been recognized by many authors (22-30).

To solve this problem we have devised a procedure to assay volatile liquids quantitatively (31). At the heart of our procedure is the incubation vessel depicted in Figure 1. The containment system is composed of Pyrex, Teflon, stainless steel and Viton components. Therefore, it is autoclaveable and highly inert toward many chemicals. The procedure we use is as follows: First, the Ames test plates are prepared exactly as described in

3

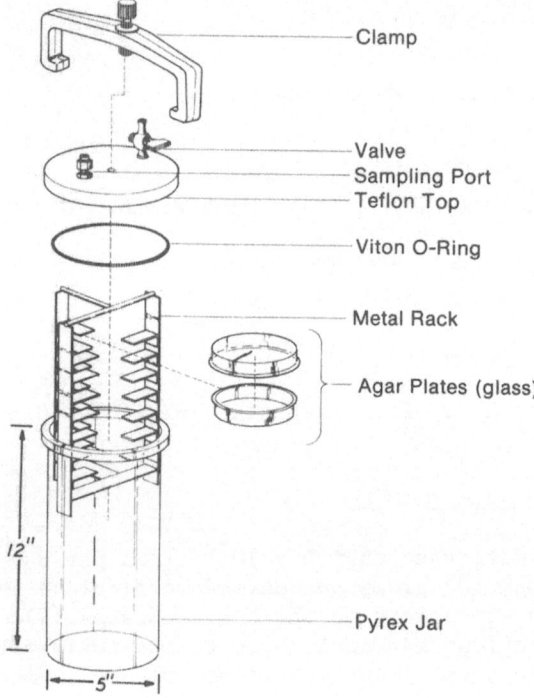

Figure 1. A schematic representation of the containment system
used in these mutagenicity studies.

reference 2 except no test chemical is added.   After solidification
of the top agar, the plates are uncovered and inserted agar side
down into the sterilized stainless steel rack.   This procedure is
carried out in a Baker class II, type II laminar flow hood which
provides a sterile work environment.   The rack is placed in the
containment vessel and the top is positioned and tightly sealed.
The container is partially evacuated by means of a vacuum pump and
an amount of the test substance is introduced into the system
through the septum (Fig 1).   When varporization of the liquid is
complete, sterile air is introduced until ambient pressure is
reached.   The containers are then incubated at 37°C for 2 days with
constant, slow magnetic agitation of the atmosphere.   After the
incubation is complete, vapor phase samples are withdrawn and
analyzed by gas-liquid chromatography (GLC).   The vessels are
opened and samples of distilled water (included in each container)
are taken and analyzed by GLC and revertant colonies on the plates
are counted.   The data are plotted as revertants per plate versus
amount of chemical per plate and the mutagenicity calculated in
units of revertants per nanomole.   This allows comparison of the
data with published values for other non-volatile materials.

Table 1. Distribution of 1,2-dibromoethane in the closed system

	^{14}C–EDB mg/plate	Percent of Water Value
Water	0.93	100
V.B. Salts Solution	0.90	96.8
V.B. Salts Solution + 2% Glucose	0.85	91.4
V.B. Agar Plate	0.70	75.6

Glass Petri dishes containing 25 mL of the fluids listed were placed in the containment system and 13.2 mg of ^{14}C labelled 1,2-dibromoethane, specific activity 6.66 x 10^5 DPM/mg, was added. After 48 hours incubation at 37°C, the vessel was opened and weighed amounts of the fluids and agar samples were immersed in scintillation fluid and ^{14}C determined by scintillation counting. V.B. = Vogel–Bonner salts mixture (see 32).

Some of the characteristics of the containment system were investigated using carbon-14 labelled 1,2-dibromoethane. Table 1 shows comparisons among various solid and liquid media incubated together in the system. Water was found to be the best absorbing medium for this chemical. Addition of the standard Vogel-Bonner salts mixture (32) depresses absorption by about 3% whereas further addition of 2% glucose decreases absorption by about 9% when compared to pure water. A standard Ames test agar plate absorbs about 75% as much chemical as pure water. Thus our GLC measurements which are performed, of necessity, using distilled water samples probably overestimate the actual amount of material in an agar plate by about 25%. Table 2 shows the results from two containers that were filled with ten agar plates each. The first contained glass Petri dishes and the second contained the more usual disposable plastic dishes. It can be seen that about 90% of the radiolabel is recovered in the plastic dishes when these are used. Therefore, all subsequent work has been performed using glass plates.

Table 3 shows the total recovery of various amonts of added C-14 dibromoethane after the standard 48-hour incubation period. In containers 1-3, about 80% of the added material is recovered in the vapor phase with about 5% present in the 0-ring and septum of the container. In containers 4-6, which each contained ten glass plates filled with 25 mL of standard bottom agar, the vapor phase recoveries drop to about 20%. The bulk of the material is recovered in the agar plates with a small amount in the 0-ring and septum. The mean recovery is 87±10% of the added chemical.

Table 2. Effect of plastic plates in closed system

Plate Material	Recovery ^{14}C-EDB - % of Dose			
	Vapor	Agar	Plastic	Total
Glass	18.3	69.5	-------	87.8
Plastic	5.7	14.5	87.4	107.6

Two containment vessels were filled with 10 Petri plates, containing 25 mL of standard bottom agar (2). Weighted amounts, 13.48 and 13.55 mg respectively, of ^{14}C-1,2-bromoethane were injected into the vessels containing the glass and plastic plates. After 48 hours incubation at 37°C, the recovery of label in the vapor, agar and plastic was determined by scintillation counting.

We conclude from these figures that the system is working well and that chemicals introduced into the system become distributed between the aqueous and vapor phases.

We have used our system to measure the mutagenicity of ten common haloalkane solvents. The data are summarized in Table 4. For each compound we list boiling point, distribution coefficient (K_D) (defined as mg per Liter in the aqueous phase divided by mg per Liter in the vapor phase at 48 hours) and mutagenicity results. Among the ten compounds, seven are active and three are inactive. In contrast, only three of these compounds can be detected in the standard system: 1-bromopropane, 1,2-dibromoethane and 1-bromo-2-chloroethane. Among the seven active compounds, all are positive for strains TA-1535 and TA-100 and only one, 1,2-dibromoethane, is active on TA-98. None of these ten is active for TA-1537 or TA-1538. In all cases, the mutagenic activity is independent of the rat liver S-9 activation system. For the three compounds found to be negative at least two concentrations that produced visible toxicity to the test organisms were included.

The statistical minimum detection limits are given in Table 5 for the same ten chemicals. For each tester strain we can calculate a minimum significant number of revertants per plate by reference to Student's t tables and assuming n= 5 replicates for both negative control and experimental plates. Then, knowing the distribution coefficient and mutagenic potency of each compound we can calculate the minimum vapor concentration of that compound required to produce that result. These figures represent the threshold vapor concentration at which mutagenicity will become evident using our system. It is interesting to note that in every case, TA-1535, because of its low background reversion frequency, is the most efficient detector strain.

Table 3. Recovery of ^{14}C, 1,2-dibromoethane by liquid scintillation counting

| Container Contents | Amount Added $(CH_2)_2Br_2$, (mg)[a] | Recovery After 48 Hr. Incubation Percent of Dose | | | |
		Vapor	Agar[b]	0-Ring + Septum	Total
1. Control, no plates	5.06	74.3	–	4.5	78.8
2. Control, no plates	9.25	71.6	–	5.2	76.8
3. Control, no plates	13.59	84.5	–	5.7	90.2
4. 10 Glass plates	4.84	22.8	77.9	1.9	102.6
5. 10 Glass plates	9.35	15.0	69.1	1.5	85.6
6. 10 Glass plates	13.48	18.3	69.5	1.7	89.5

$\overline{X} \pm$ S.D. = 87.3 $\pm$ 9.3

[a] Amounts determined by differential weighing. Specific Activity = 6.61 x 10^5 DPM/mg.

[b] Recovery based on amount found in weighed agar samples.

Table 4. Mutagenicity, distribution coefficients, boiling points, purities and structures often common haloalkanes listed in order of decreasing mutagenicity for TA-100

Chemical Name	% Purity[a]	Boiling Point °C	K (Wt/Vol)Ag/(Wt/Vol)Vap[b]	Mutagenicity, Rev/nmole[c]					
				TA-1535		TA-98		TA-100	
				-S9	+S9	-S9	+S9	-S9	+S9
1,2-Dibromoethane	99.89	132	35.2	0.146	0.165	0.018	0.018	0.173	0.172
Bromoethane	98.97	39	5.5	0.185	0.131	Neg.	Neg.	0.103	0.118
Iodoethane	99.43	71	4.2	0.134	0.191	Neg.	Neg.	0.093	0.116
1-Bromopropane	99.85	71	2.9	0.113	0.089	Neg.	Neg.	0.082	0.088
1-Bromo-2-Chloroethane	98.39	107	24.0	0.107	0.084	Neg.	Neg.	0.055	0.073
1-Bromobutane	94.12	101	2.1	0.021	0.018	Neg.	Neg.	0.019	0.024
1,2-Dichloroethane	99.98	84	26.7	0.002	0.002	Neg.	Neg.	0.001	0.001
1,1,2-Trichloroethane	98.0	112	34.1	Neg.	Neg.	Neg.	Neg.	Neg.	Neg.
1-Chlorobutane	99.7	78	1.4	Neg.	Neg.	Neg.	Neg.	Neg.	Neg.
Carbon Tetrachloride	98.5	77	0.71	Neg.	Neg.	Neg.	Neg.	Neg.	Neg.

a - Determined from GLC analysis of vapor phase samples.

b - K_D values for each compound were calculated from GLC analyses of vapor and aqueous phase concentrations after 48 hour incubation at 37°C. Each value is an average from at least five containers.

c - Mutagenicities were obtained from linear regression analysis. The data were plotted and only those on the linear portion of the response curve were included in the analysis.

Table 5. Minimum detectable vapor concentrations (β = 0.05),
ppm with N = 5 replicates

Compound	Ta-1535	TA-98	TA-100
1,2-Dibromoethane	1.7	24.8	4.3
Bromoethane	10.5	----	34.4
Iodoethane	13.4	----	59.8
1-Bromopropane	31.2	----	106.5
1-Bromo-2-chloroethane	4.0	----	17.1
1-Bromobutane	223.8	----	581.4
1,2-Dichloroethane	171.7	----	983.2
1,1,2-Trichloroethane	N.D.[a]	N.D.	N.D.
1-Chorobutane	N.D.	N.D.	N.D.
Carbon Tetrachloride	N.D.	N.D.	N.D.

$$\text{Min. Detectable ppm} = \frac{\text{Minimum Significant Rev/Plate}}{(K_D)\ (1.2)\ (\text{Rev/nmole})}$$

[a]N.D. = No detectable mutagenic activity.

Three of the compounds in Table 4 give positive results in
both the standard and containment systems. Comparison of the
results showed that the mutagenic potencies in the two systems
were widely different. 1-Bromopropane (BP = 71° C) shows an 85
times higher response in the closed system; 1-bromo-2-chloroethane
(BP = 101° C) gives a 55 fold greater response while 1,2-dibromo-
ethane (BP = 132° C) exhibits about a 35 fold elevation in response.
To better understand these differences, we studied the mutagenicity
of the dibrominated series of compounds given in Table 6. These
liquids have boiling points ranging from 132° C for 1,2-dibromoethane
to 251°C for 1,6-dibromohexane. The mutagenicity results are given
in Table 6 and it can be seen in the last column that the ratio
of the mutagenic potencies decreases with increasing boiling point.
These results presumably reflect the decreasing vapor pressure of
the compounds with increasing boiling point. The Clasius-Clapeyron
equation predicts an exponential decrease of vapor pressure as
boiling point increases and we therefore plotted our ratio data in
this format in Figure 2. The linear trend is apparent and suggests

Table 6. Mutagenicity of the dibrominated series

| Compound | B.P., °C | Mutagenicity, Rev/nmole (TA-100)[a] | | |
		Standard	Containment	Ratio[b]
1,2-Dibromethane	132	0.0076	0.172	22.6
1,3-Dibromopropane	167	0.0025	0.030	12.2
1,4-Dibromobutane	198	0.0350	0.265	7.6
1,5-Dibromopentane	227	0.0170	0.066	3.9
1,6-Dibromohexane	251	0.0083	---	---

[a]For the standard test, mutagenicity is expressed as revertants/
nmole of material added to the test plates whereas for the
containment system mutagenicity is expressed as revertants/
nmole of material found by GLC analysis.

[b]The ratio is the potency found in the containment system
divided by that determined in the standard system.

that the differences observed between the two systems are related
to the boiling point of the test compound. The regression line
reaches a value of 1 at BP = 265°, suggesting that no difference
would be observed with liquids boiling at or above this temperature.
Liquids with boiling points less than 265°C will give erroneous
results in the standard system by the factors indicated in Figure 2.

We were curious as to why highly mutagenic materials with
boiling points less than 100°C seem to give no response at all in
the standard test. The losses by vaporization surely cannot be
instantaneous and therefore the tester strains should be exposed
to the chemicals for several minutes or hours. To answer this
question we designed a pulsed dose experiment using bromoethane,
BP = 39°C. The kinetics of entry and exit from a water plate in
the containment system were studied by GLC with the results shown
in Figure 3. Bromoethane was injected and vaporized at time zero
and at 60 min the plate was removed from the container and placed
in the laminar flow hood for an additional 60 min. The concentration of bromoethane in the aqueous phase rises steadily during the
first hour to a level of about 7 mg per plate and after removal
the concentration falls rapidly until, at 60 min, less that 2%
remains in the plate. We therefore used bromoethane as a probe
to study the kinetics of reversion in the Ames Test. The results

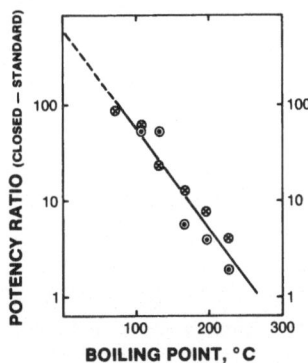

Figure 2.   The logarithm of the ratio of the mutagenic potency
observed in the containment system to that observed in the standard
system versus the boiling point of the test liquid.   Data were
obtained for both TA-1535 (-θ-) and TA-100 (-◉-).   Test compounds
were the dibrominated series given in Table 6 plus 1-bromopropane
and 1-bromo-2-chloroethane.   The line was obtained by linear regression analysis of the data and predicts a ratio of unity at boiling
point 265°C.

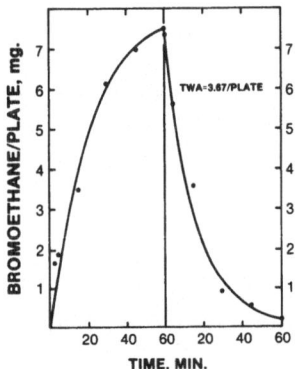

Figure 3.   The rates of entry and exit of bromoethane to and from
the aqueous phase.   A glass plate containing 25 mL of distilled
water was placed in the containment system along with ten plates
containing 25 mL of standard bottom agar (2).   200 μL of bromoethane was introduced and vaporized under reduced pressure.   The
vessel was brought to ambient pressure by introducing sterile air
and samples of the aqueous phase were withdrawn through the septum.
After 1 hour, the vessel was opened and the plates were placed in
a laminar flow hood for an additional 60 min.   Aqueous samples
were taken at the time indicated and analyzed by GLC for the presence
of bromoethane.

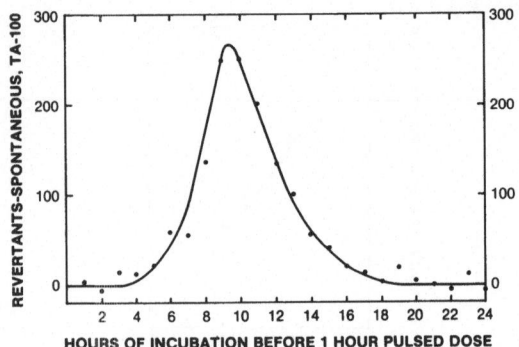

Figure 4. The kinetics of reversion in the standard plate
incorporation assay. Approximately 10^8 TA-100 cells from an
overnight culture grown in nutrient broth were plated in 2.0 mL
to top agar (2) at time zero. All plates were placed into a 37°C
incubator. At the times indicated three plates were removed and
subjected to the pulsed dose of bromoethane described in Figure 3.
Plates were returned to the 37°C incubator and revertants counted
at 48 hours. The mean value of the triplicate measurements is
plotted. Spontaneous revertants have been subtracted (114±14).

are shown in Figure 4. Approximately 10^8 TA-100 cells were plated
in top agar at time zero and placed into a 37°C incubator. At the
times indicated on the X-axis, the triplicate plates were removed
from the incubator, exposed to the pulsed dose of bromoethane depicted in Figure 3, and then returned to the 37°C incubator.
Revertants were counted at 48 hours and the results of the triplicate
plate counts are given in the figure. It can be seen that the cells
are refractory to mutagenesis for about 4 hours after plating and
that the rate of reversion is maximal at 9-10 hours after plating.
Chemicals present for only a brief time (4 hours) will therefore
have no measurable effect in the test. This finding explains
why highly volatile materials give no response and has implications
that reach beyond the measurement of volatile substances. For
example chemicals with limited stability under the conditions of
the test could disappear before the cells are "ready" for muta-
gensis. Active metabolites produced by the rat liver S-9 fraction
might also disappear before exerting any mutagenic effect. In
this regard, the duration of S-9 activity in semisolid agar has
been estimated to be only 6-7 hours at 37°C (33,34).

SUMMARY

We have designed a procedure for the analysis of volatile
liquids in the Ames Salmonella/microsome assay. The closed, inert

containment system (Fig 1) allows vapor-phase samples to be withdrawn at any time during incubation. Using the system we can
study the equilibrium distribution of the test substance and thus,
its concentration in the aqueous phase. The use of glass, rather
than plastic, Petri dishes (Table 1) is highly desirable when
assaying haloalkanes. About 90% recovery of C-14 dibromoethane is
observed after 48 hour incubation in the containment system (Table
3). We have used the system to measure the mutagenicity of ten
common haloalkane solvents (Table 4). Seven of the ten compounds
were positive with mutagenicities ranging from 0.001 revertants
per nanomole to 0.172 revertants per nanomole for strain TA-100.
All seven of the positives were mutagenic for strains TA-1535 and
TA-100 while only 1,2-dibromoethane was positive for TA-98. In
agreement with others (25,28,30), we found no significant effect of
the rat liver S-9 mix (2) on the mutagenicity data. The distribution coefficients vary from 0.71 for carbon tetrachloride to 35.2
for 1,2-dibromoethane and are independent of the mutagenicity data.
The statistical limitations of the system are given in Table 3 and
show that TA-1535 is the best detector for these compounds. We
have shown in Figure 2 that the ratio of mutagenic potency observed
in the containment system to that observed in the standard open
system varies in an inverse exponential manner with the boiling
point of the test compound. Liquids whose boiling points exceed
265°C will give similar results in both systems; those whose
boiling points lie between 100°C and 265°C may be detectable in
the standard system with diminished response and liquids boiling
at less thatn 100°C will almost certainly require containment to
be detectable.

    We have used the containment system and bromoethane, a highly
volatile (BP = 39°C) liquid mutagen to measure the kinetics of
reversion in the Salmonella plate assay. The results, given in
Figure 3, show that the maximum rate of production of revertants
occurs about 9-10 hours after the cells have been plated and is
near zero in hours 1 through 5. This result explains why liquids
such as bromoethane give no response in the standard assay system-
they are gone before the critical time during which the cells are
revertable. This observation has implications for any compound
producing a time-dependent concentration of active mutagen in the
Salmonella/microsome assay.

    We feel that our system has several important advantages when
compared to the standard system for the assay of volatile liquids.
First, the system produces mutagenicity data in units of revertants
per nanomole which are comparable to other literature data. Second,
the purity of the test substance is obtained as part of the analytical procedure. Finally, the system allows us to avoid false
negative results that can be caused by vaporization of the test
compounds in the standard plate incorporation method.

REFERENCES

1. T. J. Hughes, E. Pellizzari, L. Little, C. Sparacino, and
   A. Kolber, Ambient air pollutants: Collection, chemical
   characterization, and mutagenicity testing, Mutat. Res. 76:
   51-83 (1980).

2. B. N. Ames, J. McCann, and E. Yamasaki, Methods for detecting
   carcinogens and mutagens with the Salmonella/Mammalian
   microsome mutagenicity test, Mutat. Res. 31:347-364 (1975).

3. T. Sugimura, M. Nagaro, T. Kawachi, M. Honda, T. Yahagi, Y.
   Seino, S. Sato, N. Matsukura, T. Matsushima, A. Shirai,
   M. Sawamura, and H. Matsumoto, Mutagen-Carcinogens in food
   with special reference to highly mutagenic pyrolytic prod-
   ucts in broiled foods in origins of human cancer, in:
   "Origins of Human Cancer", H. H. Hiant, J. D. Watsun
   and J. A. Winston, eds., Cold Spring Harbor Laboratory,
   Cold Spring Harbor, NY, 1561-1578 (1977).

4. B. Commoner, A. J. Vithayathil, P. Dolara, S. Nair, P. Madyastha,
   and G. C. Cuca, Formation of mutagens in beef and beef extract during cooking, Science 201:913-916 (1978).

5. N. E. Springarm, L. A. Slocum, and J. H. Weisburger, Formation
   of mutagens in cooked foods. III. Foods with high starch
   content, Cancer Lett. 9:7-12 (1980).

6. H. U. Aeschbacher and H. P. Wurzner, An evaluation of instant
   and regular coffee in the Ames mutagenicity test, Toxicol.
   Lett. 5:139-145 (1980).

7. J. S. K. Lee and L. Y. Y. Fong, Mutagenicity of Chinese
   alcoholic spirits, Food Cosmet. Toxicol. 17:575-578 (1979).

8. B. Commoner, Carcinogens in the environment, Chemtech,
   February (1977).

9. J. N. Pitts, K. A. VanCauvenberghe, D. Grosjean, J. P. Schmid,
   D. R. Fitz, W. L. Belser, G. B. Knudson, and P. M. Hynds,
   Atmospheric reactions of polycyclic aromatic hydrocarbons:
   Facile formation of mutagenic nitro derivatives, Science
   202:515-519 (1978).

10. C. E. Chrisp, G. L. Fisher, and J. E. Lammert, Mutagenicity of
    filtrates from respirable coal fly ash, Science 199:73-75
    (1977).

11. B. A. Glatz, C. D. Chriswell, M. D. Arguello, H. J. Svec,
    J. S. Fritz, S. M. Grimm, and M. A. Thomson, Examination of
    drinking water for mutagenic activity, J. Amer. Water Works
    Asso., August, 465 (1978).

12. N. Gruener, Mutagenicity of ozonated, recycled water, Bull.
    Environ. Contam. and Toxicol. 20:522-526 (1978).

13. W. G. Honer, M. J. Ashwood-Smith, and C. Warby, Mutagenic act-
    ivity of swimming pool water, Mutat. Res. 78:137-144 (1980).

14. S. Sato, Y. Seino, T. Ohka, T. Yahagi, M. Nagao, T. Matsushima,
    and T. Sugimura, Mutagenicity of smoke condensates from
    cigarettes, cigars, and pipe tobacco, Cancer Lett. 3:1-8
    (1977).

15. J. McCann, E. Choi, E. Yamasaki, and B. N. Ames, Detection of carcinogens as mutagens in the Salmonella/microsome mutagenicity test, Proc. Nat. Acad. Sci. (U.S.A.) 72:5135-5139 (1975).

16. J. M. Muzzall and W. L. Cook, Mutagenicity test of dyes used in cosmetics with the Salmonella/mammalian microsome test, Mutat. Res. 67:1-8 (1979).

17. W. T. Speck, A. B. Stein, and H. S. Rosenkranz, Mutagenicity of metronidazole: Presence of several active metabolites in human urine, J. Nat. Cancer Inst. 56:283-284 (1976).

18. M. D. Gold, A. Blum, and B. N. Ames, Another flame retardant, tris-(1,3-dichloro-2-propyl)phosphate and its expected metabolites are mutagens, Science 200:785-787 (1978).

19. J. A. Heddle and W. R. Bruce, Comparison of tests for mutagenicity or carcinogenicity using assays for sperm abnormalities, formation of micronucleii and mutations in Salmonella, in: "Origins of Human Cancer", H. H. Hiatt, J. D. Watson, and J. A. Winsten, eds., Cold Spring Harbor Laboratory, Cold Spring Harbor, NY, 151557 (1977).

20. T. Sugimura, S. Sato, M. Nagao, T. Yahagi, T. Matsushima, Y. Seino, M. Takeuchi, and T. Kawachi, Overlapping of carcinogens and mutagens, in: "Fundamentals in Cancer Prevention", Baltimore University Park Press, 191-215 (1976).

21. S. J. Rinkus and M. Legator, Chemical characterization of 465 known or suspected carcinogens and their correlation with mutagenic activity in the Salmonella typhimurium system, Cancer Res. 39:3289-3318 (1979).

22. J. M. Baden, M. Brinkerhoff, R. S. Wharton, B. A. Hitt, V. F. Simmon, and R. I. Mazze, Mutagenicity of volatile anesthetics: Halothane, Anesthesiology 45:311-318 (1976).

23. J. M. Baden and V. F. Simmon, Mutagenic effects of inhalation anesthetics, Mutat. Res. 75:169-189 (1980).

24. H. Bartsch, C. Malaveille, R. Montesano, and L. Tomatis, Tissue mediated mutagenicity of vinyledine chloride and 2-chlorobutadiene in S. typhimurium, Nature 255:641-643 (1975).

25. H. Brem, A. B. Stein, and H. S. Rosenkranz, The mutagenicity and DNA-modifying effect of haloalkanes, Cancer Res. 34:2576-2579 (1974).

26. B. A. Bridges, On the detection of volatile mutagens with bacteria: Experiments with dichlorovos and epichlorohydrin, Mutat. Res. 54:367-371 (1978).

27. C. deMeester, M. Diverger-vanBogaert, M. Lambotte-Vandepaer, M. Roberfroid, F. Poncelet, and M. Mercier, Mutagenicity of vinyl chloride in the Ames test. Possible artifacts related to experimental conditions, Mutat. Res. 77:175-179 (1980).

28. W. M. F. Jongen, G. M. Alink, and J. H. Koeman, Mutagenic effect of dichloromethane on Salmonella typhimurium, Mutat. Res. 56:245-248 (1978).

29. J. F. Russel, M. E. Sippel, and D. F. Krahn, Methods for the
     detection of mutagenic gases and volatile liquids in the
     Salmonella/microsome assay, Environmental Mutagen Society
     11th Annual Meeting, Nashville, Abstract (1980).
30. V. F. Simmon, Structural correlations of carcinogenic and
     mutagenic alkyl halides, in: Proceedings of the Second FDA
     Office of Science Summer Symposium, I. M. Asher and
     C. Zervos, eds., (1977).
31. E. D. Barber, W. D. Donish, and K. R. Mueller, Quantitative
     measurement of the mutagenicity of volatile liquids in the
     Ames Salmonella/microsome assay, Environmental Mutagen
     Society 11th Annual Meeting, Nashville, Abstract (1980).
32. H. J. Vogel and D. M. Bonner, Acetylornithinase of E. coli:
     Partial purification and some properties, J. Biol. Chem.
     218:97-106 (1956).
33. C. Malaveille, G. Planche, and H. Bartsch, Factors for
     efficiency of the Salmonella/microsome mutagenicity assay,
     Chemical-Biological Interactions 17:129-136 (1977).
34. M. S. Kelley and J. M. Baden, Duration of S-9 Activity,
     Environmental Mutagenesis Society 11Annual Meeting,
     Nashville, Abstract (1980).

DISCUSSION

Q. Shimada (American Health Foundation): What is the cytotoxicy
of those compounds at the mutagenic concentrations?

A. Barber (Eastman Kodak): For negative compounds, the highest
concentration is visibly toxic on the plates. For the positive
ones, we normally don't check for toxicity.

Q. Shimada (American Health Foundation): Have you tried other
kinds of S9 preparations or species other than rat?

A. Barber (Eastman Kodak): No.

Q. Shimada (American Health Foundation): In the last part of
your study, when you checked kinetics, did you do it by the ordinary
conventional agar plate method?

A. Barber (Eastman Kodak): Yes, we made some standard Ames
plates with the exception that they were poured in glass plates
rather than in plastic. We took the plates, put them in the incubator and then at the times indicated, we took them out and gave
them the pulse dose of bromoethane.

Q. Borg (BNL): The standard Ames assay strains are notoriously
insensitive to haloalkanes and, for that matter, to many oxidants.
Both classes of compounds are of interest to those individuals

concerned with air pollution. I have recently heard Dr. Ames
say that the reasons for their insensitivity lies in the particular
nucleic acid sequences involved in those specific backward
mutation strains. These sequences are not sensitive to radiation-
like substances or to chemicals which autooxidize and produce
strong electrophiles of the same class. Dr. Ames and his colleagues
have already developed some Salmonella strains that are more
sensitive to that class of materials. When they are introduced
into the world of testing, we will find materials such as the
haloalkanes that now give low mutational results perhaps giving
much higher mutational frequencies.

(Editors Note: Dr. Ames presented information in the development
of these more sensitive Salmonella strains at the Annual Meeting
of the EMS Society, San Diego, CA, March 1981).

A. Barber (Eastman Kodak): One of the reasons for the insensitivity is also due to the modis operondum of the test; volatile
gases boil off. Even if you had more sensitive strains, it
would still be better if you confined the exposure to a closed
system.

Q. Gager (Phillip Morris): I like your system very much. Do
you see any problems if you were to use a more complicated and
reactive mixture when using a flow system to pass the environment
across the plates?

A. Barber (Eastman Kodak): No, we are in the process of doing
that. Up to this point, we have just used volatile liquids. I
see absolutely no problem. With just a few hardware changes, we
can put gas into one port and take it out the other port. We could
do this, either on a single dose basis--by putting in a gas,
stopping the flow, closing the system and watching for equilibration. Alternatively we could put the mixture through the system,
continuously through for the whole 48 hours, The latter approach
would offer a steady state rather than an equilibrium situation.

Q. Krahn (DuPont): In response to Dr. Gager's question, we have
also tested many gases in our Ames system in a manner similar to
the system that Gene Barber has described, passing gas through
and holding for 48 hours and we get very excellent results.

Q. Snyder (Jefferson Medical College): I have two brief questions.
First of all, have you looked at any aromatic compounds such as
benzene, toluene, etc. in your system?

A. Barber (Eastman Kodak): No.

Q. Snyder (Jefferson Medical College): Secondly, the question
of the lacke of an S9 interaction raises concern about possible
direct alkylating activity in the compounds you have investigated.
Have you looked at that possibility?

A. Barber (Eastman Kodak): There is some evidence that Salmonella
themselves can do some metabolic activation of haloalkanes. Most
people who have tested these compounds have found that the presence
of the S9 fraction makes no difference. However, there is one
report with dichloroethane that an S9 fraction enriched with
glutathione transferase and glutathione enormously raises the
mutagenic potency of that compound. We have not really attempted
too much alteration of the metabolic activation systems.

Q. Snyder (Jefferson Medical College): I'll just make the comment
that we have been looking at a system like this but for a different
purpose. Mainly, we have been looking at reductive dehalogenation
of carbon tetrachloride by running the system anaerobically. You
get less bacterial growth, but you can still measure mutaganisis
in the Salmonella. There seems to be some indication that this
reaction will go with the S9 with carbon tetrachloride.

A. Barber (Eastman Kodak): There was a paper last year at the
S.O.T. meetings which indicated that for dichloromethane, they
were able to retrieve 0-18 labeled carbon monoxide and carbon
dioxide when the agent was incubated with strain TA-100. So there
is some evidence that Salmonella are not neutral to haloalkanes and
that they can certainly do some metabolic transformation.

REVIEW OF FRACTIONATION AND BIOASSAY CHARACTERIZATION TECHNIQUES

FOR THE EVALUATION OF ORGANICS ASSOCIATED WITH AMBIENT AIR PARTICLES

Larry D. Claxton

Health Effects Research Laboratory
U.S. Environmental Protection Agency
Research Triangle Park, NC   27711   (U.S.A.)

INTRODUCTION

The primary objective of coupling chemical fractionation and
genetic bioassay techniques for particulate organics is the identification of potentially genotoxic components associated with the
collected particles. There are, however, other objectives which
can be of major importance. These objectives include: determining
the primary sources of particles with genotoxic properties, understanding chemical transformation in the atmosphere, determining
exposure levels of toxic airborne agents and assisting in the
design of control technology. The number of references integrating
the chemical analysis of air particulate organics with the bioassay
of those organics is relatively few. The purpose of this review
is to demonstrate, with these and other related references, the
approaches that one may use in identifying genotoxic substances
associated with air particles.

GENERALIZED APPROACHES

Approaches that may be used to integrate bioassay results
and chemical identification procedures are categorized in Table 1.
Within the open literature, each investigator has applied a unique
combination of literature, chemical, and/or bioassay approaches in
identifying mutagens and/or carcinogens associated with air
particles. This categorization simply supplies convenient means
of examining the different approaches.

19

Table 1.   Categories of approaches for integrating chemical
           fractionation and bioassay characterization techniques

---

1.   The Pairing Method

2.   The Chemical Search Method

3.   The Mutagen Search Method

4.   The Bioassay Identification Method

5.   The Bioassay-Directed Fractionation

The Pairing Method

The Pairing Method combines information independently obtained
from different sources. Figure 1 illustrates how the pairing of
information from the bioassay of individual chemicals and chemical
fractionation of particulate organics may identify genotoxic substances associated with air particles.

As seen in Figure 1, the pairing of data independently derived
from the bioassay of chemicals and the fractionation of particulate
organics will provide very little overlap between the two lists
(A* - A*) and the identification of very few mutagens and carcinogens (A*) associated with air particles. As seen from this illustration some genotoxicants that are not associated with air particles
are identified, and genotoxicants associated with the particles
would be missed. The Pairing Method is typified by a classical literature review. For example, Bridbord and French (1) reviewed the
carcinogenic and mutagenic risks associated with use of fossil fuels.
By relying upon the IARC list of carcinogens they identified organics
associated with fossil fuels which are carcinogens.

The Chemical Search Method

In this second approach (Fig 2), primary effort is given to
identifying genotoxicants (A*, B*, and C*) and, after identification,
performing chemical analysis for only known genotoxicants that may
be associated with air particles. When searching for specific
chemicals with this approach, chemical methods for identifying
specific chemical species may have to be developed before the
genotoxicant can be identified in complex organic mixtures. Knowing
that nitroparaffins are used as industrial solvents, Hite and Skeggs
(2) surveyed a series of nitroparaffins with the Ames and micronuc-

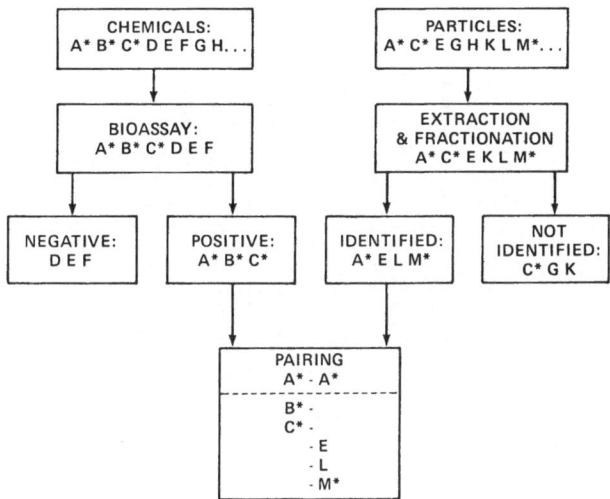

Figure 1. The pairing method

leus assays. Since 2-nitropropane was found to give a positive response in the Ames test, screening by chemical means for the compound within air particulate organics can be accomplished.

The Mutagen Search Method

The Mutagen Search Method (Fig 3) is simply the reverse situation of the Chemical Search Method. With this approach, the majority of effort is spent in identifying as many particle-associated chemicals as possible (A*, E, I and M*). Secondly, the identified chemicals are bioassayed to determine genotoxic activity (A*, M*).

An excellent example of this approach is the work of Florin et al. (3). After the identification of 239 compounds in the gaseous and semivolatile phase of tobacco smoke, the authors tested each compound with a Salmonella mutagenicity assay. The identified 9 positive compounds that included five polyaromatic hydrocarbons (PAH), two methylindoles, and two amines. A similar approach was also taken by Bjorseth et al. (4) for the particulate emissions from a gasifier demonstration project. These investigators identified eight positive PAH compounds. However, these compounds did not account for the

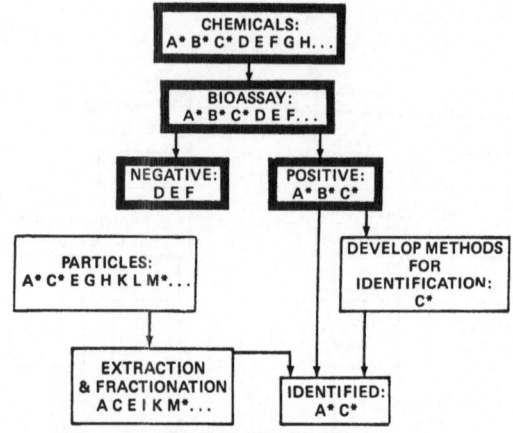

BIASED TOWARD DETECTED MUTAGENS

Figure 2.   The chemical search method

total mutagenic activity of the samples.

The Bioassay Identification Method

This highly specialized approach can provide some unique
capabilities. Bioassay identification (Fig 4) depends upon the
development of highly specialized strains of bacteria (or other
organisms) with altered metabolic pathways. The altered chemical

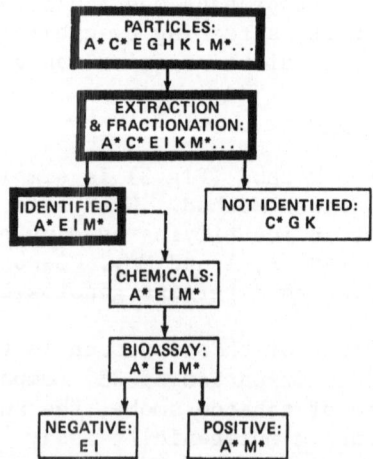

BIASED TOWARD CHEMICAL IDENTIFICATION METHODS

Figure 3.   The mutagen search method

pathway determines whether the organism endogenously activates a promutagen to an active metabolite. By comparing the response of mixtures in these highly specialized strains, specific types of chemicals can be identified by the use of bioassay alone. Rosenkranz et al. (5) recently used this approach in identifying mutagenic nitropyrenes associated with carbon black. Further expansion of this work is reported by Mermelstein et al. (6) within this symposium. In addition, by using the normal testing strains developed by Ames (7), and taking into account that exogenous activation is needed by certain compounds, potential classes of active compounds within a mixture can be identified.

The Bioassay-directed Fractionation Method

The Bioassay-Directed Fractionation Method (Fig 5) is an intimate coupling of chemical fractionation and bioassay, in which each fraction sample is bioassayed for genotoxic activity. Typically, in this approach only the bioassay-positive chemical fractions (and in some cases only the most positive fractions) are further subfractionated, bioassayed, and chemically identified. This scheme limits the identification of compounds to a relatively few genotoxic and nongenotoxic compounds within the the last subfraction step. These compounds could then be bioassayed individually.

This method can be illustrated by recent efforts involving ambient air particles within the author's laboratory. Ambient air particles were collected in two cities (Philadelphia and New York). After Soxhlet extraction with methylene chloride the organics were solvent partitioned. The five fractions produced were an acid fraction, a base fraction, a nonpolar neutral (NPN) fraction, a polynuclear aromatic (PNA) fraction, and a polar

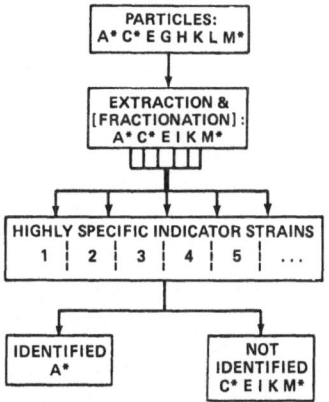

Figure 4.   The bioassay identification method

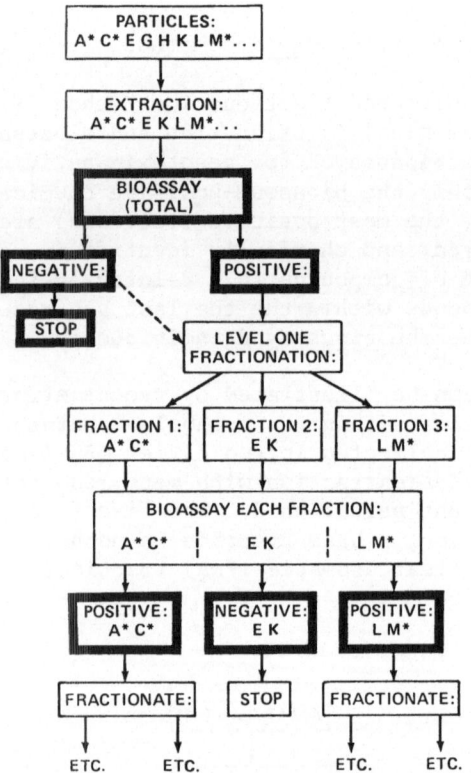

Figure 5.　The bioassay-directed fractionation method

Table 2. Examples of relative organic weights and mutagenic activity (S. typhimurium, TA98, without activation) from two ambient air particle collections

Fraction	Philadelphia			New York		
	Percent of Total Organic Weight	Slope (rev/μg)[a]	Percent of Total Mutagenic Activity	Percent of Total Organic Weight	Slope (rev/μg)[a]	Percent of Total Mutagenic Activity
ACID	4.6	0.8	41.3	5.3	0.01	30.7
BASE	0.7	0.2	1.4	4.2	0.0	0.0
NPN	53.2	0.0	0.0	69.0	0.0	0.0
PNA	8.9	0.05	4.7	6.9	0.2	69.3
PN	32.4	0.1	52.6	14.6	0.0	0.0

[a]Slope measured as revertants per microgram of organic per plate

neutral (PN) fraction. Table 2 demonstrates that the majority of
the organics are NPN's which are not mutagenic. Most of the
mutagenic activity for the Philadelphia particulate organics was
found in the acid and polar neutral fractions, whereas, in the New
York particles, the acids and PNA fractions accounted for all of
the detected mutagenic activity.

Future investigations should be based on an understanding of
the specific advantages and disadvantages associated with each
approach. Table 3 summarizes some of these. As can be seen from
Figures 1 through 5 and Table 3, each approach tends to identify
different genotoxicants. Only the bioassay-directed fractionation
type of approach has the potential of identifying all genotoxicants.
This potential is created for the following reasons: (a) all
portions of the complex mixture can be assayed, (b) as fractionation
occurs, concentration of genotoxicants occurs, (c) toxic materials
that mask genotoxic activity in the assay will be separated from
the genotoxicants, and (d) compounds not previously bioassayed will
be identified in active fractions. However, even the bioassay-
directed fractionation will miss certain genotoxicants, depending
upon the chemical and bioassay procedures used. For example, if
the Ames test is used with a solvent partitioning fractionation
scheme, azonaphthols and halogenated hydrocarbons are likely to go
unidentified (8).

The Bioassay-Directed Approach, because of the amount of
sample needed, is generally limited to using in vitro bioassay
techniques. The Pairing, Chemical Search, and Mutagen Search
Techniques are not limited by bioassay methods. The Bioassay
Identification Method cannot be used to identify many genotoxicants.
However, in some instances, this method is superior to presently
available chemical methods.

MICROBIAL BIOASSAYS

Although a wide variety of microbial bioassays are available
for the assessment of organic chemicals, the Salmonella typhimurium
plate incorporation assay (7) is the most extensively used procedure.

Basically, a specific amount of the chemical to be tested and
a mixture of the indicator bacterial strain are added to melted
soft agar, mixed, and poured into a minimal media plate that will
select for bacteria that have undergone mutation. The plate is
incubated for 48 to 72 hours at $37°C$ and then scored for the
number of colonies per plate. The procedure scores for gene
mutations that occur in the histidine locus in strains of
Salmonella typhimurium created by Ames (7). Each of these strains
has undergone a specific mutation within the histidine locus and
thus is unable to grow unless histidine is supplied.

Table 3. Comparison of generalized methods for coupling chemical fractionation and bioassay procedures

Method	Advantage	Disadvantage*
1 Pairing	simple; may rely on literature; not limited to bacterial bioassays	misses genotoxicants not common to bioassay and chemical procedures (B,C,M)
2 Chemical Search	relies on genotoxicity assay data; not limited to bacterial bioassay	misses genotoxicants on particles that are not within bioassay list (M): chemical methods may have to be developed
3 Mutagen Search	uses well established chemical procedures; not limited to bacterial bioassay	misses genotoxicants not identified by chemical procedures (C)
4 Bioassay Identification	can be accomplished without chemical procedures; occasionally more sensitive than existing chemical procedures	highly limited as to type of chemicals that can be identified; many genotoxicants are missed (C,M); limited to in vitro assays
5 Bioassay-Directed Fractionation	all genotoxicants potentially identified; efficient use of bioassay and chemical methods	amounts of material needed for successive fractionations; initially limited to in vitro assays

*Letters in parenthesis refer to lettering scheme used in Figures 1 through 5.

Since these strains are "mutating back" to a wild type condition, a reverse mutation is said to occur; however, the bio-chemical mechanisms are the same as occur in forward mutation. Systems that detect forward mutation, though, are generally not limited to a specific type of genetic event, e.g. a frameshift or a point mutation, whereas a reverse mutation system does detect a specific type of mutation. Since screening systems need to score for a variety of genetic mutational events, several different Salmonella strains have been developed. Each strain detects a specific type of mutation and incorporates other changes needed to increase its sensitivity.

By assaying for resistance to 8-azaguanine, Skopek et al. (9) developed a preincubation forward mutation system using Salmonella. This and other forward mutation systems utilizing bacteria (10) should be considered for screening samples limited in amount. Since bacterial assays are most valuable as screening systems, assays utilizing more advanced organisms can be used for further evaluation of mutagenic compounds associated with airborne particles. A review of these systems is beyond the scope of this paper, but a thorough review of genetic bioassays is being completed by expert committees through the EPA-sponsored Gene-Tox Program (11). In this effort, scientists in 22 work groups will evaluate each bioassay, formulate generalized protocols, and recommend areas of additional research. These evaluations are being published in the open literature.

CHEMICAL FRACTIONATION PROCEDURES

Even more variable than the bioassay procedures are the poten-tial chemical fractionation procedures that are available. Table 4 lists five types of commonly used preparative fractionation proce-dures and the types of partitioning materials associated with each. In addition, each of these procedures can be used in various combina-tions. The techniques utilized by individual investigators appear to depend upon cost, availability, type of organics of greatest interest, and personal preferences.

Tokiwa et al. (12) used a solvent partitioning scheme for particulate organics collected at an industrial site (Ohmuta) and a heavy traffic site in Fukouka, Japan. Most of the fractions demonstrated some mutagenic activity with the Salmonella system. These investigators also used thin layer chromatography (TLC) to identify specific PAH compounds. Pellizzari et al. (13) reported a different solvent partitioning scheme for the analysis of organics from air particles; and in examining air particulate organics from Chicago, Commoner and colleagues (14) used TLC. In examining particles other than those in ambient air, Rosenkranz et al. (5) and Lofroth et al. (15) used high pressure liquid chromatography.

Table 4. Fractionation procedures available for ambient
          air organics

Type of Method	Materials
Solvent Partitioning	Solvents of differing polarity
Thin Layer Chromatography (TLC)	Silica gel Alumina Cellulose acetate
Column Chromatography	Alumina Silica gel Ion exchange Porous polymer
Batch Chromatography	Ion exchange Porous polymer
High Performance Liquid Chromatography	Alumina Silica gel Ion exchange Porous polymer

Huisingh et al. (16) fractionated the organics from diesel emission
particles with a combination of solvent partitioning and silica gel
chromatography.

DIFFICULTIES IN INTERPRETATION

The collection, extraction, fractionation and bioassay of
complex organics associated with air particles is a multistep process
that presents many difficulties in the final interpretation of data.
Table 5 lists examples of some of these difficulties, divided into
three categories: artifactual, statistical/mathematical, and
interpretive.

Artifactual Difficulties

Each stage of the analysis provides potential artifacts. Pitts
et al. (16) have shown that when nitrogen oxides pass over particle-

Table 5. Difficulties in the interpretation of ambient
          air genotoxicity studies

Class	Stage	Example
Artifactual	Collection	NO_x, organic interaction, oxidation by ozone
	Extraction	Selective loss of compounds
	Fractionation	Recovery efficiencies
	Storage	Storage half-life of compounds
	Bioassay	Endogenous activation
Statistical/ Mathematical	Collection	Collection efficiencies, TSP calculations
	Extraction/ Fractionation	Recovery efficiencies
	Bioassay	Data summary method
Interpretive	Collection	Seasonal, daily variation
	Extraction	Classes of substances removed
	Fractionation/ identification	Correlations (bioassay, site, etc.)
	Bioassay	Chemical/strain selectivity
	Other	Particle residence time, core material, ambient interactions (ozone)

bound organics on filters, mutagenic components can be produced
during the collection processes. Extraction processes are selective
with respect to the organics that are removed from particles (19).
Pellizzari et al. (13) have shown that when artificial mixtures of
compounds are chemically fractionated, the individual compounds
are recovered with differing efficiencies. As shown by Rosenkranz
et al. (5), with the nitroreductase-deficient strains of Salmonella,
bacteria and other indicator cells and organisms may contain
endogenous metabolic capabilities that can produce false positive
and/or false negative bioassay results.

Statistical/Mathematical Difficulties

     Collection efficiencies vary with the type of collector, the
weather conditions, air particle concentrations, and other factors
(18). Therefore, it is sometimes necessary to make data transformations based on these variables. Each transformation will tend
to introduce increased variance and decreased confidence in any
final data transformation. Additional data transformations may be
considered for the chemical extraction and fractionation data as
well. For example, if only 60% of total organic mass is recovered
after fractionation, can one correct for this reduction before
calculating the genotoxic activity per mass of air particles?
Since the recovery efficiency may vary for each compound, this
correction step may introduce additional errors into the final
interpretation.

     Controversy persists regarding the statistical methods most
applicable for each of the bioassays generally available. For
example, can Salmonella plate incorporation data be summarized by
the slope of a linear regression line plotted over some portion of
the dose-response curve, or should another statistical model such
as the one developed by Stead et al. (19) be used? Many of these
statistical questions are not generally recognized. Indeed, when
examining a unified set of data handled in a single statistical/
mathematical manner, transformations may have little influence on
final interpreation. However, when examining reports from multiple
investigators, these transformations (or the lack of them) may make
comparisons very difficult.

Interpretive Difficulties

     The characteristics of ambient air samples fluctuate because
of the variability within sources, changes in ratios of types of
sources, weather conditions, alterations in collection conditions,
etc. When samples are collected over a short time period, results
may not be representative of a yearly averaged sample. In addition,
the proximity of a collector to sources, human exposure areas, etc.,
could be critical in interpretation. Extraction and fractionation
procedures must be considered for selectivity. Genotoxic agents

may be extracted, they may "spill over" into multiple fractions, or
their genotoxic activity may be masked by components toxic to the
indicator organism. Certain bioassays also are very poor detectors
of genotoxic activity for certain classes of compounds. For example,
Rinkus and Legator (8) suggest that the Ames test is a poor detector
of certain categories of carcinogens -- namely, azonphthols, benzo-
dioxoles, polychlorinated aromatics, and steroids. In addition to
the above difficulties, a complete interpretation of any hazard
associated with air particles would have to take into account the
particle residence time of the respirable particles, the distribution of particles of differing sizes, the composition of the core
material of the particles, the effect of ambient interactions,
chemical transformations and a variety of other related processes.

CONCLUSIONS

The coupling of bioassay and chemical identification presents
many difficulties, but such an effort provides some unique opportuni-
ties. First, genotoxicants associated with ambient air particulates
can be identified, a process directly useful to those engaged in
atmospheric measurements and in laboratory and theoretical studies
of relevant health factors in our environment. Secondly, comparisons can be made, although with considerable difficulty, between
sites, seasons of the year, and proximity to various types of
genotoxicant sources. This type of information will allow more
revealing epidemiological studies and will foster the development
of needed monitoring and control technologies. The integration
of bioassay and chemical methodologies, therefore, has great
utility in present environmental research.

ACKNOWLEDGEMENT

This paper has been reviewed by the Health Effects Research
Laboratory, U.S. Environmental Protection Agency, and approved for
publication. Mention of trade names or commercial products does
not constitute endorsement or recommendation for use.

REFERENCES

1. K. Bridbord and J. G. French, Carcinogenic and mutagenic risks
      associated with fossil fuels, in: "Carcinogenesis", Vol. 3,
      P. W. Jones and R. I. Freudenthal, eds., Raven Press, N.Y.,
      451-463 (1978).
2. M. Hite and H. Skeggs, Mutagenic evaluation of nitroparaffins
      in the Salmonella typhimurium/mammalian - microsome test
      and the micronucleus test, Environ. Mutagenesis 1:383-389
      (1979).

3. I. Florin, L. Rutger, M. Curvall, and C. R. Enzell, Screening
    of tobacco smoke constituents for mutagenicity using the
    Ames test, Toxicol. 18:219-232 (1980).
4. O. M. Bjorseth, C. P. Flessel, N. M. Monto, J. J. Wesolowski,
    T. R. Parker, and P. K. Ouchida, Polycyclic aromatic
    hydrocarbon (PAH) content and mutagenic activity in products
    and emissions from a gasifier demonstration project,
    California State Air Resources Board Report No. CA/DOH/AIHL/
    R-196, 26 pp, (1979).
5. H. S. Rosenkranz, E. C. McCoy, D. R. Sanders, M. Butler,
    D. K. Kiriazides, and R. Mermelstein, Nitropyrenes:
    Isolation, identification, and reduction of mutagenic
    impurities in carbon black and toners, Science 209:1039-1043
    (1980).
6. R. Mermelstein, H. S. Rosenkranz, and E. C. McCoy, The micro-
    bial mutagenicity of nitroarenes, in: "Symposium on the
    Genetic Effects of Airborne Agents" R. R. Tice, D. L. Costa
    and K. Schaich, eds., Plenum Press, New York (1981).
7. B. N. Ames, J. McCann, and E. Yamasaki, Methods for detecting
    carcinogens and mutagens with the Salmonella/mammalian -
    microsome mutagenicity test, Mutat. Res. 31:347-364 (1975).
8. S. J. Rinkus and M. S. Legator, Chemical characterization of
    465 known or suspected carcinogens and their correlation
    with mutagenic activity in the Salmonella typhimurium
    system, Cancer Res. 39:3289-3318 (1979).
9. T. R. Skopek, H. L. Liber, D. A. Kaden, and W. G. Thilly,
    Quantitative forward mutation assay in Salmonella
    typhimurium using 8-azaguanine resistance as a genetic
    marker, Proc. Natl. Acad. Sci. (U.S.A.) 75:410-414 (1978).
10. G. R. Mohn and J. Ellenberger, Appreciation of the value of
    different bacterial test systems for detecting and for
    ranking chemical mutagens, Arch. Toxicol. 46:45-60 (1980).
11. M. D. Waters, The Gene-Tox Program, in: "The Banbury Report 2:
    Mammalian Cell Mutagenesis: The Maturation of Test Systems"
    A. W. Hsie, J. P. O'Neill, and V. K. McElheny, eds., Cold
    Spring Harbor Laboratory, Cold Spring Harbor, N.Y., 449-
    457 (1979).
12. H. Tokiwa, S. Kitamori, K. Takahashi, and Y. Ohnishi, Mutagenic
    and chemical assay of extracts of airborne particulates,
    Mutat. Res. 77:99-108 (1980).
13. E. D. Pellizzari, L. W. Little, C. Sparacino, T. J. Hughes,
    L. Claxton, and M. D. Waters, Integrating microbiological
    and chemical testing into the screening of air samples for
    potential mutagenicity, in: "Application of Short-Term
    Bioassays in the Fractionation and Analysis of Complex
    Environmental Mixtures", M. D. Waters, S. Nesnow,
    J. L. Huisingh, S. S. Sandhu, and L. Claxton, eds., Plenum
    Press, New York, N.Y., 331-352 (1978).

14. B. Commoner, A. J. Vithayathil, and P. Dolara, in: "Application
     of Short-Term Bioassays in the Fractionation and Analysis
     of Complex Environmental Mixtures", M. D. Waters, S. Nesnow,
     J. L. Huisingh, S. S. Sandhu, and L. Claxton, eds., Plenum
     Press, New York, N.Y., 529-570 (1978).
15. G. Lofroth, E. Hefner, I. Alfheim, and M. Moller, Mutagenic
     activity in photocopies, Science 209:1037-1039 (1980).
16. J. Huisingh, R. Bradow, R. Jungers, L. Claxton, R. Zweidinger,
     S. Tejada, J. Bumgarner, F. Duffield, M. Waters, V. Simmon,
     C. Hare, C. Rodriguez, and L. Snow, Application of bioassay
     to the characterization of diesel particle emissions, in:
     "Application of Short-Term Bioassays in the Fractionation
     and Analysis of Complex Environmental Mixtures", M. D. Waters,
     S. Nesnow, J. L. Huisingh, S. S. Sandhu, and L. Claxton,
     eds., Plenum Press, New York, N.Y., 381-418 (1978).
17. J. N. Pitts, Jr., D. M. Lokensgard, P. S. Ripley, K. A. Van
     Cauwenberghe, L. Van Vaeck, S. D. Shaffer, and W. L. Belser,
     Jr., "Atmospheric" epoxidation of benzo(a)pyrene by ozone:
     Formation of the metabolite benzo(a)pyrene-4,5-oxide,
     Science 210:1347-1349 (1980).
18. R. Jungers and J. Huisingh, Collection and extraction techniques
     for the evaluation of airborne particles in microbial muta-
     genicity bioassays, in: "Symposium on the Genotoxic Effects
     of Airborne Agents", R. R. Tice, D. L. Costa and K. Schaich,
     eds., Plenum Press, New York, NY (1981).
19. A. G. Stead, V. Hasselblad, J. P. Creason, and L. Claxton,
     Modeling the Ames test, Mutat. Res. 85:13-27 (1981).

AIRBORNE PARTICLE COLLECTION AND EXTRACTION METHODS

APPLICABLE TO GENETIC BIOASSAYS

Robert H. Jungers and Joellen Lewtas

U.S. Environmental Protection Agency
Research Triangle Park, NC 27711 (U.S.A.)

INTRODUCTION

Particulate matter is present in the ambient air in a wide range of sizes and chemical compositions. Ambient air contains both carbonaceous soot particles and silica-alumina particles, which range in size from below 0.01 µm to over 100 µm. Carbonaceous soot particles are emitted from oil combustion, such as in residential heaters and diesel vehicle exhausts. Coal-fired power plants emit fly ash particles with a silica-alumina matrix. These and other sources of urban and industrial air particles also emit significant amounts of organic compounds which condense onto the surfaces of particles after dilution and cooling in the ambient air. Advances in the application of chemical analytical techniques have led to the identification of a number of genotoxic agents in these particle-bound organics, including polycyclic aromatic hydrocarbons (PAH), e.g. benzo(a)pyrene and benzo-flouranthene, oxygenated compounds, e.g. 7-8-benz(d,e)anthracene-2-one, and aromatic amines, e.g. aminoanthracene.

The development of the Ames Salmonella typhimurium mutagenesis bioassay (1) has provided a simple, sensitive, and rapid bioassay for potential carcinogenic activity applicable to air samples collected by conventional techniques. The initial application of this bioassay to particulate organic matter is ambient air (2-4) has stimulated research in both (1) the characterization and identification of classes of potential carcinogens and of specific carcinogens present in ambient air particles and (2) in the evaluation of emission sources and atmospheric conditions responsible for the observed mutagenicity of urban air particles. Advances in these areas, however, are dependent upon improvements in sample collection, extraction and bioassay methods.

This review addresses air particle collection and extraction
methods for application to genetic bioassays. Below are discussed
the various particle collection systems currently in use, methods
for extracting organic compounds from particulate samples, and
results from bioassays of such extracts.

COMPARISON OF AVAILABLE COLLECTION METHODS

A great number of air particle samplers have been designed
and used for field applications. Filtration and impaction are the
two collection techniques most commonly employed for the determination of particle concentration and subsequent chemical analysis.
The conventional air particle collection instrument, recommended
in the Code of Federal Regulations for determining total suspended
particles (TSP) (5), is the standard high-volume sampler (Fig 1).

The first reported studies on the mutagenicity of particulate
organic matter from ambient air used high-volume samplers to collect
air particles on fiberglass filters (2-4). This sampler collects
both respirable particles (<5 μm in diameter) and larger non-
respirable particles. In these studies, the organics were extracted

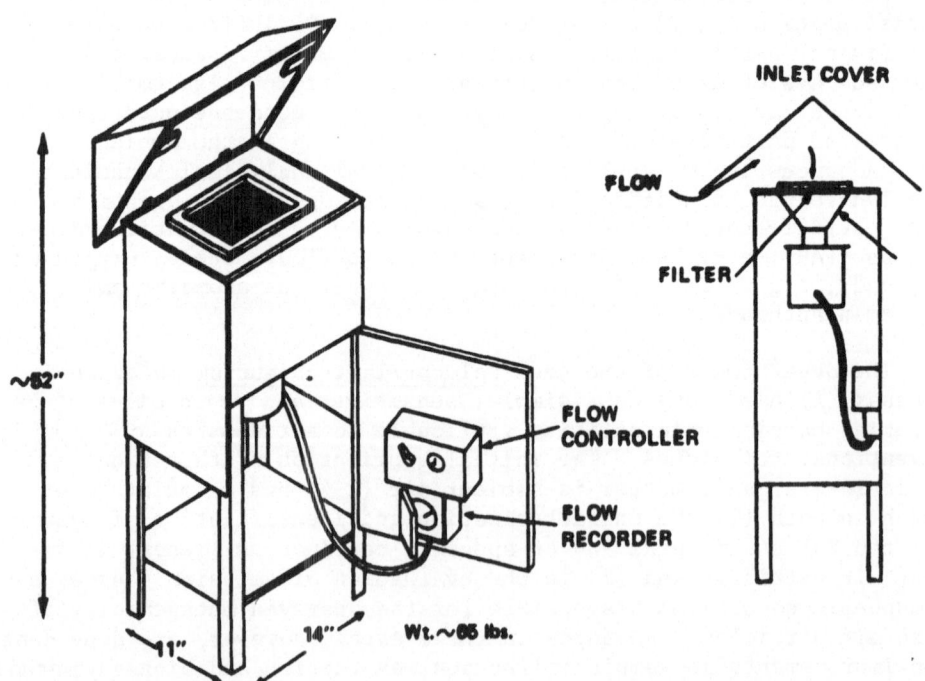

Figure 1.   Schematic diagram of a standard TSP high-volume sampler.

from the particles with methanol (2), acetone (4) or a mixture of equal amounts of methanol, benzene, and dichloromethane (3). Since high-volume samplers are widely used in air-monitoring progams to determine TSP levels, these studies provided comparative mutagenicity data for different sites.

Although high-volume samplers provide the simplest, and for many investigators the only, method available for collection of air particles, they present several serious disadvantages. The most consequential disadvantage is that respirable particles are collected simultaneously with larger particles. In cases where the smaller respirable particles are considerably more mutagenic than the larger particles, the larger particles dilute the overall mutagenicity of the sample, thereby biasing the analysis of the particle composition to which the human lung is exposed.

Other potential disadvantages may result from the large volume of air being drawn continuously over collected particles. Samples collected by this method may lose volatile organics by evaporation. The organics present on the particles are also potentially subject to reactions with nitrogen dioxide (NO_2), ozone (O_3) or peroxyacetyl nitrate (PAN), which are all present in urban air. Pitts et al. (6,7) have shown that PAH, e.g. benzo(a)pyrene and perylene, directly coated onto fiberglass filters reacted with NO_2, O_3, PAN, and ambient photochemical smog to form several direct-acting mutagens (mutagens that do not require an exogenous microsomal activation system). Although these reactions have not been shown to occur with PAH adsorbed on the surface of air particles, potential surface reactions during filtration still must be considered when interpreting studies using filtration for sample collection.

A recent modification of the standard high-volume sampler, the size-selective inlet (SSI) high-volume sampler (Figures 2,3), collects only particles <15 μm (the definition of "inhalable particles"). This sampler excludes the larger particles, which are not normally inhaled (8).

A sampling device that does collect particles in separate size fractions is the cascade impactor (9) (Figures 4,5). These samplers use a series of plates with either holes or slots offset at each stage to collect separate size fractions ranging from <2 μm to >7 μm. The cascade impactor is generally attached to a high-volume sampler that collects the smallest particles by filtration. Teranashi et al. (10), using an Andersen high-volume cascade impactor, found that the organics from particles <1.1 μm, collected on the backup filter, were significantly more mutagenic than the organics from the larger particles. Pitts et al. (7) reported similar findings using a Sierra high-volume cascade impactor in downtown Los Angeles. This method, while providing a size-fractionated sample for bioassay,

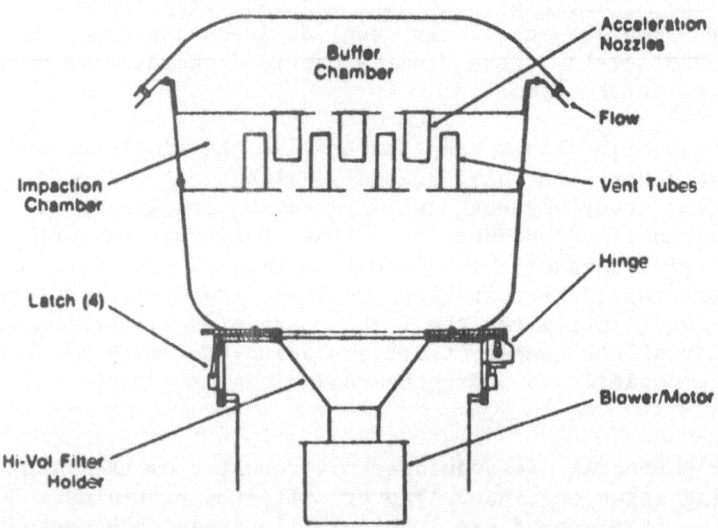

Figure 2.   Schematic diagram of a size-selective inlet (SSI) high-volume sampler.

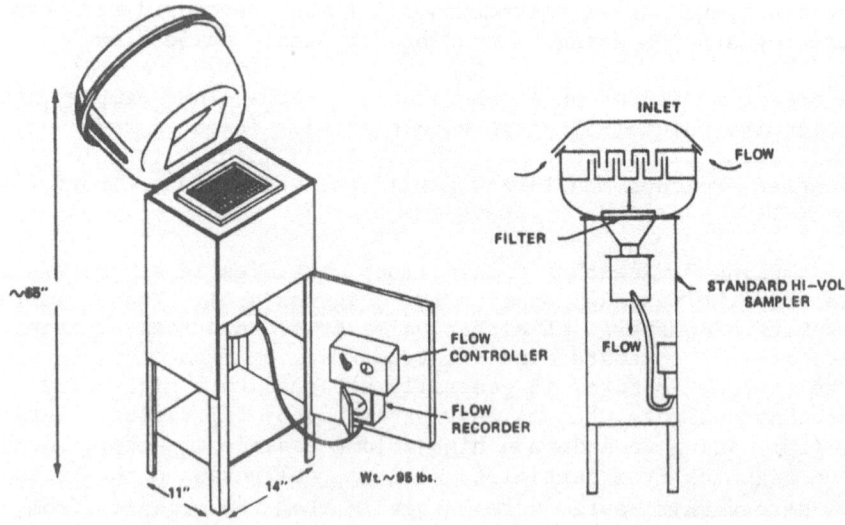

Figure 3.   Schematic diagram of the head of an SSI high-volume sampler.

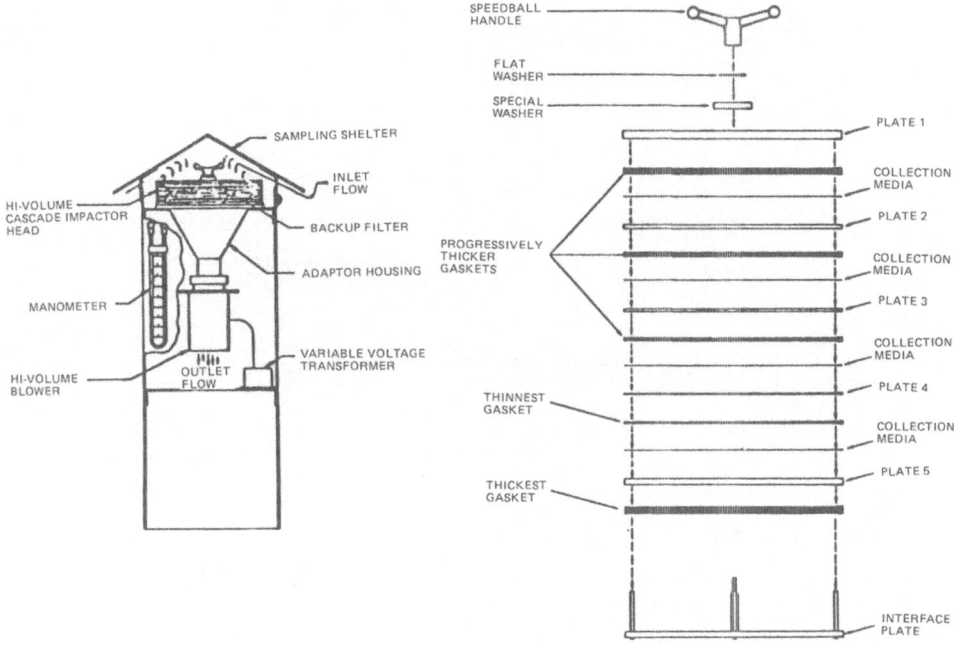

Figure 4.   Schematic diagram of a cascade impactor.

still employs filtration to collect the smallest particles, which contain most of the mutagenic components.

In addition, the dichotomous sampler (vitual impactor) (Fig 6) can be employed when elemental chemical analysis is desired (8). This sampler operates by drawing air into the sampler at the rate of 1 m^3/hr. The sampler inlet is designed to allow only particles ≤15 μm in diameter to enter. The airstream carrying the particles is divided into two streams according to size of particles (<2.5 μm; 2.5-15 μm) and each stream is collected on separate filters.

All three of these instruments employ filtration as the primary collection method. Preliminary studies have shown that these samplers do not collect sufficient amounts of particles for complete bioassay studies. Mitchell et al. (11) designed and constructed for the U.S. Environmental Protection Agency (EPA) a massive air volume sampler (MAVS) that would collect large quantities of size-fractionated air particles. The principles and operation of this samplers are described in the next section. The new sampler, MAVS, capable of collecting gram amounts of size particles in a reasonable sampling period, is currently being evaluated for collecting ambient air particles for bioassay studies.

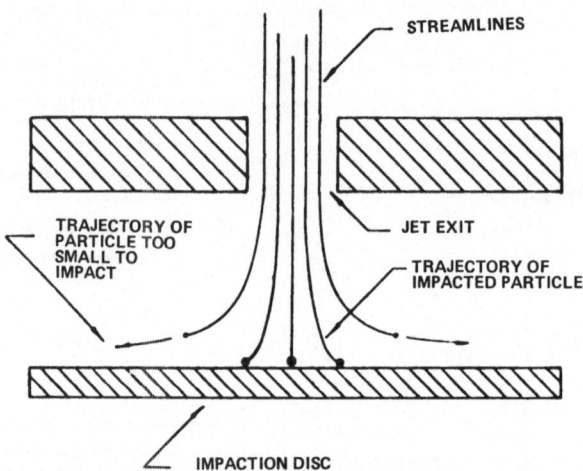

Figure 5.   Schematic diagram of a cascade impactor plate.

Table 1 shows a comparison of particle masses that can theoretically be collected by the TSP high-volume sampler, the SSI high-
volume sampler, the dichotomous sampler, and the new MAVS for
different air pollution TSP concentrations. The comparison was
made for 24-hr sampling periods in ambient air with TSP particle
concentrations of 60, 100, and 200 ug/m^3. As noted in the table,
the mass of particles collected depends on the ambient particle
concentration and, more importantly, on the flow rate of air
through the given sampler. The flow rate of the MAVS (18.5 m^3/min)
enables it to collect a greater amount of particulate matter than
the other samplers.

The MAVS employs two impactors that collect 3.5 to 20 μm and
1.7 to 3.5 μm particles, followed by an electrostatic precipitator
(ESP) that collects the particles <1.7 μm. In an initial study
with the MAVS at a Los Angeles freeway site, the particles with
mean diamters <1.7 μm were found to be significantly more mutagenic
than the larger particles. However, when the ESP was charged to
maximize collection efficiency, as much as 0.05 ppm of O_3 was measured
at the flow outlet of the sampler (11). Jungers et al. (12) evaluated
the effect of O_3 under the MAVS operating conditions on both the

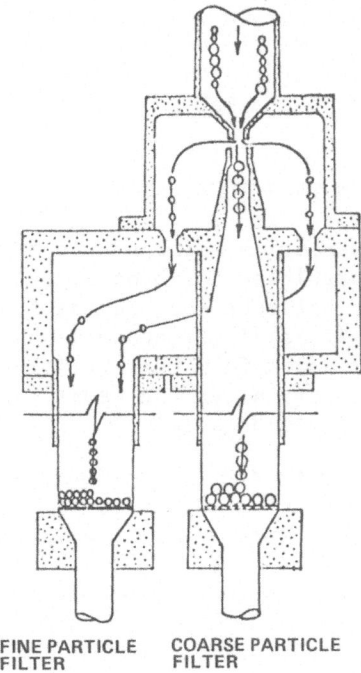

FINE PARTICLE
FILTER

COARSE PARTICLE
FILTER

Figure 6.   Schematic diagram of the head of a dichotomous sampler.

Table 1.   Comparison of amounts of particulate matter
theoretically collected by four type of samplers

Type of Sampler	Flow Rate (m^3/min)	Volume Sampled[a] (m^3)	60 µg/m³	100 µg/m³	200 µg/m³
			TSP Concentrations		
			Calculated Mass Collected[a] (g)		
Dichotomous	0.017	24.0	0.0015	0.0024	0.0048
SSI high-volume	1.13	1,627.0	0.097	0.16	0.33
TSP high-volume	1.40	2,016.0	0.12	0.20	0.40
MAVS	18.5	26,640.0	1.6	2.7	5.3

[a]During 24 hr sampling period.

mutagenicity and chemical composition of the particulate organic
matter collected in the ESP. It was concluded from these studies
that, under these operating conditions, the O_3 did not significantly
affect the mutagenicity or the PAH content of the organics.

EVALUATION OF EXTRACTION TECHNIQUES

A wide range of extraction methods have been used to remove
the organics from air particles for bioassay. Solvents employed
range from the nonpolar solvents cyclohexane (13) to the more polar
solvent methanol (2). Since different solvents will preferentially
remove different constituents from the particles, the method of
extraction can significantly alter the resulting composition and
observed mutagenicity of the organics.

To determine which extraction technique was the most effective
in removing mutagens from ambient air particles, two extraction
methods and seven solvent systems were evaluated. The two extraction
methods were Soxhlet extraction for 24 hr and sonication for both
30 min and 2 hr. The four basic single solvents were cyclohexane
(CH), dichloromethane (DCM), acetone (Ac) and methanol (MeOH),
with polarity indices of 0.0, 3.4, 5.4, and 6.6, respectively.
Three additional comparative solvent systems were: sequential
sonication in cyclohexane and methanol (CH/MeOH); equal amounts of
toluene, dichloromethane, and methanol (T:DCM:MeOH); and direct
suspension in dimethylsulfoxide (DMSO). These solvent systems
were selected for comparison with previous work (3,14). Only the
four single solvents were used in the Soxhlet extraction method,
while the four single solvent and the comparative solvent systems
were used in sonication.

The data in Table 2 indicate no significant difference for
percentage of organics extractable after 30 min or 2 hr of sonication,
but acetone and methanol extracted approximately three times as
much mass as cyclohexane and twice as much as DCM. The Soxhlet
extraction method indicated a significant increase in the amount of
organics extractable with acetone as the solvent.

An aliquot of each extract was dried under nitrogen and diluted
with a 1:1 mixture of DCM and cyclohexane. Fifty microliters were
spotted on a thin-layer chromatography plate, developed in a tank
containing a 1:2 mixture of DCM and ethanol, and analyzed for
benzo(a)pyrene (B(a)P) with a Perkin Elmer MPF-44B Fluorescence
Spectrometer. The results are shown in Table 3. Regardless of the
method of extraction or solvent used, the amount of B(a)P extracted
was relatively constant on a weight-by-weight basis.

An aliquot of 10% of the extract was dried and prepared in
aqueous solution for analysis on a Dionex Model 10 ion chromatograph

Table 2. Percentages of total particle masses
extractable by various solvents

Solvent	Soxhlet 24 hr	Sonication 30 min	Sonication 2 hr
CH	4.0	2.9	3.0
DCM	6.4	4.3	4.7
Ac	21.0	9.0	8.2
MeOH	8.9	11.4	11.5
CH/MeOH		11.6	11.6
T:DCM:MeOH		8.7	8.9

(IC). The IC was equipped with a standard column and used a 0.003
M sodium bicarbonate and 0.0024 M sodium carbonate eluent. The
sample was analyzed for fluoride, chloride, nitrate and sulfate
ions. Total anion concentrations were calculated and compared for
the four single solvents used in Soxhlet extraction and for the
two additional comparative solvent systems used in the sonication
extraction techniques. Table 4 shows that the increasing polarity
of the solvent is comparable to the increasing quantity of total
anions extracted, regardless of extraction technique, although the
increased time of sonication did result in increased amounts of
total anions extracted. Table 4 also shows that approximately
the same amounts of total anions were extracted by methanol in
both the Soxhlet and 30-min sonication extraction techniques and

Table 3. Benzo(a)pyrene extracted from fiberglass
filters by various solvents and techniques

Solvent	B(a)P (µg/g) Soxhlet 24 hr	Sonication 30 min	Sonication 2 hr
CH	11.8	8.2	8.4
DCM	12.0	10.9	9.5
Ac	11.4	11.7	10.1
MeOH	11.8	11.1	10.4
CH/MeOH		13.2	12.8
T:DCM:MeOH		13.9	11.7

Table 4.   Comparison of total anion concentrations

Total Anions (μg/g)

Solvent	Soxhlet 24 hr	Sonication 30 min	Sonication 2 hr
CH	29.12	17.09	13.80
DCM	19.32	21.77	81.89
Ac	583.71	319.66	528.49
MeOH	992.67	878.54	1124.87
CH/MeOH		949.44	997.49
T:DCM:MeOH		443.69	500.17

by the comparative solvent system CH/MeOH in the 30-min sonication technique. It appears that methanol is the basic extracting solvent of total anions in both systems.

Table 5 summarizes the anion analysis of the four single solvents in the Soxhlet extraction technique. The main ion extracted by acetone is nitrate, while methanol predominantly extracts sulfate. Neither of these ions appreciably affects the microbial mutagenicity bioassay results.

While the comparative solvent systems do not extract significantly more nitrate ion than does the single solvent acetone, they extract far less of the sulfate ion than does the single solvent methanol. Therefore, it appears that the two comparative solvent

Table 5.   Ambient air anion analysis

Anion Concentrations (μg/g)[a]

Solvent	Fl	Cl	NO_3	SO_4
CH	0.02	2.6	7.0	19.5
DCM	0.02	2.2	2.8	14.3
Ac	2.41	83.7	477.5	20.1
MeOH	0.77	18.7	56.1	917.1

[a]Soxhlet extraction for 24 h.

Table 6. Mutagenic activities in TA98 (with S-9 activation)
of solvent-extractable materials from air particles

Solvent	Soxhlet Extraction		Sonication	
	Rev/µg	95% Confidence Limits	Rev/µg	95% Confidence Limits
CH	0.11	0.05 - 0.16	0.09	0.06 - 0.13
DCM	0.52	0.43 - 0.62	0.48	0.37 - 0.58
Ac	0.28	0.23 - 0.33	0.45	0.38 - 0.53
MeOH	0.26	0.23 - 0.30	0.35	0.28 - 0.41
CH/MeOH			0.38	0.30 - 0.45
T:DCM:MeOH			0.37	0.26 - 0.48
DMSO			0.07	0.03 - 0.10

systems (CH/MeOH and T:DCM:MeOH) are not more effective than the
Soxhlet-extracted single solvents (acetone and methanol).

BIOASSAY RESULTS AND DISCUSSION

The S. typhimurium mutagenesis data for tester strains TA98
for each solvent system and extraction tecnique are shown in
Table 6. DCM extraction, either by Soxhlet or sonication, resulted
in an extractable material that was more mutagenic than any resulting
from the other solvents. The least polar solvent, cyclohexane, was
the least effective in extracting mutagens by either method. Acetone
and methanol solvent extraction generally resulted in less mutagenic
samples than DCM extraction, except when acetone was used in sonication. Acetone removed considerably more extractable mass from
air particles, including more inorganics. Therefore, when the
mutagenic activity was calculated per microgram of particle, as
shown in Table 7, acetone appeared to result in greater mutagenic
activity. Consequently, the use of acetone may be advantageous
for certain studies. Analytical problems associated with the use
of acetone plus the presence of higher concentrations of inorganic
salts resulted in selection of DCM for subsequent studies.

While DMSO can be used directly both to suspend the particles
and administer them to the bioassay without evaporation or solvent
exchange, it was the least effective method for detecting the
mutagenic activity present. Thus, DCM is the preferred solvent,
particularly for studies in which the amount of nonmutagenic mass
should be minimized.

Table 7.   Ambient air particle mutagenicity in TA98
            (with S-9 activation)

Solvent	% Extractable	Revertants/μg Extractable	Revertants/μg Particle
CH	4.0	0.11	4.4
DCM	6.4	0.52	33.3
AC	21.0	0.28	58.8
MeOH	8.9	0.26	23.1

[a]Soxhlet extraction.

REFERENCES

1. B. N. Ames, J. McCann and E. Yamasaki, Methods for detecting
    carcinogens and mutagens with the Salmonella/mammalian
    microsome mutagenicity test. Mutat. Res. 31:347-364 (1975).
2. H. Tokiwa, H. Takeyoshi, K. Morita, K. Takahashi, N. Soruta
    and Y. Ohnishi, Detection of mutagenic activity in urban
    air pollutants. Mutat. Res. 38:351-359 (1976).
3. J. N. Pitts, D. Grosjean, J. M. Mischke, V. F. Simmon and D.
    Poole, Mutagenic activity of airborne particulate organic
    pollutants. Toxicol. Lett. 1:65-70 (1977).
4. R. Talcott and E. Wei, Airborne mutagens bioassayed in
    Salmonella typhimurium. J. Natl. Cancer Inst. 58:449-451
    (1977).
5. Code of Federal Regulations. Title 40, part 50, appendix B.
    Reference Method for Determination of Suspended Particulates
    in the Atmosphere (High Volume Method). General Service
    Administration: Washington, D.C. 531-535 (1980).
6. J. N. Pitts, K. A. Van Cauwenberghe, D. Grosjean, J. P. Schmid,
    D. R. Fitz, W. L. Belser, G. B. Knudson and P. M. Hynds,
    Atmospheric reactions of polycyclic aromatic hydrocarbons:
    Facile formation of mutagenic nitro derivatives. Science
    202:515-519 (1978).
7. J. N. Pitts, K. A. Van Cauwenberghe, D. Grosjean, J. P. Schmid,
    D. R. Fitz, W. L. Belser, G. B. Knudson and P. M. Hynds,
    Chemical and microbiological studies of mutagenic pollutants
    in real and simulated atmospheres, in: "Application of
    Short-term Bioassays in the Fractionation and Analysis of
    Complex Environmental Mixtures," M. D. Waters, S. Nesnow,
    J. L. Huisingh, S. S. Sandhu and L. Claxton, eds., Plenum
    Press, New York, NY 353-379 (1979).
8. A. McFarland and C. E. Rodes, Characteristics of aerosol
    samplers used in ambient air monitoring. Presented at
    the 86th National Meeting of Chemical Engineers, Houston,
    TX (1979).

9. A. A. Andersen, A sampler for respiratory health hazard
    assessment. Am. Ind. Hyg. Assoc. J. 27:160-165 (1966).
10. K. Teranashi, K. Hamada, N. Tekeda and H. Watanabe, Muta-
    genicity of the tar in air pollutants. Proc. 4th Int.
    Clean Air Congress, Tokyo 33-36 (1977).
11. R. I. Mitchell, W. M. Henry and N. C. Henderson, Fabrication,
    optimization and evaluation of a massive air volume sampler
    of sized respirable particulate matter. EPA-600/4-78-031.
    U.S. Environmental Protection Agency; Research Triangle
    Park, NC (1978).
12. R. Jungers, R. Burton, L. Claxton and J. Lewtas Huisingh,
    Evaluation of collection and extraction methods for
    mutagenesis studies on ambient air particulate, in: "Short-
    term Bioassays in the Analysis of Complex Environmental
    Mixtures," M. D. Waters, S. S. Sandhu, J. L. Huisingh, L.
    Claxton and S. Nesnow, eds., Plenum Press, New York, NY
13. M. Moller and I. Alfheim, Mutagenicity and PAH-analysis of
    airborne particulate matter. Atmos. Environ. 14:83-88
    (1980).
14. E. D. Pellizzari, L. W. Little, C. Sparacino, T. J. Hughes,
    L. Claxton and M. D. Waters, Integrating microbiological
    and chemical testing into the screening of air samples
    for potential mutagenicity, in: "Application of Short-
    term Bioassays in the Fractionation and Analysis of
    Complex Environmental Mixtures," M. D. Waters, S. Nesnow,
    J. L. Huisingh, S. S. Sandhu and L. Claxton, eds., Plenum
    Press, New York, NY 382-418 (1979).

SESSION II: ASSAY AND EXPOSURE TECHNOLOGY OF IN VITRO

MAMMALIAN CELL SYSTEMS APPLIED TO AIRBORNE AGENTS

Donald E. Rounds, Moderator

Pasadena Foundation for Medical Research
99 North El Molino Avenue
Pasadena, CA 91101 (U.S.A.)

Welcome to the session on in vitro mammalian assay systems. I would like to thank Dr. Tice and the organizers of this symposium for the opportunity to present our particular bias regarding the use of mammalian cells for short-term bioassays of genotoxic agents. Our position is that bioassay systems using mammalian cells in vitro are competitive with those systems using microbial or plant cells with respect to speed, economy, sensitivity and reliability. In addition, mammalian cells offer greater relevance to genotoxic responses in mammals or even human subjects. Although the in vitro models do not always include the pharmacokinetic conditions which are important considerations for whole animal exposures, we are beginning to develop techniques which can apply in vitro short-term bioassays to tissues derived from experimental animals exposed to genotoxic agents in vivo.

The speakers for this session were selected because they have each been successful in developing methods for directly exposing cells in culture to genotoxic agents in the form of gases, particulates or vapors, while maintaining a biocompatible environment for the cell cultures being tested. Secondly, the four presentations were selected because they represent at least four different methods for measuring genotoxic effects.

IN VITRO ANALYSIS OF MAMMALIAN CELLS EXPOSED IN VITRO AND IN VIVO TO AIRBORNE AGENTS

R. R. Guerrero, and D. E. Rounds

Pasadena Foundation for Medical Research
99 North El Molino Avenue
Pasadena, California 91101 (U.S.A.)

INTRODUCTION

Airborne genotoxic agents consitute a major environmental
hazard. A reflection of this hazard is the fact that lung
cancer constitutes the greatest cause of cancer deaths in men
and, disturbingly, is rapidly increasing in women as well (1).
Genotoxic airborne agents such as gaseous and particulate agents
from industrial wastes, automobile exhaust, cigarette smoke and
many mining and manufacturing processes (2,3) constitute
the most important sources of exposure.

There is a great need to address the problems caused by
airborne genotoxic agents. Progress in this area has been
hampered by the technical difficulty of working with biologically
hazardous gases and airborne particulates, and by the lack of
suitable models that accurately reflect the biophysics of in vivo
exposure. Unfortunately, most mutagenic and carcinogenic bioassay systems require that the test agent be in liquid form. The
solubilized agent can then be conveniently fed to, injected into
or painted on experimental animals or can be added to the medium
used in in vitro culture systems. Useful data generated by these
systems is difficult to interpret since extracts from gases and
particulates preclude the influence of the various pulmonary
pharmacokinetic responses. Mucociliary clearance, cellular-
specific differences in metabolic activation, particulate inges-
iiuu and clearance by pulmonary macrophages, and plating out of
gases and particulates on nasopharyngeal membranes can all act to
minimize or accentuate the damage caused by the exposure.

51

Studies which use chronic exposure of experimental animals
to airborne agents are preferable in that the whole animal response
is incorporated into the final analysis. Unfortunately, this experimental approach is prohibitively expensive as a routine
screening technique.

In this presentation, we would like to review our experiences
in testing airborne genotoxic agents and will demonstrate our
progression from directly exposing mammalian cells in vitro with
gaseous agents to our current approach of applying short-term in
vitro bioassay methods to lung cells following in vivo exposures.
The current techniques are fast, inexpensive, sensitive and
incorporate the advantages of an in vivo exposure.

METHODS FOR IN VITRO EXPOSURES TO GASEOUS AGENTS

The first procedure is one in which human diploid lung
fibroblasts (4) were exposed directly in vitro to controlled
concentrations of ozone employing a "rocking table" apparatus
(5).

Monolayers of WI-38 cells in their 37th population doubling
were grown in Eagle's diploid medium supplemented with 20% fetal
calf serum, 100 units/mL penicillin, and 50 µg/mL streptomycin.
The cells were harvested with 0.05% trypsin and resuspended in
Eagle's diploid medium at a concentration of 10^5 cells/mL by
Coulter count. Ten mL of the cell suspension were dispensed into
100 mm Petri dishes (Corning Products) and were incubated for 24
hr in a 37°C, 5% CO_2 incubator.

Just prior to ozone exposure, the medium was exchanged with
12 mL of fresh serum-free Eagle's diploid medium. Four dishes
were used per exposure group, beginning with the control (0 ppm
ozone) and followed by a separate set for 0.25, 0.50, 0.75, and
1.00 ppm ozone for 1 hr. The ozone chamber used in the in vitro
study was a modified Forma CO_2 incubator set for 37°C, 5% CO and
96-99% relative humidity (Fig 1). The chamber housed an ozone
generator composed of a recirculating fan and a bank of uv lamps
in a light-tight enclosure. The ozone concentration in the chamber was controlled by feedback from the output signal of a Dasibi
ozone monitor. The autospan on the ozone monitor was set at 5491
to coincide with the calibration setting used by the Air Resources
Board of California. The chamber also housed a motor-driven,
acrylic platform designed to rock on a central axle so that at
the greatest angle of movement the medium in the four Petri
dishes would collect on one side of the dishes, exposing 50% of
the dish floor directly to the atmosphere in the chamber. The
cells were exposed for 1 min after which the platform would tilt
on its axle, and the medium would collect on the opposite side of

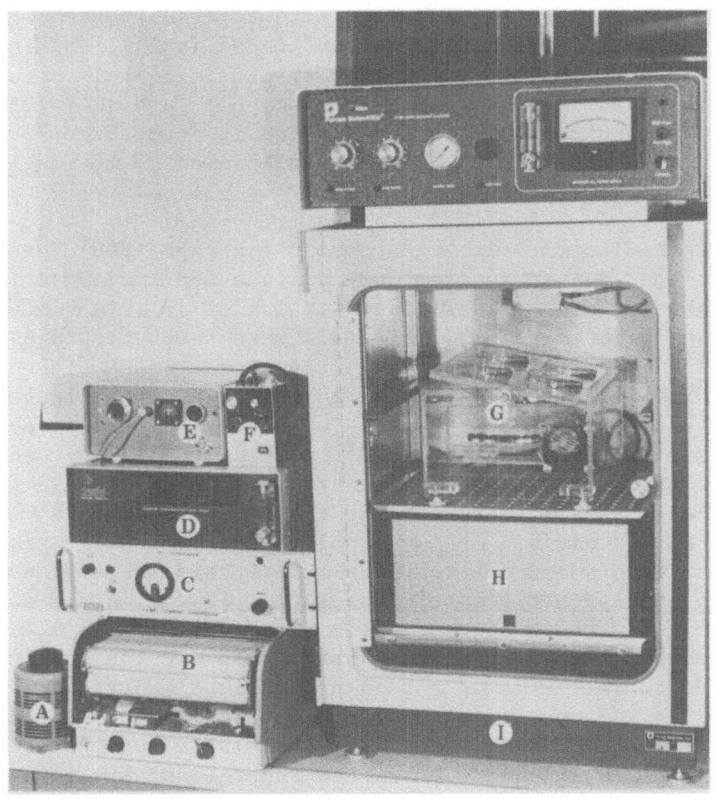

Figure 1. In vitro ozone environmental chamber with ozone generating
and monitoring systems. A. variac control for fan located inside
the ozone generator, fan used to provide uniform mixing of ozone in
the environmental chamber, B. strip chart recorder, C. UV lamp
current controller, D. Dasibi ozone monitor, E. comparator
(compares the ozone monitor's output with the desired ozone level
and sends appropriate correction signals to the lamp current
controller), F. controller for adjusting rocking table speed,
G. acrylic rocking table (the motor driven platform rocks on a
central axis, causing the nutrient medium to collect on one side
of each dish, exposing the cells on the other half of the dish's
floor to the atmosphere in the chamber, H. light tight ozone
generator containing a bank of UV lamps and a fan, I. Forma
autoregulating incubator (temperature at 37°C, relative humidity
between 96-99% and 5% CO).

the dishes, flooding the previously exposed cells with medium
while exposing the previously submerged cells to the ozone.
After treatment in this manner for 1 hr, the medium was aseptically
removed and the cells were harvested with trypsin. Each set of
exposed cells was pooled and then resuspended in 9 mL of fresh
Eagle's diploid medium supplemented with 10% fetal calf serum,
100 units/mL penicillin, 50 µg/mL streptomycin, and 10 µg/mL fungi-
zone. The cell suspension was seeded equally into three 75 cm
plastic cell culture flasks (Corning) containing 20 mL fungizone-
supplemented Eagle's diploid medium.

In preparation for sister chromatid exchange (SCE) analysis,
the flasks were wrapped in aluminum foil and 5-bromodeoxyuridine
(BrdUrd) was added to a final concentration of 10 µg/mL under
subdued lighting conditions. The cells were incubated in a 5% CO_2
incubator set at 37°C for 60 hr and then were treated with 0.05 µg/mL
Colcemid for an additional 4 hr to accumulate mitotic figures.
The medium was removed and the cells were treated for 20 sec with
0.05% trypsin supplemented with 0.05 µg/mL Colcemid and then all
but about 1 mL was removed. The flasks were incubated again
until the cells became free from the floor of the culture vessels.
Ten mL of Eagle's diploid medium supplemented with 0.05 µg/mL
Colcemid was then added to each flask. The cell suspensions were
poured into appropriately labeled conical, plastic, screw-cap
tubes (Corning) without attempting to break up the cell clumps.
The cells were pelleted at 250 x g for 10 min and the supernatant
was discarded. The cells were resuspended in 8 mL pre-warmed
(37°C) 0.075 M KCl containing 0.05 µg/mL Colcemid and were incubated
for 20 min in a 37°C water bath. The cells were sedimented for 10
min at 250 x g and all but 1 mL of the supernatant was removed.
The cell pellet was resuspended with hand agitation. The cells
were gently fixed with freshly prepared Carnoy's fixative (3
parts methanol: 1 part glacial acetic acid) and washed twice
again with Carnoy's. The cell suspensions were concentrated into
a 1 mL volume, and three drops were dropped onto microscope
slides. The chromosome spreads were differentially stained using
a modification of the fluorochrome-plus-Giemsa method of Wolff
and Perry (6) as follows: the slides were stained with Hoechst
33258 (5 µg/mL in Sorenson's buffer) for one-half hour. The
H33258 stained slides were submerged in 1/15 M Sorenson's buffer
(pH 6.8) in a clear pyrex baking dish. The buffer was overlaid
with Saran Wrap (Dow Chemical Co.) to minimize evaporation and
the tray was placed on a clear glass shelf between two banks of
plant lights (Gro and Sho F15T8PL, General Electric Co.) situated
3.8 cm above and below the slides. The fluorochromed cells
were exposed to this light source for 18 hr (overnight) and then
stained with freshly prepared 3% Giemsa in Sorenson's buffer (pH
6.8). SCE's in 30-40 chromosome spreads were scored in each set
using oil immersion bright field optics. The mean and the standard

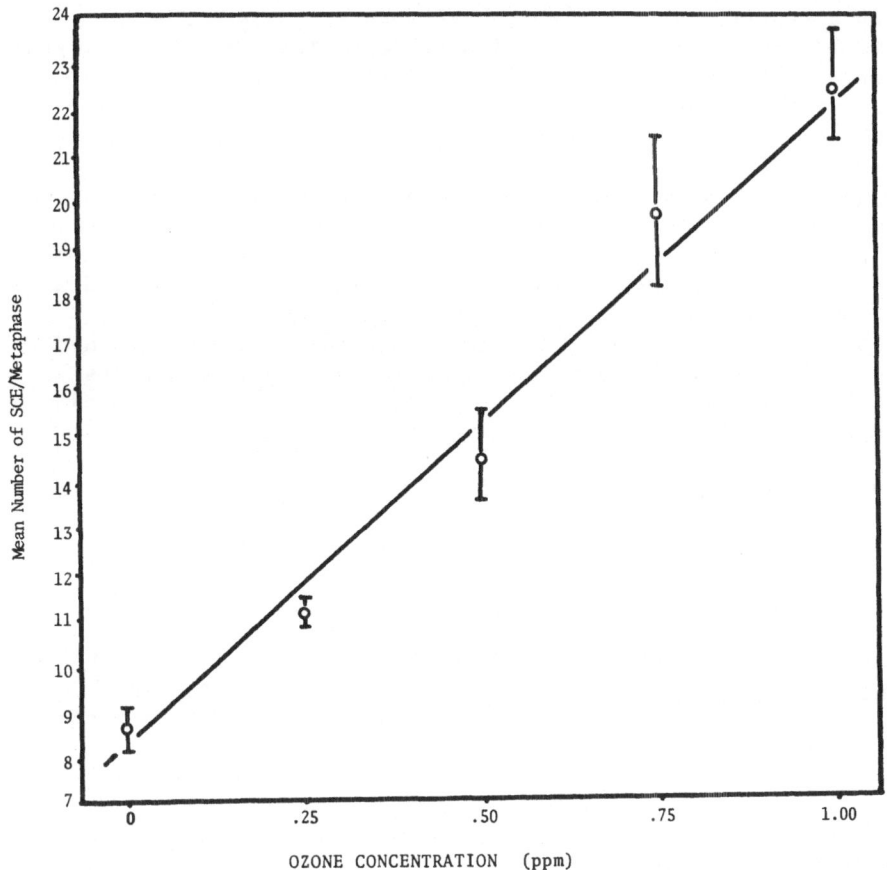

Figure 2. Sister chromatid exchange dose response curve of WI-38 cells following a 1 hr in vitro exposure to increasing concentrations of ozone. Vertical bars represent standard errors.

t-test was used to determine probability (p value). A p value of 0.05 or less was considered significant. An example of the test results is presented as a dose-response curve in Figure 2.

A second procedure, using roller tubes, was particularly useful for measuring the genotoxic effect of methyl bromide (MeBr) on cultures of Chinese hamster ovary (CHO) cells in vitro. These cells were inoculated into 15 cc glass, round-bottom tubes with screw caps at a cell concentration of 3.75 x 10^5 cells per tube. The following day three cultures per gas concentration were transferred to a fume hood and exposed to the gas by directing a steady stream of prepackaged (Matheson Gas Products) MeBr in air

through a sterile pasteur pipette attached to a tygon tube at a
flow rate of 6 L/min under 16 psi pressure. The MeBr concentrations
used were 0, 1 ppm, 6 ppm, 13 ppm and 26 ppm. The tubes were
capped quickly after exposure to trap the gas inside and then
were placed in a roller drum in a 37°C incubator. The tubes were
rotated in the drum at a rate of 30 revolutions per hr. The
cells were exposed to MeBr as they were rotated out of the culture
medium.

After 18 hr of incubation, the tubes were wrapped individually
in aluminum foil and BrdUrd added to a final concentration of
10 µg/mL in subdued light. The tubes were regassed and returned
to the roller drum. The cultures were gassed three more times
during the following 48 hr, during which time the cells entered
the second cell cycle. Colcemid was added (0.05 µg/mL) for a
final two hours, and the mitotic figures were collected in the
medium by shaking the tubes briskly 5 times. The cell suspensions
were then poured into appropriately labeled conical centrifuge
tubes and processed for SCE as previously described.

The mean SCE counts for three runs increased from 15.7±1.1
(S.E.) for the controls to 17.8±0.5 for 1 ppm, 19.9±1.3 for 6
ppm, 22.0±0.3 for 13 ppm and 26.0±0.1 for 26 ppm (Table 1). The SCE
response was dose-responsive as shown graphically in Figure 3.
Based on these data, methyl bromide was concluded to be potentially
mutagenic/carcinogenic.

IN VITRO ANALYSIS FOLLOWING IN VITRO EXPOSURES

Methyl bromide and ozone do not appear to require metabolic
activation to be mutagenic. For such agents, the two systems
just described served to answer the question about mutagenicity
in a positive or negative fashion. In studying polycyclic aromatic
hydrocarbons (PAH), which are primarily combustion by-products of
fossil fuels, the problem is much more complex. Some of these
compounds become biologically active only after metabolic activation by mixed function oxidases present in certain cells. Since
these PAH are most often encountered as gases, aerosols or
absorbed onto carbon exhaust particulates, the various biophysical
and pharmacokinetic systems which operate in an in vivo system
are crucial for accurate assessment of the net mutagenic damage
caused by any particular exposure. We have had success in solving
this problem by using an in vivo exposure coupled with an in
vitro short-term bioassay for the resultant genotoxic response.
The following are four examples of our applications:

Table 1.   The effects of methyl bromide on the frequency
of SCE in CHO cells exposed in vitro

Conc. MeBr(ppm)	# Mitosis Scored/Culture	Total SCE Observed	Average SCE Per Mitosis	Mean SCE Value±SE
0	30	526	17.9	
	30	420	14.0	15.7±1.1
	30	461	15.4	
1	30	517	17.2	
	30	526	17.5	17.8±0.5
	30	564	18.8	
6	30	540	18.0	
	30	579	19.3	19.9±1.3
	30	668	22.3	
13	30	657	21.9	
	30	650	21.7	22.0±0.3
	30	675	22.5	
26	30	783	26.1	
	30	773	25.8	26.0±0.1
	30	784	26.1	

SCE ANALYSIS ON HOST LUNG CELLS GROWN IN PRIMARY CULTURE

Syrian hamsters 4-6 weeks of age, 90-110 gm in weight (Simonsen
Laboratories, Inc.) were exposed chronically to diesel exhaust
or, by intratracheal instillation, to benzo(a)pyrene (BaP),
diesel particulates collected from glass fiber filters or to
carbon particles impregnated with dichloromethane extracts from
diesel exhaust particulates.

Intratracheal Instillation Studies

The intratracheal instillation procedure described by Saffiotti
et al. (7) was used. The animals were anesthetized with an
intraperitoneal injection of 6.7 mL/kg of a 1% solution of sodium
brevitol (Eli Lilly & Company). Each anesthetized animal was
placed with its back on a slanted board. Its mouth was kept open
by catching the lower incisors on a wire loop while the upper
incisors were held by a tight rubber band. All of the solutions
and particulate suspensions were maintained at 37°C in a water

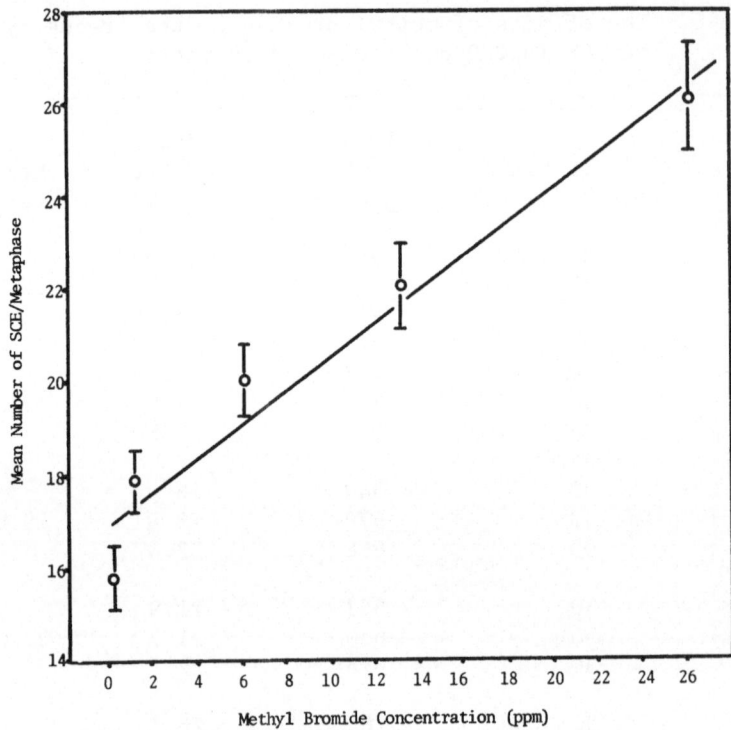

Figure 3. Sister chromatid exchange dose-response curve for CHO cells exposed to methylbromide gas for 48 hrs. Each point represents the mean from three replicate experiments. Thiry mitotic figures were scored from each culture. The bars represent standard errors.

bath. Just prior to instillation, each suspension was vortex-mixed for 5 seconds and drawn several times in-and-out of a 1.0 mL tuberculin syringe fitted with a blunt 19 gauge needle to further insure uniform distribution of the particulates. The syringe was then filled with 0.25 mL of the test suspension. The blunt needle was 60 mm long and bent at a 135° angle 45 mm from the tip. A direct focusing lamp from a dissecting microscope provided a view of the pharynx after the tongue of the hamster was gently pulled outward and laterally with forceps. The tip of the needle was inserted under the epiglottis to expose the vocal cords and then gently inserted into the trachea. The needle was pushed almost to the bottom of the trachea; the suspension gently injected and the needle withdrawn. Following instillation of the suspension into the lungs, the animals showed a brief apnea, followed by rapid recovery of regular respiration.

Twenty-four hours later, the animals were sacrificed by dislocation of the cervical vertebrae. The heart and lungs were exposed, the left atrium was removed and 20 mL Hank's Balanced Salt Solution (HBSS) was flushed through the pulmonary artery to exsanguinate the lungs. The heart and lungs were quickly excised and transferred into sterile 100 mM Petri dishes containing 10 mL HBSS with 2x penicillin/streptomycin antibiotics. The heart, trachea and bronchi were removed and the remaining lung tissue was finely minced with sterile scissors in 3 mL McCoy's 5A medium supplemented with 10% fetal bovine serum (FBS), 100 U/mL penicillin, 100 μg/mL streptomycin, and 24 mM NaHCO$_3$, 3 mM HEPES, 3 mM PIPES, 2.5 mM NaH$_2$PO$_4$ and 3 mM BES buffers. The dishes were allowed to incubate overnight in a 5% CO$_2$ incubator at 37°C. The following day unattached cells and tissue fragments were removed, and the dishes washed in HBSS to remove residual erythrocytes and cellular debris. The remaining viable cells and tissue fragments for each lung were redistributed into 4 Petri dishes in McCoy's 5A medium supplemented as above.

All cultures were incubated in a CO$_2$ incubator at 37°C. The earliest mitotic figures were observed after 60 hr of incubation at which time the medium was changed. Colonies of 50-200 cells were formed in about 5-8 days.

When the cultures showed colonies containing 50 or more cells, the medium was replaced with fresh medium supplemented with 10 μg/mL BrdUrd in subdued illumination. The dishes were wrapped in aluminum foil and incubated in the dark for an additional 60 hr, during which time the cells completed the required two-cell cycles. The cultures were then treated with 0.05 μg/mL Colcemid for 7 hr to block cell division and accumulate as many mitotic figures as possible. The cells were harvested with a rubber policeman and then processed for SCE as previously described.

The mean SCE values for animals treated with BaP-hematite suspensions are shown in Tables 2 and 3 and in Figure 4. The mean control SCE value for six animals was 11.2 ± 0.6 (S.E. between animals) and for four animals treated with 12.5 μg BaP for 24 hours was 19.5 ± 0.6. The mean SCE value for five animals given weekly instillations of 0.2 μg BaP for ten weeks was 17.2 ± 0.5 and 12.5 ± 0.5 for the control.

The intratracheal instillation technique also worked well with diesel exhaust particulates (DEP). For example, control animals treated with the carrier HBSS showed a mean SCE value of 11.4 ± 1.0 while animals exposed to DEP showed the following definitive dose responsive increase in SCE: 16.7 ± 0.6 for 6 mg/animal, 19.8 ± 1.0 for 11 mg/animal and 26.4 ± 2.5 for 20 mg/animal (Fig 5). Activated carbon containing the dichloromethane extract from DEP

Table 2.   SCE analysis of syrian hamster primary lung cell
            cultures following intratracheal instillation of
            0 or A single dose at 12.5 µg benzo(a)pyrene.*

B(a)P:FeO Conc.(µg)	# Mitosis Scored/Animal	Total SCE Observed	Average SCE Value Per Animal	Mean SCE Value±SE
0:12.5	16	162	10.1	11.2±0.6
	30	393	13.1	
	10	111	11.1	
	27	333	12.3	
	25	221	8.8	
	16	182	11.4	
12.5:12.5	17	306	18.0	19.5±0.6
	10	199	19.9	
	14	271	19.4	
	16	331	20.7	

* Exposures were for 24 hours. Iron oxide was used as the carrier
for the B(a)P in a 50:50 by weight ratio. Control animals were
treated with iron oxide only.

Table 3.   SCE analysis of syrian hamster primary lung cell
            cultures following intratracheal instillation of
            0 or 10 doses of 0.2 µg benzo(a)pyrene.*

B(a)P:FeO Conc.(µg)	# Mitosis Scored/Animal	Total SCE Observed	Average SCE Value Per Animal	Mean SCE Value SE
0:12.5	30	411	13.7	12.5±0.5
	30	361	12.0	
	20	238	11.9	
	32	381	11.9	
	26	289	11.1	
	30	423	14.1	
12.5:12.5	30	489	16.3	17.2±0.5
	30	494	16.5	
	30	575	19.2	
	10	168	16.8	
	10	174	17.4	

* Each exposure was one week apart. Iron oxide was used as the
carrier for the B(a)P in a 50:50 by weight ratio. Control
animals were treated with iron oxide only.

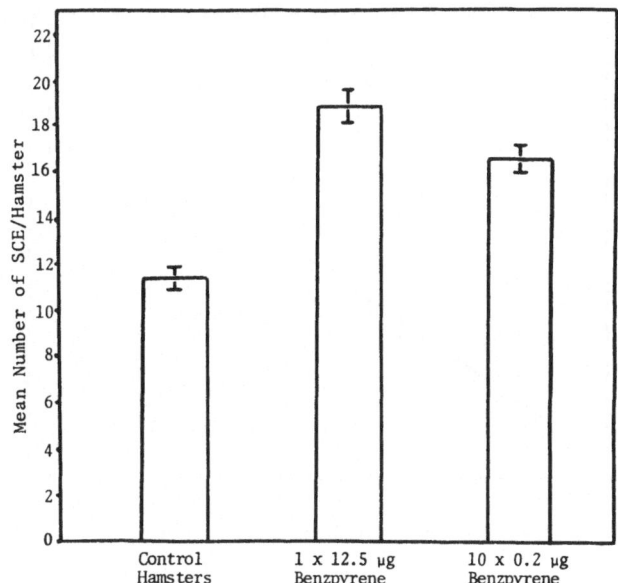

Figure 4. Histogram of the SCE frequency in control and BaP treated hamster lung tissues. The height of the bar represents the average value for each set. The bars represent standard errors.

also produced a definitive dose responsive increase in SCE. For example, the carbon controls gave 10.5±0.3 SCE while those treated with the extract gave 14.9±0.9 for 0.3 mg/animals, 16.2±0.3 for. 0.6 mg/animal and 19.1±0.4 for 1.2 mg/animal diesel extract (Table 4, Fig 6).

These data demonstrate both the feasibility of the procedure and in vivo metabolic activation of the procarcinogen.

Chronic Exposure Studies

For chronic exposure studies, the Cincinnati, Ohio branch of the EPA exposed young male Syrian hamsters in environmental chambers to either clean air, 6 mg/m^3 DEP or 12 mg/m^3 DEP for 8 hr/day, 7 days/week for 3 months. SCE analysis of the cultured lung cells from these animals provided the data shown in Table 5 and Figure 7. Eleven control animals showed a mean SCE value of 11.5±0.4. Similarly, eight hamsters exposed to 6 mg/m^3 DEP showed 11.9±0.5 SCE per cell. In contrast, four animals exposed to 12 mg/m^3 for 3 months showed a mean SCE value of 20.2±0.7. This study, although not yet complete, illustrates an in vivo exposure to an airborne genotoxic agent, followed by analysis of the cells in vitro.

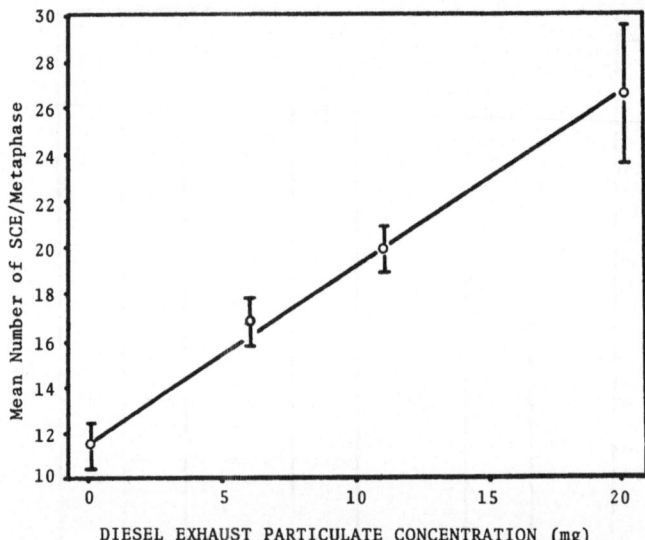

Figure 5. Dose-response curve of the SCE frequency resulting from diesel exhaust particulates administered intratracheally over a range from 0 to 20 mg per hamster. The exposure was for 24 hrs. Control animals received HBSS only. Circles represent the mean SCE values. Vertical bars represent standard errors.

A HOST-MEDIATED ASSAY FOR AIRBORNE AGENTS

Although the analysis of primary cultures of treated hamster lung cultures offers many advantages in assessing airborne genotoxic agents, it has the disadvantage of being technically difficult. Therefore, we have modified the procedure by introducing a suspension of Chinese hamster ovary (CHO) cells into the airways of Syrian hamster lungs through intratracheal instillation (8). These tester cells mirror the genotoxic response of the host tissue, since the same physiological activation systems and defense mechanisms affect all cells in the airways. The following is an example of how the system has been used to test this hypothesis: anesthetized Syrian hamsters were instilled with HBSS containing 3×10^6 CHO cells mixed with BaP: hematite over a concentration range of 0 to 20 ng BaP/animal. Scanning electromicroscopy of fixed lung tissue demonstrated that most of the CHO cells reached the terminal bronchioles (Fig 8) while some even reached the alveolar ducts and alveoli. After 4 hr of exposure in vivo the CHO cells were harvested by pulmonary lavage, washed with HBSS and incubated in roller tubes overnight in McCoy's 5A medium supplemented with 10% fetal bovine serum, 100 U/mL penicillin and 100 μg/mL streptomycin. The CHO cells were then incubated for

Table 4. SCE analysis of syrian hamster primary lung cell cultures following intratracheal instillation of activated carbon particles containing increasing concentrations of dichloromethane extract from diesel exhaust particulates.

Extract* Conc. (mg/Hamster)	# Mitosis Scored/Animal	Total SCE Observed	Average SCE Value Per Animal	Mean SCE Value SE
0	29	284	9.8	
	26	315	12.1	
	20	169	8.5	10.5 0.3
	10	106	10.6	
	16	182	11.4	
0.3	14	243	17.4	
	3**	34	11.3	14.9 0.9
	20	268	13.4	
	10	140	14.0	
	0**	–	–	
0.6	20	333	16.6	
	15	239	15.9	16.2 0.3
	13	217	16.7	
	13	203	15.6	
	0**	–	–	
1.2	20	359	18.0	
	5**	75	15.0	19.1 0.4
	20	396	19.8	
	22	429	19.5	
	12	230	19.2	

* The Exposures were for 24 hours. Control animals received activated carbon only.

** Poor yield of mitotic figures, therefore, excluded from the analysis.

Table 5.　SCE analysis of syrian hamster primary lung cells
following exposure to 0, 6 and 12mg diesel exhaust
particulates/m^3 exhaust gas for 6 hours/day,
7 days/week for 3 months.

Exposure*	# Mitosis Scored/Animal	Total SCE Observed	Average SCE Value Per Animal	Mean SCE Value±SE
Control	70	867	12.4	11.5±0.4
	40	512	12.8	
	36	446	12.4	
	8	106	13.2	
	17	164	9.6	
	24	236	9.8	
	23	332	12.3	
	24	232	11.0	
	16	162	10.1	
	18	111	11.1	
	17	393	13.1	
6 mg/m^3	26	289	11.1	11.9±0.5
	30	366	12.2	
	20	248	12.4	
	32	381	11.9	
	25	250	10.0	
	30	348	11.6	
	34	503	14.8	
	30	327	10.9	
12 mg/m^3	27	593	22.0	20.2±0.7
	13	239	18.4	
	14	280	20.0	
	6	126	20.2	

* Control animals received filtered air.

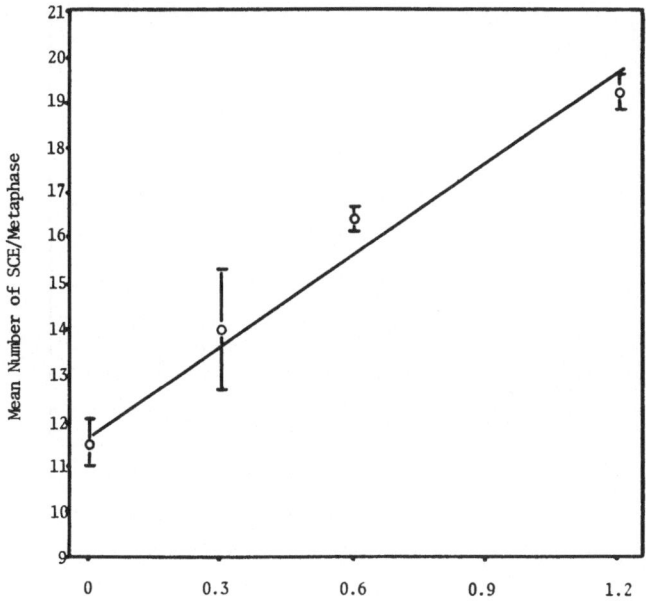

Dichloromethane Extract Concentration/Hamster (mg)

Figure 6. Dose response curve of the SCE frequency resulting
from intratracheal instillation of a dichloromethan extract
from diesel exhaust particulates. The extract was absorbed
onto activated carbon. The control animals recieved activated
carbon only. Each data point represents the mean from 5
animals and the bars represent standard errors.

two-cell cycles in medium supplemented with 10 µg/mL BrdUrd. The
dividing cells were selectively harvested using the shake-off
procedure and prepared for SCE analysis as previously described.

     The mean frequency of SCE found in the control population of
CHO cells was 12.2±1.1 (Fig 9). CHO cells from two animals
treated with 5 ng BaP each yielded 16.9±1.0 and 17.7±1.3 SCE,
while two treated with 10 ng BaP/animal showed mean SCE values of
18.4±1.3 and 20.3±1.0. CHO cells from animals treated with 20 ng
BaP each showed SCE values of 23.9±1.6 and 23.1±1.3.

     Several conclusions can be drawn from these data. Firstly,
morphological evidence (Fig 8) suggests that CHO cells concentrate
in the terminal bronchioles, probably due to the restricted size
of the airways at that site and because mucociliary clearance

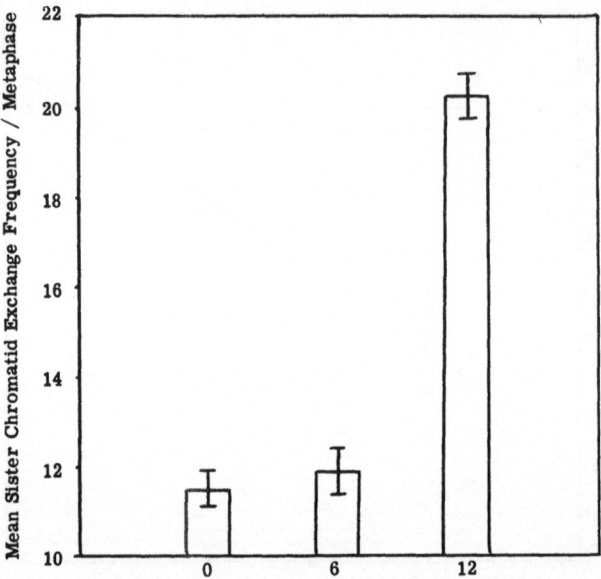

Figure 7. Histogram of the SCE frequency in cells from
control animals and experimental animals exposed to diesel
exhaust for 6 hrs./day, 7 days/week for 3 months. At least
4 animals are represented at each concentration and an
average of 25 mitotic figures were scored for each animal.
Bars represent standard errors.

eliminates the cells from the tracheal and bronchial surfaces.
Secondly, since BaP is a procarcinogen and must be metabolically
activated to become an ultimate carcinogen (9–12), the linear SCE
dose-response curve (Fig 9) confirms that localized lung tissues
contain the mixed function oxidases necessary to activate BaP in
vivo (13). Thirdly, these data illustrate that this technique is
reproducible and is sensitive to nanogram quantitates of BaP.

CHO cells constitute an easily-cultured cell line with a
fast (12 hr) doubling time, while the size of the chromosomal
complement (20–22 chromosomes) makes for easy scoring of SCE. The
shake-off procedure provides large numbers of mitotic figures which
reduces the search time during the SCE scoring procedure. These
features minimize the time, effort and cost required to perform
SCE analysis, at least for acute in vivo exposures to genotoxic agents.

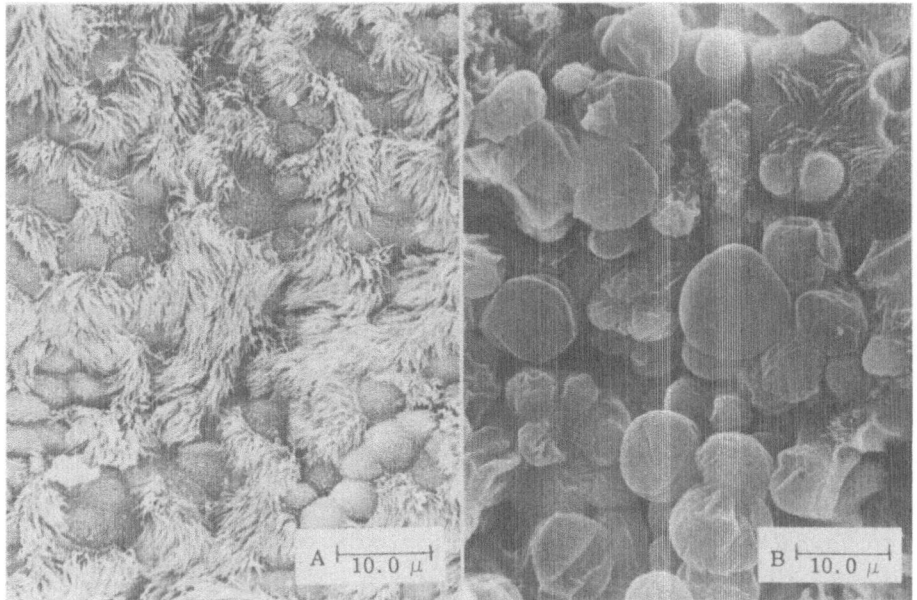

Figure 8. Scanning electronmicrographs of terminal bronchioles
in hamster lung. (a) Untreated lung showing cilia and Clara
cells. (b) Chinese hamster ovary cells adhering to the
surface of a terminal bronchiole in a hamster lung instilled
with 10^6 CHO cells. Note the comparatively large CHO cells
and the underlying smaller Clara cells.

HUMAN DIPLOID CELLS GROWN IN HUMAN URINE

The fourth technique involves growing of human diploid cells
in medium containing urine from human subjects exposed to airborne
genotoxic agents (14). SCE analysis on the cells was used to
assay for the presence of mutagenic metabolites excreted in the
urine. Urine specimens from nine smokers and seven non-smokers
were collected in late afternoon. The smokers smoked from 30-40
cigarettes per day and all smoked brands rated at 17-19 mg tar
per cigarette. A 50 mL aliquot of each urine specimen was centrifuged for 10 min at 5000 x g and the supernates sequentially passed
through Millipore filters with pore diameters of 0.80 µM, 0.45 µM
and 0.22 µM. The urine was buffered with 1.5 mg $NaHCO_3$/mL and the
pH adjusted to 7.2. The osmolality of the urine was measured
with a Wescor 5100 B vapor pressure osmometer and adjusted to
$300 \pm 4m$ osmol/kg with distilled water. The urine samples were
supplemented with 10% FBS, 100 U/mL penicillin and 100 µg/mL

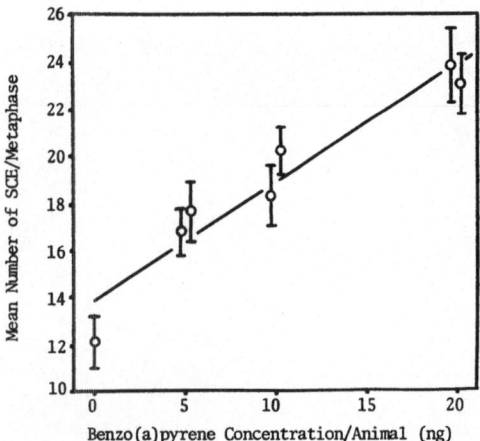

Figure 9.   Sister chromatid exchange dose-response curve
representing duplicate runs in which CHO cells instilled
intratracheally into hamster lungs, were exposed in vivo
to intratracheally instilled BaP, recovered by pulmonary
lavage and processed for SCE in vitro.   Iron oxide was
used as the carrier for the B(a)P in a 50:50 by weight
ratio.   Vertical bars represent standard errors.

streptomycin, and then pressure filtered through sterile 0.22 μM
pore size Millipore membrane filters.

Disposable T75 flasks (Corning) were inoculated with 7.5x10
WI-38 cells/flask in 20 mL Eagle's diploid medium (Grand Island
Biological Company) supplemented with 10% FBS, 100 U/mL penicillin
and 100 μg/mL streptomycin.   The cells were incubated in a 37°C
incubator for 18 hours to allow the cells to attach and spread,
then a volume of medium equal to from 5-20% was removed from each
flask and replaced with an equal volume of conditioned urine.
The flasks were wrapped in aluminum foil to exclude light, and
BrdUrd was added to a final concentration of 10 μg/mL.   The cells
were reincubated for 48 hours and then were processed for SCE as
previously described.   Figure 10 shows that the number of SCE
from smokers increased in proportion to the amount of urine
present with a much steeper slope than for the controls (2.67:1).
Figure 11 demonstrates that the system is quite reproducible; in
the first run the average SCE value was 9.9±0.2 for non-smokers
and 14.7±0.3 for smokers, while on a second run on the same people
the averages were 9.6±0.3 SCE for non-smokers and 14.2±0.2 for
smokers.   The SCE response represents the net mutagenic activity
of a complex mixture of products some of which have been activated
and others inactivated.   In the presence of a genotoxic airborne
agent, such as cigarette smoke, the mutagenic response is measurable

over the background level presumably derived from normally encountered mutagenic contaminants in such things as drinking water, food and air.

DISCUSSION:

Exposing mammalian cells in vitro directly to extracts from aerosols, particulates or gases has the advantage of convenience and enables one to eliminate any toxic or neutralizing constituent which may interfere with the measurement of the mutagenic response. While such data are important, the approach does not provide enough information for accurately assessing genetic risk. The ozone and methyl bromide data, for example, demonstrate clearly that both gases are mutagenic to mammalian cells exposed in

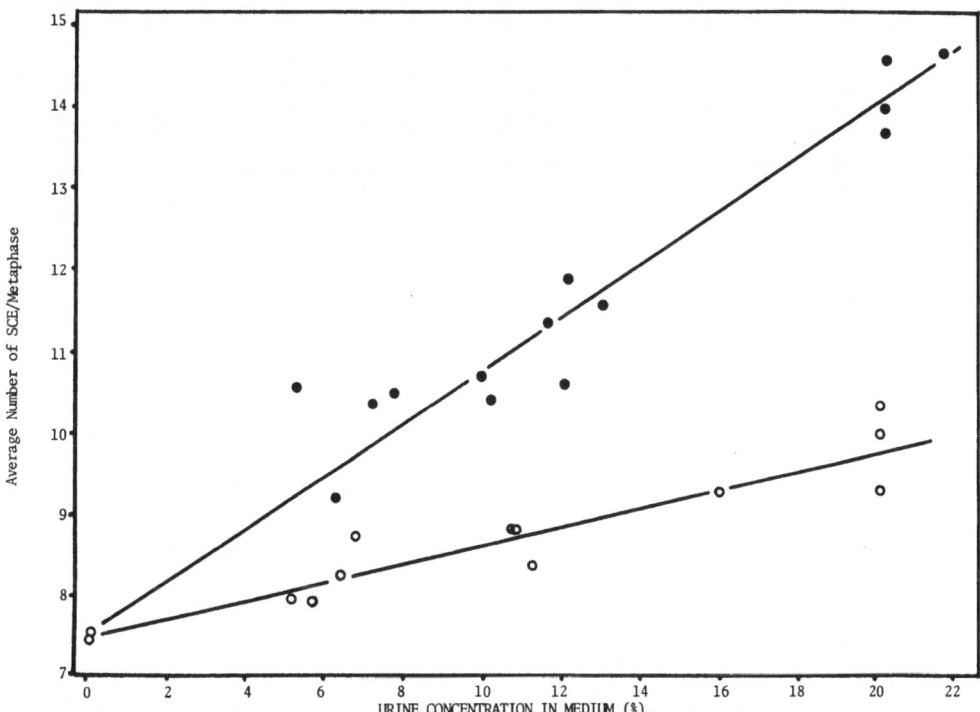

Figure 10. Sister chromatid exchange dose-response curves for WI-38 cells treated for 48 hours with urine-medium mixtures. The closed circles represent specimens derived from smokers and the open circles represent urine specimens from non-smokers. Each point represents the average from 30 metaphases.

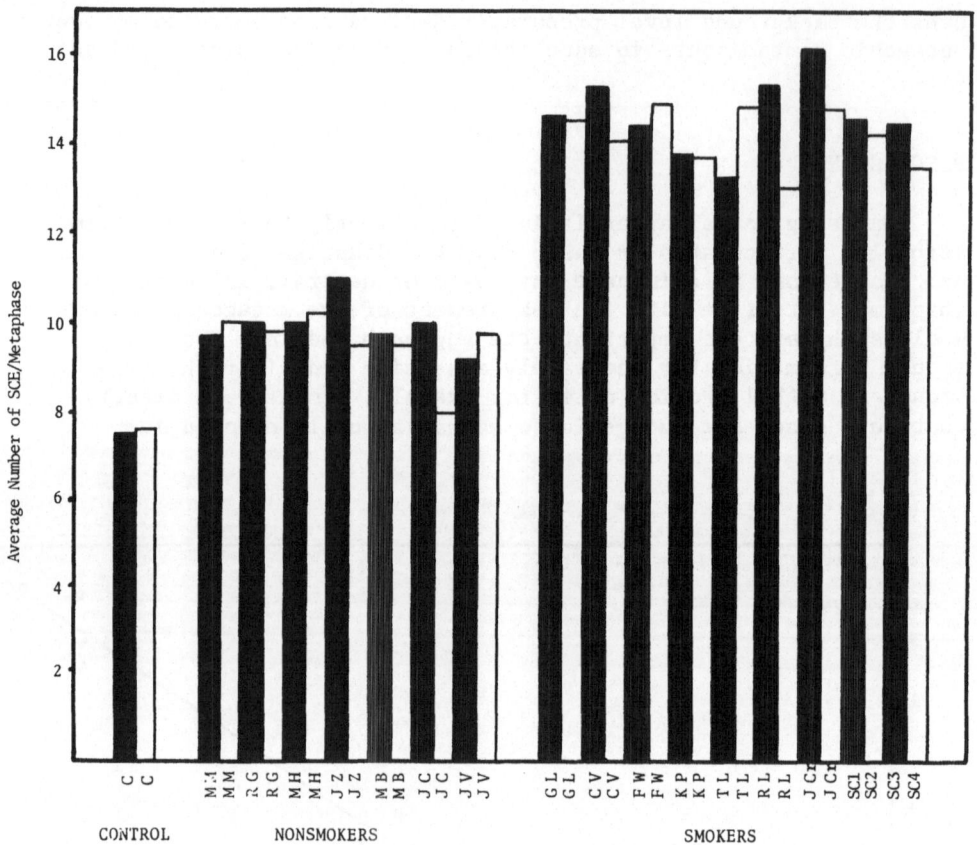

Figure 11. Average number of sister chromatid exchanges per
metaphase for WI-38 cells treated with 20% urine from 7 nonsmokers
and 8 smokers. The dark bars show the results of experiment 1
and the open bars represent the results of a replicate experiment
2. Standard errors were less than 8% from the mean for each
analysis.

vitro. The question still remains, however, as to how much of a
genetic risk these gases constitute in the in vivo situation
where all the various biophysical and pharmacokinetic mechanisms
of activation, inactivation and clearance inherent to the whole
animal come into play. The in vivo exposure to diesel exhaust is
a good example; a dichloromethane extract of the diesel particulate
was mutagenic to mammalian cells in vitro, while lung cells from

animals exposed in vivo for three months (6 mg particulates/m^3 diesel exhaust/6hr/day) showed no detectable mutagenic response as measured by SCE analysis. Presumably, clearance, metabolism and excretion contributed to this difference. The animal tissues did show a mutagenic response when the exposure was increased to 12 mg/m^3 for the three months.

The CHO cell instillation assay is a variation on the same basic concept of in vivo exposure followed by in vitro analysis. The instillation of CHO cells has the advantage of using a fast dividing cell line which, when instilled in the animals and then harvested by pulmonary lavage, reflects in vitro the net effect of the in vivo exposure. This approach provides an extremely fast and inexpensive analysis of the in vivo exposure since the CHO cells require only 24 hours for processing after being recovered from the lung, as opposed to the approximately two weeks processing time needed for the pulmonary cells from primary cultures. The disadvantage is that progressively fewer CHO cells are recoverable over time and virtually none can be recovered after three days in the animal. Consequently, this approach is valuable for studying acute exposures but not chronic exposures.

The urine assay, on the other hand, is valuable for both acute and chronic studies. This system has some distinct advantages over the other in vivo systems presented here in that it is totally human and, therefore, reflects the human response. In addition, because it is a non-invasive procedure, the same subjects can serve as their own control and can be tested repeatedly. The test can be used to monitor the buildup or elimination of genotoxic substances in individuals over time. For example, epidemiologic studies have identified numerous biologically hazardous working conditions (15). Such studies, however, cannot evaluate the variation in individual responses caused by differences in the level of exposure and inherent differences in detoxification capabilities. This system allows the collection of specimens from the same individuals at minimum and maximum occupational exposure to genotoxic substances, and measure the specific mutagenic response to each exposure. Such information may provide a means to best assess the mutagenic risk for each individual.

It is imperative that new and better assay systems be developed to measure the mutagenic properties of the many airborne genotoxic agents present in our environment. Such evaluation should translate ultimately into genetic risk assessment. Since in vivo exposures for the lifetime of the animal are impractical due to the length of time required for evaluation, the number of compounds needing testing and the overriding cost, the techniques described herein provide reasonable alternatives and in some ways definite advantages over strictly in vitro or in vivo approaches.

REFERENCES:

1. E. S. Pollack and J. W. Horn, Trends in Cancer Incidence
    and Mortality in the United States, 1969–76, Proc. Natl.
    Acad. Sci. 64(5):1091–1103 (1980).
2. S. S. Epstein and J. B. Swartz, Fallacies of Lifestyle
    Cancer Theories, Nature 289:127–130 (1981).
3. D. L. Davis and B. H. Magee, Cancer and Industrial Chemical
    Production, Science 206:1356–1358 (1979).
4. L. Hayfick and P. S. Moorhead, The Serial Cultivation of
    Human Diploid Cell Strains, Exp. Cell Res. 25:585–621 (1961).
5. R. R. Guerrero, D. E. Rounds, R. S. Olson and J. D. Hackney,
    Mutagenic Effects of Ozone on Human Cells Exposed in
    vivo and in vitro Based on Sister Chromatid Exchange
    Analysis, Environ. Res. 18:336–346 (1978).
6. S. Wolff and P. Perry, Differential Giemsa Staining of
    Sister Chromatids and the Study of Sister Chromatid
    Exchanges without Autoradiography, Chromosoma 48:341–
    353 (1974).
7. U. Saffiotti, F. Cefis and L. H. Kolb, A Method for the
    Experimental Induction of Bronchogenic Carcinoma,
    Cancer Res. 28:104–124 (1968).
8. R. R. Guerrero, D. E. Rounds and K. S. Narayan, The Use
    of CHO Cells in Intact Hamster Lungs to Test the Effect
    of In Vivo Application of Airborne Mutagens, Presented
    at the EPA Second Symposium on the Analysis of Complex
    Environment Mixtures, Williamsburg, Virginia (1980).
9. C. C. Harris, I. C. Hsu, G. D. Stoner, B. F. Trump and
    J. K. Selkirk, Human Pulmonary Alveolar Macrophages
    Metabolize Benzo(a)pyrene to Proximate and Ultimate
    Mutagens, Nature 272:633–634 (1978).
10. I. C. Hsu, G. D. Stoner, H. Autrup, B. F. Trump, J. K.
    Selkirk and C. C. Harris, Human Bronchus-Mediated
    Mutagenesis of Mamalian Cells by Carcinogenic Poly-
    nuclear Aromatic Hydrocarbons, Proc. Natl. Acad. Sci.
    (USA) 75(4):2003–2007 (1978).
11. J. M. Mass and D. G. Kaufman, [^{3}H] Benzo(a)pyrene Metabolism
    in Tracheal Epithelium Microsomes and Tracheal Organ
    Cultures, Cancer Res. 38:3861–3866 (1978).
12. S. K. Yang, H. V. Gelboin, B. F. Trump, H. Autrup and
    C. C. Harris, Metabolic Activation of Benzo(a)pyrene
    and Binding to DNA in Cultured Human Bronchus, Cancer
    Res. 37(4):1210–1215 (1977).
13. C. C. Harris, H. Autrup, R. Connor, L. A. Barrett,
    E. M. McDowell and B. F. Trump, Interindividual
    Variation in Binding of Benzo(a)pyrene to DNA in
    Cultured Human Bronchi, Science 194:1067–1069 (1976).

14. R. R. Guerrero, D. E. Rounds and T. C. Hall, Bioassay
        Procedure for the Detection of Mutagenic Metabolites
        in Human Urine with Use of Sister Chromatid Exchange
        Analysis, J. Natl. Cancer Inst. 62(4):805-809 (1979).
15. P. Cole, R. Hoover and G. H. Friedell, Occupation and
        Cancer of the Lower Urinary Tract, Cancer 29:1250-1260
        (1972).

DISCUSSION:

Q. Sivak (A. D. Little, Inc): The lung is an exquisitely
complex tissue. In your hamster lung-mincing technique, is
there any way to identify in which cells the sister chromatids
are occurring? One might expect that the frequency of SCE
would be higher in fibroblast cells derived from the lining,
since that's where most of the cancers occur, rather than in
parenchymal cells or in cells derived from supporting tissue.

A. Guerrero (Pasadena): That's right. There are ways but
we haven't tried them. You can separate specific cell types
using various kinds of gradients or cell sorters. We've just
been looking at an overal response using, at this point,
relatively crude approaches.

Q. Height (Aberdeen): I was wondering at what times post-exposure
did you sacrifice the animals and remove the lungs in the hamster
lung-mincing experiments?

A. Guerrero (Pasadena): The animals were killed one day after
the particulate instillation.

Q. Height (Aberdeen): I would imagine that the nonspecific
response in the lungs would be different at different times post-
exposure.

A. Guerrero (Pasadena): It's very possible. We haven't examined
that.

Q. Bernstein (Brookhaven): In many of your particulate exposures,
you showed dose response curves as a function of agent concentration.
Have you looked at the effect of control particles--let's say
latex spheres--to see if you get a similar response simply as a
function of the number of particles.

A. Guerrero (Pasadena): We have, with activated carbon, no
SCE were induced.

METHODS FOR DETECTING GASEOUS AND VOLATILE CARCINOGENS USING

CELL TRANSFORMATION ASSAYS

G. G. Hatch, P. D. Mamay, M. L. Ayer, B. C. Casto and
S. Nesnow*

Northrop Services, Ine.
Research Triangle Park, NC 27709 (U.S.A.)

INTRODUCTION

In order to prevent in humans cancers as a result of exposure
to environmental carcinogens, suitable methodologies and bioassays
must be developed to detect all classes of carcinogenic agents.
Quantitative cell culture systems provide procedures for detecting
potential carcinogens in a relatively short period of time compared
with the years required to complete in vivo tests. Mammalian cells
in culture are particularly appropriate because they represent an
extension of the in vivo experimental models, eliminate host
factors, afford a means to control test chemical levels, and can
be conducted in a controlled environment. Quantitative linear
dose-response relationships have been demonstrated for cell transformation using fibroblasts from hamster embryo cells (1-4). The
transformation event data are consistent with a one-hit phenomenon
indicating a direct cause and effect relationship (2,3,5).

A short-term in vitro assay for carcinogens and mutagens has
been developed that determines the ability of various chemical
carcingens to enhance the adenovirus transformation of hamster
embryo cells (6-10). The viral enhancement assay reflects the
capacity of a chemical to damage cell DNA by either direct or
indirect means, and provides a rapid mammalian cell bioassay
system for screening of suspect genotoxic environmental chemicals.
This bioassay has detected a large number of compounds from a
variety of chemical classes including alcohols and phenols, aliphatic
amines, alkyl sulfates and sulfones, aromatic amines, aryl halides,

*Genetic Toxicology Division, U. S. Environmental Protection Agency,
 Research Triangle Park, NC 27709 (U.S.A.)

carbohydrates and derivatives, hydroxylamines, mycotoxins, polycyclic hydrocarbons, hydrazines, metal salts, steroid hormones, and
chlorinated hydrocarbons (10,11). The last four classes are generally
negative in standard bacterial, e.g. Salmonella, assays (12, 13).
The majority of agents examined include: a) those chemicals more
commonly used in other short-term in vitro assays, b) those used
in various industrial applications, c) chemicals produced annually
in large volumes, d) inorganic metal salts, and e) complex environmental mixtures including diesel emissions, coke oven tars, fossil
fuel emissions, and cigarette smoke condensates (14). Data from
approximately 105 test performances in 29 different chemical classses
have been published using the enhancement of viral transformation
assay (10,11). Unpublished or preliminary data exist for approximately 100 other chemicals. Using the enhancement of adenovirus
transformation as an index of carcinogenicity, 94% of 130 chemicals
with known positive or negative activity agreed with their current
in vivo classification.

Several factors make this assay an ideal candidate for
evaluating potentially hazardous compounds. These include the
relative ease of performance, sensitivity of the system, objective
scoring of foci, quantitation, and the existence of a large data
base for a wide variety of compounds and complex mixtures.

Volatile or gaseous chlorinated compounds are ubiquitous
environmental contaminants. Many of them have annual production
volumes in excess of billions of pounds per year in the U.S.
Small amounts can be detected in municipal drinking water supplies,
some are used in certain food extraction processes, and there is
widespread use in industrial workplaces. Some have been identified
as tumorigenic in rodents. Previous attempts using viral enhancement assays to evaluate the genotoxic potential of several volatile
chlorinated organic compounds by incorporation into liquid cell
culture media were unsuccessful or produced only a weak or
inconsistent enhancement response.

This report documents the successful development and implementation of methods for treating mammalian cell cultures with
volatile organic liquids and gaseous compounds. Treatment effects
were determined by measuring the enhancement of DNA viral transformation in hamster embryo cells. The compounds tested include:
1,1,1-trichloroethane (TCE), dichloromethane (DCM), 1,2-dichloro-
ethane (DCE), and chloromethane (CM).

MATERIALS AND METHODS

Cell cultures. Primary hamster embryo cells (HEC) were
prepared by trypsinization of eviscerated and decapitated embryos
(Charles River Laboratories, Boston, MA) after 13 to 14 days of

gestation. Cells were resuspended in modified Dulbecco's medium (MDM) (6) supplemented with 10% heat-inactivated fetal bovine serum (FBS) (Sterile Systems, Inc., Logan, UT) and $NaHCO_3$ (0.22 g/100 mL). Approximately 5×10^6 cells in 5 mL of medium were plated into 60-mm plastic Petri dishes and incubated in a 5% CO_2 atmosphere at 37°C. Total cell counts after three days ranged from 4 to 6×10^6 cells/plate.

Continuous lines of Vero monkey kidney cells (ATCC:CC1-81) were subcultured weekly in 100-mm plastic Petri dishes. Growth medium consisted of Eagle's minimum essential medium with 10% heat-inactivated FBS.

Virus. Simian adenovirus SA7 was inoculuated onto Vero cell cultures in 100-mm dishes at an input multiplicity of two to three plaque-forming units per cell. After the virus had been adsorbed for 2 hr, 5 mL of medium were added to each plate. Cytopathic effects were usually complete by 72 hr, after which the infected cells were harvested and then frozen-thawed four times. Finally, the virus was separated from cell debris by low-speed centrifugation. Virus stocks for transformation assays were stored in 1-or 2-mL aliquots at -65°C.

Transformation Assays. Adenovirus transformation procedures have been presented in detail elsewhere (6,15). Briefly, the procedure was as follows. SA7 was added to HEC (3 to 4×10^7 plaque-forming units per culture) and adsorbed for 3 hr; the virus-inoculated cells were removed with trypsin (0.25% trypsin in MDM with 0.1 mM $CaCl_2$), centrifuged, and resuspended to 10^6 cells/mL in MDM with 10% FBS and $NaHCO_3$ (0.11 g/100 mL). The cells were mixed vigorously and plated into 60-mm dishes using 2×10^5 cells/dish; 3 mL of the above medium were then added to each plate. After incubation for 3 days, the medium was changed to MDM with 0.1 mM CaCl (15,16), 10% FBS, and $NaHCO_3$ (0.22 g/100 mL). After six days, transformation assay plates were overlaid with 3 mL of the above medium containing Bacto-agar (0.3 g/100 mL). At intervals of four, five and six days, 3 mL of additional agar medium were added. Final focus counts were made 25 to 30 days from the beginning of the experiment.

Survival Assays. After resuspension of the HEC to 10^6 cells/mL (see "Transformation Assays"), the cells were diluted 1:300 in medium to give 333 cells/0.1 mL. Two-tenths mL (666 cells) was then added to each of five plastic dishes, followed by 3 mL of MDM with 10% FBS and $NaHCO_3$ (0.11 g/100 mL). Five to six days later, 3 mL of medium with double the amount of $NaHCO_3$ were added to each plate, and after eight to nine days of total incubation, the cell colonies were fixed in 10% buffered formalin and stained with 0.02% crystal violet. The cloning efficiency of virus-inoculated cells under these conditions was usually 8% to 12%.

Determination of Enhancement. The fraction of cells surviving
treatment was determined from assay plates which were seeded with
666 cells/plate. The number of colonies in five plates seeded with
treated cells was divided by the number of colonies in five plates
with control cells to give the surviving fraction of treated cells.

The total number of SA7 foci in five control plates, each
receiving 2×10^5 cells, was used as the frequency of transformation
per 10^6 virus-inoculated control cells. The transformation frequency (TF) of treated cells (number of SA7 foci per 10^6 surviving
cells) was calculated by multiplying the number of SA7 foci from
five plates by the reciprocal of the surviving fraction of treated
cells. Detailed methods for determining the frequency of transformation and enhancement ratios using data from actual enhancement
experiments have been published (8,17). Similar methods for calculation have been used for the determination of mutation frequency
and for the demonstration of enhancement of SV40 transformation by
DNA base analogs (18) in mammalian cells.

Enhancement was expressed as the ratio between the TF of treated,
surviving cells and the TF of control cells. Statistical significance was determined using a table of ratios (9) derived from the
Lorenz table (19) that is based upon the Poisson distribution. The
increased TF was considered statistically significant at the 5% or
1% confidence level if the enhancement ratio exceeded the appropriate
value obtained from the Lorenz table.

EXPOSURE TO GASES AND VOLATILE COMPOUNDS

Treatment of Cells in Modular Chambers. Polycarbonate treatment chambers of 4.6-liter volume (Billups-Rothenberg, Flow Labs,
McLean, VA) are equipped with inlet and outlet ports for easy
introduction and purging of test gases. The chambers could be
opened for insertion of dishes of cultured cells and then resealed
with spring-steel clamps. After exposure, the test gases or vapors
were exhausted through charcoal under negative pressure. Before
reuse of test chambers, any residual chemical was removed by heating
at 100°C under negative pressure. Individual chambers were used
for exposure to each concentration of volatile organic chemical or
gas.

Cells were cultured in 60-mm glass (Corning Pyrex) Petri
dishes. Fresh medium was added to the cells 8 hr to 24 hr prior to
treatment. This medium was replaced with 1 mL of medium supplemented with 7.5 mM HEPES (N-2-hydroxyethyl piperazine) buffer in
place of sodium bicarbonate. Ten milliliters of sterile water were
added to the bottom of each chamber to ensure adequate humidification
and to prevent desiccation of the small volume of cell culture
medium. Following introduction of test chemicals, the sealed

chambers and cells were placed onto a rocker platform at 37°C. The chambers were tilted to provide a sufficient inclination (10°) and rocked at a rate appropriate (4X/min) for maintaining a portion of the cell monolayer in contact with test gases or vapors at all times. Treatment times ranged from 2 hr to 20 hr after which the gases were exhausted and the cells assayed for viability and enhancement of viral transformation.

   Treatment with Volatile Liquids. Test compounds, pipettes, and Petri dishes were prechilled to -20°C. Appropriate volumes of volatile compounds were mechanically dispensed with a glass pipette into an open glass Petri dish. Dishes containing the test chemicals were quickly placed into the exposure chambers containing dishes of cells, and the chambers were rapidly sealed. Depending on the density of the test chemical vapor, relative to air, the dish containing the chemical was placed above or below the test cells. Most volumes (generally 0.05 mL to 2.0 mL) of test compounds were completely volatilized within 10 min to 30 min at 37°C.

   Treatment with Chloromethane. Two-to-five liter cylinders of certified, analytical grade test gases were used without additional purification. A gas delivery system with a mass flow meter was used for quantitative distribution of test gases to exposure chambers. Regulators (two-stage) for each hazardous test gas were all stainless steel and Teflon, with specified Compressed Gas Association (CGA) fittings. All tubing, connectors, and valves were either stainless steel, brass, Teflon, or other suitable inert material(s). Oxygen (supplied at 20% final concentration) and nitrogen (used as an inert diluent gas) were dispensed and metered through standard regulators and rotameters. The accuracy of all flow rates was confirmed by use of a standardized bubble chamber. Flow rates were chosen: a) to deliver accurately five changes of test and diluent gases to each test chamber in 2 min to 6 min, and b) not to exceed the vaporization rate of gases with a low vapor pressure and/or under low pressurization. Test chambers were purged of residual air with five changes of test gas that has been documented to give an approximately 99% accurate atmosphere concentration. All gas lines were purged of test gas with nitrogen for 1 min following delivery of the final concentration of each test gas. To confirm the accurate dilution and delivery of calculated gas concentrations by this system, nitrogen oxide was dispensed similarly and a nitrogen oxide analyzer was used to determine the actual gas concentrations.

   Chemicals. Chloromethane gas (certified, analytical grade, Matheson Gas, East Rutherford, NJ) was prepared immediately prior to testing. 1,1,1-Trichloroethane (97%) and 1,2-dichloroethane (99%) were obtained from Aldrich Chemical, Milwaukee, WI. Dichloromethane (99%) was obtained from Burdick & Jackson Labs, Muskegon, MI.

Analytical Determinations. Samples of chamber atmospheres
were obtained with separate 50-μL Pressure-Lok A-2 gas syringes,
equipped with an all-Teflon shutoff valve between the syringe barrel
and needle (Supelco, Inc., Bellefonte, PA). Gas chromatographic
analyses were performed using a Hewlett-Packard 5840 gas chromatograph with flame ionization detection (FID) and a capillary inlet
system (Hewlett-Packard, Palo Alto, CA). Retention times and peak
areas for each compound were computed and recorded by the microprocessor which controls the chromatographic variables.

Gas chromatographic analysis operating conditions were as
follows:

Oven Temperature: Isothermal @ 50-70°C

Injector Temperature: 300°C

Detector Temperature: 325°C

Carrier: He

Column Flow: 1.2 m^3/min

Column: 26-meter SE-54 fused silica capillary
                           column

RESULTS

DCE and TCE were tested in a hamster cell system for cytotoxicity and enhancement of viral transformation using two-fold
chemical dilutions extending over an approximately 20-fold range.
The results of exposure to DCE and TCE in liquid cell culture
media for approximately 20 hr are shown in Table 1. Under these
conditions, neither compound was overly toxic to mass cultures of
hamster embryo cells as demonstrated by subsequent cloning assays
for viability. TCE reduced cell viability only at the highest
concentration (1,000 μg/mL) tested. Cells exposed to DCE and TCE
with this treatment protocol produced no increased enhancement of
transformation by SA7 at any dose level; no SA7 foci were observed
at 1,000 μg/mL of TCE.

Table 2 shows the effect of pretreatment of mass culture of
hamster embryo cells with the vapor phase of DCE and TCE for 20
hr (see Table 1). DCM, a volatile chlorinated organic liquid, as
well as the gas, CM, were also evaluated in this manner. All
compounds were tested using five two-fold chemical or gas dilutions
extending over approximately a 20-fold dose range. Dose ranges
were selected from preliminary toxicity screens (data not shown).
For brevity, some 100% lethal dose levels have been omitted.

Table 1. Effect of 20-hr pretreatment of HEC with
          1,2-Dichloroethane of 1,1,1-Trichloroethane
          on transformation by SA7: exposure of cells
          to chemical in culture medium

Chemical[a]	Final Concentration (μg/mL)	Surviving Fraction[b]	SA7 Foci[c]	Enhancement Ratio[d]
1,2-Dichloroethane	1,000	0.90	26	1.0
	500	1.04	31	1.1
	250	1.00	37	1.3
	125	0.98	24	0.9
	62	1.00	30	1.0
	0	1.00	29	1.0
	A[e]	1.01	21	0.7
1,1,1-Trichloroethane	1,000	0.31	0	0
	500	1.06	49	0.9
	250	0.96	43	0.9
	125	1.04	35	0.7
	62	0.99	36	0.7
	0	1.00	49	1.0
	D[e]	0.96	42	0.9

[a]Chemical dilutions in culture medium were added to mass cultures of HEC 20 hr before SA7 addition. Virus was absorbed 3 hr and the cells were transferred for survival (500-700 cells/dish) and for transformation assays (200,000-300,000 cells/dish).

[b]Determined from plates receiving 500-700 cells. The number of colonies from virus and chemically treated cells was divided by the number of colonies from virus-inoculated control cells to give the surviving fraction. Cloning efficiency of control cells was from 10% to 15%.

[c]Number of foci from 10⁶ plated cells.

[d]Enhancement ratio was determined by dividing the transformation frequency (TF) of treated cells (TF=SA7 foci x reciprocal of the surviving fraction) by that obtained from control cells.

[e]Solvent controls, A = acetone, D = dimethylsulfoxide at an 0.5% final concentration.

Table 2.   Effect of 20-hr pretreatment of HEC with
            volatilized liquids of gaseous chlorinated
            hydrocarbons on transformation by SA7:
            exposure of cells to gases or vapors in a
            sealed chamber

Chemical[a]	ml of liquid added parts/10^3 gas[b]	Surviving Fraction[c]	SA7 Foci[d]	Enhancement Ratio[e]
Chloromethane	50	0	0	NA [f]
	25	12	64	8.3
	13	80	143	2.8
	6	86	134	2.4
	3	89	86	1.5
	0	100	64	1.0
Dichloromethane	5.0	0	0	NA
	2.5	26	19	2.6
	1.3	60	50	3.0
	0.6	93	44	1.7
	0.3	122	39	1.1
	0	100	28	1.0
1,2-Dichloroethane	0.8	0	0	NA
	0.4	78	197	2.5
	0.2	64	283	4.3
	0	100	103	1.0
1,1,1-Trichloroethane	1.0	0	0	NA
	0.5	41	57	1.5
	0.3	63	146	2.5
	0.1	95	137	1.6
	0.05	147	103	0.8
	0	100	42	1.0
Acetone	1.6	0	0	NA
	0.8	66	67	0.9
	0.4	75	85	1.0
	0.2	61	66	1.0
	0.1	84	102	1.1
	0	100	109	1.0

[a]Mass cultures of hamster embryo cells were treated for approximately
20 hr with volatilized liquids or gaseous compounds before addition
of SA7. Virus was absorbed 3 hr and the cells were transferred for
survival (500-700 cells/dish) and tranformation assays (200,000-
300,000 cells/dish).

Table 2.   (continued)

---

[b]See Table 3 for the actual amounts of volatilized liquid vapors
present in the treatment chamber atmosphere as determined by gas
chromatography.

[c]Determined from plates receiving 500–700 cells. The number of
colonies from virus and chemically treated cells was divided by
the number of colonies from virus-inoculated control cells to
give the surviving fraction. Cloning efficiency of control cells
was from 10% to 15%.

[d]Number of foci from 2×10^6 plated cells.

[e]Enhancement ratio was determined by dividing the transformation
frequency (TF) of treated cells (TF = SA7 foci x reciprocal of
the surviving fraction) by that obtained from control cells.
Underlined values are statistically significant at the 1% level.
Dash underlined values are significant at the 5% level.

[f]NA = not applicable.

     All compounds were tested from 100% lethal to non-lethal doses
that are generally found within a narrow range (Table 2).   Short
treatment period (2 hr) required two- to four-fold increases in
concentration, when compared to the 20 hr exposure but demonstrated
similar dose responses (data not shown).

     All four chlorinated compounds tested, either volatilized
liquids (DCM, DCE, or TCE) or gas (CM), significantly enhanced the
viral transformation frequency in a dose-dependent manner. An
enhanced frequency of transformation was shown for at least two or
more dose levels tested. All the chlorinated organics also produced a significant absolute increase in the number of viral foci
at several doses. Therefore, increased enhancement ratios were not
dependent upon cell killing for significance. The commerical
preparation of TCE tested was allegedly stabilized with 3% 1,4-dioxane
(manufacturer's specifications) itself a suspected cancer agent
(20). However, independent analytical analysis by mass spectro-
metry of this commercial sample found no detectable 1,4-dioxane,
but did identify 2-methyl-2-butanol (data not shown). This latter
compound has no known carcinogenie activity. These results indicate
TCE, not the stabilizer, was responsible for the enhancement response.

     A noncarcinogenic, nonchlorinated volatile solvent control,
acetone, was tested from lethal to nonlethal concentrations and
produced no enhancement of viral transformation (Table 2).

The concentrations of chlorinated chemicals present in sealed treatment chambers at different time intervals were determined by standard gas chromatographic techniques (Table 3). Triplicate samples from individual chambers with three doubling volumes of DCM show an approximate two-fold decrease in detectable DCM at 2 hr (120, 73, and 30 $\mu g/cm^3$, respectively). Concentrations of DCM detected at 2 hr are approximately 20% of calculated theoretical values, and decrease to approximately 10% by 20 hr to 24 hr. The concentration of TCE detected at 2 hr is related to the volume of TCE initially volatilized in the treatment chamber and is, as with DCM, a fraction of calculated theoretical values. TCE values do no significantly decrease over a 24 hr period. Two hours was selected as the initial sampling period since this was the minimal time necessary for equilibration of the system, and produced maximal values. Samples taken earlier (data not shown) showed low values and were not reproducible or consistent.

DISCUSSION

Volatile liquids and gases constitute a significant portion of our exposure to potential carcinogenic agents. For example, a significant number of the chemicals used in the industrial workplace are gases or may volatilize under ambient temperature. Petrochemical refining and energy-generating plants produce large amounts of gaseous pollutants (21). Some gases and organic liquids are slightly soluble in water and may contaminate municipal drinking water supplies (22).

Closed treatment chambers have been utilized by others for treatment of prokaryotic or eukaryotic cell systems with volatilized liquids or gases. Cultured cells treated in this manner have subsequently been assayed for toxicity or mutation (23-25; see Session III, these proceedings). However, previous attempts to assay for transformational events following this type of treatment have not been published.

A sensitive and well validated bioassay (enhancement of viral transformation in hamster cells) that closely correlates with in vivo carcinogenesis was selected as an indicator system for developing and applying methods appropriate for testing volatile liquids and gases. The treatment procedures increased the sensitivity of hamster embryo cells to viral transformation. Additionally, the treatment resulted in the consistent and reproducible detection of some compounds that previously produced weak or no enhancement responses when incorporated into liquid cell culture media.

Concentrations of chemicals detected by gas chromatographic analysis after relatively short volatilization times (2 hr) were related to the volume of chemical added. However, analytical determinations confirmed that the concentrations of gaseous chemicals

Table 3. Concentrations of volatilized chlorinated compounds present in treatment chamber atmospheres as determined by gas chromatography

Chemical	Chemical Volatilized per Chamber (mL)	Theoretical Concentration ($\mu g/cm^3$)	Concentration (Gas Chromatograph) ($\mu g/cm^3$)		
			2–4 hr	12 hr	20–24 hr
1,1,1-Trichloroethane	2.00	540	58	NT	59
	1.00	270	25	7	10
	0.50	134	26	27	23
	0.25	67	19	14	15
	0.12	32	11	NT	8
	0	0	0	0	0
Dichloromethane	2.50	670	120	NT	67
	1.25	330	73	NT	17
	0.63	170	30	13	12
	0	0	<2	<2	<2
1,2-Dichloroethane	0.75	190	14	2	6
	0.38	94	4		2
	0.19	47	3	2	2
	0.10	24	3	<1	<1
	0.05	12	<1		<1
	0	0	0		<1

decreased with time in the sealed treatment chambers; this decrease
was not unexpected. Absorption and/or penetration of vapors or
gases into cells, media, and the treatment chamber walls seem the
most likely explanation and is consistent with results reported
by others (23; see Krahn, these proceedings).

Although the smaller volumes of chemicals added produce low
concentrations that approach the limits of detectability by gas
chromatography, and larger volumes saturate the treatment chamber,
the volumes needed to produce cell killing and/or enhanced viral
transformation can be accurately determined during the initial
treatment period. It is evident that the concentrations of some
compounds do not remain constant for the entire 20 to 24 hr treatment period. Therefore, some chemicals may need to be re-added to
treatment chambers if first results are negative. Conversely,
shorter treatment times may be necessary if the chemical concentrations fall low enough that cell repair processes are effective
prior to subsequent bioassay (17).

The technical problems involved in accurately determining
the amount of chemical interacting with critical cellular target
macromolecules, i.e. DNA, are considerable and need not be addressed
for preliminary screening assays. The approach taken in these
studies was to use a range of exposure concentrations and confirm
the interaction of test chemical with cellular macromolecules by
assaying for the ability of the cells to attach and divide.

The positive effects observed, i.e., enhanced viral transformation, in cultured cells treated with the volatile liquids or gases
is in agreement with their positive carcinogenic response, i.e.,
tumor formation, observed in rodents treated with these same chlorinated compounds. These latter responses have been summarized by
Infante (see these proceedings) from ongoing experiments with
animals exposed to chloroethane and chloromethanes from the NCl/NTP
Bioassay program and from privately supported industrial studies.
1,2-Dichloroethane produced tumors in several tissue sites of both
sexes of mice and rats treated by gavage. Male rats treated by
inhalation with dichloromethane developed salivary gland tumors.
Male mice similarly exposed to chloromethane had kidney tumors.
Animal bioassays (NCl/NTP) have been initiated with 1,1,1-trichloro-
ethane, but no data are available at this time. However, 1,1,1,-tri-
chloroethane was shown to transform cultured rate embryo cells, if
treated cells were passaged a sufficient number of times following
exposure (26). Its close structural analog, 1,1,2-trichloroethane,
given by gavage has produced liver and adrenal gland tumors in
male and female mice.

In summary, a treatment protocol suitable for testing diverse
volatile liquids and gases in a well validated mammalian embryo
cell system has been reported. Results of tests with a series

of chlorinated compounds of industrial and environmental importance
document that this system is sensitive, quantitative, and reproducible.
The relative ease and simplicity of the treatment protocol described
should allow evaluation of these and additional compounds in other
validated in vitro bioassay systems for carcinogenesis and muta-
gensis.

ACKNOWLEDGEMENTS

The helpful discussions and support of Dr. Michael D. Waters
of the Genetic Toxicology Division, Health Effects Research Laboratory, U. S. Environmental Protection Agency, Research Triangle
Park, NC, were invaluable to these studies. Analytical Serices
(gas chromatography) were provided by the Northrop Services, Inc.
laboratory of Dr. John Windsor, through the Aerosol Research
Branch, U.S. Environmental Protection Agency, Research Triangle
Park, NC.

REFERENCES

1. Y. Berwald and L. Sachs, In vitro transformation of normal cells
    to tumor cells by carcinogenic hydrocarbons, J. Natl. Cancer
    Inst. 35:641-661 (1965).
2. J. A. DiPaolo, P. J. Donovan, and R. L. Nelson, Quantitative
    studies of in vitro transformation by chemical carcinogens,
    J. Natl. Cancer Inst. 42:867-874 (1969).
3. J. A. DiPaolo, P. J. Donovan, and R. L. Nelson, Quantitative
    studies of in vitro transformation of hamster cells by
    polycyclic hydrocarbons: Factors influencing the number of
    cells transformed, Nature New Biol. 236:240-241 (1971).
4. T. Kuroki and H. Sato, Transformation and neoplastic development
    in vitro of hamster embryonic cells by 4-nitroquinolino-1-
    oxide and its derivatives, J. Natl. Cancer Inst. 41:53-71
    (1968).
5. T. T. Chen and C. Heidelberger, Quantitative studies on the
    malignant transformation of mouse prostate cells by carcino-
    genic hydrocarbons in vitro, Intern. J. Cancer 4:166-178
    (1969).
6. B. C. Casto, Enhancement of adenovirus transformation by
    treatment of hamster cells with ultraviolet irradiation,
    DNA base analogs, and dibenz(a,h)anthracene, Cancer Res.
    33:402-407 (1973).
7. B. C. Casto, Enhancement of viral oncogenesis by chemical
    carcinogens, in: "Chemical Carcinogenesis. The Biochemistry
    of Disease", P. O. P. Ts'o and J. A. DiPaolo, eds., Vol. 4,
    Marcel Dekker, Inc., N.Y., 607-618 (1974).
8. B. C. Casto, W. J. Pieczynski, and J. A. DiPaolo, Enhancement
    of adenovirus transformation by treatment of hamster embryo

cells with diverse chemical carcinogens, Cancer Res. 34: 72-78 (1974).

9. B. C. Casto, W. J. Pieczynski, and J. A. DiPaolo, Enhancement of adenovirus transformation by pretreatment of hamster cells with carcinogenic polycyclic hydrocarbons, Cancer Res. 33:819-824 (1978).

10. B. C. Casto, Detection of chemical carcinogens and mutagens in hamster cells by enhancement of adenovirus transformation in: "Advances in Modern Environmental Toxicology", Vol. 1, N. M. Mishra, V. Dunkel, and M. Mehlman, eds., Senate Press, Princeton, N.J., 241-271 (1981).

11. B. C. Casto, Mutagenesis and Carcinogenesis testing in vitro, Second Chemical Congress of the North American Continent, Biotechnology and Bioengineering, Las Vegas, NV, August 24-28, 1980 (in press) (1981).

12. J. McCann, E. Choi, E. Yamasaki, and B. N. Ames, Detection of carcinogens as mutagens in the Salmonella/microsome test: Assay of 300 chemicals, Proc. Natl. Acad. Sci. (U.S.A.) 72: 5135-5139 (1975).

13. J. McCann and B. N. Ames, Detection of carcinogens as mutagens in the Salmonella/microsome test: Assay of 300 chemicals, Part II, Proc. Natl. Acad. Sci. (U.S.A.) 73:950-954 (1976).

14. B. C. Casto, G. G. Hatch, S. L. Huang, J. L. Huisingh, S. Nesnow, and M. D. Waters, Mutagenic and carcinogenic potency of extracts of diesel and related environmental emissions: In Vitro mutagenesis and oncogenic transformation, in: "Health Effects of Diesel Engine Emissions", Proceedings of an International Symposium, Vol. 2, USEPA-600/9-80-057a 843-860 (1980).

15. B. C. Casto, Adenovirus transformation of hamster embryo cells, J. Virol. 2:376-383 (1968).

16. A. E. Freeman, P. H. Black, R. Wolford, and R. J. Huebner, Adenovirus type 12-rat embryo transformation system, J. Virol. 1:362-367 (1967).

17. B. C. Casto, W. J. Peiczynski, N. Janosko, and J. A. DiPaolo, Significance of treatment interval and DNA repair in the enhancement of viral transformation by chemical carcinogens and mutagens, Chem.-Biol. Interactions 13:105-125 (1976).

18. G. J. Todaro and H. Green, Enhancement by thymidine analogs of susceptibility of cells to transformation by SV40, Virology 24:393-400 (1964).

19. R. J. Lorenz, Zur statistik des plaque-Testes, Archiv. Gesamte Virus forschung 12:108-137 (1962).

20. IARC Mongraphs 11, 247, Lyon, France (1976).

21. E. Sawicki, Atmospheric genotoxicants - What numbers do we collect?, in: "Application of Short-term Bioassays in the Fractionation and Analysis of Complex Environmental Mixtures", M. D. Waters, S. Nesnow, J. L. Huisingh, S. S. Sandhu, and L. Claxton, eds., Environmental Science Research, Vol. 15, Plenum Press, N.Y., 171-194 (1978).

22. T. Page, R. H. Harris, and S. S. Epstein, Drinking water and
     cancer mortality in Louisiana, Science 193:55-57 (1976).
23. D. F. Krahn, Utilization of the CHO/HGPRT System: metabolic
     activation and method for testing gases, in: "Banbury Report
     2, Mammalian Cell Mutagenesis: The Maturation of Test
     Systems", W. H. Abraham, J. P. O'Neill, and V. R. McElheny,
     eds., Cold Spring Harbor Laboratory, Cold Spring Harbor,
     N.Y., 249-276 (1979).
24. V. F. Simmon, K. Kauhanen, and R. G. Tardiff, Mutagenic activity
     of chemicals identified in drinking water, in: "Progress
     in Genetic Toxicology", D. Scott, B. A. Bridges, and
     F. H. Sobels, eds., Elsevier/North Holland Biomedical Press,
     Amsterdam, 249-258 (1977).
25. G. S. Samuelson, R. E. Rasmussen, B. K. Nair, and T. T. Crocker,
     Pollutants on mammalian cells, Environ. Sci. and Tech. 12:
     426-430 (1978).
26. P. J. Price, C. M. Hassett, and J. I. Mansfield, Transforming
     activities of trichloroethylene and proposed industrial
     alternatives, In Vitro 14:290-293 (1978).

DISCUSSION

Q. Sivic (Arthur D. Little): You do your actual assay in low
calcium medium and you do your survival, I assume, in regular
medium. I'm wondering whether this will effect the quantitation
of the system when you measure one parameter in one type of medium
and then use a different medium for the other parameter?

A. Hatch (Northrop): We allow for that. The treated cells
are not shifted to low calcium medium until about eight or ten
days afterwards, which should provide enough time for toxicity
to be expressed.

Q. Sivak (Arthur D. Little): You did some analytical chemistry
on the gaseous phase of your chambers. Is there anything else
besides the parent compound? Chemicals like dichloromethane are
very easily oxygenated without stabilizers around and you could
get some pretty active materials in there.

A. Hatch (Northrop): I'll comment on that and it's certainly a
legitimate question. There are chlorinated and other compounds
present in the treatment chambers, as you would expect since they're
plastic. The levels of those compounds are generally at parts
per million to parts per billion and that puts us about 1,000 to
10,000 times lower than the lowest dose needed to produce any
type of biological effect including toxicity. If you're worried
about residual carryover, we heat these chambers under vacuum after
each treatment exposure. I would caution you that if you have

your analytical chemist look at the treatment chambers when they
arrive from the commercial manufacturer, you will pick up the
presence of some compounds. However, the levels are so low that
while they can be detected analytically, they are certainly of no
biological significance.

CHO/HGPRT MUTATION ASSAY: EVALUATION OF

GASES AND VOLATILE LIQUIDS

David F. Krahn, Frances C. Barsky and Kevin T. McCooey

Haskett Laboratory for Toxicology and
Industrial Medicine
E. I. du Pont de Nemours & Co.
Newark, DE 19711 U.S.A.

INTRODUCTION

Over the last several years, attention has been focused on
the development and use of rapid and reliable assays to identify
potentially carcinogenic and mutagenic chemicals. These assays,
if understood and properly evaluated, can be valuable tools for
gathering information to aid in deciding whether chemicals may
be hazardous to human health.

The Chinese hamster ovary (CHO) cell mutation assay as
developed by Dr. A. W. Hsie and coworkers at Oak Ridge National
Laboratory (1,2) is one of the in vitro assays used at Du Pont
to evaluate the mutagenic activity of chemicals. This cell
culture system was designed to detect mutations in the gene
coding for hypoxanthine-guanine phosphoribosyl transferase (HGPRT).
Mutants deficient in HGPRT grow in the presence of the purine
analog, 6-thioguanine (6TG), but normal cells convert 6TG to its
toxic nucleotides and die.

The CHO/HGPRT assay has been used to test several different
types of chemicals. The chemical classes from which representative
compounds have been tested include: nitrosamines (3,4), nitrosamides and nitrosamidines (5), alkanesulfonates and alkylsulfates
(6), nitrogen heterocyclics (7), metal containing compounds (8),
aromatic amines (4) and others (4, 9-12).

There are only a few published reports of gases or highly
volatile chemicals being tested for mutagenic activity in

91

cultured mammalian cell mutation assays (4, 13). It is important
to develop reproducible and quantitative methods for testing compounds exhibiting these physical properties, because they are a
substantial portion of the samples encountered in a routine test
program. The subject of this article is the development and use of
methods for testing gases and volatile liquids in the CHO/HGPRT
assay.

Materials and Methods

The protocol currently used in the laboratory is described
below and differs slightly from the protocol used to conduct the
experiments reported in this article. The differences are identified
in the Results Section.

Test Chemicals. Chlorofluoromethane (fluorocarbon-31): b.p.
-9.1°C @ 760 mm Hg; source, Du Pont. Dichlorodifluoromethane
(Freon 12): b.p. -28°C @ 760 mm Hg; source, Du Pont. Trichloro-
fluoromethane (Freon 11): b.p., 23.7°C @ 760 mm Hg; source, Du
Pont. Vinyl chloride (chloroethene): b.p. -14°C @ 760 mm Hg; source,
Matheson.

Cell Culture. Experiments are performed with cells from the
BH4 clone of the CHO-K1 cell line. The growth medium for stock
cells is Ham's F12 medium without hypoxanthine containing 5-10%
dialyzed heat-inactivated fetal or newborn calf serum. The
same growth medium containing penicillin (50 U/mL) and streptomycin
(50 µg/mL) is used for mutagenesis experiments (culture medium).
An aliquot of cells stored in liquid nitrogen is thawed each week
for the experiments scheduled for the following week. Cells are
tested for mycoplasma.

Activation System. Livers are obtained from at least six,
8- to 9- week old male rats (Charles River CD) injected with
Aroclor 1254 (500 mg/kg) five days before sacrifice. The livers
are homogenized (Waring Blender and Dounce homogenizer) in 0.15 M
KCl and centrifuged at 9,000 x g for 10 min. The supernatant
fraction (S-9) is centrifuged again under the same conditions
and the resulting S-9 is stored in liquid nitrogen. All steps are
performed at 4°C. The BIO-RAD Assay is used to determine the
protein content.

Treatment Medium. For experiments not involving metabolic
activation, the culture medium (3 mL) is buffered to a pH of 7.2
by the addition of HEPES buffer (2.5×10^{-2} M). For experiments
involving metabolic activation, the culture medium (3mL) is
supplemented with HEPES buffer (2.5×10^{-2} M), $MgCl_2$
(5.6×10^{-3} M), glucose-6-phosphate (5.0×10^{-3} M), NADP (1.5×10^{-3} M), NADH (1×10^{-3} M) and S-9 (0.6-3.0 mg protein).

Treatment and Analysis. Gases are metered into treatment flasks with a flowmeter system calibrated with air and corrected for gas density as described by Nelson (14). The treatment flasks containing cells and treatment medium are flushed with about 10 volumes of the desired test gas-air mixture in one min. Each flask is closed with a cap containing a rubber septum. A gas tight syringe is used to take headspace samples from the treatment flasks at the beginning and end of the exposure period. These samples are injected into a gas chromatograph and the peak heights are compared to the peak height resulting from a sample taken from a standard chamber charged with 20% or 50% test gas in air, but containing no culture or treatment medium.

Liquids are introduced directly into the treatment flasks containing cells and treatment medium. The concentration of chemical in the flask headspace is measured as described for gases.

Experimental Design. Preliminary cytotoxicity experiments, with and without activation, are conducted to establish appropriate test sample and S-9 concentrations for the mutagenesis experiments. An attempt is made to attain about 10% relative survival at the highest concentration. In each mutagenesis experiment, all treatments are made in duplicate. Two or more experiments are performed with and without activation. Each experiment includes a positive and a solvent or air control and at least four concentrations of the test chemical.

S-9 Evaluation. After a preliminary cytotoxicity experiment with 1.0 mg of S-9 protein per mL of treatment medium, a second cytotoxicity experiment is conducted using 0, 0.2, 0.5 and 1.0 mg of S-9 protein per mL of treatment medium at two concentrations of test chemical giving about 70% and 30% survival relative to control. The S-9 level resulting in the greatest cytotoxicity is chosen for mutagenesis experiments.

Mutagenesis Assay. Five x 10^5 cells are plated per 25 cm T-flask in 5 mL of culture medium. The next day the culture medium is removed and the treatment medium is added. The test chemical is added to the treatment medium or introduced into the treatment flask. The cells are incubated for 18-19 hr (without activation) or for 5 hr (with activation). Treatment flasks were incubated on a rocker panel in a 37°C incubator. After incubation, the flasks are flushed with air, the treatment medium is removed and the flasks were washed with culture medium. Cells treated without activation are subcultured immediately after treatment while the cells treated with activation are incubated in culture medium for 21-26 hr after treatment before subculturing.

At the time of the initial subculturing, one portion of the cells is plated to assess cytotoxicity and another portion is

plated to allow expression of mutants resistant to 6-thioguanine
(6-TG). To assess cytotoxicity, 200 cells are plated per 60 mm
dish (6 dishes) from each flask of cells in the experiment. These
dishes are incubated for 7 days and the colonies are stained and
counted. Cell survival is determined by dividing the total number
of colonies by the total number of cells plated and is expressed
both as the percent plated and as the percent of the solvent or
air control survival.

To allow for expression of the 6-TG resistant phenotype, cells
from each treatment flask are plated at 1×10 /100 mm dish (1 dish).
These cells are maintained in exponential growth for 7 days by
subculturing twice (2 days after the initial subculturing and then
3 days later or vice versa). On the 7th day the cells are plated
to assess cell survival and the frequency of 6-TG-resistant
cells. Cell survival is determined as described above. To
identify 6-TG-resistant cells, 2×10^5 cells are plated per 100
mm dish (5 dishes) in culture medium containing 1×10^{-5} M 6-TG.
The cells are incubated for 7 days and the colonies are stained and
counted. The mutation frequency is expressed as the number of
mutant colonies/10^6 surviving cells at the time of mutant selection.

Data Analysis. The mutation data are analyzed by the method
of Irr and Snee (12). The data are transformed using the formula
$Y = (\text{mutation frequency} + 1)^{0.15}$ because this power transformation
provides data which satisfies assumptions required for performing
parametric statistical analyses. In one analysis, the data at each
concentration of chemical are compared to the solvent or air control
response by a t-test to determine whether any of these test chemical
concentrations cause a significant increase in the mutation frequency.
A second analysis evaluates a dose response relationship with an
analysis of variance technique.

RESULTS

In all experiments about to be described, the growth medium
was hypoxanthine-free Ham's F12 supplemented with 10% dialyzed heat-
inactivated fetal calf serum. Experiments with an activation system
were conducted with 2 mg of S-9 protein per mL of treatment medium.
An "S-9 evalution" was not performed.

Flask headspace samples were taken for analysis at the beginning
and the end of the treatment period in most experiments. The concentrations of test compound in the headspace at the end of the
treatment period were similar to or slightly less than the initial
concentrations. Only the initial measurements are reported.

Chlorofluoromethane increased the reversion frequency of
Salmonella typhimurium strains TA 1535 and TA 100 with and without

an activation system (15). In the CHO/HGPRT assay, this chemical was cytotoxic and mutagenic without activation when cells were exposed to chlorofluoromethane in air for different lengths of time (Fig 1). In Trial 1, the average measured concentration of chloro-fluoromethane in the headspace of all treatment flasks was 41±3.1%. The zero time control is an average of untreated controls incubated for 0, 1, 2, and 3 hrs. The statistical analysis of the data shown in Figure 1 indicates that the mutagenic activity at each time point was significantly greater than the activity observed at zero exposure, that the time-response exhibited a significant linear component, and that the duplicate experiments were not significantly different (Table 1).

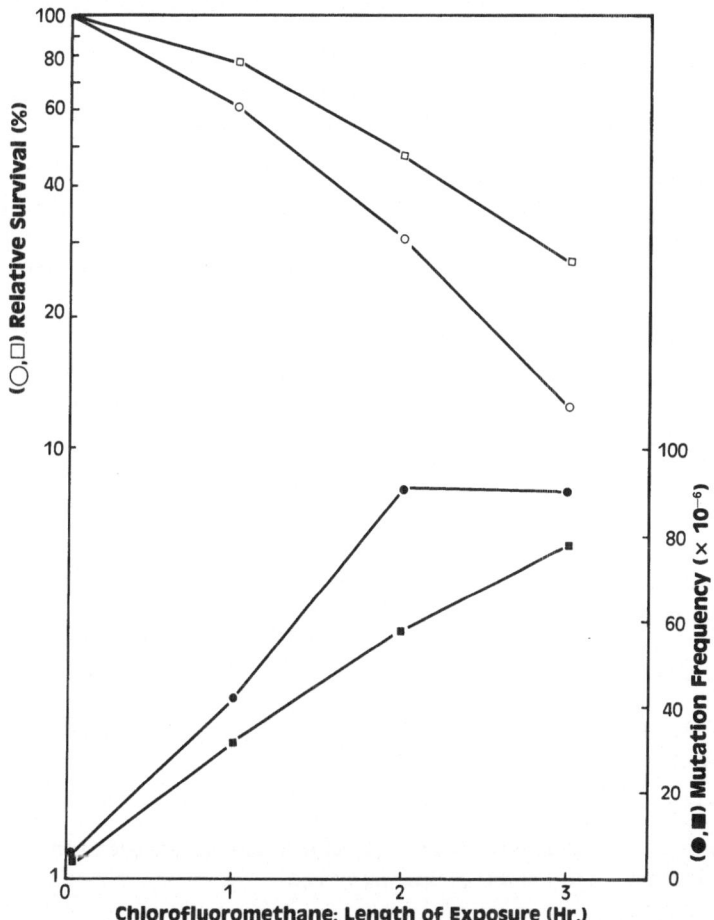

Figure 1. Cytotoxicity and mutagenic activity of chlorofluoro-methane without activation. Average measured concentration: 41±3.1% (SD, standard deviation), Trial 1 (□), Trial 2 (0).

Chlorofluoromethane also caused significant concentration-dependent cytotoxicity and mutagenicity in CHO cells exposed for 4 hours in the absence of an activation system (Fig 2). The concentrations were determined by gas chromatographic analyses of headspace samples. The mutagenic activity exhibited by chlorofluoromethane in the presence of the activation system was similar to that observed in experiments without activation.

Cytotoxicity and significant mutagenic activity were observed in the presence of the activation system when cells were exposed to a single concentration of vinyl chloride for different times

Table 1.   Statistical Analysis of Chlorofluoromethane without Activation

A.   Comparison of mutation frequency at each concentration to the negative control by Student's t test

Exposure (hrs)	No. of Results	Average Frequency[a]	Transformed Average[b]	t[c]	'p' value
0	4	2.1	1.1875		
1	4	35.9	1.7181	5.46	0.0002
2	4	68.0	1.8872	7.20	0.0000
3	4	82.5	1.9421	7.76	0.0000

B.   Dose response analysis of variance

The dose response relationship exhibited a significant linear component ($p < 0.0001$)

C.   Dose X trial analysis

The dose response relationships of each experiment were similar (p = 0.9746).

[a]Mutants per 10^6 cells surviving at the time of selection. Calculated from the Transformed Average.

[b]The average of the individual transformed values. The transformed mutation frequency equals the (mutation frequency + 1)$^{0.15}$

[c]Value calculated by Student's t test.

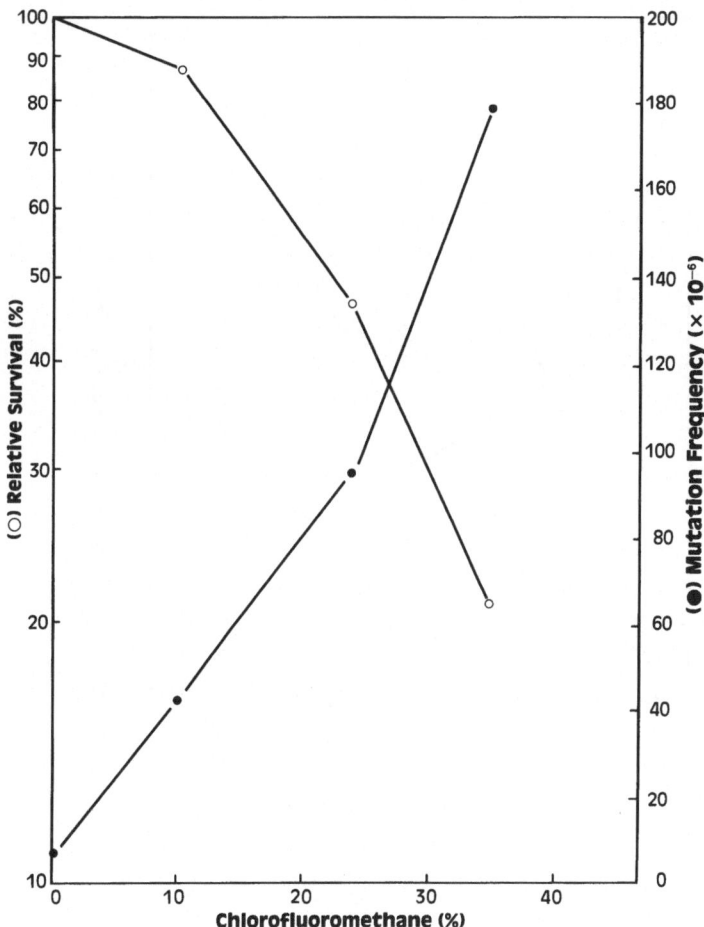

Figure 2.   Cytotoxicity and mutagenic activity of chlorofluoromethane with activation.   The points are the averages of triplicate treatments in one experiment.   Exposure: 4 hr.

(Fig 3).   Activity was directly related to the length of exposure. In this experiment, the average measured concentration of vinyl chloride in the headspace of the treatment flasks was 34±5.2% (SD). Significant concentration-dependent responses were also observed in the presence of activation when cells were exposed to different concentrations of vinyl chloride for 3 hours (Fig 4).   Vinyl chloride (50% in flask headspace) was not mutagenic after a 5 hour exposure with treatment medium lacking S-9 and cofactors.

    Trichlorofluoromethane was cytotoxic at the high test concentration but did not exhibit any mutagenic activity in the presence

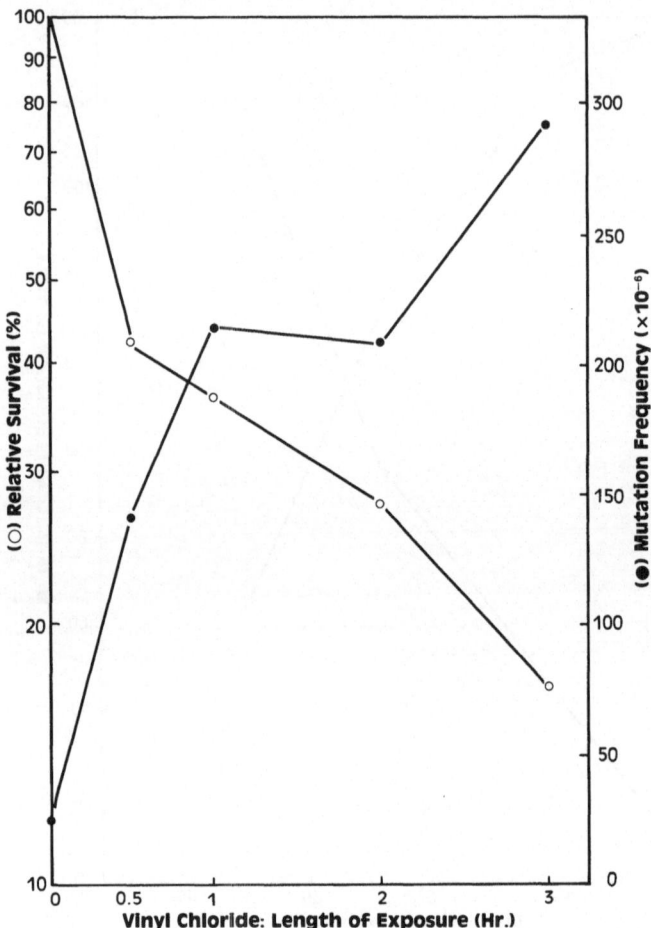

Figure 3. Effect of length of exposure on cytotoxicity and mutagenic activity of vinyl chloride with activation. Average concentration: 34±5.2% (SD).

of an activation system (Table 2). This compound was also non-mutagenic in the absence of an activation system.

Dichlorodifluoromethane exhibited neither cytotoxicity nor mutagenic activity without the activation system (Table 3). The concentrations in the treatment flasks with and without activation were 43.4±5.3% (SD) and 42.7±5.5% (SD), respectively. This compound was also nonmutagenic in the presence of an activation system.

Trichlorofluoromethane and dichlorofluoromethane are also nonmutagenic for the S. typhimurium strains TA 1535, TA 1537, TA 98, and TA 100 either with or without and activation system (15).

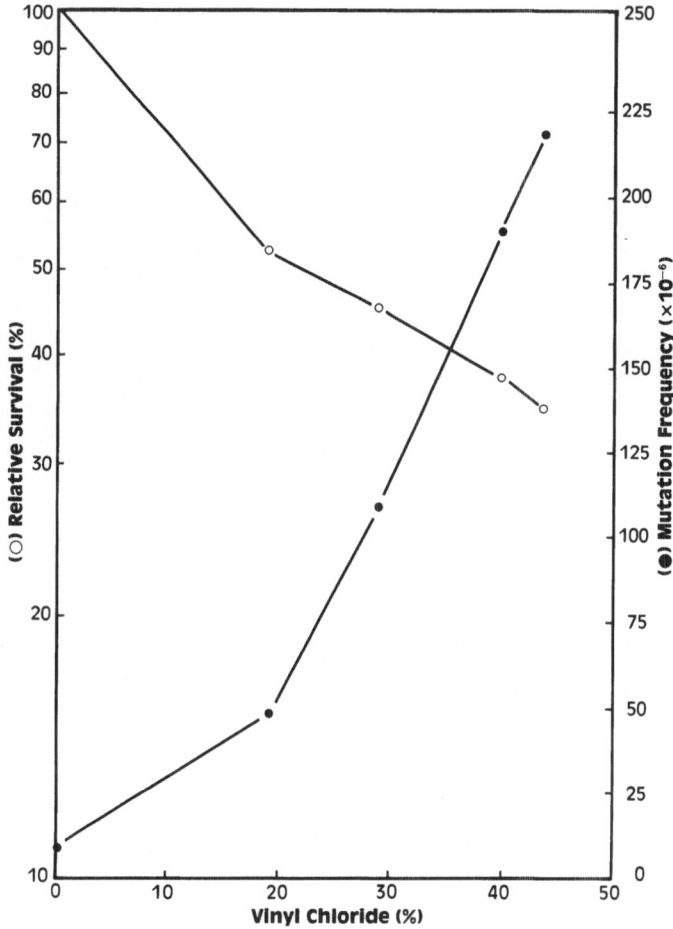

Figure 4. Cytotoxicity and mutagenic activity of different
concentrations of vinyl chloride with activation. Exposure: 3 hr.

DISCUSSION

    Reliable methods for evaluating the mutagenic activity of gases
are important because a substantial portion of chemicals used in
industry are gases or highly volatile liquids. The qualitative
responses observed in the CHO/HGPRT assay with chlorofluoromethane,
dichlorodifluoromethane, and trichlorofluoromethane were identical
to the activities exhibited by these chemicals in the Ames strains
of Salmonella typhimurium (15).

Table 2. Trichlorofluoromethane: Cytotoxicty and Mutagenic Activity with Activation

Concentration[a] (μL)	Cytotoxicity[b] (% Control) I	II	Mutagenic Activity Total Mutants I	II	Mutation Frequency[c] I	II
0	100	100	0	10	0	11.3
			1	8	1.3	9.5
20	99.0	105	0	23	0	26.7
	88.5	89.9	4	4	4.6	5.2
50	101	92.5	5	2	6.5	2.3
	102	107	4	2	5.0	2.3
100	89.8	111	3	9	3.4	10.1
	95.0	88.6	6	20	7.5	23.6
200	36.0	35.8	0	5	0	5.0
	31.5	5.9	0	0	0	0
PC[d]	21.2	33.5	180	115	262	160
	20.3	29.4	96	138	162	196

[a]Amount added to 25 cm^2 T-flasks containing cells and treatment medium. The average and standard deviation of the measured concentrations (% v/v in headspace) were 7±2.4 (20 μL), 25±1.7 (50 μL), 42±7.9 (100 μL), 68±7.7 (200 μL).

[b]The absolute control plating efficiencies (cells plated the day after treatment) for the duplicate experiments I and II were 70% and 64%, respectively.

[c]Mutants per 10^6 cells surviving at the time of selection with 6TG.

[d]Positive control, vinyl chloride. Three hr exposure in flasks treated with undiluted gas.

Table 3. Dichlorodifluoromethane: Cytotoxicity and Mutagenic Activity without Activation

| Length of Exposure (Hr)[a] | Cytotoxicity[b] (% Control) | | Mutagenic Activity | | | |
| | | | Total Mutants | | Mutation Frequency[c] | |
	I	II	I	II	I	II
0	100	100	3.6[e]	5.2[e]	4.9	6.0
1	92.8	112	4	2	5.2	2.3
	95.4	93.6	1	3	1.2	3.4
5	92.8	87.1	3	21	4.4	23.5
	108	94.9	0	1	0	1.1
18	102	103	4	11	5.9	14.3
	98.2	114	4	4	5.7	5.5
1[d]	80.1	61.4	42	21	61.9	35.1

[a]The average of measured concentrations of dichlorodifluoromethane for all treatments was 45±3.6% (SD).

[b]The absolute control plating efficiencies (cells plated the day after treatment) for the duplicate experiments, I and II, were 62.7% and 64.2%, respectively.

[c]Mutants per 10^6 cells surviving at the time of selection with 6TG.

[d]Positive control, chlorofluoromethane, at 68% (average calculated from measurements).

[e]Average of the 1, 5 and 18 hour controls.

Vinyl chloride is mutagenic for <u>Salmonella</u> in both the presence and absence of an activation system (15-17). However, in the CHO/HGPRT assay, vinyl chloride was mutagenic only when the treatment was conducted with activation. Vinyl chloride in the presence of an activation system is reported to cause mutations in V79 cells (13) and to transform BHK_{21} cells (18).

Methods are available for testing gases and volatile liquids for mutagenic activity in cultured mammalian cells. Concentration- and time-related effects can be demonstrated for direct-acting chemicals, as well as for those requiring an activation system. Refinements of the treatment methods will provide more quantitative and reproducible information. Treatment chambers would provide a means to treat replicates identically. Also, while headspace analysis did allow the observation of concentration-related effects, the use of a method for measuring the concentration of a test chemical in the treatment medium would provide a more accurate indication of the true exposure.

REFERENCES

1. A. W. Hsie, P. A. Brimer, T. J. Mitchell, and D. G. Gosslee, The dose-response relationship for ethyl methanesulfonate-induced mutations at the hypoxanthine-guanine phosphoribosyl transferase locus in Chinese hamster ovary cells, <u>Somatic Cell Genet.</u> 1:247-261 (1975).
2. J. P. O'Neill, P. A. Brimer, R. Machanoff, G. P. Hirsch, and A. W. Hsie, A quantitative assay of mutation induction at the hypoxanthine-guanine phosphoribosyl transferase locus in Chinese hamster ovary cells (CHO/HGPRT system): Development and definition of the system, <u>Mutat. Res.</u> 45:91-101 (1977).
3. J. P. O'Neill, D. B. Couch, R. Machanoff, J. R. San Sebastian, P. A. Brimer, and A. W. Hsie, A quantitative assay of mutation induction at the hypoxanthine-guanine phosphoribosyl transferase locus in Chinese hamster ovary cells (CHO/HGPRT system): Utilization with a variety of mutagenic agents, <u>Mutat. Res.</u> 45:103-109 (1977).
4. D. F. Krahn, Utilization of the CHO/HGPRT System: Metabolic activation and method for testing gases, <u>in</u>: "Banbury Report 2 - Mammalian Cell Mutagenesis: The Maturation of Test Systems", A. W. Hsie, J. P. O'Neill, and V. K. McElheny, eds., Cold Spring Harbor Laboratory, Cold Spring Harbor, N.Y., 251-261 (1979).
5. D. B. Couch and A. W. Hsie, Mutagenicity and cytotoxicity of congeners of two classes of nitroso compounds in Chinese hamster ovary cells, <u>Mutat. Res.</u> 57:209-216 (1978).
6. D. B. Couch, N. L. Forbes, and A. W. Hsie, Comparative mutagenicity of alkylsulfate and alkane-sulfonate derivatives in Chinese hamster ovary cells, <u>Mutat. Res.</u> 57:217-224 (1978).

7. J. C. Fuscoe, J. P. O'Neill, R. M. Peck, and A. W. Hsie, Muta-
    genicity and cytotoxicity of nineteen heterocyclic mustards
    (ICR compounds) in cultured mammalian cells, Cancer Res. 39:
    4875-4881 (1979).
8. N. P. Johnson, J. D. Hoeschele, R. O. Rahn, J. P. O'Neill,
    and A. W. Hsie, Mutagenicity, cytotoxicity and DNA binding
    of platinum (11)-chloramines in Chinese hamster ovary cells,
    Cancer Res. 40:1463-1468 (1980).
9. D. B. Couch, E. Bermudez, G. M. Decad, and J. G. Dent, The
    influence of activation systems on the metabolism of 2,4-
    dinitrotoluene and its mutagenicity to CHO cells, in:
    "Banbury Report 2 - Mammalian Cell Mutagenesis: The Matur-
    ation of Test Systems", A. W. Hsie, J. P. O'Neill, and
    V. K. McElheny, eds., Cold Spring Harbor Laboratory, Cold
    Spring Harbor, N.Y., 303-309 (1979).
10. E. J. Greene, M. A. Friedman, J. A. Sherrod, and A. J. Salerno,
    In Vitro mutagenicity and cell transformation screening of
    phenylglycidyl ether, Mutat. Res. 67:9-19 (1979).
11. E. J. Greene, M. A. Friedman, J. A. Sherrod, and A. J. Salerno,
    In Vitro mutagenicity and cell-transformation screening of
    N-(2,3-epoxy-propyl)phthalimide, Mutat. Res. 68:251-257
    (1979).
12. R. C. Snee and J. D. Irr, Design of a statistical method for
    the analysis of mutagenesis at the hypoxanthine-quanine
    phosphoribosyl transferase locus of cultured Chinese hamster
    ovary cells, Mutat. Res. 85:77-93 (1981).
13. C. Drevon and T. Kuroki, Mutagenicity of vinyl chloride, vinyl-
    idene chloride, and chloroprene in V79 Chinese hamster cells,
    Mutat. Res. 67:173-182 (1979).
14. G. O. Nelson, "Controlled Test Atmospheres, Principles and
    Techniques", Ann Arbor Science Publishers, Inc., Ann Arbor,
    MI, 21-57 (1972).
15. Du Pont, unpublished data.
16. A. W. Andrews, E. S. Zawistowski, and C. R. Valentine, A
    comparison of mutagenic properties of vinyl chloride and
    methyl chloride, Mutat. Res. 40:273-276 (1976).
17. H. Bartsch, C. Malaveille, A. Barbin, H. Bresil, L. Tomatis,
    and R. Montesano, Mutagenicity and metabolism of vinyl
    chloride and related compounds, Environ. Health Perspect.
    17:193-198 (1976).
18. J. A. Styles, A method for detecting carcinogenic organic
    chemicals using mammalian cells in culture, Br. J. Cancer
    36:558-563 (1977).

LUNG CELLS GROWN ON CELLULOSE MEMBRANE FILTERS AS AN IN VITRO MODEL OF THE RESPIRATORY EPITHELIUM

Ronald E. Rasmussen and T. Timothy Crocker

Department of Community and Environmental Medicine
University of California
Irvine, CA 92717 (U.S.A.)

INTRODUCTION

Airborne agents enter the body for the most part by first contacting some part of the respiratory epithelium. Particulates and gases may cross the epithelium by solution in the cell membranes or pinocytosis. Insoluble particles may be engulfed and removed by macrophages and mucociliary clearance mechanisms, but macrophages engorged with particulate materials may remain in the lung for prolonged times and serve as a source for continued exposure of other cells to potential toxins.

Because the cells of the epithelium are in nearly direct contact with the inhaled air, they may be especially vulnerable to intoxication by air pollutants. This is evident from studies of the lungs of experimental animals exposed to low levels of oxidant gases. Chronic exposure to less than 1 ppm of ozone (O_3) or 2 ppm of nitrogen dioxide (NO_2) is sufficient to cause clearly observable lesions in the epithelium of the terminal bronchioles (1,2). Biochemical studies of the lungs of rats exposed to these gases have shown changes in enzymatic activities (3) and peroxidation of lipids found in the cell membranes (4). The relationship of these effects to lung disease, especially the degenerative diseases such as emphysema and lung cancer, is of great interest.

Recent approaches to the problem of environmentally induced cancer have involved preparation of cell cultures from fetal tissue of experimental animals, and treatment of these cultures in vitro with the chemical of interest. The criteria for neoplastic transformation are alteration in growth habit coupled with the ability of the treated cells to form a malignant tumor upon inoculation into a suitable host animal (5,6). Application of this method to specific organs, especially to the respiratory tissues, is complicated by the difficulty in preparation of pure cultures of the cell types thought to be the target for neoplastic transformation. However, methods are rapidly being developed for the preparation of relatively pure cultures of lung cells of specific types, such as Type 2 alveolar cells (7,8) and mucus-secreting Clara cells (9).

Methods for studying the effects of airborne pollutants on cultured cells have been limited to the solution or suspension of the materials of interest in the cell culture medium. Exposure of cells to gaseous materials has required that the cells be immersed in liquid medium between periods of exposure in order that the cells not dry out and die. The requirement for intimate contact between the cells and gases such as NO_2 or O_3 in order to observe any effect is well illustrated by the work of Pace et al. (10). They found that a few millimeters of culture medium protected the cells against very high concentrations of oxidant gases; the cells being rapidly killed upon direct exposure to these gases by inverting the culture flasks.

Attempts to provide significant periods of exposure of cells to gases have usually employed some variation of a rocker platform arrangement in which a layer of cells is alternately exposed and immersed in the culture medium (11). Other methods have been to manually invert culture flasks to expose the cell layer (10) or to allow cells to attach to membrane filters and float the filters on culture medium with the cells facing upwards (12). In the latter case, passive diffusion was relied upon to maintain the viability of the cell culture. The methods which use rocker platforms or roller bottles have the disadvantage that the moving meniscus may dislodge cells or may remove and dilute solubilized gases or their reaction products. Evaporation of water from the membrane filter arrangement would lead to the concentration of salts from the growth medium with consequent harmful effects on the cells.

The recent development of cell culture dishes with gas-permeable surfaces seemed to offer the possibility of being able to study the effects of gases on cells while maintaining them beneath the culture medium. Using such a system, Alink et al. (13) showed cytotoxic effects and alterations in enzyme activities in V-79 Chinese hamster lung cells. However, the concentrations of O_3 required to produce measurable effects were several hundred parts per million, well above the level encountered in ordinary situations.

Our aim has been to provide more realistic exposure conditions for cultured cells. We have considered the structure of the in vivo respiratory epithelium as a model, and have been attempting to construct an in vitro replica. The principle of the system at its present stage of development consists of allowing cells to attach

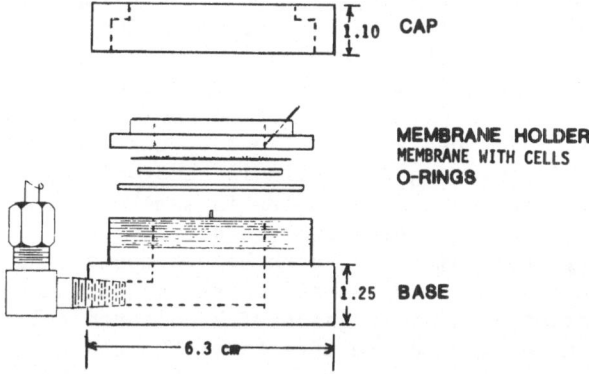

Figure 1. Diagram of filter holder for holding Millipore membrane filters during exposure to gases or vapors. Holders have been fabricated from Lexan plastic and stainless steel.

Figure 2. Stainless steel filter holders, assembled with filters in place, ready for placement in the gas exposure chamber.

to a cellulose membrane filter, and then assembling the filter into
a specially designed holder that allows perfusion of growth medium
through the filter from the side opposite the cells. For exposure
to gases, the medium perfusing through the filter is removed by a
pump so that a layer of fluid does not accumulate over the cells.
The cells remain moistened and nourished by the continuing perfusion.
The layer of fluid covering the cells is thus only as much as is
held by surface tension, and resembles in some ways the mucus layer
covering the cells of the respiratory tract in vivo. Toxic gases
have only a very short diffusion path before contacting the cell
surface. Therefore, this system offers the possibility of studying
the direct interaction of test gases with cultured cells. Using
this system we have studied the effects of NO_2 and O_3 on a number
of cellular properties. Cells exposed in this system are uniquely
sensitive to the effects of these gases, even when the gases are
present at concentrations of less than 1 ppm. The exposure system
seems well suited to studies of cellular effects at concentrations
of pollutant gases that are frequently encountered in urban environments.

MATERIALS AND METHODS

Cell Culture Methods

     The cell line used in this work was obtained from Dr. E. H. Y.
Chu (Chinese hamster lung fibroblast line V-79). Cell line V-79
was grown routinely in Dulbecco's minimal essential medium supplemented with 10% fetal bovine serum and antibiotics. All cell culture
medium and reagents were obtained from Grand Island Biological Co.,
Grand Island, NY.

Development of Cell Culture Exposure System

     With the respiratory epithelium as a model, the cell culture
system was designed to allow continuous feeding of the cells during
exposure to the atmosphere of choice. Millipore filters, type HAWP,
37 mm in diameter, were chosen as the cell substrate after many trials
of other filters and sizes. These filters have a nominal pore size
of 0.45 μm; smaller or larger pore sizes were less conducive to cell
growth. Filters of larger diameter tended to distort excessively
under the pressure of the perfusion pumps that supplied medium to
the rear of the filters. Following the seeding of the filters with
the cell line to be used, the filters were assembled into specially
designed holders (Fig 1). These holders were then connected to
a perfusion pump that supplied medium to the rear of the filter at
a rate of approximately 0.02 mL/mm^2/hr. This very low flow rate
was sufficient to maintain viability of the cells, but not high
enough to dislodge the cells. Figure 2 shows two of the stainless
steel filter holders assembled with filters, ready for placement
inside the gas exposure chambers (see below).

Figure 3. Stainless steel cell filter holders in place inside a stainless steel lined gas exposure chamber. Nutrient cell growth medium enters through tubes at the top of the chamber. Medium which perfuses through the filters is drawn off through the tubes at the rear wall of the chamber. The test gas enters at the left end of the chamber and exits at the right end.

Procedure for Exposure of Cells to Test Atmospheres

Gas exposure chambers have been constructed of Lexan plastic and wood lined with stainless steel foil (Figure 3). The chambers provide exposure capability for oxidant gases or corrosive atmospheres as well as for materials such as hydrazine which are not compatible with metals. Control and monitoring of exposure to oxidant gases is accomplished by means of a microprocessor-controlled solenoid valve arrangement that permits the use of up to six exposure chambers simultaneously. Figure 4 is a diagram of the system, showing inter-connections for use with any or all of three gases, (NO_2, O_3, or SO_2). Individual ozone generators are provided for each chamber, while NO_2 and SO are obtained by dilution of cylinder gas. Spent gas leaving the chambers is vented to the outside atmosphere; this does not constitute an environmental hazard because the gas concentrations used are only a fraction of a part per million, and are frequently

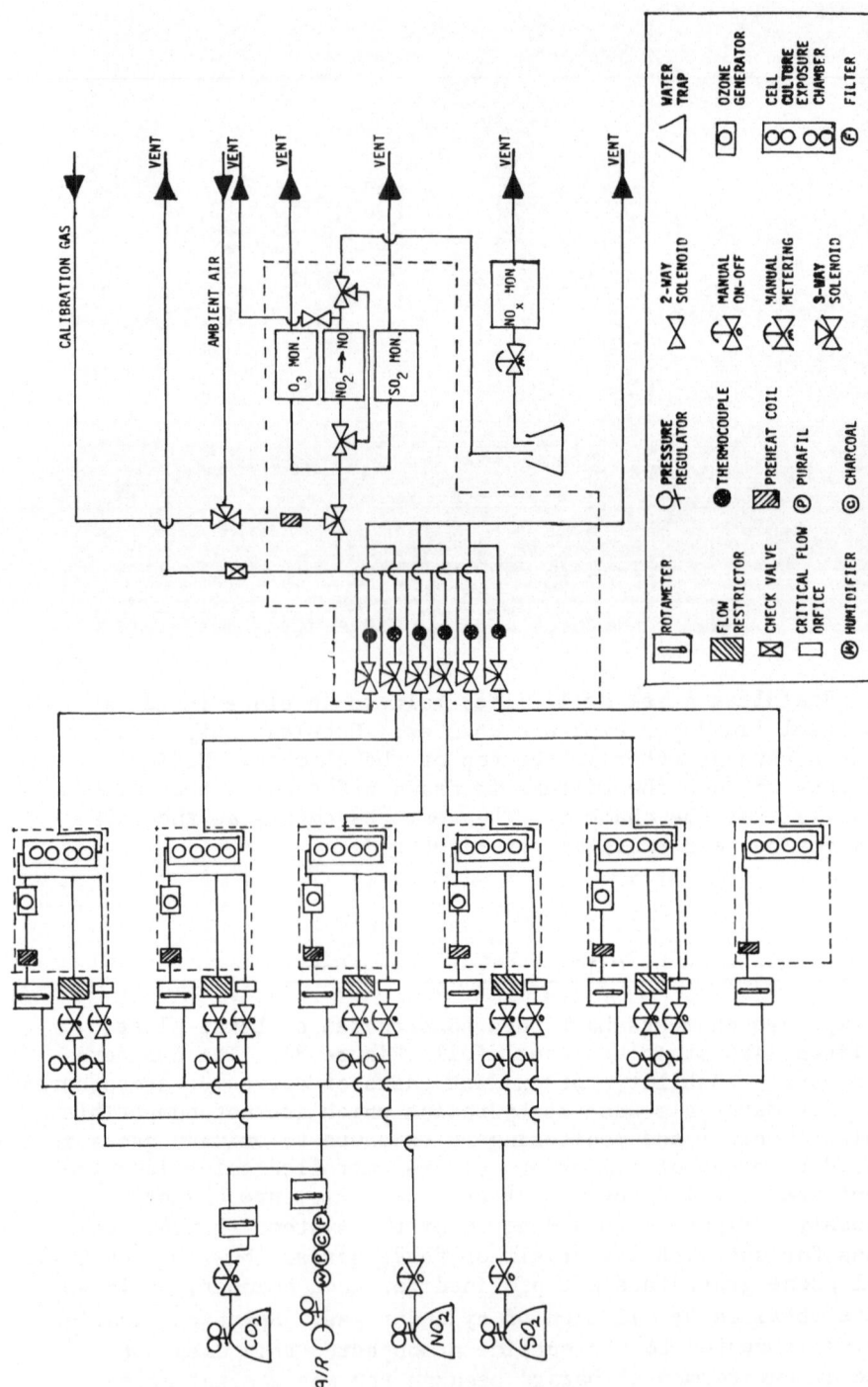

Figure 4. Six-chamber system for exposing cell cultures to NO_2, O_3, or SO_2.

lower than the ambient concentrations in the environment. This means that the laboratory air used as carrier gas must be purified before use. The system employs a sequence of Purafil (Purafil, Inc., Napa, CA), activated charcoal, and bacteriologic filters to remove the major contaminants. With this filtration system, the concentration of NO_2 in the air supply is reduced to approximately 0.01 ppm, and O_3 is below the level of detectability (1 ppb).

The gas-analysing instruments were chosen to be specific for the gases to be measured, and are a Beckman 952A $NO/NO_2/NO_x$ analyser and a Dasibi 1003AH ozone monitor. An SO_2 monitor has not been installed at this writing.

For exposure to gases or vapors, cell suspensions were seeded onto the filters, either with the filters in plastic petri dishes, or with the filters already assembled into the holders. In the latter case the cell suspension was placed in the top well of the holder (Figure 2) and the cells were allowed to settle and attach to the filters over a period of several hours. After assembly, the filter holders were placed in the exposure chambers, the chambers sealed, and gas flow through the chambers was initiated. When the desired gas concentration was reached within the chambers, the medium overlying the cells was withdrawn by a peristaltic pump and the exposure begun. Perfusion of medium through the filters was started at the same time. At the end of the exposure period, the chambers were opened, the holders removed, and the cell-bearing filters treated as appropriate to the experiment.

For measurements of cytotoxic effects of O_3 or NO_2, the filters were transferred to petri dishes containing culture medium, and returned to a 37°C, 5% CO_2 incubator. Cell survival, or colony forming ability (CFA), was measured as the ability of cells on the filter to form macroscopic colonies after 6–7 days of incubation.

The effects of oxidant gases on DNA synthesis in V-79 cells was measured as described by Painter (14). Cellular DNA was pre-labeled with ^{14}C-thymidine (^{14}C-TdR) by incubating the cells for at least two cell generations in the presence of the tracer. After exposure to gas, the cultures were pulse-labeled (10 min) at intervals with ^{3}H-thymidine (H-TdR 5 μCi/mL, 50–60 Ci/mmole), and immediately fixed with cold 5% trichloracetic acid. Measurement of the $^{3}H/^{14}C$ ratio by scintillation counting gave an index of the rate of DNA synthesis at the instant of labeling with ^{3}H-thymidine.

The incidence of sister chromatid exchanges (SCE) in V-79 cells was determined essentially by the method of Perry and Wolff (15). The cells, which have a cycle time of about 12 hr, were planted on filters and grown for approximately 18 hr in medium containing $10^{-5}M$ 5-bromodeoxyuridine (BrdUrd). At this point, the cells were exposed to O_3 at 0.035 ppm for 1 hr, and then returned to

medium containing BrdUrd and colcemid (0.1 µg/mL) for an additional
3-4 hr. Thus, SCE induced during the second replication cycle in
the presence of BrdUrd would be detected.

EXPERIMENTAL RESULTS

Cytotoxic Effect of NO$_2$ on V-79 Cells

Figure 5 shows the results of studies on the survival of
V-79 cells after exposure to 0.15 ppm of NO$_2$ for the times indicated.
These data confirm previous findings (16) of the great sensitivity
of V-79 cells to inhibition of colony formation by NO$_2$. The basis
for this inhibition is probably the same as for the effects seen
with O$_3$ (see below), i.e., cells are being destroyed or otherwise
lost from the filters as a consequence of the NO$_2$ exposure.

Effects of NO   on DNA Synthesis in V-79 Cells

Preliminary studies suggested that NO$_2$ may have had an
inhibitory effect on DNA synthesis in V-79 cells. Further studies
have shown that, at NO$_2$ doses which do not result in extensive cell
loss, there is a transient inhibition of DNA synthesis followed by
recovery. In some cases, the rate of DNA synthesis in the NO$_2$- ex-

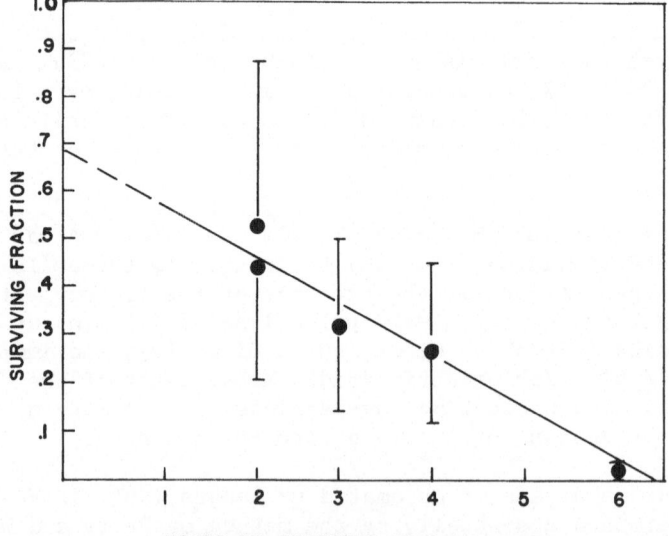

HOURS OF EXPOSURE TO 0.15 PPM NOx

Figure 5. Cytotoxic effect of NO$_2$ and V-79 cells. The data points
indicate the survival of colony-forming ability relative to cells
exposed to clean air for the same time period.

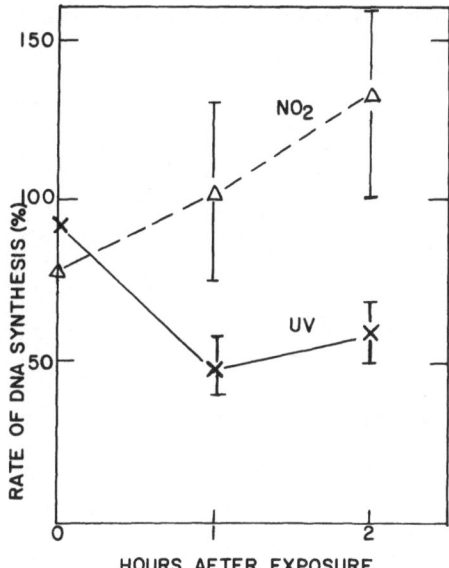

Figure 6.   Effect of ultraviolet light or NO_2 on the rate of DNA synthesis in V-79 cells measured immediately after exposure and at 1 and 2 hours later.   The values are plotted as a percentage of the rate of DNA synthesis in corresponding control cultures.

posed cultures exceeded that in the air-exposed.   The data from these studies are summarized in Figure 6.   The results of positive-control studies with UV light were as expected in that the rate of DNA synthesis continued to decline after the short exposure to UV light (14).   In the cultures exposed to 0.15 ppm of NO_2 for 1 hr, there was no evidence of significant cell loss from the filters as judged by comparison of the ^{14}C label remaining on the filters between NO_2 and air-exposed cultures.   The conclusion from these studies is that a one hour exposure to 0.15 ppm of NO_2 did not cause DNA damage sufficient to inhibit DNA synthesis.

Cytotoxic Effects of O_3

Cell suspensions were seeded onto filters previously assembled into the holders at 2.5 x 10^5 cells per filter.   Twenty-four hr later, the cell-bearing filters were exposed to 0.05 ppm of O_3 for periods of 1-4 hr.   The filters were then removed from the holders, and treated with 0.25% trypsin to remove attached cells.   The cells were collected by centrifugation, counted, an aliquot taken for viability testing with trypan blue (17), and a second aliquot seeded into dishes to determine CFA.   The results are shown in Table 1. It is evident that exposure to O_3 caused a substantial loss in the

Table 1.   Cytotoxic effect of 0.05 ppm O_3 on V-79 cells

Expt. No.	Treatment	Fraction of Cells Recovered[a]	Fraction Viable[b]	CFA[c]
35-79	Immersed	1.00	1.00	0.40
	Air, 2 hr	0.70	0.95	0.31
	Air, 4 hr	0.44	0.84	0.30
	O_3, 2 hr	0.12	0.92	0.26
	O_3, 4 hr	0.06	0.89	0.42
37-79	Immersed	1.00	0.99	0.28
	Air, 1 hr	0.22	0.87	0.44
	Air, 3 hr	0.36	0.83	0.26
	O_3, 1 hr	0.07	0.78	0.22
	O_3, 3 hr	0.08	0.94	0.25

[a] Compared to immersed control.
[b] Fraction of recovered cells indicated as viable by trypan blue exclusion (8).
[c] Colony-forming ability of cells indicated as viable by trypan blue exclusion expressed as a fraction of cells plated. Values are based on five replicate plates.

fraction of cells recovered from the filters. Exposure to air also caused some loss, but not to the same extent. The cells recovered were nearly all viable as judged by trypan blue exclusion and the CFA were comparable to cells held in immersed culture. The state of viability of the cells which detached from the filters during gas exposure could not be determined since they would presumably be carried off with the medium perfusing through the filters. These results support the conclusion that O_3 more than air, affects the cells in a way that causes them to detach from the filters.

Effect of O_3 on DNA Synthesis in V-79 Cells

Chemicals and physical agents which damage cellular DNA also tend to inhibit DNA replication. If the DNA damage is not rapidly repaired, the rate of DNA replication will continue to decline with time (14). There have been reports that O_3 may cause chromosomal damage (18), and may therefore cause DNA damage. Experiments to examine the effect of low levels of O_3 on DNA sythesis were done in order to obtain evidence regarding possible DNA damage by O_3.

The method was as described above. Cells were prelabeled with ^{14}C-TdR, planted on filters, and after allowing time for attachment,

exposed to O_3, air, or to UV light as a positive control agent known
to damage DNA.   Cultures were pulse-labeled with ^{3}H-TdR at intervals
following the gas or UV exposure.

The results of three experiments are summarized in Table 2.

One effect of the O_3 exposure was to cause a loss of ^{14}C-labeled
cells from the filters.   This finding confirms previous observations
which indicated cells were being lost after O_3 exposure.

In most experiments, O_3 produced a decrease in the rate of DNA
synthesis.   This decrease persisted for some time after exposure, but
eventually returned to the control level or above.   Exposure to UV
light also caused a reduction in the rate of DNA sythesis, which did
not recover to control levels during the time period studied.   In some
experiments with UV, the rate of DNA sythesis continued to decline.
The conclusion from these studies is that exposure of cells to O_3
under the conditions described did not produce DNA damage to an
extent that would inhibit DNA sythesis.   The pattern of inhibition
of DNA synthesis by O_3 resembled that seen after exposure of cell
cultures to toxins that are metabolic poisons which do not damage
DNA (14).   The data also suggest that if cells are not lost from the
filters, they will recover from the effects of the O_3 exposure.
This latter conclusion is supported by the experiments in which
cytotoxic effects were examined using trypan blue dye exclusion
tests and CFA (Table 1).

A Test for Sister Chromatid Exchange (SCE) Induction by O_3

The results of tests for SCE induction in V-79 cells by O_3
showed no significant difference between the SCE frequency in air-
exposed and O_3-exposed cells.   The data in Table 3 are from two
independent counts of the SCE incidence.   The graphs of one of
these sets of data are shown in Figure 7.   Although the data are
suggestive of a higher incidence of SCE in the O_3-exposed cells,
it does not appear that O_3 induces SCE.

DISCUSSION

The cell exposure system used in these studies was originally
designed to be used with the oxidant gases NO_2, O_3 and possibly
with SO_2.   The materials used in construction were therefore selected
to be compatible with these gases.   Exposure chambers were fabricated
from Lexan plastic; gas lines and other plumbing were of stainless
steel or teflon; the special cell culture holders were made of Lexan
plastic.   Physically, the exposure system is very sturdy.   As
development of the system has proceeded, it has taken on a modular
character.   Lexan plastic and stainless steel chambers and cell
culture holders are readily interchangeable.   Additional chambers
have been developed and used to exposure of organ cultures or rodent

Table 2. Effect of O_3 exposure on the rate of DNA synthesis in V-79 cells

Expt.	O_3 Dose	Ratio O_3/Air[a]	Ratio UV/PBS[a]	^{14}C-DPM on O_3-exposed Filter[b]
1-80	0.03 ppm, 1 hr	0 hr = 0.27 4 hr = 0.34	0 hr = 0.95 4 hr = 0.84	Air, 0 hr = 6756 O_3, 0 hr = 3810 Air, 4 hr = 6403 O_3, 4 hr = 3329
2-80	0.03 ppm, 1 hr	0 hr = 0.35 3 hr = 0.84 4 hr = 1.57	0 hr = 0.68 3 hr = 0.71 4 hr = 0.44	Air, 0 hr = 3048 O_3, 0 hr = 2299 Air, 3 hr = 1635 O_3, 3 hr = 1642 Air, 4 hr = 510 O_3, 4 hr = 2016
4-18-80	0.035 ppm, 1 hr	0 hr = 0.87 0.5 hr = 0.89 1.5 hr = 0.54 3.5 hr = 1.91	0 hr = 0.89 0.5 hr = 0.90 1.5 hr = 0.69 2.5 hr = 0.59 3.5 hr = 0.69	O_3, 0 hr = 2663 O_3, 0.5 hr = 2061 O_3, 1.5 hr = 598 O_3, 3.5 hr = 408

[a]Calculated as the ratio of ^{3}Hdpm/^{14}Cdpm in the O_3 or UV exposed cells divided by the ^{3}Hdpm/^{14}Cdpm ratio in the corresponding control.
[b]Total ^{14}Cdpm remaining on filter at the time of measurement. Values are the mean of 3 or 4 filters.

Table 3.   Test for sister chromatid exchange induction
            in V-79 cells by ozone

Cell cultures were exposed on filters to 0.03 ppm of O_3 for 2
hours.   The cells had been grown in medium containg BrdUrd (10^{-5} M)
for 18 hours prior to exposure to O_3 .   Following exposure, the
cell-bearing filters were transferred to culture dishes with
medium containg BrdUrd for an additional 4 hours at which time
the cells were harvested and stained for SCE analysis as described
(7).   The values are SCE per cell± S.D.   The numbers in parentheses are the number of cells counted.   Only complete metaphases
were scored.   The slides were scored twice with different observers.

	O_3-Exposed	Air-Exposed
Count #1	6.6 ± 2.2(53)	5.1 ± 1.7(60)
Count #2	4.9 ± 2.0(60)	4.0 ± 1.6(60)

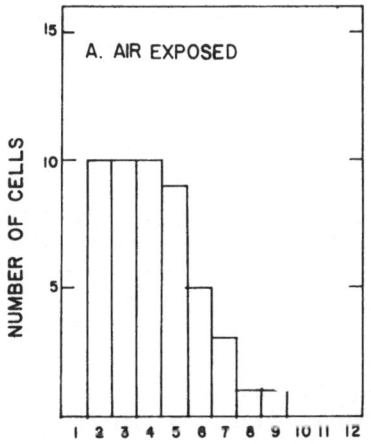 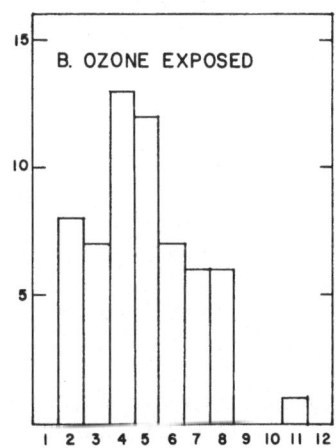

NUMBER OF EXCHANGES

Figure 7.   Sister chromatid exchange incidence in cell line V-79
after exposure to ozone for 1 hour at 0.035 ppm.

tracheal explants. The gas flow capacity of the system (4 L/min
per chamber) is sufficient for acute exposure studies with small
numbers of mice or rats. This may enable direct comparison of
in vivo and in vitro effects in the same species. The stainless
steel chambers and culture holders have been used to study the effect
of fuel vapors (JP-5) on cell cultures. Methods have been developed
for vaporizing JP-5 or other similar hydrocarbons and mixing them
with air or other gas mixtures. Further methods have been developed
for condensing the fuel vapors or allowing the vapors to contact
the cell cultures. These studies, to be described elsewhere, have
demonstrated the flexibility of the system. With minor modifications,
a variety of potential airborne pollutants can be vaporized and
their toxic effects on cell cultures measured.

The results found with the cell exposure studies suggest
that some modes of cell exposure may exaggerate the sensitivity of
cells well beyond that which is present in vivo. This may indicate
an important application of the exposure system in studies of
pollutant-cell membrane interactions on a morphological level, perhaps
using electron microscopy. However, studies of mutagenesis, or other
events requiring cellular metabolism following exposure may be
overshadowed by the toxic effects resulting from the impact of
airborne pollutants on the cell surface. This tentative conclusion
is based on studies with established mammalian cell lines, and may
not apply to primary cultures of cells which retain their in vivo
characteristics.

In summary, the exposure system is now at a stage of development
that allows generation and monitoring of a variety of materials,
mixing them with O_3 or NO_2, and exposure of cell cultures to the
resulting vapor or condensed material. The system is probably best
suited for laboratory research studies rather than as a device for
screening of possible biological effects is test materials.

ACKNOWLEDGEMENTS

The development of the cell exposure system was a collaborative
effort between the Department of Community and Environmental Medicine
and the School of Engineering, University of California, Irvine.
We are very grateful to Dr. G. S. Samuelsen and J. T. Taylor for
their participation.

Support from the Environmental Protection Agency (Contract No.
68-02-1204), the National Cancer Institute (Grant No. CA 15079), the
National Institute for Environmental Health Sciences (Grant No.
ES-01835), and Air Force Office of Scientific Research (Grant No.
77-3343) at various stages of this project is gratefully acknowledged.

We also thank D. S. Swedberg, M. E. Witte, and E. R. Fox for excellent technical assistance.

REFERENCES

1. C. G. Plopper, D. L. Dungworth and W. S. Tyler, Morphometric evaluation of pulmonary lesions in rats exposed to ozone. Am. J. Pathol. 71:395-408 (1973).
2. M. J. Evans, R. J. Stephens, L. J. Cabral and G. Freeman, Cell renewal in the lungs of rats exposed to low levels of NO_2. Arch. Environ. Health 27:180-188 (1972).
3. C. E. Cross, A. J. de Lucia, A. K. Reddy, M. Z. Hussain, C. K. Chow and M. G. Mustafa, Ozone interactions with tissue: Biochemical approaches. Am. J. Med. 60:929-935 (1976).
4. B. D. Goldstein, C. Lodi, C. Collinson and O. J. Balchum, Ozone and lipid peroxidation. Arch. Environ. Health 18:631-635 (1969).
5. Y. Berwald and L. Sachs, In vitro transformation of normal cells to tumor cells by carcinogenic hydrocarbons. J. Natl. Cancer Inst. 35:641-661 (1965).
6. J. DiPaolo and P. Donovan, Properties of Syrian hamster cells transformed in the presence of carcinogenic hydrocarbons. Exptl. Cell Res. 48:361-377 (1967).
7. R. J. Mason, M. C. Williams, R. D. Greenleaf and J. A. Clements, Isolation and properties of Type II alveolar cells. Am. Rev. Resp. Dis. 115:1015-1026 (1977).
8. Y. Kikkawa and K. Yoneda, The Type II epithelial cell of the lung. Lab. Invest. 30:76-84 (1974).
9. T. R. Devereaux and J. R. Fouts, The nonciliated cell bronchiolar epithelial (Clara) cell of the rabbit lung: Isolation and identification. Fed. Proc. 1092 (1980).
10. D. M. Pace, J. R. Thompson, B. T. Aftonomos and H. G. O. Holck, The effects of NO_2 and salts of NO_2 upon established cell lines. Can. J. Biochem. Physiol. 39:1247-1255 (1961).
11. R. H. Fetner, Ozone-induced chromosome breakage in human cell cultures. Nature 194:793-794 (1962).
12. C. Voisin, C. Aerts, E. Jakubczak and A. B. Tonnel, La culture cellulaire en phase gazeuse. Bull. Europ. phisopath. Resp. 13:69-82 (1977).
13. G. M. Alink, J. C. M. van der Hoeven, F. M. H. Debets, W. S. M. ven de Ven and J. H. Koeman, A new exposure model for in vitro testing of effects of gaseous pollutants on mammalian cells by means of gas diffusion through plastic films. Chemosphere 2:63-73 (1979).
14. R. B. Painter, Rapid test to detect agents that damage human DNA. Nature 265:650-651 (1977).
15. P. Perry and S. Wolff, A new giemsa method for the differential staining of sister chromatids. Nature 251:156-158 (1974).

16.   G. S. Samuelsen, R. E. Rasmussen, B. K. Nair and T. T. Crocker,
      Novel culture and exposure system for measurement of
      effects of airborne pollutants on mammalian cells. Env.
      Sci. Technol. 12:426-430 (1978).
17.   H. J. Phillips, Dye exclusion tests for cell viability, in:
      Tissue Culture Methods and Applications, P. F. Kruse, Jr.,
      and M. K. Patterson, Jr., Eds., Academic Press, New York,
      406-408 (1973).
18.   P. C. Gooch, D. A. Creasia and J. G. Brewen, The cytogenetic
      effects of ozone: Inhalation and in vitro exposures.
      Environ. Res. 12:188-195 (1976).

DISCUSSION

Q. Hatch (Northrop): I'd like to make one comment. Its' been our
experience that if you want to look at genotoxic effects, as opposed
to cytotoxic effects, you might want to increase your innoculum
density and you might also want to go to shorter treatment periods
at slightly higher doses.

Q. Rounds (Pasadena): Do you see any difference in the response
of different cell types to these gases? What cell type do you
normally use?

A. Rasmussen (Irvine): The data that I've shown today was obtained
with Chinese hamster V-79 cells. We've done similar studies with
CHO cells. We have a couple of strains of rat lung cells, presumably
derived from Type II cells, which show similar toxic effects and
again no genotoxic effects. Therefore, the results are identical
for a number of different kinds of cells. I should mention that it
is possible to establish cultures of almost any kind of cell that
you can grow on a filter and expose them in this system.

Q, Rounds (Pasadena): Also, could you comment on what type of
filter you are using and whether or not you think there is any
interaction between the test gases and the filter itself.

A. Rasmussen (Irvine): Yes. We have tested this. The filter
that we use is a Millipore HAWP filter, with a 0.47 micron pore
size. We've tried three or four different manufactures and different
pore sizes and by trial and error arrived at this filter. We did
specific experiments to look for the formation of toxic materials
by simply exposing filters to the gas without cells and then looking
at the plating efficiency of cells on those exposed filters and
found no indication of toxic material.

SESSION III. PLANT BIOASSAYS FOR THE DETECTION OF

AIRBORNE GENOTOXIC AGENTS

L. A. Schairer, Moderator

Biology Department
Brookhaven National Laboratory
Upton, NY 11973 (U.S.A.)

The first two sessions of this Symposium were devoted to a
description of the state-of-the-art technology for in vitro
carcinogen/mutagen testing using microorganisms and mammalian cells.
Those systems are very amenable to assessment of aqueous and particulate chemical exposure. The emphasis in this session (now) shifts
to in vivo assay technology utilizing higher plants. Several
higher plants, and _Tradescantia_ in particular, have not only
demonstrated the potential for assessing the biological activity
of aqueous genotoxic agents but, and more importantly, they have
demonstrated the potential for detecting the mutagenicity of
gaseous or airborne agents. In most cases the whole organism is
exposed to chemical vapors under either laboratory or field conditions. Constraints of sterile culture are not a factor. These
eukaryotic plant systems can be exposed for periods of a few
minutes to days or weeks, thus simulating acute workplace exposures
as well as chronic exposure to ambient atmospheres.

Assessment of potential hazards from airborne environmental
contaminants must include a battery of short-term bioassays. The
number of false negatives and false positives reported for various
test systems underscores the fact that no single assay, whether
microbial, plant or animal, will tell the whole story.

Effects of genotoxic agents on plants are important for several
reasons: plants make up a large portion of our biosphere, have
a major aesthetic and economic impact on our society and constitute
a vital link in man's food chain. Promutagens have been shown to be
activated by plants and, in the food chain, could become a direct
threat to human health. In situ exposures of plants to industrial
effluents have yielded positive mutagenic responses which support

121

the use of plant assays as early warning systems for the detection and assessment of hazardous atmospheres.

The three papers in this session will present the state-of-the-art for some of the plant assays and put their role as monitors for airborne genotoxic agents into proper perspective.

MONITORING AMBIENT AIR FOR MUTAGENICITY USING THE HIGHER

PLANT TRADESCANTIA

L. A. Schairer, R. C. Sautkulis and N. R. Tempel

Biology Department
Brookhaven National Laboratory
Upton, NY 11973 (U.S.A.)

INTRODUCTION

The theme of the conference on "Genotoxic Effects of Airborne
Agents" concerned the state of the art of bioassay systems from
cultured microbes to laboratory animals and their application to
the assessment of human health effects of airborne environmental
contaminents. The major emphasis for short-term bioassays has
been placed on bacterial and mammalian cell lines. However, for
increased perspective on the state-of-the-art of specific in vitro
assays it is important to consider the environmental impact on
whole organisms by reviewing the contributions made by in vivo
assays. The more classical non-mammalian in vivo systems such as
Drosophila, Zea mays and Tradescantia are characterized by well
defined genetic bases, versatility in mode of treatment, relatively
low cost and short term and/or high sensitivity to both physical
(radiation) and chemical mutagens (1-5). This paper will deal
exclusively with somatic mutation in the Tradescantia stamen hair:
describing the system briefly, demonstrating its relevance to
environmental mutagen assessment and discussing its adaptation for
in situ ambient atmosphere monitoring.

As a result of very great increases in world population and
expansion in industrial and agricultural development, man is being
exposed to an ever increasing level of airborne environmental
pollutants. The sources of these chemical environmental contaminants
include automotive and industrial combustion products (6), pesticides (7,8) and commercially used chemical additives, solvents
or catalysts (9,10). In the past few years there has been
an accelerating effort to detect and identify the most hazardous
pollutants (11-16). It is to this end that we are exploiting

123

the very sensitive Tradescantia plant somatic mutation test system
to study the genetic effects of physical and known or suspected
chemical mutagens.

THE TRADESCANTIA STAMEN HAIR SYSTEM

The stamen hair system has been described in detail elsewhere
(1,5,17,18) so only certain features will be reviewed here. The
plant used exclusively in the field studies to be described here
is clone 4430, a diploid interspecific hybrid (T. subacaulis x T.
hirsutiflora) bred at Brookhaven National Laboratory (Fig 1). This
clone is a hybrid between pink and blue flowering parents with blue
being dominant over pink. The visible marker used in this test
system is the phenotypic change in pigmentation from blue to pink
in mature flowers. The pigmentation change (hereafter called mutational or pink event) is induced in young developing floral tissue
and is expressed 5 to 18 days later as isolated pink cells or groups
of pink cells in the stamen hairs of mature flowers (Fig 1 c,d).
The pink events are essentially nonlethal; large mutant sectors
indicate genetic injury early in the development of that tissue.

Conventional plant breeding tests were made with clone 4430 and
appropriate parental stocks to confirm the genetic basis for the
pink color locus. Classical Mendelian segregation ratios of 3:1
and 1:1 (blue to pink) were obtained indicating single gene control
of flower color (19,20). Other endpoints documented by radiobiolog-
cal studies and available for environmental chemical assessment in
clone 4430 include colorless (non-pigmented) stamen hair cells,
stunted hairs (cell death or growth inhibition) and aborted pollen
grains (Fig 2). Figure 2a shows one and two-celled hairs; such
hairs and others less than 12 cells in length are called stunted
and are scored as nonsurvivors. Thus, it is possible to construct
survival curves as well as mutation dose-response curves for stamen
hairs after mutagen treatments (1,17). Hence, environmental stress
expressed as both genetic and toxic effects can be measured with this
system.

Figure 1. (a) Normal stock plant of Tradescantia clone 4430 showing
several mature inflorescences. (b) Floral cuttings are shown
prepared for exposure. Sixty cuttings per dish are grown in aerated
Hoagland's nutrient solution during and after treatment. (c) Somatic
mutation, indicated by dotted area, at the blue color locus results
in expression of the recessive pink allele in petal sectors; the
large size of the sector shown indicates induction early in the
development of the bud. (d) Enlargement of stamen hairs with pink
mutant cells, indicated by dotted area.

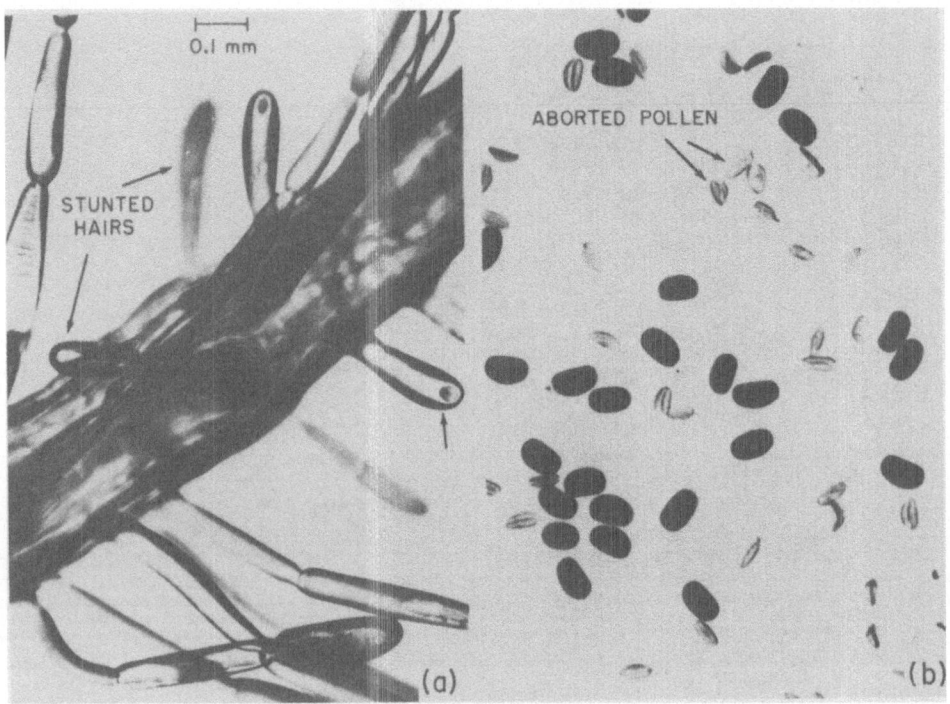

Figure 2.   (a) Enlargement of filament showing several stunted stamen
hairs typical of those seen after exposure to radiation or chemical
mutagens.   (b) Mutagen-induced pollen abortion in Tradescantia as
indicated by unstained (cotton blue) or collapsed grains.

The stock plants are easily maintained by vegetative propagation and flower continuously throughout the year in controlled-
environment growth chambers. The material treated consists of
unrooted, fresh cuttings each with a young inflorescence containing
flower buds in a range of developmental stages (see Fig 1b,c).
Following exposure to either chemical or physical mutagens, the
cuttings are grown in aerated Hoagland's nutrient solution under
standard conditions (1) and the flowers are analyzed each day
as they bloom for approximately two weeks after treatment. Induced
pink-event rates are expressed as the mean of the rates for several
consecutive peak response days, usually days 11 to 15 for acute x
rays and 7 to 12 for acute chemical exposures. Detailed descriptions
of laboratory techniques for radiation and chemical exposures and
methods for calculating mutation rates are given elsewhere (1,17,18,
21).

Radiobiology

Extensive use of Tradescantia in radiobiological studies has provided much information about somatic mutation and cell death produced by x, gamma, neutron and heavy ion radiations (1,22). Many data exist on induced chromosome aberrations in both somatic and gametic tissues (23,24). The classical linear, non-threshold, dose-response curves established for several endpoints in Tradescantia serve as models for chemical mutagen responses; these include prediction and extrapolation to very low dose effects (Fig 3,4). Recent work in our laboratory has demonstrated that the methods used to study the genetic or cytogenetic effects of physical mutagens can be applied directly to chemical mutagen studies (1).

Chemical exposures under laboratory conditions

The techniques developed for vapor phase chemical exposures in the laboratory have been described in detail elsewhere (17,25) so comments here will be limited to a few pertinent observations.

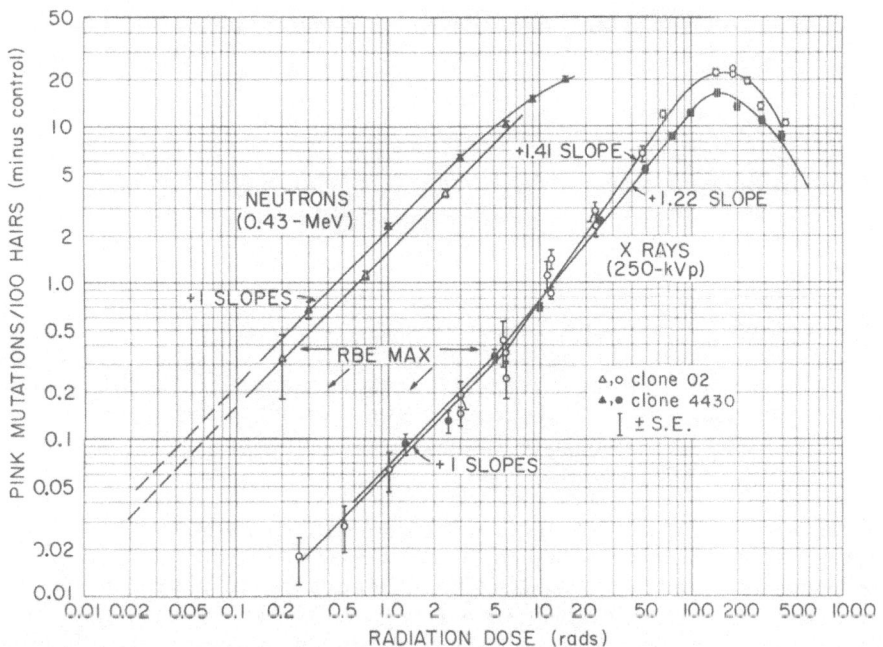

Figure 3. Neutron and x-ray dose-response curves for pink mutant events in stamen hairs of Tradescantia clones 4430 and 02. Note slope change from plus 1 below 6 rads x rays to a 2-hit component resulting in 1.22 and 1.41 slopes at higher doses.

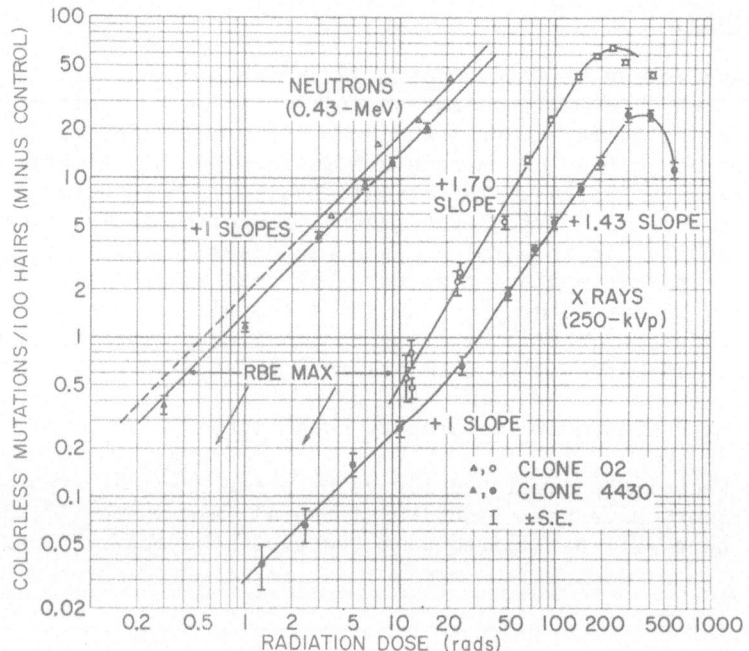

Figure 4.   Neutron and x-ray dose-response curves for colorless
mutant events in stamen hairs of <u>Tradescantia</u> clones 4430 and 02.

Young inflorescences of <u>Tradescantia</u> clone 4430 were exposed
to gaseous 1,2-dibromoethane (DBE), an alkylating agent used as a
standard chemical mutagen in our studies.   The mutation frequency
increased linearly with both increasing chemical concentration
(0.1 to 100 ppm) and duration of exposure (2-144 hours).   These data
may be expressed in terms of total dose by plotting induced mutation frequency against the product of concentration (ppm) and
duration of exposure (hours) (Fig 5).   For purposes of comparison,
a standard curve for x-ray effect is shown in rads.   Slope and
shape of the curve for DBE induction of color change resemble
those for radiation injury.

Of particular interest are the shapes of the dose response
curves following DBE exposures of 6 and 144 hours (Fig 6).   Data
from 6-hour exposures show a significant decrease in slope in the
low dose range.   Although the curve for the 144-hour exposure shows
a higher level of response, the data are consistent with the shallow,
low dose slope shown for the 6-hour exposures.   Differences in
mechanisms responsible for mutation at high and low doses are unknown for chemicals, but as with radiation results, the response
is greater at low doses (Fig 3,4,6).   Extrapolation from high to

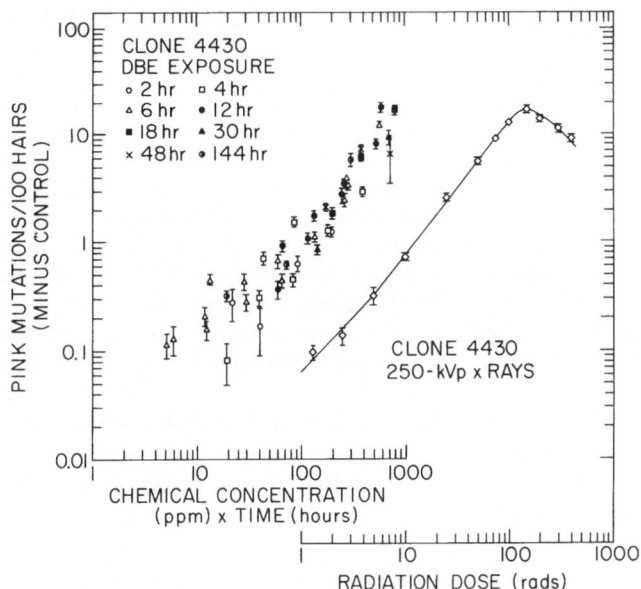

Figure 5. DBE-induced mutation response plotted against total dose (ppm X hours of exposure) results in a linear response curve. The standard x-ray curve is shown for comparison. (From Schairer et al. (5)).

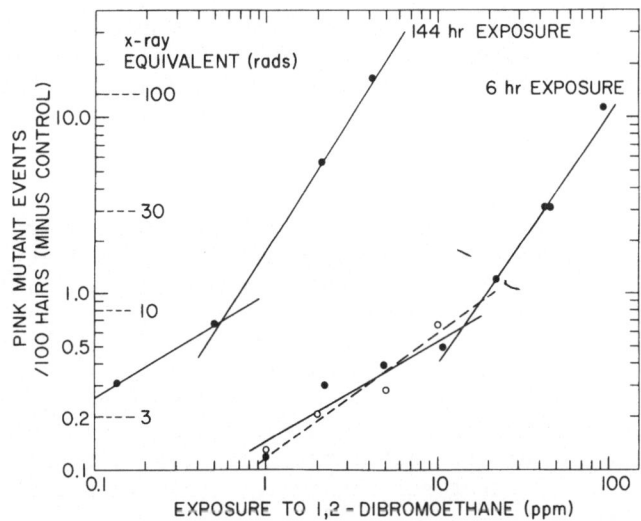

Figure 6. Response curves for pink stamen hair mutations in Tradescantia clone 4430 show the results of 6- and 144-hour exposure to DBE. Slopes less than +1 were obtained at low doses. Open and closed symbols (6-hr) represent separate experiments. (From Schairer (34)).

low levels of mutagen exposures would lead to an underestimate of
the hazard (26,27).

Although a large percentage of the effort of this group has
been spent on the development of a mobile monitoring vehicle, a
number of chemicals have been tested in the laboratory to validate
the system as a monitor for gaseous mutagens. Chemicals such as
the gasoline additives 1,2-dibromoethane (DBE) and trimethyl phosphate (TMP) were found to be potent mutagens whiles SO_2, NO_2, vinyl
chloride, and freon-22 were weak mutagens in this test system. These
and other chemicals or air pollutants tested are listed in Table 1.
The concentration listed is the lowest value tested that showed a
statistically significant mutagenic response.

Benzo(a)pyrene and other mutagens (numbers 24-34 in Table 1)
were applied topically in 1% DMSO solution. The aqueous results
should be considered preliminary but are generally consistant
with published mutagenic responses.

Chemical exposures under field conditions (ambient air

Laboratory studies with chemicals indicated that the stamen
hair system was highly sensitive to gaseous chemicals (17,28) and
that the system should respond to in situ exposures to industrial
pollution at ambient levels. The criteria for monitoring air
pollution for mutagenicity in the field include: (a) a roadworthy
vehicle to house test organisms during exposure; (b) exposure of
the test organisms under suitable culture conditions; (c) a constant
flow of untempered ambient air; and (d) a chronic exposure capability
simulating natural exposures of plants and animals. A detailed
description of the design and operation of the mobile monitoring
vehicle (MMV) has been given elsewhere (5,18). The mobile laboratory (Fig 7) was insulated and air conditioned to permit year-
round operation of the laboratory. Three Model M-13 growth chambers
(Environmental Growth Chambers, Chagrin Falls, Ohio) were installed.
One of the chambers serves as a clean air control and the second is
used for ambient air exposures. The chambers are designed to maintain any desired standard laboratory condition or, if desired, to
simulate fluctuations in the temperature and relative humidity of
the ambient air outside. Ambient air is drawn into the exposure
chamber through a 10.16 - cm glass duct at continuous flow rates up
to about .5 cubic meters per minute, a maximum of one air change
every two minutes. An air filter train is used to scrub the air
continually in the chamber serving as the concurrent control.

Field exposures are accomplished in the following manner: fresh
cuttings of Tradescantia clone 4430 are made from stock plants
grown in controlled environment chambers at Brookhaven National
Laboratory; they are hand carried to the test site by car or air-
plane; cuttings are placed in the chambers in glass containers
filled with Hoagland's nutrient solution (Fig 1b); and the cuttings

Table 1. A list of chemicals with which Tradescantia was treated, the conditions of the treatment and their mutagenic effect given in terms of the level of statistical significance (%).

No.	Chemical	Phase	Exposure Time(hr)	Minimum[a] Conc.	Maximum[b] Conc. for Effect	Statistical Signific. %
1	Ozone	Vapor	6	5 ppm		2
2	Sulfur Dioxide	Vapor	6	40 ppm		1
3	Nitrogen Dioxide	Vapor	6	50 ppm		5
4	Nitrous Oxide	Vapor	6	250 ppm		1
5	Ethyl Methanesulfonate	Vapor	6	5 ppm		1
6	1,2-Dibromoethane	Vapor	6	0.14 ppm		1
			144	0.14 ppm		1
7	Trimethylphosphate	Vapor	13	2 ppm		2
8	Trichloroethylene	Vapor	6	0.5 ppm		1
9	Vinyl Chloride	Vapor	6	75 ppm		2
10	Vinylidene Chloride	Vapor	6	—	1288 ppm	Insig.
11	Vinyl Bromide	Vapor	24	22 ppm		5
12	2-Bromoethanol	Vapor	24	50 ppm		1
13	Dichlorodifluoromethane (Freon-12)	Vapor	6	—	392 ppm	Insig.
14	Chlorodifluoromethane (Freon-22)	Vapor	6	194 ppm		2
15	Hexamethylphosphoramide	Vapor	6	Saturated		1
16	Benzene	Vapor	6	4000 ppm		1
17	Caffeine	Liquid	Chronic	10^{-3} M		Insig.
18	Atrazine	Liquid	Chronic	0.045 gm/pot		Insig.
19	Sodium Azide	Liquid	3	10^{-4} M		1
20	1,1-Dibromoethane	Vapor	6	—	250 ppm	Insig.
21	Dimethylamine Hydrochloride	Liquid	2	10^{-2} M		10

Table 1. (cont'd).

No.	Chemical	Phase	Exposure Time(hr)	Minimum[a] Conc.	Maximum[b] Conc. for Effect	Statistical Signific. %
22	Ethyl Alcohol	Vapor	24	1000 ppm		1
23	Hydrazoic Acid	Vapor	6	3.5 ppm		1
24	N-Methyl-N-nitro-N-nitroso-guanidine	Liquid	1.5	10^{-3} M		1
25	Safrole	Liquid	24	6.17×10^{-6} M		1
26	3,3',5,5'-tetramethyl benzidine	Liquid	24	4.16×10^{-6} M		1
27	Alpha-naphtylamine	Liquid	24	6.75×10^{-5} M		1
28	Beta-naphthylamine	Liquid	24	6.98×10^{-5} M		1
29	Benzidine	Liquid	24	5.43×10^{-7} M		1
30	Hexamethylphosphoramide	Liquid	24	5.58×10^{-5} M		1
31	Pyrene	Liquid	24	4.94×10^{-6} M		1
32	Diethylstilbestrol	Liquid	24	3.73×10^{-4} M		1
33	Benzo(a)pyrene	Liquid	24	3.96×10^{-5} M		1
34	N,N'-Ethylenethiourea (2-Imidazolidinethione)	Liquid	24	9.79×10^{-5} M		1
35	2-nitrofluorene	Liquid	24	4.7×10^{-6} M		2
36	Tetrachlorethylene	Vapor	6	1035 ppm		5
37	Diallate	Vapor	24	800 ppm		1

[a] the lowest concentration at which an effect was observed
[b] the highest concentration at which no effect was observed

Figure 7. View of the front of the Mobile Monitoring Vehicle (MMV) which shows remote mounting of the air conditioning and filter train for the three heat exchangers for the growth chambers. Lower right, 220 volt 100 amp electrical power cable. (From Schairer et al. (18)).

are exposed continuously for a ten-day period. At the end of the exposure the cuttings were taken back to Brookhaven National Laboratory for posttreatment analysis of the flowers as they bloom each cay. Exposures of a few hours to several weeks could be made but a ten-day period was chosen for the Tradescantia plants because it was long enough to maximize the sensitivity of the system, yet short enough to permit analysis of a sufficient number of flowers over the peak response period back at Brookhaven. The peak mutation response period following a ten-day exposure is 11 to 17 days after the start of the exposure. The mean of the mutation rates for a 5-to 7- day scoring period results in an observed rate for a given test site based on an average stamen hair population between 200,000 and 300,000. A population of 300 cuttings in each ambient air and control chamber will yield enough data to resolve as small as a 10% increase in pink events over the background frequency.

Results of in situ exposures to ambient air

The first field trials for the MMV were conducted in the summer
of 1976 in Elizabeth, NJ. Over the next four years a total of 17
additional sites throughout the United States were monitored in a
preliminary study. This study had two objectives: (i) to demonstrate
the adaptability of the stamen hair system to ambient air monitoring
and (ii) if mutagenicity were observed, to look for causative agents
common to positive sites (Fig 8). These sites were selected because
of known high levels of human cancer incidence or the presence of
high levels of suspect compounds in the atmosphere. Two exceptions
were the sites at Grand Canyon, AZ, and Pittsboro, NC. It was
deemed essential to conduct a "clean ambient air" exposure to verify
the efficiency of the filter on the concurrent control chamber and to
eliminate the possibility of an artifact generated in the ambient
air chamber. Two exposures at Grand Canyon produced similar results
with no significant difference in background mutation frequency
between the control and ambient air samples. Since the Grand Canyon
test represented a good baseline exposure to clean air in situ and
included all of the stress of shipping plant material, field
handling, etc., the weighted mean for four replicated ambient air
samples ($3.35\pm0.09/10^3$ hairs) was determined and this mutation frequency was established as the standard for comparison for all other
field sites monitored.* The results of mutagenicity monitoring at
all sites are shown in Figure 8 as the net induced mutation frequency
following subtraction of the Grand Canyon rate.

Clearly significant increases in mutation frequency at the flower
color lucus were observed at many industrial sites. The most consistent mutagenic response was that associated with processing of petroleum products. Mutation rate varied not only from site to site, but
also with repeated exposures at the same site. The induced mutation
frequency ranged from 16.6% at Elizabeth, NJ in January to 90.6% in
September. Some of this variation in response was undoubtedly due
to seasonal change in effluent production and varying wind direction;
prevailing winds in the summer and fall are southwest while winter
winds are northwest. The fixed location of the Elizabeth site
placed petroleum refining operations directly upwind during the summer.
It is important to note that atmospheric monitoring at fixed sites
is very dependent upon wind directtion and speed. Although
trailer locations were selected downwind from the desired source of

---

*In January, 1980 a second clean air site was selected at Pittsboro,
NC, and the results were in good agreement with those of the Grand
Canyon, AZ, study.

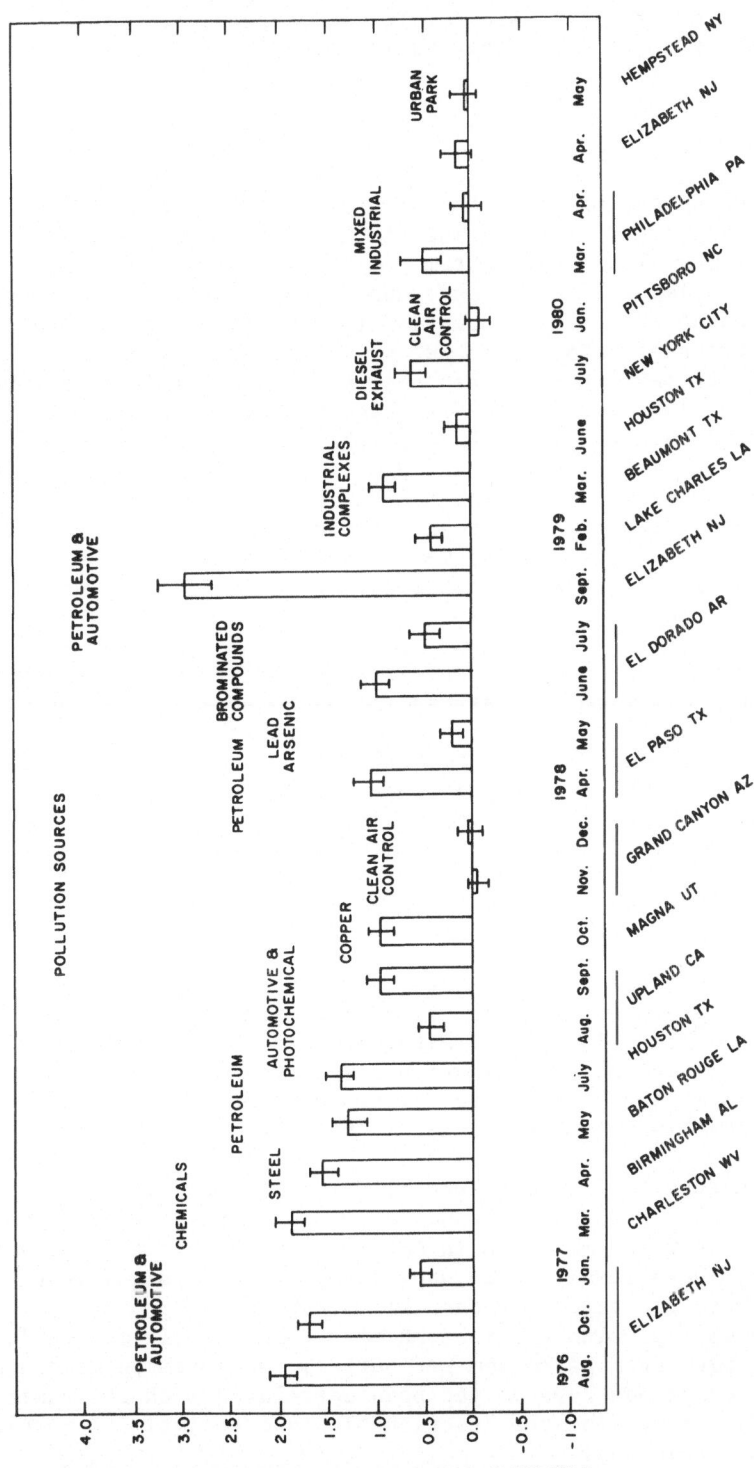

Figure 8. The mutagenicity of ambient air as measured by _Tradescantia_ in the MMV is summarized for eighteen test sites visited.

pollution, a false negative may result from unpredicted wind changes, extended periods of rain or other environmental factors.

SUMMARY AND CONCLUSIONS

Final assessment of human health effects resulting either directly or indirectly from exposure to harmful environmental agents may rest with mammalian test system results. In vitro systems, including mammalian cell and tissue cultures are short-term assays used most frequently for extrapolation to humans. However, the present consensus of opinion is that no single assay system is adequate. Discrepencies have been observed between false negatives, false positives and metabolic activation of promutagens in established genetic systems such as Salmonella, Saccharomyces, Drosophila and Zea mays. The more expensive long-term tests must be augmented by multiple assays designed for mere redundancy or to uniquely fill gaps in present state-of-the-art of environmental monitoring. The Tradescantia stamen hair test system is one such assay which can offer redundancy as well as fill the gap of monitoring ambient air for mutagenic agents.

The flower color lucus in heterozygous clones of Trandescantia has been shown to mutate when exposed to various categories of agents from fumigants (ethylene dibromide and trimethylphosphate), solvents (hexamethylphosphoramid and trichlorethylene), chemical additives or catalysts (vinyl chloride) and to compounds requiring activation such as benzo(a)pyrene. This laboratory validation of the stamen hair system supports its use as a short-term bioassay for chemical mutagens.

The most unique application of this plant system is its ability to respond to low levels of airborne compounds. Without the constraints of sterile culture and other complicated experimental procedures the Tradescantia plant is very adaptable to in situ monitoring at suspect industrial sites. Collection of data on the mutagenicity of ambient air pollution is an area of testing not amenable to mammalian or in vitro systems and hence an area in which Trandescantia can make a unique contribution.

The high sensitivity and versatility of the Tradescantia stamen hair system were put to good use by employing the plant as an in situ monitor for mutagens in ambient air in polluted industrial sites. Preliminary results from many sites showed a significant increase in mutation rate with one (Elizabeth, NJ) approaching a background doubling. The environment most consistantly mutagenic was that downwind from petroleum refineries. No specific compounds or groups of compounds have as yet been correlated with the positive sites. Studies involving the fractionation and analysis of complex mixtures are continuing.

FUTURE POSSIBILITIES

As an established short-term bioassay, the stamen hair system can be used as one of a battery or tier of tests for chemical mutagen assessment. The most unique contribution will be in the area of atmospheric monitoring. A collaborative study with Research Triangle Institute, Inc., under contract with the U.S. Environmental Protection Agency, has shown good correlation between mutagenicity of airborne particulates (Ames/Salmonella) and vapors (Tradescantia) at the same industrial sites (29). Fractionation, quantitative analysis and bioassessment of environmental mixtures using the above and other organisms may identify specific hazardous compounds which can be tested by the more rigorous mammalian systems. Fractionation techniques have the obvious advantage of simplfying atmosphere clean-up and regulation procedures for specific hazardous compounds.

Greater use of basic radiobiological data should be encouraged. There is a very large background of radiation data describing dose response curve patterns (27), dose rate effects (30,31), oxygen enhancement ratios in tissue (32) and nuclear factors influencing sensitivity (33). Dose response curve patterns have been shown to be similar in Tradescantia for both physical and chemical agents. Other phenomena established by radiation studies should be explored for chemicals. Predictive potential for effects of environmental mutagen agents is of equal importance to earlier studies related to nuclear fallout and its impact on ecology in general but food chain in particular.

ACKNOWLEDGEMENTS

This work was supported jointly by the U. S. Department of Energy, National Institute of Environmental Health Sciences, and U. S. Environmental Protection Agency. The authors acknowledge with thanks the special efforts of Dr. E. Pellizzari and staff of Research Triangle Institute for organic vapor analysis and Dr. J. Van't Hof for assistance in the preparation of this paper. The many hours of flower analysis by Mr. E. E. Klug, Ms. A. F. Nauman, Ms. M. M. Nawrocky, Ms. V. Pond, and Ms. R. C. Sparrow are also gratefully acknowledged.

REFERENCES

1. A. G. Underbrink, L. A. Schairer, and A. H. Sparrow, Tradescantia stamen hairs: a radiobiological test system applicable to chemical mutagenesis, in: "Chemical Mutagens: Principles and Methods for their Detection", Vol. 3, A. Hollaender, ed., Plenum Press, New York, N.Y., 171-207 (1973).

2. E. Vogel and F. H. Sobels, The function of Drosophila in genetic toxicity testing, in: "Chemical Mutagens: Principles and Methods for their Detection", Vol. 4, A. Hollaender, ed., Plenum Press, New York, N.Y., 93-142 (1976).

3. C. E. Nix and B. Brewen, The role of Drosophila in chemical mutagenesis testing, EPA-600/9-78-027, 111-123 (1978).

4. M. J. Plewa, Activation of chemicals into mutagens by green plants: A preliminary discussion, Environ. Health Perspectives 27:45-50 (1978).

5. L. A. Schairer, J. Van't Hof, C. G. Hayes, R. M. Burton, and F. J. de Serres, Exploratory monitoring of air pollutants for mutagenicity activity with Tradescantia stamen hair system, Environ. Health Perspectives 27:51-60 (1978).

6. R. G. Bond and C. P. Straub, eds., "Handbook of Environmental Control", Vol. 1, "Air Pollution", CRC Press, Cleveland, OH, (1973).

7. R. White-Stevens, ed., "Pesticides in the Environment", Vol. 1, Part II, Marcel-Dekker, N.Y. (1971).

8. S. S. Epstein and M. S. Legator, eds., "The Mutagenicity of Pesticides, Concepts and Evaluation", MIT Press, Cambridge, MA (1971).

9. L. Fishbein, W. G. Flamm, and H. L. Falk, "Chemical Mutagens, Environmental Effects on Biological Systems", Academic Press, N.Y. (1970).

10. W. G. Flamm and M. A. Mehlman, eds., "Advances in Modern Toxicology", Vol. 5, "Mutagenesis", Hemisphere Publishing Corporation, London (1978).

11. I. J. Hindawi, "Air Pollution Injury to Vegetation", National Air Pollution Control Administration Publ. No. AP-71, U.S. Dept. of Health, Education, and Welfare, Washington, D.C. (1970).

12. F. Vogel and G. Röhrborn, eds., "Chemical Mutagenesis in Mammals and in Man", Springer-Verlag, New York, N.Y. (1970).

13. B. L. Oser, Toxicology of pesticides to establish proof of safety, in: "Pesticides in the Environment", Vol. 1, Part II, R. White-Stevens, ed., Marcel-Dekker, New York, N.Y., 411-456 (1971).

14. W. G. Flamm, A tier system approach to mutagen testing, Mutat. Res. 26:329-333 (1974).

15. J. V. Neel, Developments in monitoring human populations for mutation rates, Mutat. Res. 26:319-328 (1974).

16. M. D. Waters, S. Nesnow, J. L. Huisingh, S. S. Sandhu, and L. Claxton, eds., "Application of Short-term Bioassays in the Fractionation and Analysis of Complex Environmental Mixtures", Plenum Press, New York, N.Y. (1978).

17. A. H. Sparrow, L. A. Schairer, and R. Villalobos-Pietrini, Comparison of somatic mutation rates induced in Tradescantia by chemical and physical mutagens, Mutat. Res. 26:265-276 (1974).

18. L. A. Schairer, J. Van't Hof, C. G. Hayes, R. M. Burton, and
    F. J. de Serres, Measurement of biological activity of
    ambient air mixtures using a mobile laboratory for in situ
    exposures: preliminary results from the Tradescantia plant
    test system, in: "Application of Short-term Bioassays in the
    Fractionation and Analysis of Complex Environmental Mix-
    tures", M. D. Waters, S. Nesnow, J. L. Huisingh, S. S. Sandhu
    and L. Claxton, eds., Plenum Press, New York, N.Y., 419-440
    (1978).

19. M. Emmerling-Thompson, and M. M. Nawrocky, Genetic studies
    of flower color in Tradescantia, J. Heredity 70:115-122
    (1979).

20. M. Emmerling-Thompson, and M. M. Nawrocky, Genetic basis for
    using Tradescantia clone 4430 as an environmental monitor of
    mutagens, J. Heredity 71:261-265 (1980).

21. J. Van't Hof and L. A. Schairer, Tradescantia assay system for
    gaseous mutagens, Mutat. Res., in press (1981).

22. A. H. Sparrow, A. G. Underbrink, and H. H. Rossi, Mutations
    induced in Tradescantia by small doses of x-rays and
    neutrons: analysis of dose-response curves, Science 176:
    916-918 (1972).

23. J. R. K. Savage, Dose-response curves for chromosome type
    aberrations in Tradescantia. I. Preliminary considerations,
    Radiation Botany 15:87-140 (1975).

24. T.-H. Ma, A. H. Sparrow, L. A. Schairer, and A. F. Nauman,
    Effect of 1,2-dibromoethane (DBE) on meiotic chromosomes of
    Tradescantia, Mutat. Res. 58:251-258 (1978).

25. C. H. Nauman, P. J. Klotz, and L. A. Schairer, Uptake of
    tritiated 1,2-dibromoethane by Tradescantia floral tissues:
    relation to induced mutation frequency in stamen hair cells,
    Environ. and Exp. Botany 19:201-215 (1979).

26. C. H. Nauman, A. H. Sparrow, A. G. Underbrink, and L. A.
    Schairer, Low-dose mutation response relationships in
    Tradescantia; principles and comparison to mutagenesis
    following low dose gaseous chemical mutagen exposure, in:
    "Radiobiological Protection, First European Symposium on
    Rad-equivalence", Commission of the European Community,
    Luxemburg, EUR 5725e, 13-23 (1977).

27. C. H. Nauman and A. H. Sparrow, Problems of extrapolation from
    high dose to low dose in Tradescantia mutation studies,
    Environ. Health Perspectives 22:161-162 (1978).

28. C. H. Nauman, A. H. Sparrow, and L. A. Schairer, Comparative
    effects of ionizing radiation and two gaseous chemical muta-
    gens on somatic mutation induction in one mutable and two
    non-mutable clones of Tradescantia, Mutat. Res. 38:53-70
    (1976).

29. T. J. Hughes, E. Pellizzari, L. Little, C. Sparacino, and
    A. Kolber, Ambient air pollution: Collection, chemical
    characterization, and mutagenicity testing, Mutat. Res. 76:
    51-83 (1980).

30. A. G. Underbrink and A. H. Sparrow, The influence of experi-
    mental end points, dose, dose rate, neutron energy, nitro-
    gen ions, hypoxia, chromosome volume, and ploidy level on
    RBE in Tradescantia stamen hairs and pollen, in: "Biological
    Effects of Neutron Irradiation", International Atomic Energy
    Agency, Vienna, Austria, 185-214 (1974).

31. C. H. Nauman, A. G. Underbrink, and A. H. Sparrow, Influence
    of radiation dose rate on somatic mutation induction in
    Tradescantia stamen hairs, Radiat. Res. 62:79-96 (1975).

32. A. G. Underbrink, A. H. Sparrow, D. Sautkulis, and R. E.
    Mills, Oxygen enhancement ratios (OERs) for somatic muta-
    tions in Tradescantia stamen hairs, Radiation Botany
    15:161-168 (1975).

33. A. H. Sparrow, K. P. Baetke, D. L. Shaver, and V. Pond, The
    relationship of mutation rate per roentgen to DNA content
    per chromosome and to interphase chromosome volume,
    Genetics 59:65-78 (1968).

34. L. A. Schairer, Mutagenicity of ambient air at selected sites
    in the United States using Tradescantia as a monitor, in:
    "Proceedings of Symposium on In Vitro Testing of Environ-
    mental Agents: Current and Future Possibilities", A. R.
    Kolber, ed., September 22-29, 1979, Monaco, France,
    Elsevier Press, New York, N.Y., in press (1981).

DISCUSSION

Q. Loveday (Bioassay Systems): Could you tell me again what the
control value for stamen hair mutations was in the Grand Canyon,
how many pink hairs per thousand?

A. Schairer (BNL): 3.31 mutant events per thousand hairs.

Q. Loveday (Bioassay Systems): After subtracting that control
value you're getting from 1 to 3 pink hairs per thousand in
Elizabeth, NJ and other industrial sites. Do you feel that this
difference is significant? How many thousands of hairs would you
have to count to have confidence that a three hair per thousand
increase was significant?

A. Schairer (BNL): We score something like 300,000 hairs, so that
the total number of mutant events for the whole experiment was on
the order of 1,000. The per thousand hairs is just used as a unit
value.

ENVIRONMENTAL CLASTOGENS DETECTED BY MEIOTIC POLLEN MOTHER CELLS

OF TRADESCANTIA

Te-Hsiu Ma, Van A. Anderson and Iftikharuddin Ahmed

Western Illinois University
Macomb, IL 61455 (U.S.A.)

INTRODUCTION

Many airborne agents are routinely produced in large quantities from industrial effluents, vehicular emissions, agricultural fumigations and sprays, and from natural or man-made explosions. These common air pollutants, in both gaseous and particulate forms, are dissipated into the open air of large geographic areas of the globe. Other airborne agents, e.g. cigarette smoke, volatile cosmetics and common household pesticides, are more commonly release into small confined areas more intimately related to daily life. Humans are frequently acutely exposed to these latter pollutants at relatively high concentrations. The possible harmful effects of each of these agents and the possible synergistic interactions between them are essential topics in environmental studies.

Although studies on mutagenicity, cellular biochemical reactions and preferred metabolic pathways in vivo are important, in situ studies of environmental effects on living systems are essential for policy making in reference to human health. For in situ studies, plant bioassays are more efficient and less expensive than are animal bioassays (1,2). In fact, no animal system is presently known which can be used efficiently for in situ air monitoring and provide short-term reliable quantitative data.

In addition to the Salmonella and other microorganism mutation tests which have been extensively used to test the mutagenicity of diesel engine emission condensates (3-5), several higher plant systems have been used for in situ monitoring. These include waxy mutation in maize (6,7), somatic mutation in Tradescantia (8,9), and Tradescantia-micronucleus (Trad-MCN) tests (10,11). Both

141

<u>Tradescantia</u> (12) and Maize (7) are reliable for testing agents
requiring metabolic activation, while extracts of these plants
can be used to activate promutagens in vitro. Of these plant
systems, the Trad-MCN assay is highly efficient for in situ air
pollution monitoring.

The Trad-MCN test was first tested in studies of 1,2-dibro-
moethane mutagenesis (10) and was further improved in trials with
ethyl methanesulfonate, sodium azide and X-rays (13,14).
The ability for <u>Tradescantia</u> floral tissue to absorb gaseous
agents was demonstrated by using radioisotope labelled dibromoethane (15). Dose-response curves were established for this
system using both dibromethane and X-rays. To date, 32 chemicals
of gaseous and liquid form have been tested, and 15 sites in the
United States and the People's Republic of China have been monitored
with Trad-MCN (11). This assay was designed to take advantage of
the high sensitivity (14,16,17) and high degree of synchrony in
the early meiotic prophase I chromosomes. At this stage, the
acentric fragments of the damaged chromosomes become micronuclei
(MCN) in the tetrad stage of the same meiotic cycle (Fig 1). The
sensitivity of this assay was further enhanced by the high
degree of synchrony of the early tetrads in which the MCN

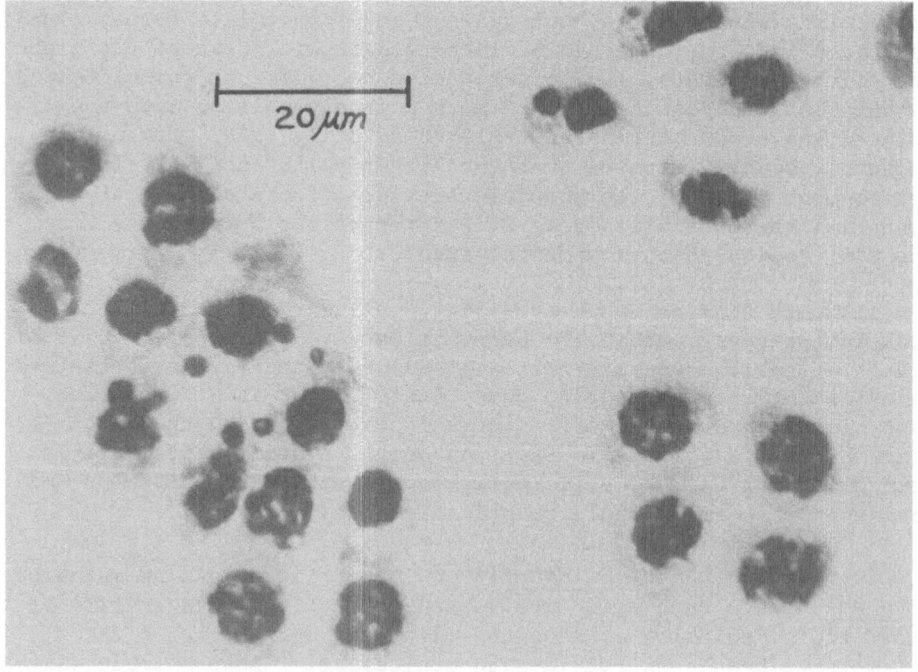

Figure 1. Normal tetrad (N) and tetrads containing micronuclei
(MCN) in <u>Tradescantia</u> pollen mother cells.

are scored. These various aspects all make this test highly
efficient. This paper is a consolidated report of the results of
Trad-MCN test on several airborne agents.

MATERIALS AND METHODS

     The Trad-MCN bioassay was used to examine the genotoxic
potency of airborne agents in three different ways: in laboratory
tests of gases, in laboratory tests of liquid suspensions of airborne particulates and during in situ monitoring of various chosen
locations. A detailed account of the Trad-MCN bioassay is described
in earlier publications (10,13).

     The general scheme of the present study is outlined below.
Tradescantia paludosa, clone #03, was utilized in all tests except
in an early study of 1,2-dibromoethane (10). For laboratory tests,
plant cuttings were collected from greenhouse-or phytotron (Duke
University Phytotron)-grown plants. They were maintained in tap
water in cups covered with perforated aluminum foil and exposed
continuously to gaseous agents of known concentration during the
treatment period (Fig 2). The plant cutting were illuminated
with incandescent or fluorescent light.

     Gaseous treatments were also accomplished by keeping the
plant cuttings in an air-tight bell jar containing a known concentration of gas in a single application without replenishments of
gas during the total treatment time.

     For treatment with particulate suspensions, the plant cuttings
absorbed the dissolved components through the stem, peduncle and
pedicel to the pollen mother cells (Fig 3).

     For in situ monitoring, the cuttings in the tapwater cups
were carried in a "clean-air box" to and from the testing site
and exposed to ambient air for 0.5 - 6.0 hr (Fig 4).

     The treated inflorescences were then allowed to carry on
meiotic processes for 24 - 30 hr, referred to as the recovery
time. This recovery time allows the damaged chromosomes in the
synchronized early prophase I pollen mother cells to arrive at
the synchronized tetrad stage where the MCN were scored. At the
end of the recovery time, the whole inflorescences of each experimental group (around 15 - 20) were fixed in aceto-alcohol (1:3)
for 24 hr and microslides prepared. The fixed material can also
be stored in 70% ethanol and slides prepared at a later date.
Only the early tetrad stage of the pollen mother cells were used
for scoring MCN fequencies (Fig 2). Generally 300 tetrads from
each slide were scored in order to derive the MCN frequency (MCN/
100 tetrads) for each sample population. Five to nine sample

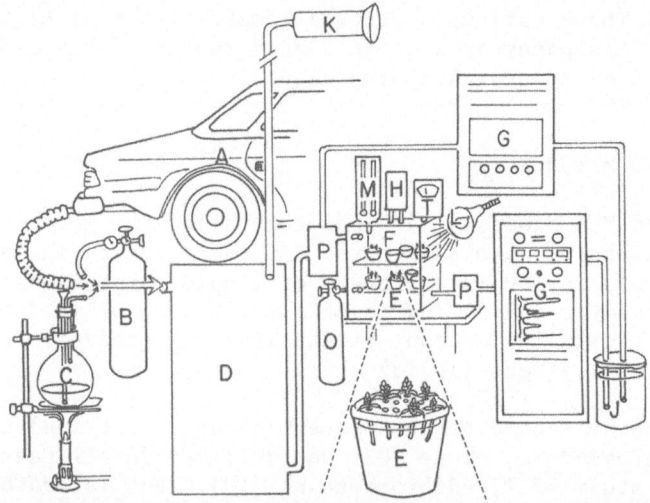

Figure 2. Diagram showing the scheme of gaseous treatment from
three different sources. A = Automobile exhaust fumes, B = Bottled
gases, C = Gases generated from chemical reactions; D = Dilution
chamber, E = Experimental material, Trandescantia plant cuttings,
F = Fumigation chamber for treatment, G = Gas chromatographic
system for gas analysis and measurement, H = Hygrometer, J = Jug
for waste gas disposal, K = Clean filtered air from outside, L =
Light for illumination during treatment, M = Monometer, O = Compressed air or pure oxygen, P = Pumps for gas flow, T = Thermometer.

populations were scored from each experimental group in order to
derive the means and variances of each of the experimental groups.
Statistical analyses were applied to determine if the means of the
treated and control groups were different at a statistically
significant level of $p < 0.01$.

RESULTS AND DISCUSSION

Public parking garages are usually built underground and/or
are partially enclosed. Occasionally, these structures can
accumulate relatively high concentrations of automobile exhaust
fumes. People are often subjected to high concentrations of
these fumes, which are composed of both gaseous and particulate
material. Major mutagens present in these fumes are sulfur and
nitrogen oxides, hydrocarbons and organo-metallic compounds. To
date, three different parking garages have been monitored with the
Trad-MCN bioassay (Table 1). The magnitude of the clastogenic

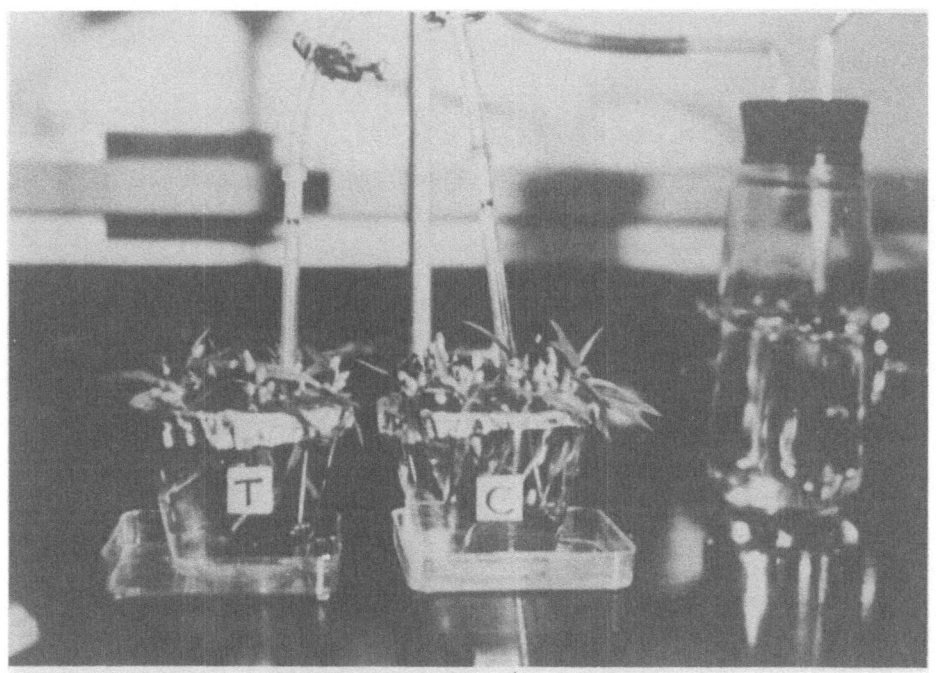

Figure 3. Set-up treatment with liquid or water mixable agents, use of an aerator to pump air into the solution.

response in these parking garages was largely dependent upon the weather conditions, volume of traffic and monitoring period. The most critical monitoring periods were during rush-hour traffic.

Three sites of truck or bus stops were monitored and the results are shown in Table 2. Both truck stops had an average of 20 large diesel trucks plus 50 - 100 automobiles parked on the lot, with frequent arrivals and departures. The negative response in the repeated tests at the Dixie Truck Stop was primarily due to the prevailing wind (around 10 mile per hour) during the testing period. The only bus stop monitored was located at the intersection of two major streets in a downtown area. The major pollutants were similar to that of the truck stops except that the fumes were partially confined between the tall buildings rather than dissipated in the open field. This bus stop showed a positive response (at the $p < 0.05$ significant level) even with a one hour exposure. Longer exposure at this site was not possible due to an onset of rain during this monitoring trip.

Figure 4.   In situ monitoring of air pollutants near a truck stop.

        In situ monitoring of industrial sites has been previously
carried out using Tradescantia stamen hair test (8,9) and by
using signs of damage to plant leaves. However, few related
studies are currently in process, mainly because of the lack
of an efficient bioassay with distinct end points. In situ
monitoring on the industrial sites with Trad-MCN bioassay has been
carried out in United States since 1978 and recently in People's
Republic of China during the summer of 1980.

        The results of these monitoring tests and a few in situ tests
of specific laboratory and field conditions of non-industrial
nature are shown in Table 3.

        Positive response was obtained from a three hour exposure at
the WRI site, presumably as a result of pollutants generated from a
group of local petroleum refineries. The repeated positive
responses obtained from the PRC-IS-AC site in Qingdao, People's
Republic of China (PRC), were presumably due to gaseous agents
released from an Agriculture Chemical (factory) plant. The PRC-
IS-AC site, a Rubber Company, did not give conclusive results.
P-Dichlorobenzene fumes from a herbarium also gave inconclusive
results. Further studies of these sites are underway.

Table 1. Results of in situ monitoring of public parking garages

Site location (code numbers)	Expt. group	Dose (hr)	MCN/100 tetrads mean ± S.E.		Response 0.01 Signif.	Remarks
Chicago, IL	HC*	-	6.65 ±	1.60		
(CMP-1)	FC*	-	6.60 ±	0.93		
	T-1	4	8.69 ±	1.32	-	Clear, wind from lake
	T-2	6	8.34 ±	0.85	-	Clear, wind from lake
Decatur, IL	HC	-	7.91 ±	1.60		
(DMP-1)	FC	-	9.00 ±	2.00		
	T-1	1	9.90 ±	1.00	-	
	T-2	2	11.00 ±	1.70	-	
	T-3	3	10.00 ±	1.60	-	
	T-4	4.5	12.30 ±	2.50	-	
Peoria, IL	HC	-	7.80 ±	1.18		
(SRP-1)	FC	-	9.04 ±	0.43	-	
	T-1	2	24.19 ±	1.41	+	
	T-2	4	27.26 ±	1.46	+	
	T-3	6	28.33 ±	2.89	+	
Peoria, IL	HC	-	9.04 ±	0.77		
(SRP-2)	FC	-	10.50 ±	1.34		
	T-1	2.5	16.28 ±	2.73	+/-	Few cars parked
	T-2	4	12.20 ±	1.07	+/-	Few cars parked

*HC = Home Control   FC = Field Control

Table 2.   Results of in situ monitoring of the bus and truck stops

Site location (code numbers)	Expt. group	Dose (hr)	MCN/100 tetrads mean ± S.E.	Response 0.01 Signif.	Remarks
Dixie Truck	HC*	–	6.85 ± 0.86		
(DTS-1)	FC*	–	7.81 ± 1.07		
	T-1	2	10.13 ± 3.44	–	
	T-2	4	6.29 ± 0.56	–	
Dixie Truck	HC	–	9.70 ± 2.26		
(DTS-2)	FC	–	9.53 ± 1.64		
	T-1	2	6.87 ± 0.87	–	
	T-2	3	12.22 ± 1.32	–	
Woodhull Truck	HC	–	5.12 ± 0.69		
(WTS-1)	FC	–	5.62 ± 0.49		
	T-1	2.5	10.54 ± 1.05	+	
	T-2	5	10.84 ± 0.80	+	
Peoria Bus stop	HC	–	9.04 ± 0.77		
(PBS-1)	FC	–	10.50 ± 1.34		
	T-1	0.5	9.73 ± 0.85	–	
	T-2	1	13.53 ± 1.57	+/–	Interrupted by rain

*HC = Home Control   FC = Field Control

Table 3. Results of in situ monitoring of industrial sites, farm and laboratory atmosphere

Site location (Code Numbers)	Expt. groups	Dose (hr)	MCN/100 tetrads means	S. E.	Response 0.01 Signif.	Remarks
Industrial sites						
Granite city (GCI-1)	HC*	-	4.33	0.53		
	FC*	-	7.70	1.19		
	T-1	month	9.12	1.20	+/-	clone 4430
Wood River (WRI-1)	T-1	3.5	15.89	2.45	+	clone 03
Qingdao, PRC (PRC-IS-1)	HC	-	6.34	0.89		
	FC	-	9.90	0.58		
Residential RI-	T-1	4.5	11.30	1.16	-	
Agri-Chem. AC-	T-1	4.0	14.30	0.98	+	
Bus Station BS- (PRC-IS-2)	T-1	5.0	9.20	1.34	-	
	HC	-	8.40	0.44		
	FC	-	12.30	1.07		
Agri-Chem. AC-	T-2	6	15.20	1.27	+	
Rubber Co. RC-	T-1	6	12.80	0.93	+/-	Positive with HC
Laboratory and field atmosphere						
Office (JTC-1)	HC	-	7.04	1.28		
	T-1	4	9.70	1.41	-	
Herbarium PDB (PDB-1)	HC	-	5.20	0.44		
	T-1	3	9.94	1.95	+/-	
	T-2	6	10.34	1.27	+/-	
Animal Farm (UFO-1)	HC	-	6.40	2.00		
	T-1	6	6.26	1.47	-	Exhaust from swine house

*HC = Home Control   FC = Field Control

Under laboratory conditions, diesel exhaust fumes generated from a Nissan automobile were tested, and the detailed account of this study is now in press (18). The relative concentrations of the fumes were represented by the concentrations of hydrocarbons. The dosages were determined by both the concentration of the fumes and the duration of exposure (23 minutes per driving cycle X number of cycles). The overall results are summarized in Table 4.

Significant (p<0.01) positive responses were obtained at around 2 ppm of hydrocarbons for three driving cycles (69 minutes) of exposure and 4.5 ppm of hydrocarbons for two driving cycles (46 minutes). If the dosages were expressed in terms of concentration (ppm) X time (driving cycles), the frequencies of MCN/100 tetrads increased in a positive dose-response manner.

These findings from the in situ air monitoring and laboratory tests with the Trad-MCN bioassay could be utilized to substantiate some of the epidemiological studies made on the correlation between automobile emissions and cancer incidence in humans. Positive correlations were found from the epidemiological studies in Europe (19,20), in the United States (21) and in Japan (22). When laboratory animals such as cats, rats, mice, guinea pigs and golden hamsters were exposed to total exhaust fumes (5,23) or treated with exhaust condensate (24,25), an increase in the frequency of tumors, sarcomas and pulmonary adenomas was observed.

Laboratory tests of gases in the enclosed chambers were carried out in two different ways. The first method was to allow a given concentration of gas and air mixture to flow through the plexiglass chamber at a given rate, and expose the plant cuttings in the chamber for a given length of time. The second way was to keep the plant cuttings in an air-tight bell jar containing a known concentration of gas. No flow or gas or replenishment of gas was allowed during the total exposure time. The results of the Trad-MCN tests on the clastogenicity of gases in the enclosed chambers are shown in Table 5.

These tests showed a positive Trad-MCN response after 24 hr exposure to 5 ppm NO_2 and a 22 hr exposure to 1 ppm SO_2. Gas flow was constant in the treatment chamber throughout the exposure periods. In contrast, it requires a life-long (2 year) exposure of 20 ppm NO_2 to induce tumors (lymphosarcomas of testis and gut) in rats (23). This comparison clearly indicates the extraordinary sensitivity of the Trad-MCN test to environmental gases. Well-known mutagens such as hydrazoic acid (HN_3) and ethyl methanesulfonate (EMS) gave positive responses when treatments were applied in an air-tight bell jar without replenishment of gases.

Similar tests conducted in the laboratory with 1,2-dibromoethane (10) demonstrated a linear dose response curve at concentations

Table 4. Results of Trad-MCN tests on diesel exhaust fumes (DEF).

Expt. groups (Code numbers)	Concentrations* hydrocarbon ppm	Dose equivalent Cycle X ppm		MCN/100 tetrads Mean ± S. E.			Response 0.01 Signif.
DEF-1 HC**	-	-		8.32	±	1.63	
FC **	-	-		5.67	±	0.61	
T-1	0.3	X 1	0.3	5.45	±	0.68	-
T-2	0.3	X 2	0.6	5.97	±	0.67	-
T-3	0.3	X 3	0.9	7.75	±	0.78	-
DEF-2&3 HC	-	-		4.76	±	0.90	
FC	-	-		5.04	±	0.70	
DEF-3 T-1	1.8 - 2.1	X 1	1.8	7.35	±	0.32	-
T-2	1.8 - 2.1	X 2	4.2	7.54	±	1.62	-
T-3	1.8 - 2.1	X 3	6.3	14.80	±	0.44	+
DEF-2 T-1	4.4 - 4.7	X 1	4.4	8.56	±	2.07	-
T-2	4.4 - 4.7	X 2	9.4	20.81	±	3.13	+
T-3	4.4 - 4.7	X 3	13.2	10.90	±	1.68	-
Positive control (40 R X-rays)				62.06	±	3.10	+

*Fume concentration represented by hydrocarbon concentrations.

**HC = Home Control: FC = Field Control

Table 5. Results of Trad-MCN tests on gases in chambers

Experimental groups (code)	Concentration (ppm)	Duration (hr)	MCN/100 tetrads Mean ± S. E.		Response 0.01 Signif.	Remarks
NO$_2$ –1 C	–	–	5.76	± 0.44		
T–1	5	6	4.56	± 0.53	–	
NO$_2$ –2 C	–	–	3.68	± 1.11		
T–1	5	24	7.53	± 0.23	+	
SO$_2$ –1 C	–	–	7.29	± 0.84		
T–1	1	6	3.74	± 0.41	–	
SO$_2$ –2 C	–	–	6.17	± 0.66		
T–1	1	22	10.01	± 1.03	+	
O$_3$ –1 C	–	–	5.80	± 0.44		
T–1	5	5.5	4.20	± 0.95	–	
NO$_2$+ O$_3$–1 C	–	–	6.18	± 0.66		
T–1	5/each	3	5.60	± 0.60	–	
HN$_3$ –1 C	–	–	5.20	± 0.58		
T–1	136*	6	21.20	± 3.66	++	
T–2	272*	6	14.30	± 0.89	++	
EMS–1 C	–	–	5.20	± 0.58		
T–1	1000*	6	17.40	± 1.49	+	

*Single application of gas in the confined chamber without replenishment in 6 hr of treatment.

ranging from 4.6 to 77.7 ppm (correlation coeffecient=0.88, slope= 0.754). This demonstrated the reliability of the Trad-MCN test for detecting cause-effect relationships.

The Tradescantia-micronucleus bioassay is a simple, efficient and inexpensive test system. Since the test materials are the selected inflorescences of plant cuttings, they can be transported to and from the monitoring sites at ease. While the plant cuttings are maintained in water or nutrient solution, the normal life processes of the cuttings occur under adequate illumination in the open field or in a confined chamber. Therefore, these experiments can be considered as an in vivo test with in vitro convenience. Although results of a test can be obtained as early as 24-48 hours after treatment, the fixed experimental material can also be stored for up to several years before cytological observation for their MCN frequencies.

In addition to its application to the testing of airborne agents for mutagenicity, the Trad-MCN test can also be effectively used to monitor river and lake waters for their pollutants.

SUMMARY

Tradescantia-Micronucleus (Trad-MCN) bioassay was utilized for in situ monitoring of air pollutants and laboratory tests on gaseous agents. This efficient, short-term test involves exposing plant cuttings which bear young influorescences of Tradescantia (spiderwort) to gaseous agents and observing chromosomal damage as reflected by micronucleus formation in the meiotic pollen mother cells. In situ monitoring was conducted on three public parking garages, two truck stops and one bus stop. Test results showed positive responses at one garage and one truck stop. The bus stop also showed a positive response for a one hour exposure, but at a lower level of significance. Six industrial sites were monitored. Among them, one site near an agricultural chemical plant in the People's Republic of China and one site in the midst of a number of petroleum refineries in the United States showed positive responses. Fumes from diesel automobile exhaust gave increased positive responses at increased concentrations of fumes or exposure time. Five gaseous agents including NO_2, SO_2, O_3, HN_3 and EMS were tested in confined chambers. Exposure to the continuous flow of NO_2 and SO_2 at 1 and 5 ppm concentrations (22 - 24 hour exposure) gave positive responses. A single application of hydrazoic acid (HN_3) at 136 and 272 ppm and of EMS at 1,000 ppm concentrations gave positive responses when a 6-hour exposure was applied in an air-tight chamber without replenishment of the gas.

ACKNOWLEDGEMENT

The authors wish to express their gratitude to Drs. Michael Waters, and Shahbeg S. Sandhu of the Research and Development Division of U. S. EPA, Research Triangle Park, NC, for their encouragement and assistance to carry out the studies on the diesel exhaust fumes and the gaseous agents at the Health Effect Research Laboratory, U.S. EPA, to Dr. Roy Zweidinger and the staff of Northrop Services, Inc. for their technical assistance in providing an analysis of the diesel exhaust fumes. We also wish to acknowledge the assistance of Dr. William Lower of the Environmental Trace Substance Research Center of the University of Missouri, Columbia, MO, and Dr. Tsungci Fang of Sahndong College of Oceanology, Qingdao, People's Republic of China in conducting the in situ monitoring of air pollutants. The collaboration of Dr. Heny Helmyers of the Duke Phytotron, Duke University, Durham, NC, is also appreciated.

Research projects were supported by the U.S. Environmental Protection Agency Contract #DA8-5095J and Research Grant #R806422-01 and partially funded by the Program of the Committee of Scholarly Communication with the People's Republic of China, co-sponsored by the Council of Learned Societies, National Academy of Sciences, and Research Council of Social Science.

REFERENCES

1. F. J. de Serres, Introduction: Utilization of higher plant systems as monitors of environmental mutagens, Environ. Health Perspectives 27:3-6 (1978).
2. R. A. Nilan, Potential of plant genetic system for monitoring and screening mutagens, Environ. Health Perspectives 27:45-50 (1978).
3. J. Huisingh, S. Nesnow, R. Bradow, and M. Waters, Application of a battery of short term mutagenesis and carcinogenesis bioassay to the evaluation of soluble organics from diesel particulates, Inter. Symp. on Health Effects of Diesel Engine Emissions, Session II, Part 1, Cincinnati, OH Abstract (1979).
4. Y. Y. Wang, S. M. Rappaport, R. F. Sawyer, R. E. Talcott, and E. T. Wei, Direct-acting mutagens in automobile exhaust, Cancer Lett. 5:39-48 (1978).
5. D. S. Barth and S. M. Blacker, The EPA program to assess the public health significance of diesel emissions, J. Air Pollut. Control Assoc. 28:269-271 (1978).
6. M. Freeling, Maize Adh-1 as a monitor of environmental mutagens, Environ. Health Perspectives 27:91-97 (1978).
7. M. J. Plewa, Activation of chemicals into mutagens by green plants: A preliminary discussion, Environ. Health

Perspectives 27:45-50 (1978).

8. L. A. Schairer, J. Van T Hof, C. G. Hayes, R. M. Burton, and
    F. J. de Serres, Exploratory monitoring of air pollutants
    for mutagenicity activity with the Tradescantia stamen hair
    system, Environ. Health Perspectives 27:51-60 (1978).

9. W. R. Lower, P. S. Rose, and V. K. Drobney, In situ mutagenic
    and other effects associated with lead smelting, Mutat. Res.
    54:83-93 (1978).

10. T. H. Ma, A. H. Sparrow, L. A. Schairer, and A. F. Nauman, Effect
    of 1,2-dibromoethane (DBE) on meiotic chromosomes of
    Tradescantia, Mutat. Res. 58:251-258 (1978).

11. T. H. Ma, V. A. Anderson, and I. Ahmed, In situ monitoring of
    air pollutants and screening of chemical mutagens using
    Tradescantia micronucleus bioassay, 11th Ann. Meet. Environ.
    Mutagen Soc., Abstract (1980).

12. B. R. Scott, A. H. Sparrow, S. S. Schwemmer, and L. A. Schairer,
    Plant metabolic activation of 1,2-dibromoethane (DBE) to a
    mutagen of greater potency, Mutat. Res. 49:203-212 (1977).

13. T. H. Ma, Micronuclei induced by X-rays and chemical mutagens
    in meiotic pollen mother cells of Tradescantia - A promising
    mutagen test system, Mutat. Res. 64:307-313 (1979).

14. T. H. Ma, G. J. Kontos, Jr., and V. A. Anderson, Stage sensiti-
    vity and dose response of meiotic chromosomes of pollen
    mother cells of Tradescantia to X-rays, Environ. Exp. Bot.
    20:169-174 (1980).

15. C. H. Nauman, P. J. Klotz, and L. A. Schairer, Uptake of
    tritiated 1,2-dibromoethane by Tradescantia floral tissues:
    Relation to induced mutation frequency in stamen hair cells,
    Environ. Exp. Bot. 19:201-205 (1979).

16. A. H. Sparrow, Radiation sensitivity of cells during mitotic
    and meiotic cycles with emphasis on possible cytochemical
    changes, Ann. New York Acad. Sci. 51:1508-1540 (1951).

17. D. Lingren, Sensitivity of premeiotic and meiotic stages to
    spontaneous and induced mutations in barley and maize,
    Hereditas 79:227-238 (1975).

18. T. H. Ma, V. A. Anderson, and S. S. Sandhu, A preliminary study
    on the clastogenic effects of diesel exhaust fumes using
    Tradescantia Micronucleus Bioassay, in: Proc. EPA Second
    Symposium on Application of Short-term Bioassays in the
    Analysis of Complex Environmental Mixtures, Williamsburg,
    VA (in press) (1981).

19. M. Blumer, W. Blumer, and T. Reich, Polycyclic aromatic
    hydrocarbons in soil of a mountain valley: Correlation with
    highway traffic and cancer incidence, Environ. Sci. Technol.
    11:1082-1084 (1977).

20. W. Blumer, M. Blumer, and T. Reich, Carcinogenic hydrocarbons
    and the incidence of cancer mortality among residents near
    an automobile highway, Fortschr. Med. 95:1497-1498, 1551-1552
    (1977).

21. J. M. Colucci and C. R. Begeman, Carcinogenic air pollutants
    in relation to automotive traffic in New York, Environ. Sci.

Technol. 5:145–150 (1971).

22. H. Shimizu, K. Aoki, and T. Huroishi, An epidemiological study
    of lung cancer in relation to exhaust gas from cars, Lung
    Cancer 17:103–112 (1977).
23. M. Stupfel, F. Romary, M. H. Tran, and J. P. Moutet, Lifelong
    exposure of SPF rats to automotive exhaust gas, dilution
    containing 20 ppm of nitrogen oxides, Arch. Environ. Health
    26:264–269 (1973).
24. U. Mohr, Toxicologic and pathologic data on polycyclic aromatic
    hydrocarbons and automobile exhaust condensate, Ecotoxicol.
    Environ. Safety 2:267–276 (1978).
25. H. Reznik-Schuller and U. Mohr, Pulmonary tumorigenesis in
    Syrian golden hamster after intratracheal instillations with
    automobile exhaust condensate, Cancer 40:203–210 (1977).

DISCUSSION

Q. Lewtas (EPA): I think you said that you did simultaneous experiments with the micronucleus test and the stamen hair assay for
dibromoethane. Could you say which assay was more sensitive?

A. Ma (Illinois): I should say that the experiments weren't done
simultaneously, but in two different time periods. The micronucleus test was about 36 times more sensitive than the pink
mutation assay.

Q. Lewtas (EPA): What do you think that means?

A. Ma (Illinois): I think one factor is the degree of cell
synchrony. The cells in the stamen hair are not dividing at the
same time. In the micronucleus test the cells are exposed while in
the same stage of division. Furthermore, the synchrony of tetrads
for scoring gives you the additional advantage of observing all
of the cells at the same time. I don't know of any other system
that has that high degree of synchrony.

Q. Lewtas (EPA): One comment I'd like to make is that with a
system that's so sensitive and that can respond to SO_2 and NO_2 ,
it's important to do in situ outdoor air studies in such a way that
you know the concentrations of those agents. Knowing the lowest
detectable dose for these criteria pollutants from laboratory
studies, you can take their genotoxic effects into account in
making an interpretation about what other agents might be present.

A. Ma (Illinois): Yes, I hope we can have that kind of analysis
at every site we test.

Q. Schairer (BNL): I might make one other addition of comment.
The micronucleus test would also have an apparent increased

sensitivity due to the fact that any lesion in the chromosomes
would result in a micronucleus whereas the somatic mutation we
measure in <u>Tradascantia</u> occurs in a single chromosome. The
probability of getting a higher frequency of micronuclei would
be greater. We have room on the mobile laboratory for simultaneous
mutation and micronuclei experiments. This could be done by
sending Dr. Ma fixed material from routine in situ exposures.

PLANT GENETIC SYSTEMS WITH POTENTIAL FOR THE DETECTION OF

ATMOSPHERIC MUTAGENS

Milton J. Constantin

Comparative Animal Research Laboratory
Oak Ridge, TN 37830 (U.S.A.)

INTRODUCTION

Higher plant genetic systems for the detection and monitoring
of environmental mutagens have been described (1-5). In contrast
to microbes, insects and mammals, few plant genetic systems have
been developed primarily for research in environmental mutagenesis.
Although other plant genetic systems have been used, the Tradescantia
stamen hair and micronucleus systems are the only ones for which an
appreciable amount of data on atmospheric mutagens is available
(see Schairer et al. and Ma et al., these Proceedings).

I shall discuss briefly the advantages and disadvantages of
plant genetic systems relative to environmental mutagenesis research
and list examples of systems for which information on their response
to mutagens is available. The waxy locus in maize (Zea mays L.)
and barley (Hordeum vulgare L. emend. Lam.) will be discussed as
examples, in addition to the Tradescantia stamen hair and micronucleus systems, that might be useful for the detection and monitoring of atmospheric mutagens.

ADVANTAGES AND DISADVANTAGES OF PLANT GENETIC SYSTEMS FOR
ENVIRONMENTAL MUTAGENESIS RESEARCH

Inherent Value. Aside from the relevance of higher plant
genetic and cytogenetic systems for environmental mutagenesis
research directed toward risk assessment to humans, the integral
role of plants in the biosphere merits their study in their own
right (5). It should also be emphasized that plants can play an
important role in environmental mutagenesis relevant to man through

their contributions to the human food chain (5). A good example
of this is the activation of the herbicide atrazine into mutagenic
metabolites by maize (6).

Advantages. Since the advantages of plant genetic systems have
been discussed in detail (1-5), only a brief discussion will be
presented here.

Higher plants are eukaryotic organisms whose genetic complexity
(in terms of total amount of DNA per nucleus, structural genes,
organelle DNA, chromosome morphology, etc.) is similar to that of
man. Also, plants are multicellular; like animals, they display
the classical embryological phenomena of determination, competence,
and cell heredity (7). Genetic and organizational complexity argue
in favor of using plants as well as animals as surrogates for humans
in studies of environmental mutagenesis.

In comparison with mammalian genetic systems, the culture of
higher plants as experimental organisms is relatively inexpensive.
Furthermore, the regeneration of whole plants from in vitro cultured
cells makes it possible to do genetic studies of isolated mutant
cells. Certain plant species have relatively short life cycles,
others exhibit endpoints that can be scored within a short time
(from several hours to a few days) following mutagenic treatment,
and mutagenic treatment can be accomplished under a wide range of
environmental conditions, especially if seeds are used. In
some cases, endpoints that reflect toxic and clastogenic as well as
mutagenic effects in somatic and germinal cell populations can be
studied in the same treated individuals. The logistics of doing
experiments with higher plant genetic systems also argue in favor
of using them in studies of environmental mutagenesis.

Disadvantages. The advantages cited above apply individually
to select genetic systems, i.e. no single plant system exhibits
all of them. Differences in anatomy, morphology, physiology and
mobility among animals and plants are well known; however, how
these differences affect mutagenesis is not known. Knowledge of
the molecular biology and mutation process in higher plants is
limited compared to that of mammals and microbes.

Perhaps the disadvantages alluded to above should instead be
viewed as incentives for more research. Some of the shortcomings
now associated with plant genetic systems could be overcome through
an increase in research effort.

Sources of Additional Information. In addition to the publications cited earlier, more information on plant genetic and cytogenetic systems can be obtained from other publications in the
"Proceedings of the Workshop on Higher Plants as Monitors of
Environmental Mutagens" (8), "Proceedings of the Conference on

Pollen Systems to Detect Biological Activity of Environmental
Pollutants" (9), and individual reports from the Gene-Tox Program
Workgroup on "Plant Genetic and Cytogenetic Assays" (10).

Status. Genetic systems in higher plants can play a useful
role in the determination of the mutagenicity of chemicals under
laboratory conditions. In this regard, it is logical that higher
plants be included among those organisms that are generally used to
screen chemicals for mutagenicity. Furthermore, higher plants are
uniquely suited for the detection and monitoring of mutagens in our
environment under "real world" conditions. Plants can be grown
easily in situ to determine whether mutagens are present in the soil,
water or atmosphere at particular geographic locations. This can
be accomplished with minimum disturbance to the environment being
tested.

PLANT GENETIC SYSTEMS

Table 1 lists examples of genetic systems available in various
plant genera and species, the locus or loci involved, and the kinds
of phenotypic endpoints that are involved. This list indicates the
kinds of genetic systems that are available rather than all the
genetic systems in higher plants that have been used in mutagenesis
research. Although not listed in Table 1, limited information is
available concerning auxotrophy in Arabidopsis (11) and Lycopersicon
(12). An auxotrophic genetic system is especially promising because
the in vitro culture of plant cells could make it feasible to isolate
mutants affecting certain key points in metabolic pathways.

THE WAXY LOCUS SYSTEM IN POLLEN

Background Information. According to Eriksson (13), it was
reported in 1860 that the starch grains of certain varieties of
rice stained a reddish color when reacted with an iodine solution
while those of other varieties stained a blue-black color. Later
research showed that the difference in staining was the result of
a single gene that governs starch composition. Wild-type (normal)
starch, which consists of amylose and amylopectin, stains a blue-
black color when reacted with I_2KI solution. In contrast, waxy
mutant starch consists solely of amylopectin and stains a reddish
color when reacted with I_2KI solution (14). The waxy allele is
recessive to the starchy allele. The waxy character expressed in
endosperm has been used in research to better understand the effects
of physical and chemical mutagens in higher plants (13).

The waxy character is also expressed in pollen of various
species. It was demonstrated during the mid-1920s that pollen
grains from a heterozygous (Wx wx) maize plant segregated in a 1:1

Table 1. Examples of genetic systems in higher plants that can be used in environmental mutagenesis research

Genus	Species	Locus	End point
Somatic sectors			
Pisum	sativum	Numerous	Light-green to whitish spots on leaves
Glycine	max	Y_9, Y_{11}	Yellow and green (single and twin) spots on leaves
Trifolium	repens	Vby	Alteration of color pattern on leaves
Zea	mays	Yg_2	Yellowish streaks on leaves
Tradescantia	hybrids		Pink stamen hair and/or petal cells
Embryo and seedling lethals			
Hordeum	vulgare	Hundreds	Chlorophyll-deficient seedlings
Oryza	sativa	"	"
Zea	mays	"	"
Arabidopsis	thaliana	"	Chlorophyll-deficient immature embryos and seedlings
Endosperm mutants			
Zea	mays	Wx	Starch composition
		Sh_1, Sh_2	Shrunken endosperm
		Su_1	Sugary
Hordeum	vulgare	Wx	Starch composition
Oryza	sativa	Wx	"

Genus	Species	Locus	End point
		Pollen mutants	
Hordeum	vulgare	$\underline{Wx}$	Starch composition
Zea	mays	"	"
Oryza	sativa	"	"
Zea	mays	$\underline{Adh-1}$	Alcohol dehydrogenase activity
Petunia	sp.	$\underline{S}$ alleles	Sexual compatibility
Nicotiana	alata	"	"
Beta	vulgaris	"	"
Lycopersicon	sp.	"	"
Oenothera	organensis	"	"

ratio according to Mendel's first law (Brink and MacGillivray, 1924, and Demerec, 1928, as cited in 15). Maize and barley, the two species for which the most information is available, are both diploid plants; thus, their pollen grains (microgametophytes) are functional haploids. The genotype of the individual pollen grain governs the kind of starch that it will accumulate and, thereby, its color reaction to I_2KI solution. These characteristics of pollen grains are the basis for using the waxy locus system of maize and barley pollen for studies in environmental mutagenesis.

Pollen as a Genetic Test System: Factors to Be Considered. Pollen systems exhibit the following advantageous characteristics for research in environmental mutagenesis:

1. Pollen grains are produced in large numbers by most plants; therefore, they provide a high degree of genetic resolving power.
2. Pollen grains are functional haploids; therefore, the genotype of the pollen grain governs its phenotypic expression for traits such as starch composition.
3. Pollen grains are relatively small and can be handled easily in large numbers; therefore, the acquisition of data from large populations of individuals is easy.
4. Pollen grains are products of meiosis; therefore they exhibit mutations induced in germline cells of the sporophyte as well as mutations induced in post-meiotic cells.
5. Pollen grains exhibit a dose-dependent response to physical and chemical mutagens for traits such as waxy.
6. Pollen grains can be killed, fixed, and stored in ethanol for long periods of time; this facilitates the use of pollen traits, such as waxy, for studies of mutagenesis.
7. Considerable information on the biology of pollen is available in the literature.

On the other hand, pollen systems exhibit the following disadvantages:

1. Pollen is derived from germline cells of the sporophyte. The number of germline cells varies from one developmental stage to another; and it is difficult, if not impossible, to determine the exact number of germline cells for any stage. Therefore, unless tetrads are treated, there is not a 1:1 relationship between the treated cells and the scored cells for the expression of the induced mutation.
2. Morphogenetic processes that occur after mutagenic treatment can lead to mutant clusters or sectors in the tassel or spike. The number, size, and location of these clusters or sectors depend on the developmental stage involved, the fraction of the germline cells that is mutated, and the toxic effects of the mutagen dose.

3. In species where the regeneration of plants from pollen is not feasible, it is difficult to determine the genetic event that leads to the phenotypic expression scored.

4. Different developmental stages of microsporogenesis and microgametogenesis vary in susceptibility to mutagens. These processes also vary somewhat in timing among the plants of a population; in fact, they do not occur synchronously in all parts of a tassel or spike in maize and barley, respectively. This leads, therefore, to experimental variability and difficulties in the calculation of mutants per unit of mutagen dose in both chronic and acute exposures.

Overall, pollen genetic systems such as the waxy locus in maize and barley (15) and the Adh-1 locus in maize (7) are useful for studies in environmental mutagenesis (see Nilan et al. (16) for further information concerning pollen systems for studies in environmental mutagenesis).

Phenotypic Endpoint. As mentioned earlier, the phenotypic endpoint scored in the waxy locus system in maize and barley is the starch composition of the pollen grain. The starch composition can be determined simply by reacting the pollen grains with an I_2KI solution and determining the color response of individual pollen grains. In a forward mutation test, one treats a homozygous starchy (Wx Wx) wild-type genotype and determines the frequency of recessive waxy (wx) pollen grains. The mutant appears reddish-brown among the normal blue-black colored pollen grains. In a reversion mutation test, one treats a homozygous waxy (wx wx) genotype and determines the frequency of dominant starchy (Wx) pollen grains. The reverse mutant appears blue-black among the normal tan to reddish-brown pollen grains.

Procedure. The large number of experimental designs and objectives in environmental mutagenesis studies precludes the description of a standard procedure for each aspect of the waxy locus system in maize and barley. General comments will be made regarding aspects of the system that can vary appreciably in environmental mutagenesis studies; standard procedures will be described for aspects that do not vary appreciably. Generally, the same procedures can be used for maize and barley. Whenever different procedures are required, these will be pointed out.

Either physical or chemical mutagens can be used to treat dry seeds, germinating seeds or growing plants at specific developmental stages prior to the tetrad stage or continuously during the life cycle of the plant up to the time the tassel is collected. In all cases, it is important that the experimental design provide adequate control populations, sufficient numbers of individuals per treatment and control population, and unambiguous comparisons of treatment effects. For experiments involving chemical agents, proper control populations include treatment with water, a solvent (if the chemical has to be dissolved in a solvent other than water), and a buffer (if pH has to be controlled). The test agent should be used in a concentration gradient with a sufficient number of doses to permit inferences to be drawn regarding the nature of the dose-response curve. When the environment is to be tested in situ, it is extremely important to have control populations in a "clean" environment to provide baseline data.

Certain procedures must be altered to accommodate differences in physical characteristics of the agent to be tested. For example, the solubility of the chemical will limit the upper level of concentration that can be used, while the rate of hydrolysis will influence the duration of treatment, (for a discussion of this subject, refer to Konzak et al. (17)). The developmental stage of the plant being treated will influence the treatment procedure, i.e. seeds can tolerate a wide range of environmental conditions such as pH, temperature, and atmospheric composition and pressure before, during, and after treatment. Simple factors, e.g. the volume of solution per seed, aeration during treatment, post-treatment rinsing, etc., should be held constant. Frequently, preliminary trials are required to determine the range in concentration, duration of treatment, etc., that can be used effectively. Chemicals vary in their toxicity relative to mutagenicity; empirical data often constitute the only source of information in this respect.

The plant material must be selected carefully to ensure that experimental objectives can be fulfilled. In maize, for example, a genotype with the proper heteroalleles must be used (incidentally the seed source should be genetically pure). A high degree of seed viability and vigor is essential.

To have pollen grains for analysis, both maize and barley plants have to be grown until the time of anthesis. Regardless of the developmental stage treated, optimum conditions for growth and development must be provided uniformly for all populations. For example, high temperatures induce a high incidence of pollen sterility, which can seriously distort the results of an experiment. For laboratory experiments, the plant populations should be grown under controlled environmental conditions. For in situ studies of ambient environmental conditions, a control population should be grown in a clean environment for comparison with a running control population grown in situ. This type of information will establish the validity of the test.

The procedure used in processing the pollen grains for analysis is described in detail by Plewa (6) for maize and by Rosichan et al. (18) for barley. Maize tassels and barley spikes are collected, killed, and fixed in 70% ethanol on the day of anthesis or 2-4 days prior to anthesis, respectively. The material can be stored in

ethanol with or without refrigeration for a long time (many months).
Pollen grains are extracted from anthers taken at random from the
tassel or spike and reacted with I_2KI in a gelatin or glycerol
solution. The stained pollen grains are examined with the aid of
a dissecting scope. An estimate is made of the total number of
pollen grains examined (see Plewa (6), for details for maize, and
Rosichan et al. (18), for barley). The number of forward mutant
or reverse mutant grains per slide is determined and used along
with the estimated total pollen grains examined to determine the
mutant frequency on a per plant basis.

Data from approximately 10 tassels or spikes are used to
compute each treatment mean and standard deviation. A graphic presentation of the treatment means is useful in determining the nature
of the dose-response curve. An analysis of variance, with or without
a transformation of data, can be used to provide a statistical
basis for inferences to be drawn regarding the effects of various
treatments.

Genetic Events Involved. The gene locus that governs the
composition of starch accumulated in maize pollen grains is located
near the centromere on the short arm of chromosome 9 along with
other marker genes including C, B_2, SH_1, and Yg_2 (19). At least
31 heteroalleles at the waxy locus are known in maize; five of these
are controlling elements and some of the induced mutants are the
result of deletions (20). In barley, the gene locus is located
near the distal end of the short arm of chromosome 1 (4). A number
of mutant lines induced by sodium azide and γ-radiation are being
characterized (16); at least six mutant alleles have been mapped
(21).

The resolving power of the waxy locus system is sufficiently
great that both forward and reversion mutation tests can be done.
The forward mutants result from: recessive point mutations;
minute deletions that involve only the Wx locus; chromosome segmental
losses that involve a number of loci including Wx; chromosomal
aberrations that lead to position effects that repress the expression
of Wx; and/or altered regulatory genes that affect either the synthesis or the accumulation of amylose. The reverse mutation test is
based on the induction of dominant point mutations at the wx locus.
Waxy mutants derived by induced deletions will not revert. Therefore,
deletion mutant waxy lines cannot be used in reversion mutation
tests.

Validation Data. The waxy locus system in maize and barley
has been used extensively in mutagenesis research. Both acute and
chronic exposures to either physical or chemical mutagens have been
used (10,22). Acute and chronic mutagenic treatments can affect
the results.

Experimental results show a dose-dependent response for waxy
mutants in both maize and barley following acute and chronic exposure to external sources of ionizing radiation and to chronic
exposures from ^{90}Sr and ^{35}S incorporated in the plant tissue.
The nature of the dose-response curves differ somewhat among experiments, as would be expected on the basis of different genotypes,
environmental conditions, etc.

In experiments that involve chronic treatments, the sporophyte
is subjected to mutagens during the major portion of its life cycle.
Usually, the dose rates and, consequently, the total doses are low
to avoid disrupting the maturation of the sporophyte. The results
of an experiment reported by Eriksson (23) will be used to show
the exposure resonse for waxy mutants in maize when the plants were
exposed to γ-radiation for 50 days. Plants of an early flint-type
maize of Russian origin were grown for a few weeks and then placed
on the gamma field at exposure rates from 5 to 200 r/day. The
results are shown in Figure 1. The spontaneous mutant frequency
was 2.9×10^{-4}. Although the scale used in the figure does not
show this clearly, at 5 r/day the mutant frequency increased to
7.9×10^{-4}; at 200 r/day the frequency was 755×10^{-4}. The exposure
response was curvilinear over the range of exposure rates used in
this experiment.

Figure 2 shows results obtained in an experiment with barley
in which the plants were grown in a medium contaminated with ^{90}Sr
in sufficient quantity to provide the doses shown (24). In this
case, γ-radiation from the decay of ^{90}Sr and its daughter ^{90}γ
incorporated in the plant tissue was involved. The frequency of
mutation events in plants subjected to 0.03 rad/day or more was
significantly higher than the spontaneous rate. At the lower dose
rates, i.e. <0.1 rad/day, the increase in mutation event frequency
per unit of dose was greater than that at the higher dose rates,
i.e. 0.1>rad/day.

Figure 3 shows the exposure-response curves for the waxy locus
in maize when two-month-old plants were exposed to γ -radiation over
a 2 hr period (23). Exposure rates ranged from 1.5 to 200 r/hr,
and total exposures were 3, 6, 12, 25, 50, 100, 200, and 400 rad.
All anthers were collected, killed, and fixed on the same day following treatment. The spontaneous mutant frequency was 3.7×10^{-4}.
In comparison, an exposure to 3 rad increased the mutant frequency
to 5.7×10^{-4} and an exposure to 400 rad increased it to $60.1 \times
10^{-4}$. A linear response is shown for exposures through 200 rad
(the linear regression equation was $Y = 4.542 + 0.1056X$), and a
curvilinear response is shown for all exposures through 400 rad
(the regression equation was $Y = 4.4335 + 0.09288X^1 + 0.00010779X^2$).

Eriksson (25) used a homozygous recessive waxy strain of barley
subjected to acute γ-radiation from a ^{137}Cs source to study the

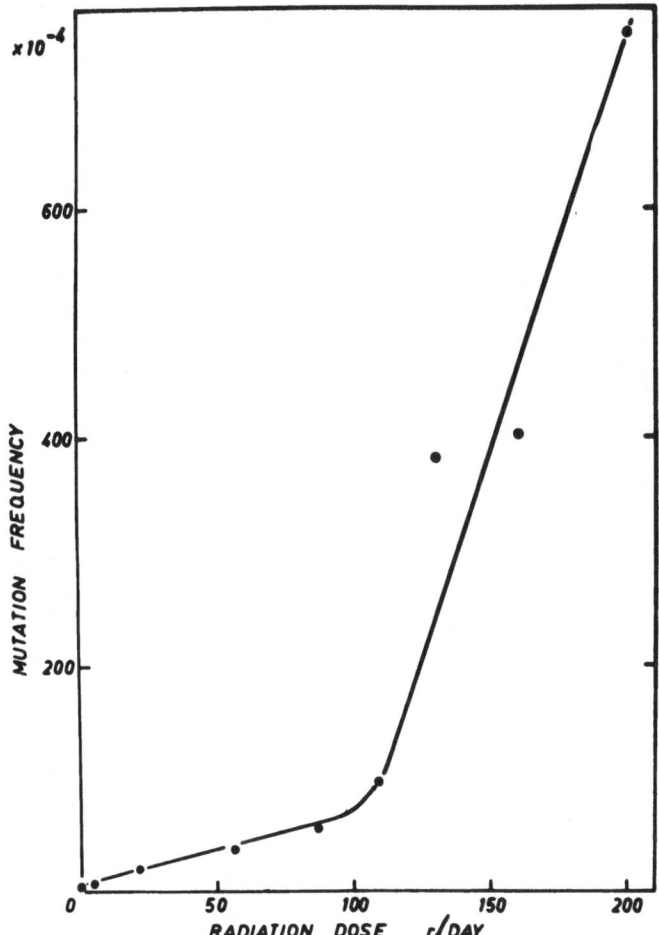

Figure 1.   Expoure response for forward mutants at the waxy locus
in maize following the exposure of plants to   γ-radiation over a
50-day period (Reproduced with permission from Hereditas, ref 23).

effects of radiation on the frequency of reversion mutants.   Exposures
were 25, 100, and 400 rad delivered over a 2 hr period.   Pollen
grains from central and lateral spikelets were compared (Fig 4).
The dose-response curve was exponential in both cases, and a difference in sensitivity probably associated with stage of development
was suggested as the cause of the difference in response between
central and lateral spikelets.

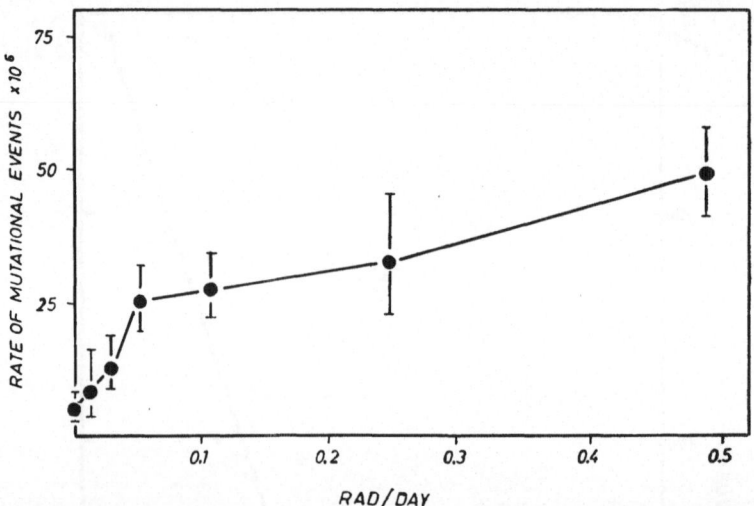

Figure 2. Dose response for forward mutants at the waxy locus in barley when the plants were grown in medium contaminated with [90]Sr (Reproduced with permission from Acta Radiological, ref 24).

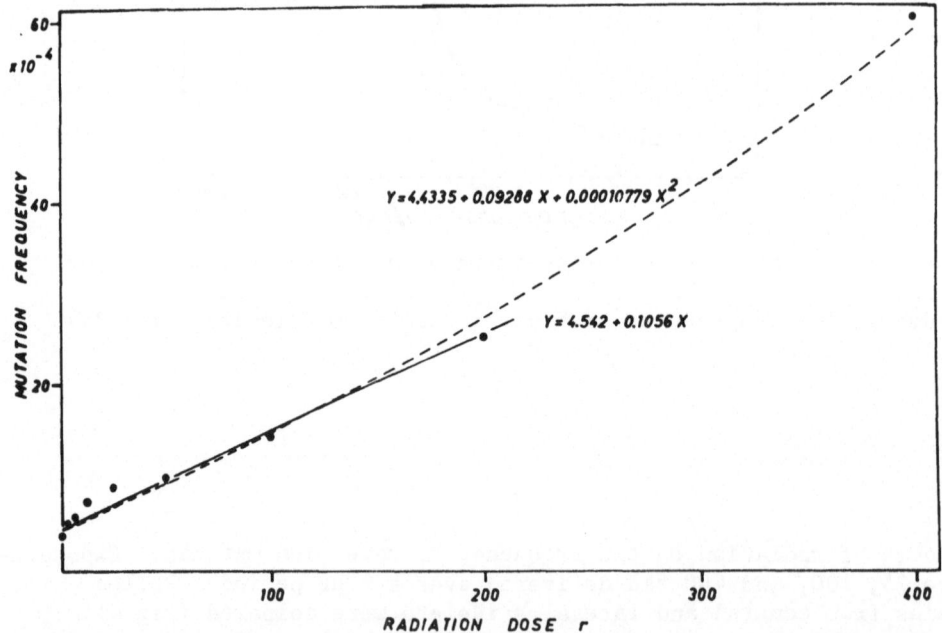

Figure 3. Exposure response for forward mutants at the waxy locus in maize following the exposure of plants to γ-radiation over a 2 hr period (Reproduced with permission from Hereditas, ref 23).

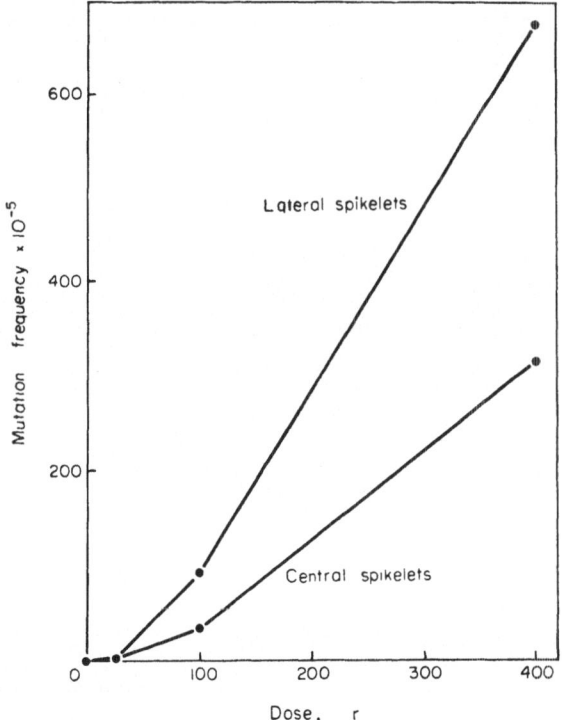

Figure 4. Exposure response for reversion mutants at the waxy locus in barley following the exposure of plants to γ-radiation over a 2 hr period (Reproduced with permission from Pergamon Press Ltd., ref 25).

Although dose-response data were not presented, ethyl methane-sulfonate (26) and isopropyl methanesulfonate (27) have been used to induce waxy mutations in barley. Sector size was smaller in barley spikes following treatment of seeds with either ethyl methanesulfonate or isopropyl methanesulfonate as compared to ^{60}Co γ-radiation. Sodium azide has induced waxy mutants in barley (16).

Thus, sufficient evidence is available to prove that the waxy locus in maize and barley is responsive to mutagenic treatment. The genetic events that govern the phenotypic response observed are not known precisely, and there are problems regarding the number of target cells at risk at different developmental stages. These, however, do not negate the use of the system in studies of environmental mutagenesis.

Pesticide Data. Plewa (6) used the wx-C allele in maize to study the effects of agricultural pesticides on the frequency of

Table 2. Effects of pesticides used alone and in combination on the frequency of reversion mutants at the wx locus in maize (reproduced from M. J. Plewa (6)).

Pesticide	Active agent	Application rate of active agent (Kg/ha)	Number of pollen grains analyzed (x10⁵)	Reversion frequency (x10⁻⁵)
Control			25.89	5.56 ± 0.98
Cyanazine	2-(4-chloro-6-ethylamino-s-triazine-2-yl amino)-2-methyl-proprionitrile	2.86	4.71	28.23 ± 3.36
Procyazine	2 [4-chloro-6-(cyclopropylamino)-s-triazine 2-yl]amino -2-methyl propanenitrile	2.86	2.41	4.65 ± 1.76
Control			19.91	4.16 ± 1.10
Heptachlor	1,4,5,6,7,8-heptachloro-3a,4,7,7a-tetra-hydro-4,7-methanoidene	0.22	7.30	10.87 ± 3.06
Chlordane	1,2,4,5,6,7,8,8-octachlor-2,3,3a,4,7,7a-hexahydro-4,-7-methanoindane	0.75	8.55	13.69 ± 1.10
Control			9.75	3.19 ± 0.67
Eradicane	S-ethyl dipropylthiocarbamate	7.20	8.09	5.31 ± 1.64
Metholaclor	2-chloro-N-(2-ethyl-6-methylphenyl-N-(2-methoxyl-1-methylethyl) acetamide	6.00	9.59	4.32 ± 1.14
Atrazine	2-chloro-4-ethylamino-6-isopropylamino-s-triazine	3.07	9.50	8.92 ± 1.29
Simazine	2-chloro-4,6-bis(ethylamino)-s-triazine	3.07	8.19	12.00 ± 2.31
Eradicane + cyanazine		3.60 ± 2.59	8.83	9.18 ± 1.46
Metolachlor + cyanazine		2.24 ± 1.79	5.17	11.08 ± 3.11
Eradicane + atrazine		3.60 ± 1.54	8.86	10.77 ± 1.59
Metolachlor + atrazine		3.00 ± 1.79	7.19	12.27 ± 2.34

reversion mutants. The results (Table 2) indicated that the s-triazine herbicides (viz. atrazine, simazine, and cyanazine) were mutagenic whereas procyazine was not. Eradicane and metolachlor, two other herbicides, were not mutagenic. However, when these were combined with the mutagenic s-triazines, the mutant frequency was increased. Both chlordane and heptachlor insecticides increased the frequency of mutants. Recently, Plewa (unpublished) has observed an increase in mutant frequency that is dose dependent when ethyl methanesulfonate or maleic hydrazide is applied periodically during a 3-week period to the growth medium in which maize plants are grown.

Results of Studies with Atmospheric Mutagens. Although the number of reports in the literature is limited, the information indicates that the waxy locus can be used to detect atmospheric mutagens. Sulovska et al. (28) exposed plants of Bonus barley (Wx Wx) for 24 hr to various concentrations of ethylene oxide in a gas chamber. The plants were exposed at about the time of meiosis in the pollen mother cells. The mutant frequency observed was as follows: 0 ppm = 1.8×10^{-6}, 100 ppm = 21.5×10^{-6}, 200 ppm = 31.8×10^{-6}, and 400 ppm = 454×10^{-6}. A later report (29) shows that 50 ppm of ethylene oxide had a significant mutagenic effect. In fact, the lowest concentration used (12.5 ppm) increased the mutant frequency. The waxy locus in barley was used to test the mutagenicity of tobacco smoke, car exhaust and city air; no strong mutagenic effects were observed (30).

The waxy locus in maize has been used along with the Tradescantia stamen hair system to test for ambient mutagenic activity next to a lead smelter in southeastern Missouri (31). Table 3 shows data on soil content of Pb, Cd, Cu, and Zn at various distances from the smelter and mutant frequencies in maize and Tradescantia grown in situ at those distances. Mutant frequency in both systems was higher at the distances tested than that of control populations grown in a clean environment. In neither system was mutant frequency a linear function of distance from the lead smelter. The increase in mutant frequency might have been induced by mutagens in the soil, in the atmosphere, or both. When Tradescantia plants were grown in soil from the smelter site in a clean atmosphere, their mutant frequency approximated that observed in situ. All four elements reportedly are mutagenic; however, the lead concentration was the highest at each location tested.

Lower et al. (personal communication) have also used maize and Tradescantia to test the environment at various geographic locations for the presence of mutagens. Table 4 lists the sites tested, the environmental conditions at each site, and the mutant frequency observed for maize waxy. All locations induced a substantial increase in mutant frequency as compared to that of a control population grown under "clean" environmental conditions.

Table 3. Soil content of Pb, Cd, Cu, and Zn, and the frequency of
          waxy mutants in maize and pink stamen hair cells in
          Tradescantia plants grown in situ at the indicated dis-
          tances from a lead smelter (Data from W. B. Lower, P. S.
          Rose and V. K. Drobney, (31)).

Distance from lead smelter	Soil content (ppm)				Mutant frequency	
					Zea	Tradescantia
(Km)	Pb	Cd	Cu	Zn	($\overline{X}$±SE) x10^{-5}	($\overline{X}$±SE) x 10^{-3}
0.3	4701	15	100	303	10.7 ± 5.01	2.084 ± 0.143
1.7	881	4	28	57	15.7 ± 5.48	2.629 ± 0.179
11.4	18	0.2	11	20	4.58 ± 0.84	1.982 ± 0.164
Control					2.93 ± 1.25	1.744 ± 0.138

CONCLUSION

     The waxy locus system in maize and barley is effective in de-
tecting and monitoring the environment for the presence of mutagens.
The following facts support this conclusion:

     1. Genetic studies have demonstrated Mendelian inheritance
patterns for the waxy character.
     2. Plants respond to mutagens at various developmental stages
prior to tetrad formation; however, specific developmental stages
vary in sensitivity to mutagens.
     3. The genetic resolving power of the system is high enough
to detect forward and reversion mutations.
     4. The endpoint scored is readily detectable with a minimum
of effort, and the conduct of a test is relatively simple and inexpensive.
     5. Mutant frequency increases as a function of dose for ionizing
radiations and known chemical mutagens; however, more than one genetic
event can give rise to the mutant pollen grains.
     6. The system responds positively when used in environmental
conditions suspected of having mutagenic activity.

ACKNOWLEDGEMENTS

     Thanks are extended to Drs. W. B. Lower, R. A. Nilan, and M. J.
Plewa, and their associates, who are doing research in environmental

Table 4. Frequency of mutants at the waxy locus in maize plants grown at various geographic locations under different environmental conditions (W. B. Lower et al., personal communication)

Location	Mutant freqeuncy $(x \times 10^{-5})$	Environment
Edwardsville, IL (SIU campus)	45.77	Petrochemical
Wood River, IL Olive and First St.	45.53	Petrochemical
Wood River, IL Olive and Date St.	39.27	Petrochemical
Wood River, IL Benred Meat Co.	47.69	Petrochemical
Granite City, IL Madison Avenue	4.33	Industrial/petro-chemical
Granite City, IL Tarpoff Meat Co.	23.38	Industrial/petro-chemical
Columbia, MO Control (GH3)	2.88 ± 0.22	

mutagenesis with the waxy locus system in maize and barley, for their contributions to this manuscript.

REFERENCES

1. L. Ehrenberg, Higher plants, in: "Chemical Mutagens: Principles and Methods for their Detection", Vol. 2, A. Hollaender, ed., Plenum Press, New York, N.Y., 365-386 (1971).
2. R. A. Nilan and B. K. Vig, Plant test systems for detection of chemical mutagens, in: "Chemical Mutagens: Principles and Methods for their Detection", Vol. 4, A. Hollaender, ed., Plenum Press, N.Y., 143-170 (1976).
3. M. J. Constantin, Utility of specific locus systems, Environ. Health Perspectives 27:69-75 (1978).
4. R. A. Nilan, Potential of plant genetic systems for monitoring and screening mutagens, Environ. Health Perspectives 27: 181-196 (1978).

5. F. J. de Serres, Higher plant systems as monitors of environ-
    mental mutagens, in: "Application of Short-term Bioassays in
    the Fractionation and Analysis of Complex Environmental
    Mixtures", M. D. Waters, S. Nesnow, J. L. Huisingh, S. S.
    Sandhu, and L. Claxton, eds., Proceedings USEPA Symposium,
    EPA-600/9-78-027, Plenum Press, New York, N.Y., 99-110 (1978).

6. M. J. Plewa, Activation of chemicals into mutagens by green
    plants: a preliminary discussion, Environ. Health
    Perspectives 27:45-50 (1978).

7. M. Freeling, Maize Adh1 as a monitor of environmental mutagens,
    Environ. Health Perspectives 27:91-97 (1978).

8. U. S. Department of Health, Education, and Welfare, Higher
    Plants as Monitors of Environmental Mutagens, Proceedings
    of Workshop, January 16-18, 1978, Marineland, FL, Environ.
    Health Perspectives 27:1-206, USDHEW Pub. No. (NIH) 79-218
    (1978).

9. U. S. Department of Health and Human Services and U. S.
    Environmental Protection Agency, Pollen Systems to Detect
    Biological Activity of Environmental Pollutants, Proceedings
    of Conference, May 5-8, 1980, The University of Tennessee,
    Knoxville, TN, Environ. Health Perspectives (in press) (1981).

10. U. S. Environmental Protection Agency, USEPA Gene-Tox Program,
    Reports by members of Workgroup on Plant Genetic and
    Cytogenetic Assays, M. J. Constantin, Chairman, Mutat. Res.,
    in preparation, (1981).

11. G. P. Redei, Arabidopsis, in: "Handbook of Genetics", Vol. 2,
    "Plants, Plant Viruses, and Protists", R. C. King, ed.,
    Plenum Press, New York, N.Y., 151-180 (1974).

12. C. M. Rick, The Tomato, in: "Handbook of Genetics", Vol. 2,
    "Plants, Plant Viruses, and Protists", R. C. King, ed.,
    Plenum Press, New York, N.Y., 247-280 (1974).

13. G. Eriksson, The waxy character, Hereditas 63:180-204 (1969).

14. G. F. Sprague, B. Brimhall, and R. M. Hixon, Some effects of the
    waxy gene in corn on properties of the endosperm starch, J.
    Am. Soc. Agron. 35:817-822 (1943).

15. M. J. Plewa, Specific locus assays in Zea mays, USEPA Gene-Tox
    Report, Mutation Research, in press, (1981).

16. R. A. Nilan, J. L. Rosichan, P. Arenaz, A. L. Hodgdon, and
    A. Kleinhofs, Pollen genetic markers for detection of muta-
    gens in the environment, Environ. Health Perspectives, in
    press, (1981).

17. C. F. Konzak, R. A. Nilan, J. Wagner, and R. J. Foster, Efficient
    chemical mutagenesis, in: "The Use of Induced Mutations in
    Plant Breeding", FAO/IAEA Tech. Mtg. Rome (Supplement to
    Radiation Botany, Vol. 5), Pergamon Press, Oxford, 49-70,
    (1965).

18. J. L. Rosichan, P. Arenaz, N. Blake, A. Hodgdon, A. Kleinhofs,
    and R.A. Nilan, An improved method for the detection of
    mutants at the waxy locus in Hordeum vulgare, Environ.
    Mutagenesis, in press, (1981).

19. M. G. Neuffer and E. H. Coe, Jr., Corn (maize), in: "Handbook of Genetics", Vol. 2, "Plants, Plant Viruses, and Protists", R. C. King, ed., Plenum Press, New York, N.Y., 3-30 (1974).

20. O. E. Nelson, Previously unreported wx heteroalleles, Maize Genet. Coop. Newsletter 50:109 (1976).

21. J. L. Rosichan, Genetic fine structure analysis of the waxy locus in barley (Hordeum vulgare), M.S. Thesis, Washington State University, Pullman, WA (1979).

22. D. deNettantcourt, G. Eriksson, D. Lindgren, and K. Puite, Effects of low doses by different types of radiation on the waxy locus in barley and maize, Hereditas 85:89-100 (1977).

23. G. Eriksson, Induction of waxy mutants in maize by acute and chronic gamma irradiation, Hereditas 50:161-178 (1963).

24. L. Ehrenberg and G. Eriksson, The dose dependence of mutation rates in the rad range, in the light of experiments with higher plants, Acta. Radiol. Suppl. 254:73-81 (1966).

25. G. Eriksson, Radiation induced reversions of a waxy allele in barley, Radiat. Bot. 2:35-39 (1962).

26. D. Lindgren, G. Eriksson, and K. Sulovska, The size and appearance of the mutated sector in barley spikes, Hereditas 65:107-132 (1970).

27. D. Lindgren and G. Ericksson, The mutated sector in barley spikes following isopropyl methanesulphonate (iPMS) treatment period. Hereditas 69:129-134 (1971).

28. K. Sulovska, D. Lindgren, G. Eriksson, and L. Ehrenberg, The mutagenic effect of low concentrations of ethylene oxide in air, Hereditas, 62:264-266 (1969).

29. D. Lindgren and K. Sulovska, The mutagenic effect of low concentrations of ethylene oxide in air, Proc. Fifth Meeting Scand. Assoc. Geneticists, Reykjavik, Hereditas 63 (Supplement) 460, Abstract, (1969).

30. D. Lindgren and K. Lindgren, Investigations of environmental mutagens by the waxy method, EMS Newsletter 6:22 Abstract, (1972).

31. W. B. Lower, P. S. Rose, and V. K. Drobney, In situ mutagenic and other effects associated with lead smelting, Mutat. Res. 54:83-93 (1978).

SESSION IV: ANIMAL EXPOSURE TECHNOLOGY FOR THE DETECTION OF

GENOTOXIC AGENTS

Robert T. Drew, Moderator

Medical Department
Brookhaven National Laboratory
Upton, New York 11973 (U.S.A.)

INTRODUCTION

In organizing this conference we felt that the first day
should be devoted to defining systems and technologies that are
available for the measurement of genotoxic effects of airborne
agents. This session covers in vivo exposure of animals considering
both whole body and nose exposure techniques; techniques for the
generation and monitoring of vapors; and finally the question of
assessing particulates for mutagenicity. It is presumed that most
of the audience have a good background in genetics, but may not
necessarily be aware of the technologies associated with the generation and monitoring of aerosols and exposure of animals. The first
paper will describe standard inhalation chambers and how they are
operated, the build-up and removal of pollutants in chambers, and
the function, applicability, advantages and disadvantages of nose
exposure systems. The second paper will discuss the methods whereby gas and mixtures of gases are generated and standardized. It
will also describe a number of techniques that can be used to monitor
the concentration of the test atmospheres. The third paper of this
series will address particles. Mutagenic particles usually exist
as complex and heterogeneous mixtures of both organic and inorganic
compounds. Collection methods that guarantee the relevance, fidelity,
and representativeness of the particles must be used. This paper
will cover sample collection, particle size considerations, sample
fidelity, and finally, procedures to test the mutagenic capabilities
of the aerosol.

179

SYSTEMS FOR EXPOSURE OF ANIMALS TO AIRBORNE AGENTS

Robert T. Drew

Medical Department
Brookhaven National Laboratory
Upton, New York 11973

INTRODUCTION

This paper will introduce the reader to the inhalation exposure of animals. It is by no means complete and for more information the reader should consult more thorough reviews (1-5). This chapter will: (a) provide basic definitions for some common terms; (b) describe standard inhalation chambers and how they are operated; (c) describe the build-up and removal of pollutants in chambers; (d) briefly discuss isolation exposure systems; and (e) describe the function, applicability, advantages, and disadvantages of nose exposure systems. Other techniques which may be useful in assessing mutagenicity but which are not described here include intratracheal instillation, partial lung exposures, and exposures during bronchoscopy.

DEFINITIONS

Atmospheric concentrations are usually expressed in one of two different units, mg/m^3 or ppm (on a volume/volume basis). While the term mg/m^3 is easily understood, the term "ppm" has historically caused much confusion. The ideal gas law can be used to convert ppm to mg/m^3. The following example shows the conversion of 10 ppm SO_2 to mg/m^3 at 760 mm H_2 and 20°C:

10 ppm SO_2 =

$$\frac{10 \text{ } \mu L \text{ } SO_2}{\text{Liter air}} \times \frac{\mu mole \text{ } SO_2}{22.4 \text{ } \mu L \text{ } SO_2} \times \frac{64 \text{ } \mu g \text{ } SO_2}{\mu mole \text{ } SO_2} \times \frac{273}{293} = 26.6 \frac{\mu g}{L} = 26.6 \frac{mg}{m^3}$$

Note that only vapors and gases can be expressed in ppm. Aerosols
(defined elsewhere in this volume) cannot be expressed in terms of
ppm.

STANDARD INHALATION CHAMBERS

     The basic concept of inhalation exposure is simple; one needs
to put animals and the test material in a closed volume or chamber
at the same time. Historically, static systems were often used
for acute exposures, but today they are seldom, if ever, used.
The regulatory guidelines require dynamic systems in which the air
and the pollutant are continuously introduced into the chamber and
the mixture is removed at the same rate. Figure 1 summarizes the
basic features of modern exposure chambers. Most chambers today
are square in cross section with a pyradmidal top and bottom. Air
is introduced at the top, either vertically or tangentially. The
pollutant is usually introduced perpendicular to the airstream with
mixing occurring in the top cone. The air passes turbulently through
the animals and is removed at the bottom, either through a Y or an
inverted U. The exhaust air should be cleaned by suitable devices
(filters, charcoal adsorbers, scrubbers, electrostatic precipitators,
etc.) before passing through the exhaust fan and being discharged
to the outside. The liquid and solid waste can be routed through
a trap and plumbed to the waste water system.

     The chambers are operated at about minus 2 cm H_2O pressure
with respect to room pressure. The temperature should be 22 ± 2°C
(68-75°F). The largest single factor governing the chamber temperature is the room temperature. Relative humidity should range from
40 to 60%, although I personally see no problem if allowed to go
as high as 70%, unless hydroscopic particles are being studied.

     During an exposure, it is necessary to record air flow, static
pressure, temperature, and humidity. It is also very important
to measure the concentration in the chamber several times during
the exposure. Because there are many features which can cause
pollutant loss during exposure it is not good technique to report
concentrations derived from dividing the amount of pollutant ex-
pended by the total volume of air passing through the system.
Ideally, chamber concentration should be recorded continuously,
but this can be expensive. We cycle our detector through several
chambers, the room, and the exhaust line, recording the concentration
at each location at half-hour intervals. Particle size should also
be measured, but not with the same frequency as concentration.
Once a generator has been shown to produce a stable aerosol, the
frequency of measuring particle size can be decreased.

     Historically, the air flow in a chamber has varied from 8 to
60 air changes per hour. The term "air change" is a carry-over from

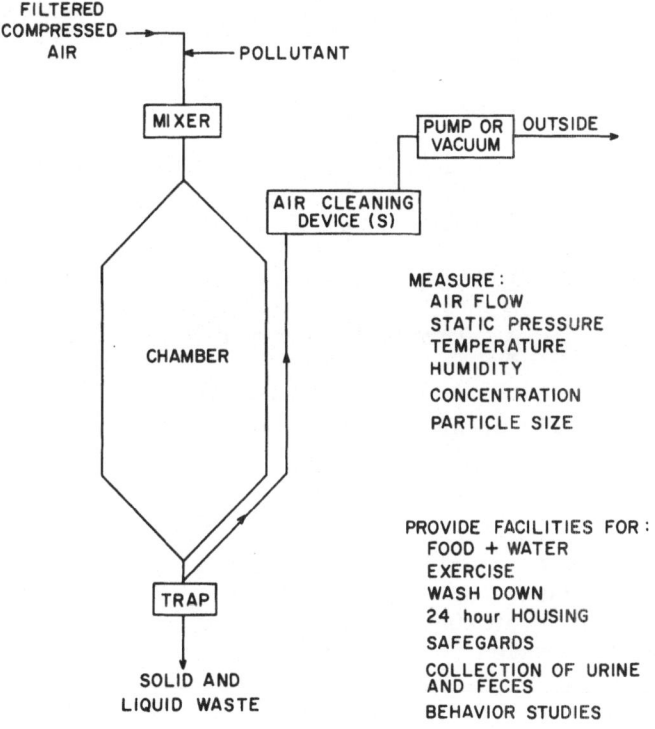

FILTERED
COMPRESSED
AIR → POLLUTANT

MIXER

PUMP OR VACUUM   OUTSIDE

AIR CLEANING
DEVICE (S)

CHAMBER

MEASURE:
  AIR FLOW
  STATIC PRESSURE
  TEMPERATURE
  HUMIDITY
  CONCENTRATION
  PARTICLE SIZE

PROVIDE FACILITIES FOR:
  FOOD + WATER
  EXERCISE
  WASH DOWN
  24 hour HOUSING
  SAFEGARDS
  COLLECTION OF URINE
  AND FECES
  BEHAVIOR STUDIES

TRAP

SOLID AND
LIQUID WASTE

Figure 1. Essential features of inhalation exposure systems.

the heating, ventilation, and air conditioning engineering termino-
logy and is an unfortunate choice. One air change is defined as
the addition and withdrawal of a volume of air equal to the volume
of the room or chamber being considered. Because the incoming air
mixes with the air already present in the chamber, a complete change
of air has not occurred. The dynamics of air mixing in inhalation
chambers was described and confirmed by Silver (8) who stated 35
years ago that the term "air change" was confusing and misleading.
Unfortunately, the term has remained in use, and it is still confusing. The equilibrium concentration of any material in a chamber
is a function of its concentration in the air entering the chamber
and the flow of air through the chamber relative to the chamber
volume. The build-up and removal of material (Figure 2) is a
logarithmic function of the air turnover rate. The build-up and
decay equation has the general form:

$$t_x = K \, a/b$$

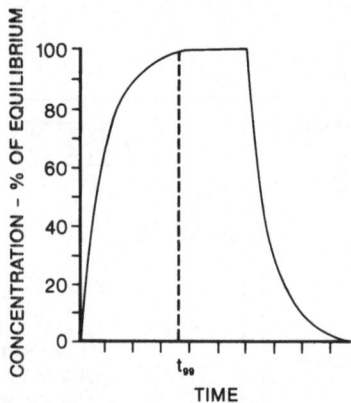

Figure 2. Build up and decay of pollutants in a chamber.

where,

    x is the percent equilibrium attained in time t
    K is a constant
    a is the volume of the chamber
    b is the flow rate

For t_{99}, the time to reach 99% of the equilibrium concentration
"K" equals 4.6; for t_{95}, "K" equals 3. If a chamber is operating
at 15 air changes per hour:

$$a/b = ab^{-1} = a\ (15a/hr)^{-1} \times 60\ min/hr = 4\ min$$

The time to reach 99% of the equilibrium concentration is:

$$t_{99} = (4.6)4\ min = 18.4\ min$$

The time to reach 95% of the equilibrium concentration is:

$$t_{95} = (3)4\ min = 12\ min$$

Various values for t_{99} and t_{95} are summarized in Table 1.

    The concentrations during a typical exposure is shown in Figure
3. Notice that the duration of the exposure is long compared to
t_{99}. Notice also that after t_b the concentration falls to zero.
The duration of the exposure is usally the time from t_a to t_b, the
exposure after t_b is minimal and equal in area (assuming constant
air flow) to that missing during the build-up of the material in
the chamber. If exposures of short duration are contemplated, the
build-up and decay effect may be important. One way to minimize
build-up and decay effects is to construct drawer units to insert
into a chamber after equilibrium has been established. Another way
would be to increase air and pollutant flow to shorten t_{99}. This

Table 1. Equilibrium values at different air change rates.

Air Changes/Hr	a/b Min/Air Change	t_{99}(min)	t_{99}(min)
60	1	4.6	3
30	2	9.2	6
15	4	18.4	12
10	6	27.6	18

has been accomplished at NIEHS where programmable calculators are used to establish concentration-time curves that approach square waves (9).

Figures 4 and 5 show schematic drawings of the NYU chamber and the Rochester chamber. Chambers similar to these are in use in several government, industrial, university and contract laboratories throughout the country. Figure 6 shows walk-in chambers in use at Brookhaven National Laboratory. Most chambers are constructed of stainless steel with either lucite or glass windows. However, they need not all be expensive. In 1965, Leonard Leach (10) published a schematic of a battery or pickle jar exposure chamber which has been in wide use ever since. A current version with aluminum faces also used at Brookhaven is shown in Figure 7. At Battelle Northwest Laboratories they used four foot lucite hemispheres coupled together in the shape of a sphere with the animals exposed on a single tier

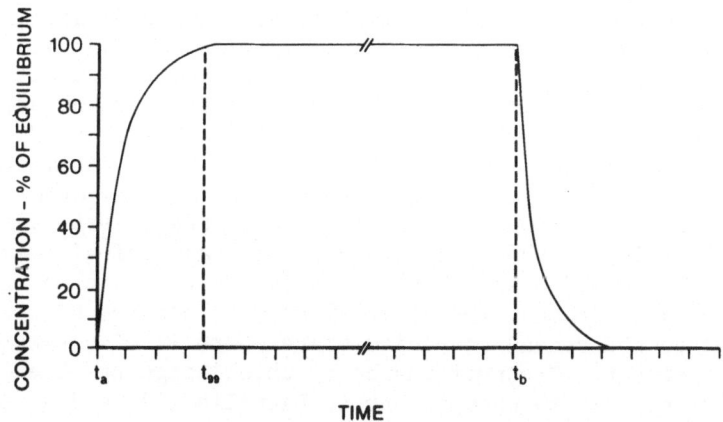

Figure 3. Concentration vs. Time during a typical exposure.

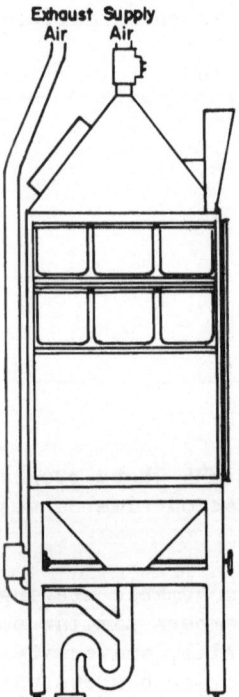

Figure 4.   The NYU chamber.

(11). Sid Laskin and I described a small plastic chamber we built
from lucite and aluminum channel (12). It could be used to expose
up to 25 rats at one time.

One problem with such chambers was the question of including
catch pans underneath each tier of cages. Catch pans interfere
with aerosol distribution. However, if not included, animals on
lower levels can get quite dirty. In systems where animals are
housed 24 hours/day, catch pans must be included if animals are on
several tiers. Recently at Battelle Northwest Laboratories, Dr.
Owen Moss and his co-workers (13) designed and tested a chamber
which allows catch pans to be used under each tier. Asymmetrical
pyramids offset from each other form the top and the bottom
of the chamber. The cages on the right side are offset from those
on the left side, as shown in Figure 8. The aerosol-air mixture
streams down the sides and swirls into the center at each catch
pan (Figure 9). Some of the material stays along the side wall
and passes to the second and third tiers where it also swirls in
toward the center. The performance of this design has been
studied extensively by groups both at Battelle (13) and at the
Lovelace Inhalation Toxicology Research Institute (14). The local

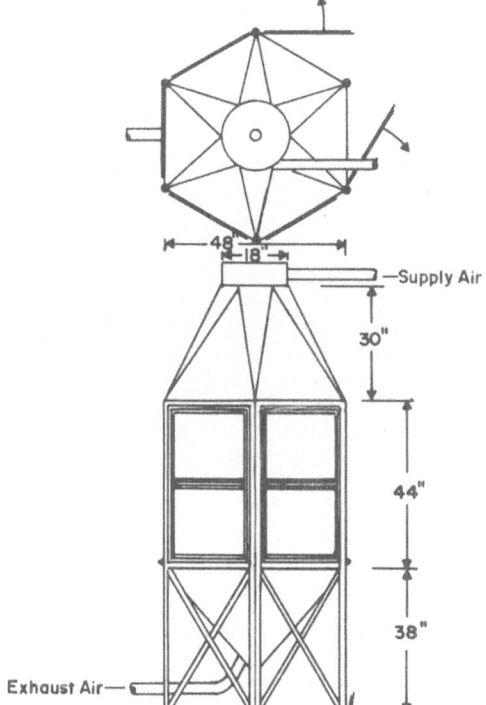

Figure 5. Rochester chamber.

concentrations vary by less than 10% from the mean. The concentration on one side is consistently higher than the other side, however, and they recommend rotating the cages within the system.
Moss also reported a variation whereby they can maintain two
concentrations in one chamber (15).

Isolation Systems

The problems associated with handling animals covered with
known particulate carcinogens mandate that isolation systems be
used when working with these agents. One such system (16) allows
for exposure in a chamber within a chamber and provides several
other glove boxes for aerosol generation, air cleaning, and
rodent housing (Figure 10). Another system similar in concept has
been built for NIOSH in Cincinnati (17). In this system, the animals
are housed in a glove box and mechanically moved into the carcinogen
exposure chamber by turning an outside crank.

Figure 6.   Walk-in chambers at Brookhaven National Laboratory.

Figure 7.   Pickle jar systems.

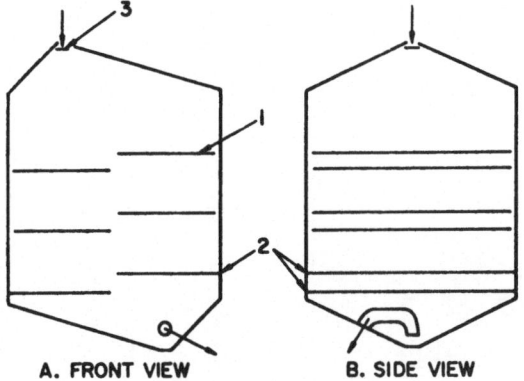

A. FRONT VIEW   B. SIDE VIEW

Figure 8. Sketch of Battelle chamber with catch pans in position.

Nose Exposures

When test agents are in short supply or very expensive, or when they are highly toxic, it may be necessary to expose only the head or nose of the animal to the test material. Head and nose exposure systems have been described by many investigators (2,18-21). This technology has grown from three diverse applications: the exposure of animals to radionuclides, the exposure of animals to biologic aerosols, and the development of systems to expose animals to cigarette smoke in a manner analogous to human smoking behavior. The advantages and disadvantages of nose exposure are outlined in Table 2.

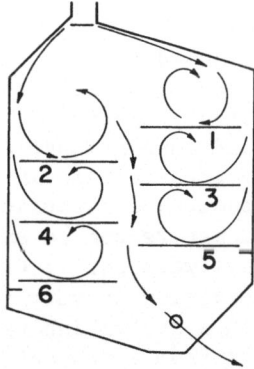

Figure 9. Air flow patterns in Battelle chamber.

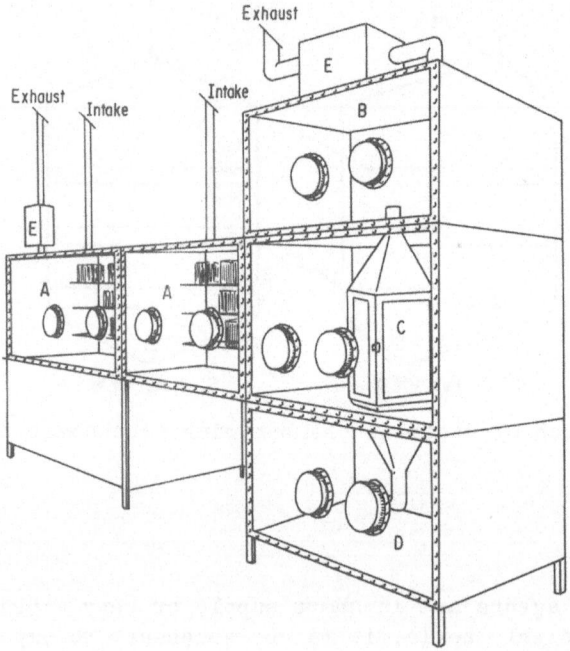

Figure 10. An isolation inhalation exposure system.

        One approach developed at the Lovelace Inhalation Toxicology
Research Institute was to fashion latex masks to fit over the noses
of animals and then insert the nose into a chamber through another
piece of soft rubber. This creates an effective seal, preventing
material loss. Another system involved the use of an inflatable
collar to make a neck seal. These systems have been used effectively
to expose dogs, rabbits (Figure 11) and rodents to highly toxic
aerosols while keeping the animal's skin virtually free of the test
agent.

        It generally has been assumed that confining rodents in nose
tubes is stressful and that such exposures should be limited to
one or at most, two hours per day. However, recently at Los Alamos
(22), scientists have modified conventional chambers to accept an
insert capable of holding several animals in tubes (Figure 12).
Their studies, as well as studies at Battelle in Geneva (23,24)
suggest that both rats and hamsters can be exposed in nose exposure
tubes six hours/day, five days/week for several months with no
untoward effects. This observation is very important since the
handling of animals from such a system inherently would be much
cleaner than handling animals from a conventional chamber. Because
of difficulties in handling animals exposed to particulate carcino-

Table 2. A comparison of nose vs. whole body exposure

Nose Exposure	Whole Body Exposure
ADVANTAGES	
No ingestion from preening	Easier
Uses less material	Large numbers of animals
Clean animals	Stable concentrations
Air clean-up simple	Can do repetitively
Material easier to contain	Minimal animal handling
DISADVANTAGES	
Laborious	Animals contaminated
Stressful	Ingestion from preening
May need cage controls	Consumes large amounts of material
	Air clean-up cumbersome
	Expensive equipment

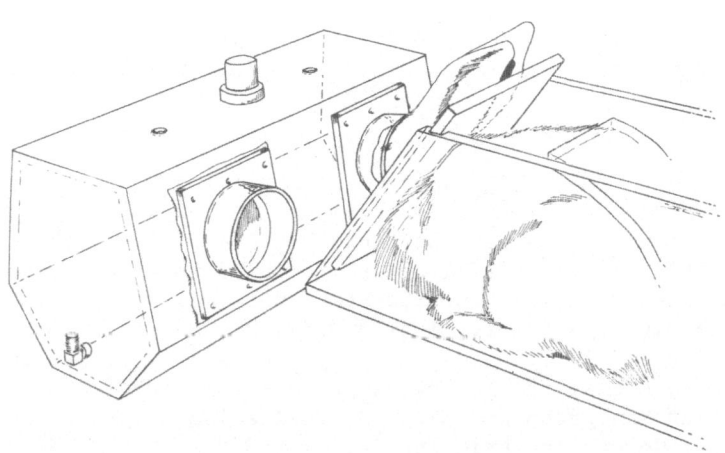

Figure 11. A nose exposure system for a rabbit.

Figure 12.   A 27" chamber modified for nose exposures.

gens, this approach may be necessary to expose animals to such
agents.

     In summary, exposure of animals to airborne agents is techni-
cally more difficult and usually more expensive than other routes
of administration. However, inhalation is necessary to mimic man's
exposure to many agents. Techniques are available to perform inhalation studies with almost any animal model. However, there are
many pitfalls in this technology and scientists are urged to seek
advice of experienced investigators before commencing such studies.

REFERENCES

 1. R. T. Drew, ed., "Inhalation Chamber Technology", BNL 51318,
       Brookhaven National Laboratory, Upton, N.Y. (1981).
 2. R. T. Drew and S. Laskin, Environmental inhalation chambers, in:
       "Methods of Animal Experimentation", Vol. 4, W. I. Gay, ed.,
       Academic Press, N.Y., 1-41 (1973).
 3. D. A. Fraser, R. E. Bales, M. Lippmann, and H. E. Stokinger,
       Exposure chambers for research in animal inhalation, Public
       Health Monograph No. 57, PHS publication No. 662, Washington,
       D.C., Government Printing Office (1959).
 4. R. G. Hinners, J. K. Burkart, and C. L. Punte, Animal inhalation
       exposure chambers, Arch. Environ. Health 16:194-206 (1968).

5. H. N. MacFarland, Respiratory Toxicology, in: "Essays in Toxicology", Vol. 7, W. J. Hayes, Jr., ed., Academic Press, N.Y. (1978).

6. U. Saffiotti, F. Cefis, and L. H. Kolb, A method for the experimental induction of bronchiogenic carcinoma, Cancer Research 28:104-107 (1968).

7. R. F. Phalen, Inhalation Exposure of Animals, Environ. Health Perspectives 16:17-24 (1976).

8. S. D. Silver, Constant flow gassing chambers: Principles influencing design and operations, J. Lab. Clin. Med. 31: 1153-1161 (1946).

9. L. J. Leach, A laboratory test chamber for studying airborne materials, AEC Progress Report, UR-629, University of Rochester, Rochester, N.Y. (1965).

10. B. O. Stuart, D. H. Willard, and E. B. Howard, Uranium mine air contaminants in dogs and hamsters, in: "Inhalation Carcinogenesis"M. G. Hanna, Jr., P. Nettesheim, and J. R. Gilbert, eds., AEC Symposium Series 18, CONF-691001, CFSTI, NBS, U.S. Dept. of Commerce, 413-427 (1970).

11. S. Laskin and R. T. Drew, An inexpensive portable inhalation chamber, Amer. Indust. Hyg. Assoc. J. 31:645-646 (1970).

12. O. R. Moss, A chamber providing uniform concentrations of particulates for exposure of animals on tiers separated by catch pans, in: "Inhalation Chamber Technology", R. T. Drew, ed., BNL 51318, Brookhaven National Laboratory, Upton, N.Y., 31-38 (1981).

13. R. L. Beethe, R. K. Wolff, L. C. Griffis, C. H. Hobbs, and R. McClellan, Evaluation of a recently designed multi-tiered exposure chamber, Inhalation Toxicology Research Institute Lovelace Biomedical and Environmental Research Institute, Publication LF-67, UC-48, November, 1979.

14. O. R. Moss, E. J. Rossignol, and M. L. Clark, Technique for simultaneously maintaining two aerosol concentrations in an inhalation exposure chamber, presented at the Annual Meeting of the American Industrial Hygiene Association, Portland, OR, May, 1981.

15. S. Laskin, M. Kuschner, and R. T. Drew, Studies in Pulmonary Carcinogenesis, in: "Inhalation Carcinogenesis", M. G. Hanna, Jr., P. Nettesheim, and J. R. Gilbert, eds., AEC Symposium Series 18, CONF-691001, CFSTI, NBS, U. S. Dept. of Commerce, 321-351 (1970).

16. E. W. Van Stee and M. P. Moorman, Monitoring for Temperature, Humidity, and Concentration, in: "Inhalation Chamber Technology", R. T. Drew, ed., BNL 51318, Brookhaven National Laboratory, Upton, N.Y., 81-88 (1981).

17. H. E. Stokinger, Toxicity following inhalation, in: "Pharmacology and Toxicology of Uranium Compounds", C. Voegtlin and H. Hodge, eds., Vol. III, McGraw-Hill, N.Y., 423-445 (1949).

18. D. W. Henderson, Apparatus for study of airborne infection, J. Hyg. 50:53-68 (1952).

19. R. G. Thomas and R. Lie, Procedures and equipment used in
      inhalation studies on small animals, AEC Research and Develop-
      ment Report, LF-11, Lovelace Foundation for Medical Education
      and Research, TID-4500 (1963).
20. W. Dontenwill, Experimental investigations on the effect of
      cigarette smoke inhalation on small laboratory animals, in:
      "Inhalation Carcinogenesis", M. G. Hanna, Jr., P. Nettesheim,
      and J. R. Gilbert, eds., AEC Symposium Series 18, CONF-691001,
      CFSTI, NBS, U.S. Dept. of Commerce 389-412 (1970).
21. L. W. Ortiz, R. F. Archuleta, J. F. Spalding, and R. G. Thomas,
      Modifications of classical whole body mode inhalation chambers
      for chronic 'nose only' exposure, presented at Amer. Indust.
      Hyg. Assoc. Conf., Portland, OR (1981).
22. C. R. E. Coggins, F. Duchosal, C. Musy, and R. Ventrone,
      Measurement of respiratory patterns in rodents using whole-
      body plethysmography and a pneumotachograph, Lab. Anim. 15:
      137-140 (1981).
23. H. Baumgartner and C. R. E. Coggins, Description of a continuous
      smoking inhalation machine for exposing small animals to
      tobacco smoke, Beitrage Zur Tabakforschung International 10:
      169-174 (1980).

DISCUSSION

Q. Height (Aberdeen): How do you feel about mixing species in
exposure chambers? What I'm referring to is a potential synergistic
effect due to Sendai virus or to mouse pneumonia virus with rats.

A. Drew (BNL): That problem is, of course, an important one.
You pay your money and you take your chances. I think the Good
Laboratory Practice Regulations are going to make it very difficult
to do multi-species exposures in one chamber. It's my guess that
large chambers are going to become less and less used. Not only
do you have the problem of interspecies communication, but in many
of your testing schemes the acceptable test concentrations may be
different for different species. One species may be much more
sensitive to the same agent concentration. On the other hand,
depending upon what endpoint and what agent you're working with,
I can think of some reasons where I would put more than one species
in a chamber.

Q. Maltoni (Italy): How many hours daily and for how many days
or weeks or months are you able to go on in your system for inhaling
fibers?

A. Drew (BNL): We've exposed animals intertracheally to fibers
for one hour a week for ten weeks.

Q. Maltoni (Italy): One hour weekly?

A. Drew (BNL): One hour weekly for ten weeks.

Q. Maltoni (Italy): How long do the exposures go in the system
where the animals were put into exposure tubes?

A. Drew (BNL): Six hours a day, five days a week for nine months.

Q. Maltoni (Italy): Nine months, do you have a problem with
urine or feces?

A. Drew (BNL): After the first four or five days, the animals
no longer seem to urinate and defecate in the tube. I'm very
excited about this system because it minimizes the problems in
handling animals that have been contaminated with compounds such
as benzopyrene.

TECHNIQUES FOR THE GENERATION

AND MONITORING OF VAPORS

Gary O. Nelson

Lawrence Livermore National Laboratory
Livermore, CA   94550 (U.S.A.)

INTRODUCTION

For a long time the preparation and measurement of test atmospheres have been of concern to those involved with animal exposure
experimentation. Test gas and vapor mixtures must be produced continuously and dynamically to prevent a build-up of unwanted contaminants
in the exposure chamber. We will discuss the general production
techniques of gas and vapor mixtures (1). We will also discuss several
methods that can be used to monitor the concentration of the test
atmospheres. In addition, we will describe an ideal system for the
production of controlled test atmospheres.

GAS AND VAPOR MIXTURE GENERATION TECHNIQUES

Flow Dilution

The most common method of producing dynamic gas mixtures is the
dilution of gases with one another after measuring their flow rates.
Gases can be metered as pure components or be previously diluted to
facilitate production of low concentrations.

A basic element of a flow dilution system is the variable-area
flowmeter (rotometer) which measures the flow of air and contaminant
gas. Flowmeters should always be calibrated against a primary or
intermediate gas standard at the pressure and temperature conditions
of use. Electronic-mass flowmeters (i.e., Kurz, Tylan and Brooks)
and controllers, though normally insensitive to pressure and temperature variations, should be calibrated with the gas of interest.

197

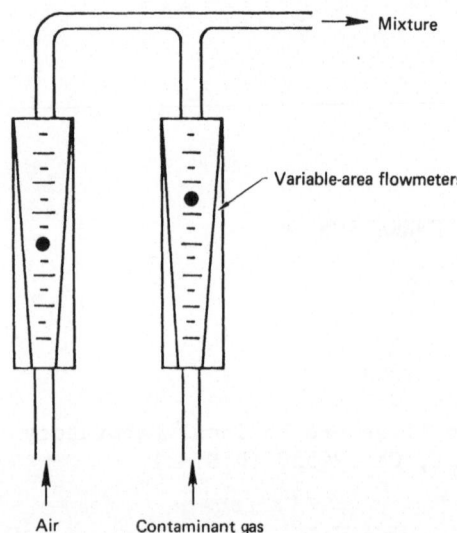

Figure 1.   Typical flow dilution using rotometers.

Figure 1 shows a typical flow dilution system in which contaminant gas is continously blended with air. The range of variable-area flowmeters extends from 5 mL/min to greater than of
300 L/min. Flows as low as 0.1 mL/min can be obtained if electronic-
flow devices are used.

Double-flow dilution techniques have been used to obtain low
gas concentrations; however, such techniques are not recommended
because the resulting pressure changes interfere with the validity
of flowmeter readings.

Solvent Injection

Vapors can be added to a moving airstream by the solvent injection method shown in Figures 2. In this technique, contaminant

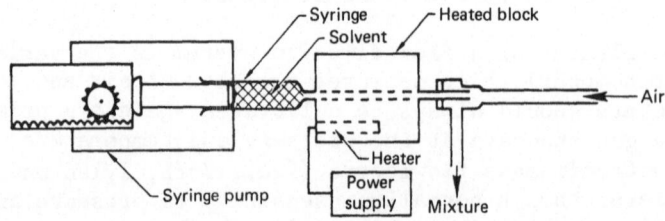

Figure 2.   Solvent injection method to produce vapors in a moving
air stream.

liquid is placed in a gas-tight syringe and pumped (i.e. Sage Model 355) through a stainless-steel needle surrounded by a heated block. As the liquid encounters the hot needle walls, it evaporates, and the vapors enter the system countercurrent to the moving airstream. If the vaporization rate is too irregular, the vapor-air mixture is passed through a large mixing chamber to dampen vapor pulsations. A cartridge of activated carbon can also be used to reduce concentration irregularities.

Saturation

High concentrations of air-vapor mixtures can be prepared by the saturation method shown in Figure 3. In this method, air is passed over a boiling, refluxing liquid, and the saturation concentration is determined from the refluxing temperature. The air-vapor mixture is generally diluted to produce the desired concentration.

Although the saturation method appears straightforward, it can yield false concentration information. This is because (i) the solvent vapor volume must be known and added to the air being saturated, (ii) the saturation temperature must be known to within 0.5°C, and (iii) the air must reside in the flask long enough to come in contact and mix with vaporized solvent. For these reasons, we suggest that a dynamic system using the saturation method be checked by an alternate chemical technique.

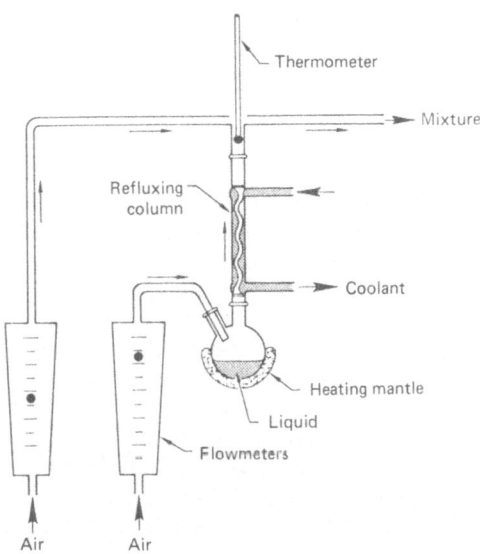

Figure 3. Air saturation method for producing high concentrations of vapors in air.

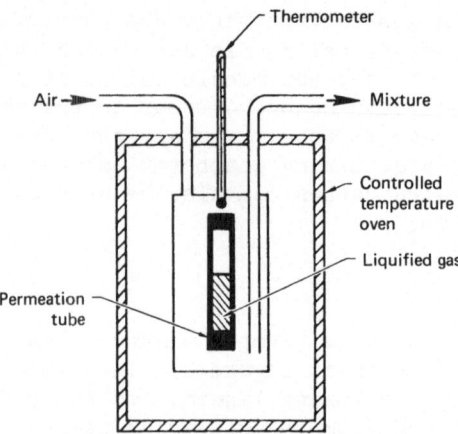

Figure 4.   Gas blending system using a permeation tube.

Permeation Tubes

A dynamic system using a permeation tube (Fig 4) can produce contaminant concentrations in the low ppm range. Commercial
examples include Metronics, Analytical Instrument Development
Company, and Dynacal. Gas is compressed to the point of liquefac-
tion and sealed inside a type of polymeric container. Contaminant
gas in the permeation tube dissolves in and permeates through the
tube walls at a constant rate, mixing with the passing, diluent gas.
The permeation rate, determined gravimetrically, is highly dependent
upon temperature. A difference of 1°C can cause a rate change of
10%. It is imperative, therefore, that proper temperature control
is maintained.

Diffusion Tubes

Vapors can diffuse through tubes at uniform rates if their
temperatures and concentration gradients remain unchanged and if
tube geometry remains fixed. Figure 5 illustrates a dynamic system
using a diffusion tube. The solvent is contained in a reservoir
beneath the diffusion tube. Air passes in and out of a controlled-
temperature oven and carries away the vapors which diffuse up through
the tube. The diffusion rate, like the permeation rate, is highly
dependent upon temperature. The rate is also influenced by the
tube's cross-sectional area, its length, and its diffusion coefficient. The diffusion concentration should not only be calculated
but also checked by a proper analytical method. As in the permeation
devices, the diffusion technique is most useful in the low (<10 ppm)
concentration range.

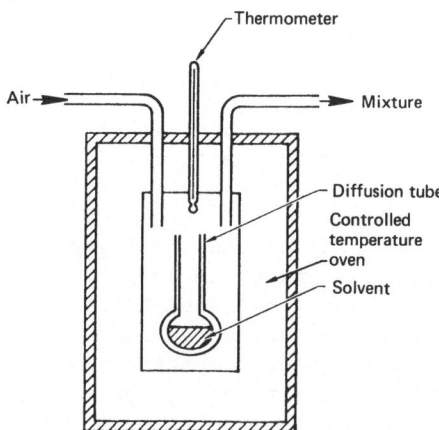

Figure 5.   Vapor blending system using a diffusion tube.

GAS DETECTION AND MONITORING TECHNIQUES

     Gas and vapor dynamic generation methods can be easily monitored
by a number of commercially available analyzers. Although a number
of specific monitors exist, only general techniques, which respond
to a wide variety of potential test contaminants, will be discussed.

     Hydrocarbon Analyzers. Hydrocarbon analyzers generally employ
the flame ionization (FID) method of detection. Commercial sources
include Beckman Instruments, Bendix and Gow-Mac. The sample is
first mixed with hydrogen and burned in air; ions and electrons
formed in the flame enter a gap between two electrodes and decrease
the gap resistance permitting a current to flow in the external
circuit.

     The magnitude of the signal is in direct proportion to the
number of carbon atoms in the flame. The small current is amplified
and converted to a millivolt signal on a meter or recorder.

     This analyzer measures total hydrocarbons in a range of
0.1–2000 ppm methane with a response time of about one second.
The FID method is insensitive to water vapor and shows diminished
response to molecules which contain no hydrogen atoms (i.e., carbon
tetrachloride). Although FID systems are relatively sensitive, they
show no specificity unless used in concert with a gas chromatograph.

     Photoionization Detectors. The process of photoionization
occurs when an organic molecule is ionized with ultraviolet

light via the reation:

$$R + h\nu \rightarrow R^+ + e^-$$

where R^- is the ionized species and $h\nu$ is a photon which has an energy equal to or greater than the ionization potential of the molecule.

The sensor consists of a sealed, UV light source that emits photons which are energetic enough to ionize many trace species (particularly organics) but do not ionize the major components of air such as oxygen, nitrogen, carbon dioxide, or water. A chamber adjacent to the ultraviolet source contains a pair of electrodes. When a positive potential is applied to one electrode, the created field drives any ions formed by absorption of the UV light to the collector electrode where the current (proportional to concentration) is measured.

The instrument is similar in response to the flame ionization detector but the photoionization detector requires no gases to produce a flame source. This process is nondestructive, and relatively nonspecific. A partial list of species is shown in Table 1 (note that three source lamps are available to fit specific detection needs) (2).

Electrochemical Analyzers. Electrochemical analyzers are a sealed module contacting the sample stream through a semipermeable membrane. The gas to be analyzed diffuses through the membrane into the module where it reacts with the electrolyte film on the surface of the sensing electrode. The electrochemical oxidation of the gas releases electrons that cause a current flow from the sensing electrode to the counter electrode. The current is proportional to the concentration. Response specificity is determined by the semipermeable membrane, the electrolyte, the electrode materials, and the retarding potential. This potential is used to retard oxidation of those sample components that are less readily oxidized than the gas of interest.

Electrochemical analyzers are typically single-gas detectors and are used for carbon monoxide, hydrogen sulfide, sulfur dioxide, oxides of nitrogen, ammonia and chlorine. Typical concentration ranges vary from 0.1 to 1000 ppm. Commercial sources include Theta Sensor, Interscan and Energetics Science.

Infrared Analyzers. Infrared analyzers offer a fast, relatively simple, nondestructive, functional-group specific method of gas and vapor detection. This method involves passing infrared radiation through a cell containing the sample and measuring the attenuation of the incident radiation. The light absorbed is proporational to some degree, to the gas concentration in the cell.

Table 1. Solvent vapor response to
photoionization detection techniques

| Class Species | Photoionization Response[a] | | |
	9.5 eV 1 amp	10.2 eV 1 amp	11.7 eV 1 amp
Paraffins and unsaturated hydrocarbons			
methane	NR	NR	NR
ethylene	NR	L	H
acetylene	NR	NR	H
1-butene	H	H	H
hexane	NR	L	H
Chlorinated hydrocarbons			
methyl chloride	NR	NR	H
carbon tetrachloride	NR	NR	H
chloroform	NR	NR	H
dichloroethane	NR	NR	H
vinylidene chloride	L	H	H
vinyl chloride	L	H	H
trichloroethylene	H	H	H
Heterocyclics & aromatics			
phenol	H	H	H
pryidine	H	H	H
benzene	H	H	H
toluene	H	H	H
xylene	H	H	H
styrene	H	H	H
aniline	H	H	H
chlorobenzene	H	H	H
nitrobenzene	NR	L	H
Nitrogen compounds			
formamide	NR	H	H
ammonia	NR	L	H
hydrazine	H	H	H
methyl amine	H	H	H
acetonitrile	NR	NR	NR
acrylonitrile	NR	NR	H

Table 1. (continued)

Class Species	Photoionization Response[a]		
	9.5 eV 1 amp	10.2 eV 1 amp	11.7 eV 1 amp
Sulfur compounds			
sulfur dioxide	NR	NR	NR
hydrogen sulfide	NR	H	H
carbonyl sulfide	NR	NR	H
carbon disulfide	H	H	H
methyl mercaptan	H	H	H
dimethyl sulfide	H	H	H
dimethyl disulfide	H	H	H
Aldehydes, ketones, alcohols, acids, esters			
formaldehyde	NR	NR	H
acetaldehyde	NR	H	H
propionaldehyde	L	H	H
acrolein	L	H	H
crotonaldehyde	L	H	H
acetone	L	H	H
methanol	NR	NR	H
ethanol	NR	L	H
formic acid	NR	NR	H
acetic acid	NR	L	H
methyl methacrylate	L	H	H
Others			
ethylene dibromide	R	H	H
ethylene oxide	R	L	H
tetraethyl lead	H	H	H
phosphine	R	H	H
arsine	R	H	H
iodine	H	H	H

[a]NR-no response; L-low response; H-high response.

The advent of the microprocessor has had a profound effect on the traditional single-beam, single-component analysis. For example, the Foxboro MIRAN, Model 80 can monitor for 10 compounds simultaneously. The Miran 801 offers multiple-station sampling, signal-averaging capability as well as interference and compensation for water vapor.

The sensitivity of the infrared method varies widely; all organic molecules are absorbed to some degree and detection limits of 0.1 ppm are not uncommon. In addition, oxygen and nitrogen are inactive infrared absorbers but water vapor and carbon dioxide can be significant interference problems in certain infrared regions.

AN IDEAL CONTAMINANT GENERATION AND MEASUREMENT SYSTEM

A high quality, exposure-test system must not only produce and measure the contaminant concentration but also control the temperature, humidity and air flow parameters. Such a system can be constructed using four, commercially available components: a temperature-humidity indicator, a dual flow-temperature-humidity control system, a series of mass-flow controllers for the contaminant gas and an infrared analyzer to monitor the concentration (see Fig 6).

The flow temperature-humidity control system and temperature-humidity indicator are linked electronically to process the air stream. Direct-reading potentiometers, mounted on the control

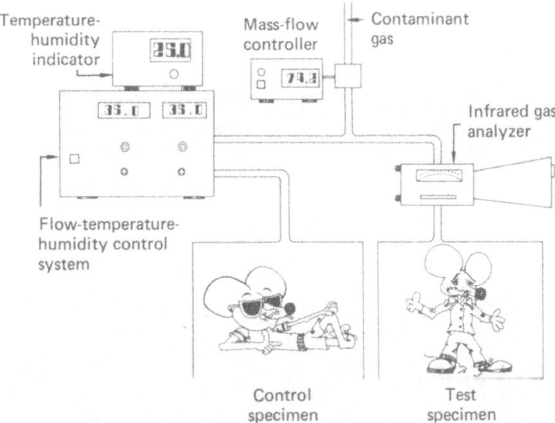

Figure 6. Schematic of the fully automated system for producing gas or vapor mixtures at a known flow, humidity, concentration, and temperature.

module, are used to select parameters. The control module shows
actual flows, while the indicator displays temperature and humidity.

Basically, the system consists of three separate controlling
mechanisms, sensors, and various safety features. Air and water
are supplied to the control module; the air then passes through a
humidification vessel in which water is maintained at a constant
level. The desired humidity is achieved by activating the heater
in the water reservoir. The air stream is then split and passes
through flow controllers where the airflows are established. These
controllers maintain and display the flow at standard conditions,
regardless of temperature and pressure variations in the flow system.
The humidified air is then heated to the proper level just before
it leaves the control module. Sensors placed in the airstream outside the control module measure the actual temperature and humidity.

Once the system parameters are set, the variation in indicated
airflow is 1% full scale. The temperature variation, after the
initial overshoot stabilizes, is 0.1°C, with an accuracy of 0.2°C.
The humidity typically varies 0.5% and is accurate to within 2%, if
the indicator is properly calibrated. System conditions are upset
slightly when cool water enters the reservoir during the refilling
process. However, the heaters react quickly to compensate for such
excursions, and conditions return to normal in less than a minute.

After the air flows exit the control system, one flow goes
directly to the control animals. The contaminants of interest are
added to the other flow. This can be done using mass-flow controllers,
diffusion or permeation devices, or some type of solvent-injection
technique.

The resulting gas mixture then passes into a Miran Model 80
infrared gas analyzer. All the contaminants are measured periodically
using the programmed averaging and wavelength-cycling process. Background concentrations of carbon dioxide and carbon monoxide are
monitored as well as any other suspected interferences. The test
gas mixture then proceeds to the animal species in the exposure
test chambers.

SUMMARY

Controlled test atmospheres can be produced using a variety
of techniques. High concentrations of gases can be generated using
flow dilution methods. High concentration of vapors can be produced
using solvent injection or saturation techniques. Low concentrations
of gases or vapors (<10 ppm) can be generated using permeation or
diffusion tubes.

The resulting concentrations of gas mixtures can be monitored and measured using flame ionization or photoionization techniques. If greater specificity is desired, then electrochemical and infrared analytical systems can be used.

ACKNOWLEDGEMENT

This work was performed under the auspices of the U.S. Department of Energy by Lawrence Livermore Laboratory under contract No. W-7405-Eng-48.

REFERENCES

1.   G. O. Nelson, Controlled Test Atmospheres-Principles and
         Techniques , Ann Arbor Science Publishers, Inc., Michigan
         1971.
2.   HNU Systems, Inc., Newton Upper Falls, MA, Bulletin PI-101/375,
         1975.

PROBLEMS ASSOCIATED WITH ASSESSING THE

MUTAGENICITY OF INHALABLE PARTICULATE MATTER

Otto G. Raabe

Laboratory for Energy-Related Health Research
University of California
Davis, CA   95616   (U.S.A.)

INTRODUCTION

Release of airborne mutagens associated with inhalable particles may occur in various industrial, governmental, and public
activities. These activities include transportation, manufacturing,
mining, refining, processing of materials, agricultural procedures,
electric power generation, and other processes involving combustion
and vaporization. People may be exposed to mutagenic aerosols in
occupational exposure to process or fugitive emissions, or during
use of aerosolizable or volatile mutagenic materials. The general
public may be exposed to environmental concentrations (released
airborne materials) or in the course of using various product
materials or other personal activities.

Particles (and related vapors), which usually exist as complex
and heterogenous mixtures of both organic and inorganic constituents, need to be studied to evaluate potentially mutagenic and
carcinogenic properties. Samples need to be collected using
methods that ensure relevance. These samples need to be given
appropriate handling and care to avoid artifacts. Testing methods
must be devised with consideration of the complex character of
these samples and with the aim of avoiding interferences and
possible alterations in active components. Finally, the mutagenic
constituents need to be identified to facilitate future analyses
and evaluations.

Chemical identification of the mutagenic constituents in a
complex and heterogeneous sample of particulate matter can be
exceedingly formidable. The mutagenic species or moieties may be
difficult to distinguish from more plentiful, less biologically

active components. Strategies for partitioning or fractionating
samples for study are required to provide data which properly
reflect the original form of the particles.

This report considers specifically those problems and considerations encountered in the assessment of the mutagenicity of
complex and heterogeneous particulate matter that may be inhaled
by people. These factors include relevance, fidelity, handling
and care of samples, extraction criteria and process interferences,
such as cell toxicity of nonmutagenic components. The goal is to
identify and quantify the mutagenic materials associated with
airborne particles and provide such information as mutagen median
physical or aerodynamic diameter. This latter information can be
used to evaluate the regional deposition of the particle in the
respiratory tract during inhalation by people. Prediction of
biological fate and carcinogenic or other untoward response will
proceed from this information.

SAMPLE COLLECTION

Airborne particle (and related vapor) samples can be collected
using a variety of established methods (Table 1) (1). The chosen
methods depend on the aerosol concentration, type, temperature and
sample size requirements, as well as consideration of the relative
importance of particle size distribution and vapor constituents.
Sample filtration using appropriate filter media will provide
samples of the total aerosol, but suffer from loading problems
with concomitant increases in flow resistance. Aerodynamic size
classification can be accomplished using inertial samplers.
Relatively large samples of size classified particles are particularly useful for comparison of other biological tests, ranging
from inhalation studies to mutagenicity (2).

A simple two-stage sampler for collecting size-classified
respirable particles at 7.5 ft/min is shown in Figure 1.* This
sampler employs a cyclone to aerodynamically separate the larger
particles from the smaller respirable particles. The smaller
particles are collected on the inside of a cylindrical filter
designed to be periodically shaken with a short pulse of high
pressure air released on the downstream side of the filter. This
pulsing shakes the layer of dust collected on the filter so that
it gradually slides down into a hopper. Provision is made for
collection of the released dust from the filter hopper and the

*Two-stage sampler designed under contract to the University of
    California by A. R. McFarland, Texas A and M University, College
    Station, Texas, 1980, as part of a study being conducted for the
    California Air Resources Board.

Table 1.   Summary of aerosol (and associated vapor)
           sampling methods

I.   AEROSOL SAMPLES

     SIZE CLASSIFICATION OF CONSTITUENTS

          Impactors
          Centripeters
          Virtual Impactors
          Cyclones
          Aerosol Centrifuges
          Diffusion Batteries
          Electrostatic Classifiers

     FILTRATION

     IMPINGERS

     BUBBLERS

     ELECTROSTATIC PRECIPITATORS

II. GAS AND VAPOR SAMPLES

     ABSORBERS

          Molecular Adsorbers
          Charcoal
          Species-Specific Reactants

     BUBBLERS-SCRUBBERS

     COLD TRAPS

     VOLUMETRIC OR COMPRESSED GASES

larger particles from the cyclone hopper. The complete system is
enclosed in an insulated and heated enclosure so that high temperature effluent streams can be sampled without condensation problems.

PARTICLE SIZE CONSIDERATIONS

     Gross samples collected by sample filtration without size
discrimination may provide adequately for screening of airborne

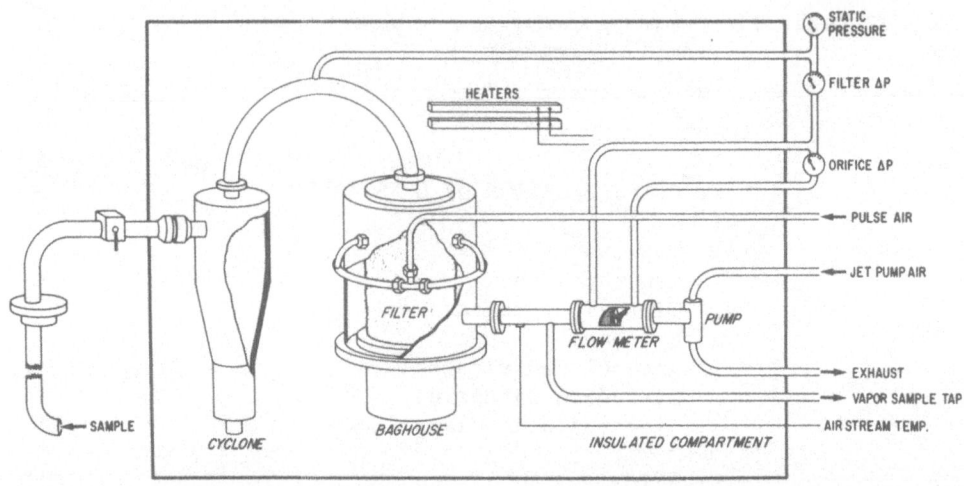

Figure 1. Two-stage aerosol sampler for collecting respirable samples of power plant fly ash (system designed by A. R. McFarland under contract to the University of California as part of a study of oil fly ash being conducted for the California Air Resources Board).

particles for mutagenic potential. Two important particle size effects tend to support the greater relevance of collection of samples using aerodynamic size separation prior to collection of particles. Both the mutagenic properties of aerosols and their fate during inhalation are a function of particle size. Failure to appreciate the significance of this fact may yield misleading results in the case where a major portion of gross samples consist of larger particles whose mutagenic potential, as well as inhalation deposition, probably are low.

The mutagenic constituents of complex aerosol-vapor mixtures are usually associated with polynuclear organic matter, certain inorganic constituents, or possibly organometallic species that tend to be absorbed on the surfaces of airborne particles. This association may be due to condensation during cooling or to gas-particle interactions or conversion. In either case the resulting surface dependence results in the small particles in the respirable size range being disproportionately more important with respect to adsorbed species than the large particles (3-4). One example is the relative size distribution of certain metals on coal fly ash particles (Table 2).

The deposition of inhaled materials in various regions of the respiratory airways depends in an important and critical way on the aerodynamic size distribution of the aerosols, as well as

Table 2.   Distribution of Size-Dependent Elements on Size-Classified
Stack-Collected Coal Fly Ash (4)

Element	Cut 1 (VMD = 20 μm)	Cut 2 (VMD = 6.3 μm)	Cut 3 (VMD = 3.2 μm)	Cut 4 (VMD = 2.2 μm)	Enhancement Factor Cut 4/Cut 1
		Concentration in μg/g (unless indicated)			
Cd	0.4(0.2)*	1.6(0.3)	2.8(0.4)	4.6(0.2)	11.5
Zn	68(1)	189(4)	301(10)	746(218)	11.0
Se	19(2)	59(2)	78(2)	198(20)	10.4
As	13.7(1.3)	56(14)	87(9)	132(22)	9.6
Sb	2.6(0.1)	8.3(0.4)	13.0(0.7)	20.6(0.7)	7.9
W	3.4(0.2)	8.6(1.6)	16(2)	24(2)	7.1
Mo	9.1(2.5)	28(1.4)	40(5)	50(9)	5.5
Ga	43(12)	116(52)	140(23)	178(90)	4.1
Pb	73(3)	169(2)	226(4)	278(3)	3.8
V	86(44)	178(17)	244(18)	327(40)	3.8
U	8.8(1.9)	16(3)	22(4)	29(4)	3.3
Cr	28(3)	54(3)	66(3)	71(4)	2.5
Ba (%)	0.168(0.001)	0.245(0.002)	0.320(0.013)	0.409(0.018)	2.4
Cu	56(1)	89(1)	107(4)	137(1)	2.4
Be	6.3(0.2)	8.5(0.2)	9.5(0.3)	10.3(0.5)	1.6
Mn	209(7)	231(5)	273(7)	309(3)	1.5
Si (%)	29.6(0.7)	28.0(0.1)	27.5(0.3)	26.8(0.1)	0.90
F⁻	60	420(40)	840(60)	2400(300)	40
SO₄=(%)	0.15	0.67(0.02)	0.99(0.15)	1.62(0.07)	10.8

*Value in parentheses is either the range of two values or standard deviation.

other properties such as hygroscopicity or delinquescence (5).
During both nose breathing and mouth breathing only particles
below 10 μm in aerodynamic diameter are subject to significant
deposition in the pulmonary region of the lung (6). Figure 2
shows the calculated regional deposition fractions for nose breathing based on the recommendations of the Task Group on Lung Dynamics
of the International Commission on Radiological Protection (7-8).

The use of the accepted criteria for respirable dust sampling
(Fig 3) provides a useful compromise between representing nose
breathing or mouth breathing. This simple two-stage separation of
particles provides two samples; one for particles that may reach the
lung tissue and one for those that are primarily deposited in the
nasal-pharynx. Separation of the collected particles into several
size groups based on aerodynamic properties allows the development
of mutagenic size distribution profiles such as shown in Figure 4
and Table 3.

SAMPLE FIDELITY

Obtaining representative aerosol samples of complex gas-
particle mixtures for mutagenesis testing requires consideration
of the ultimate goal of the assessment in each case. Several
problems can be encountered, such as the need to collect samples that
are representative of the particles to which people are exposed,
consideration of temperature and environmental effects, sample
collection artifacts, time and duration of samples with respect to
changes that may occur in the phenomena under study, and sample
handling problems.

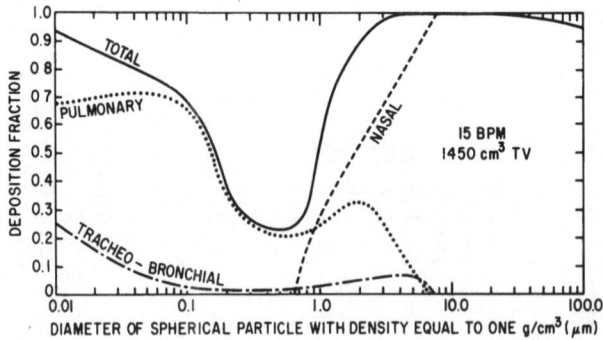

Figure 2. Regional deposition in the human respiratory tract for
airborne spherical particles with physical density equal to one
g/cm³ as calculated by the ICRP Task Group on Lung Dynamics (7)
for nose breathing at a rate of fifteen breaths per minutes (BPM)
and tidal volume of 1450 cm³ (reprinted by permission from Raabe,
1979 (7))

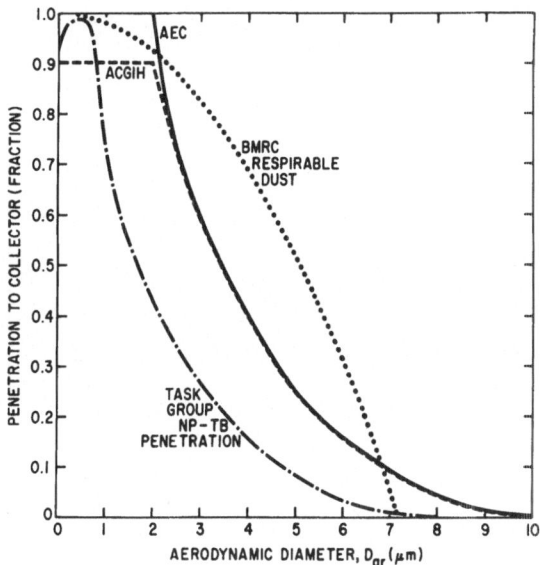

Figure 3. Respirable two-stage dust sampling criteria showing penetration to collector (fraction) versus aerodynamic resistance diameter (D_{ar}) based on recommendations of the British Medical Research Council (BMRC), the US Atomic Energy Commission (AEC), and the American Conference of Governmental Industrial Hygienists (ACGIH) (reprinted by permission from Raabe, 1981 (6)).

The kinetics associated with the aerosol stream from which samples are to be collected must be considered. In ducts operating under highly turbulent conditions, isokinetic sampling is not strictly possible since there are no streamlines of flow. Most samples are collected with probes designed to match the average bulk flow velocity recognizing the limitations of turbulent stream sampling. For respirable particles less than about 5 μm in aerodynamic diameter, particle behavior does not diverge markedly from gas flow even under anisokinetic conditions.

The sampler materials can affect the losses that may occur during collection and may lead to contamination of samples. Stainless steel and Teflon are frequently chosen for sampler components to avoid artifacts, losses and inadvertant sample contamination. Careful chemical cleaning of the components of the sampler system is essential to prevent both organic and inorganic contamination of samples. Agents such as gasket greases, silicone sealants, and the like are best avoided.

Containers for samples should also be chosen to avoid contamination. Precleaned Teflon containers provide safe, contaminant-free holders for handling and care of samples.

Table 3. Mutagenesis of Serum Filtrates Derived From Stack-Collected and ESP-Collected Fly Ash Fractions As Described by Fisher et al. (11)

Sample	Particle Size		S-9 Not Added		S-9 Added	
	VMD(μm)	σ_g	$\overline{X}$	S.D.	$\overline{X}$	S.D.
Stack-collected						
Fraction 1	20.0	1.8	15	± 8	17	± 8
Fraction 2	6.3	1.8	107	± 24	143	± 52
Fraction 3	3.2	1.8	379	± 67	447	± 11
Fraction 4	2.2	1.9	197	± 44	218	± 58
ESP-collected						
Unfractionated	27.0	2.5	4.3	± 6.1	-1.2	± 5.2
Fine fraction	2.3	1.4	2.8	± 2.2	3.7	± 8.4

*The values are observed numbers of TA 1538 his+ revertants per plate for horse serum filtrates prepared from 10 mL serum mixed with 0.8 g fly ash and maintained at 37°C for 7 days. Filtrate (100 μL) was added to 2 mL of soft top agar before plating.

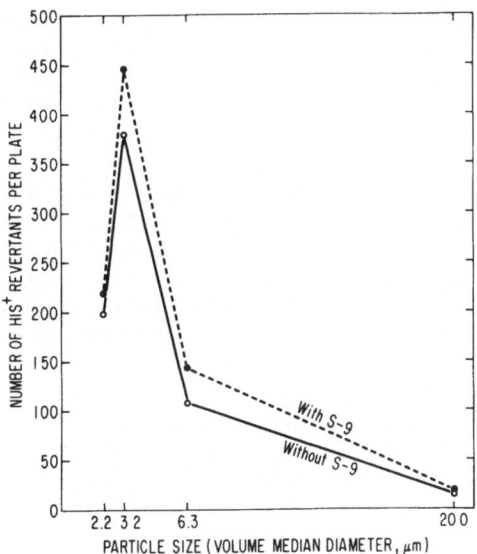

Figure 4. Distribution of mutagenic activity in the Ames
Salmonella typhimurium his+ revertant assay with TA1538 for four
size fractions of coal fly ash. Serum extracts were prepared from
10 mL horse serum mixed with 0.8 g fly ash at 37°C for 7 days.
Filtrate (100 μL) was added to 2 mL soft top agar, and ten
replicates of each test were made both with and without S-9
microsomal enzyme activation (reprinted by permission from Fisher
et al. (11)).

    Choice of filter media must be consistent with the needs of
aerosol collection, temperature extremes and avoidance of artifacts. If the filter media must be extracted because of adherence
of the sample, proper choice of filters can avoid extraction
problems. Precleaning is essential, and unused duplicate filters
should be designated as controls.

    When samples are collected on filters, the gases continuously
passing through the filter (and contacting the collected particulate material) can lead to undesirable alteration and changes of
the mutagenic constituents. Pitts et al. (9) have shown that
directly mutagenic nitro-derivatives of perylene and benzo(a)pyrene
were produced on filters after collection of polynuclear organic
matter. Pitts et al. (10) showed that benzo(a)pyrene deposited on
glass fiber filters reacts rapidly with ozone in the gas stream to
form directly mutagenic species.

    Sometimes it is necessary to collect samples at elevated
temperatures as may be found in ducts in process streams, in power

plant flue lines or in smokestacks. Potential mutagenic species may
be formed during cooling or temperature-related adsorption-conden-
sation processes and may not be associated with particles collected
at the in situ elevated temperatures. Two possibilities exist:
(a) separate collection of the vapor phase materials by condensation
or with chemical adsorbers, and (b) cooling of the sample aerosol
stream during collection. The latter process may involve undesirable
condensation of water and acid so that suitable dilution with
clean air is required. The collection of strongly acidic particles
or condensate may alter or degrade mutagenic agents during or soon
after collection. It is desirable to collect samples of aerosols
at the lowest temperature practicable if normal ambient temperatures
are not achievable.

After collection, samples need to be carefully managed to
prevent alterations and artifacts. Samples should be protected
from light, heat, and high humidity. Fisher et al. (11) showed
that heat treatment reduced the mutagenic activity of coal fly ash
samples and all activity was lost above about 350°C (Fig 5).

Finally, the unknowns in sample collection can befuddle
interpretation of some observations. Particularly, a given process
may yield mutagenic agents only under certain ideal (or nonideal)
conditions. When respirable particles of a grab sample of ash
collected from the electrostatic precipitator of a coal burning
power plant were compared to stack collected ash in the same size
range collected over twelve days, the inorganic chemical constituents were found to be nearly identical (Fig 6) (12). However,

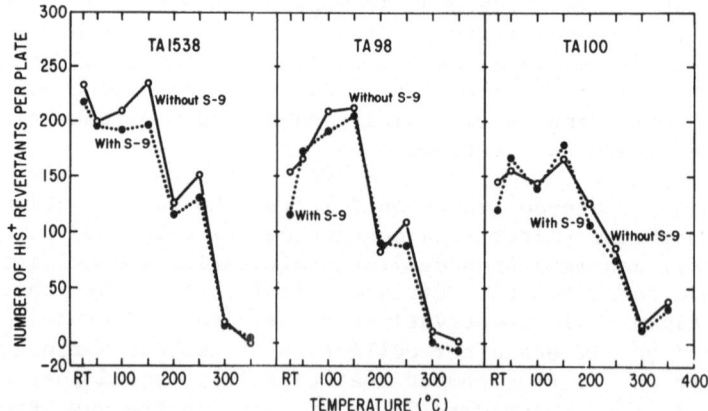

Figure 5.   Effect of temperature mutagenic activity in the Ames
Salmonella typhimurium his+ revertant assay using TA 1538, TA 98
and TA 100 as described for Figure 4 (reprinted by permission
from Fisher et al. (11)).

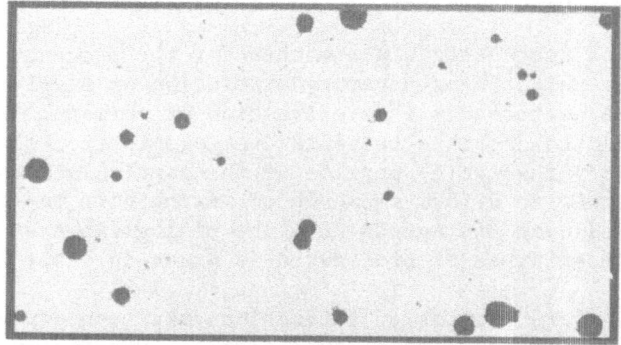

STACK FLY ASH

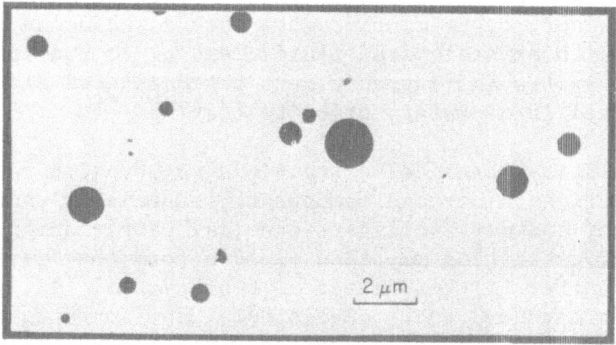

ESP HOPPER FLY ASH

Figure 6. Comparison of reaerosolized samples of (a) ESP hopper
collected size-classified coal fly ash (nonmutagenic) and (b)
stack-collected size-classified coal fly ash (mutagenic) (reprinted
by permission from Raabe (12)).

the stack collected ash was found to be mutagenic, while the
hopper ash was not (11). The temperatures of collection were very
similar, about 105°C for the hopper ash and about 95°C for the stack
ash. It is possible that certain unique conditions during one
part of the 12-day collection of stack ash yielded the mutagenic
agents and that these conditions did not exist at the time of
collection of the grab sample of hopper ash. The value of duplicate samples and extended sample times is evident. It is safe to
suggest that mutagenic coal fly ash collected from flue gas streams
are only encountered under certain yet-to-be-explained operating
conditions and possibly only when sample collection temperature is
100°C or less (11).

MUTAGENESIS TESTING PROCEDURES

In order to test particulate matter for the presence of mutagenic agents, a suitable preparatory extraction or sample partitioning must be performed. This extraction or chemical treatment
must be designed to separate the mutagenic agents from the inert
(and perhaps very insoluble) portion of the particulate matter and
convert the sample to a form amenable to mixing with top agar or
otherwise introducing the agents into the biological test system.
A summary of these types of procedures is given in Table 4.

Several different types of extractions may be necessary to
ensure adequate evaluation of a given sample. Use of physiological
extraction media such as blood serum facilitates correlating tests
to biological availability of potentially mutagenic agents. In
some cases, however, other constituents such as acids in samples
can interfere with physiological fluids leading to precipitation.
The pH of the samples and extracts must be considered and appropriately adjusted if necessary prior to testing.

Nonphysiological chemical extractions may provide for more
complete extraction of certain components, especially under ultra-
sonic agitation, and may facilitate chemical partitioning and
analysis. Such extraction may also yield a form that is not
amenable to transfer to the biological test system. Intermediate
transfer fluids may need to be developed. Many investigators use
dimethyl sulfoxide (DMSO) for extractions and for transfer to test

Table 4.   Treatments Preparatory to Mutagenesis Testing

LEACHING

DISSOLUTION

ACID NEUTRALIZATION

EXTRACTIONS (SHAKING, REFLUXING, ULTRASONICS)

SEPARATIONS AND PARTITIONING

PHOTO AND THERMAL TREATMENT

ENZYME ACTIVATION

VAPORIZATION-CONDENSATION

RE-AEROSOLIZATION FOR ANIMAL EXPOSURES

---

media. Limitations of transfer media may prevent adequate uptake
of extracted agents.

Chan et al. (13) successfully partitioned the mutagenic
components of diesel engine exhaust particles using a solvent extraction scheme beginning with dichloromethane using the partitioning scheme of Novotny et al. (14). They prepared and tested
nine chemically different fractions from the original particulate
matter and showed the mutagenic activity to be associated with a
neutral, nonpolar organic fraction.

Alternative methods that have been used include high pressure
liquid chromatographic separation of chemical constituents with
subsequent measurement of mutagenic components. Also, gas chroma-
togaphy and mass spectrometric identification of the species
present has been followed with separate testing of the mutagenic
properties of each species.

Because of the complex nature of particulate samples, many
methods are needed to test and identify mutagenic agents. The
fact that the mutagens may be of an unknown species may lead to
unexpected alterations, transfer limitations, extraction problems,
testing artifacts and toxicity effects; all of which need to be
given consideration in planning and executing mutagenicity tests
with samples of particulate matter.

SUMMARY

The collection of inhalable particulate matter for assessing
mutagenicity involves several phases, each with inherent problems.
Background contamination must be prevented. The samples must be
relevant, representative of the desirable types, and of adequate
quantity. Duplicates and controls are desirable. Stability
during collection and handling need to be protected. Analysis
methodologies need to consider detection limits, essential and
gross features, data reduction methodologies, key constituents and
overall analysis goals. Possible treatment and testing artifacts
need to be considered. Finally, chemical identification should be
made of the mutagenic agents present in samples and in the sampled
atmospheres.

ACKNOWLEDGMENTS

The author is indebted to Drs. B. Kimble, G. Fisher, C.
Chrisp, C. Wei, and A. McFarland for useful discussions of many
aspects of the mutagenesis, sampling, and chemical identification
problems discussed in this review. The illustrations were prepared by K. Shiomoto and S. Coffelt, and the manuscript was pre-

pared by N. Hardaker. This work was supported by the Office of Health and Environmental Research (OHER) of the US Department of Energy (DOE) under Contract No. DE-AM03-76SF00472 with the University of California, Davis.

REFERENCES

1. O. G. Raabe, Generation and Characterization of Aerosols, in: "Inhalation Carcinogenesis," M. G. Hanna, Jr., P. Nettesheim and J. R. Gilbert, eds., U.S. Atomic Energy Commission, Oak Ridge, TN, 123-172 (1970).

2. A. R. McFarland, R. W. Bertsch, G. L. Fisher and B. A. Prentice, A fractionator for size-classification of aerosalized solid particulate matter. Environ. Sci. Technol. 11:781-784 (1977).

3. D. F. S. Natusch, J. R. Wallace and C. A. Evans, Jr., Toxic trace elements: a perferential concentration in respirable particles. Science 183:202-204 (1974).

4. G. L. Fisher and C. E. Chrisp, Physical and Biological Studies of Coal Fly Ash, in: "Applications of Short-Term Bioassay in the Fractionation of Complex Environmental Mixture," M. D. Waters, S. Nesnow, J. L. Huisingh, S. S. Sandhu, and L. Claxton, eds., Plenum Publishing Co., New York, NY, 443-462, (1979).

5. O. G. Raabe, Physical Properties of Aerosols Affecting Inhalation Toxicology, in: "Pulmonary Toxicology of Respirable Particles," CONF-791002, C. L. Sanders et al., eds., Technical Information Center, US Department of Energy, 1-28 (1980), (Available from National Technical Information Service, Springfield, VA 22161).

6. O. G. Raabe, Comparison of the criteria for Sampling 'Inhalable' and 'Respirable' Aerosols, in: "Inhaled Particles V," Pergamon Press, 1981, in press.

7. O. G. Raabe, Deposition and Clearance of Inhaled Aerosols, Laboratory for Energy-Related Health Research, University of California, Davis, CA, UCD 472-503, 1-60 (1979). (Available from National Technical Information Service, Springfield, VA 22161).

8. P. E. Morrow, D. V. Bates, B. R. Fish, T. E. Hatch and T. T. Mercer, (International Commission on Radiological Protection Task Group on Lung Dynamics) - Deposition and retention models for internal dosimetry of the human respiratory tract. Health Physics 12:173 (1966).

9. J. N. Pitts, Jr., K. A. Van Cauwenberghe, D. Grosjean J. P. Schmid, D. R. Fitz, W. L. Belser, Jr., G. B. Knudson, P. M. Hynds, Atmospheric reactions of polycyclic aeromatic hydrocarbons: facile formations of mutagenic nitro derivatives. Science 202:515-519 (1978).

10. J. N. Pitts, Jr., D. M. Lokensgard, P. S. Ripley, K. A. Van Cauwenberghe, L. Van Vaeck, S. D. Schaffer, A. J. Thill and W. L. Belser, Jr., "Atmospheric expoxidations of benzo(a)pyrene by ozone: formation of the metabolite benzo(a)pyrene-4,5-oxide. Science 210:1347-1349 (1980).

11. G. L. Fisher, C. E. Chrisp and O. G. Raabe, Physical factor affecting the mutagenecity of fly ash from a coal-fired power plant. Science 204:879-881 (1979).

12. O. G. Raabe, Generation and Characterization of Power Plant Fly Ash, in: "Generation of Aerosol," K. Willeke, ed., Ann Arbor Science Publishers, Inc., Ann Arbor, MI 215-234 (1980).

13. T. L. Chan, P. S. Lee and J. S. Siak, Diesel Particulate Collection for Biological Testing: Comparison of Electrostatic Precipitation and Filtration, GMR-3182, General Motors Research Laboratories, Warraen, MI 48090.

14. M. Novotny, M. L. Lee and K. D. J. Bartle, The methods for fractionation, analytical separation and identification of polynuclear aromatic hydrocarbons and complex mixtures. Chromat. Sci. 12:606-612 (1974).

DISCUSSION

Q. Drew (BNL): Do you have any idea why one method of collection gives you mutagenic substances while another method doesn't?

A. Raabe (UC, Davis): We have many theories but no concrete data to suggest why.

AN OVERVIEW OF THE PROBLEM OF BENZENE TOXICITY AND SOME RECENT

DATA ON THE RELATIONSHIP OF BENZENE METABOLISM TO BENZENE TOXICITY

Robert Snyder, David Sammett, Charlotte Witmer and
James J. Kocsis

Department of Pharmacology, Thomas Jefferson University
Philadelphia, PA 19107 (U.S.A.)

INTRODUCTION

Historically, both bone marrow depression and leukemogenesis
have been associated with exposure to benzene in the industrial
setting. As early as 1897, Santesson (1) described chronic benzene
poisoning in Sweden and, around the time of World War I, Selling (2),
at Johns Hopkins, described benzene-induced aplastic anemia in
both man and animals. Impaired immunological mechanisms in benzene-
treated animals were first reported before World War I (3-8).
Prior to the development of modern industrial hygiene, the uncontrolled use of benzene in industry expanded despite the growing
appreciation of the dangers of working in an atmosphere laden
with this highly volatile solvent.

Alice Hamilton (9), one of the pioneers in occupational
health reported in 1922 on "The growing menace of benzene...
poisoning in American industry". Apparently, her warning was
at least partially successful because 6 years later she published
a paper entitled "The lessening menace of benzol poisoning in
industry" (10). However, despite her optimism, uncontrolled
benzene use continued.

Benzene was used as a solvent for rubber, paint inks, dry
cleaning, and in other industrial processes. The advent of high
speed printing presses in the 1930's created a demand for an excellent ink solvent which would dry rapidly. Benzene was ideally
suited for this purpose and by the late 1930's, several investigators noted that the rotograbure printing industry in New York was
a natural laboratory for studying benzene toxicity (11-13). With
the onset of World War II and the subsequent development of the

225

plastics industry, benzene usage and the concurrent exposure of
workers to benzene increased dramatically. During this time,
benzene became of use, not only as a solvent but as a starting
material for the synthesis of polystyrene plastics. Human exposure
to benzene eventually became regulated, at first on a voluntary
basis and then by law. Starting with a voluntary workplace level
of 100 ppm, the increasing recognition of the serious nature of
the benzene problem led to stepwise decreases in permitted exposure
levels to the present legal limit of 10 ppm. Recently, OSHA has
unsuccessfully attempted to reduce this level still further to one
ppm (14).

The classical description of benzene toxicity includes a depressed function of bone marrow, leading to a decrease in the levels of
circulating erythrocytes, platelets and the various types of
leukocytes. There is considerable variability among reports in
the literature regarding the sequence of depression of each of the
cell types in both animal and human studies. Several factors may
account for this variability: the dose of benzene and length of
exposure, the time between onset of exposure and hematological
evaluation as well as the species under investigation. In general,
each species studied to date has shown reductions in each of the
three circulating cell types, with the eventual pancytopenia,
in all likelihood, resulting from an aplastic marrow (15). The
results of exposure to benzene are similar to those of chloramphenicol, phenylbutazone, cytarabin, or methotrexate.

The problem that has plagued agencies attempting to regulate
worker exposure to benzene has been the paucity of data demonstrating bone marrow depression at air levels below 40-50 ppm.
Thus, the present OSHA standard is based not on an extrapolation
of high dose effects to low dose ones but on an estimate of a
safety factor.

Although benzene-induced depression of bone marrow can still
be observed in uncontrolled settings, the leukemia resulting from
exposure to benzene has become the disease of primary interest.
The recognition that benzene was leukemogenic was slow to develop.
It wasn't until Browning (16) and then Vigliani (17) began
collecting epidemiological data and reports of case histories
that an association of leukemia with benzene poisoning began
to be appreciated. They reported many cases of various types of
leukemia apparently associated with exposure to benzene, and, as
the data were reviewed, it became apparent that acute myelogenous
leukemia was probably the predominant type. There was recollection
of early anomalous reports in which pancytopenia had been
observed but in which bone marrow biopsies had demonstrated
large numbers of white cells. It was presumed that many cases
previously reported as aplastic anemia may have been "aleukemic

leukemia". Vigliani (18) went on to suggest that of all individuals who died of chronic benzene exposure about half died of aplastic anemia and the rest of leukemia.

The recent impetus to reduce exposure to benzene grew out of the reports of Vigliani and co-workers (17,18) and were further stimulated by the prevalence of both forms of the disease in benzene-exposed Turkish shoe makers (19). Furthermore, epidemiological research initiated by OSHA disclosed excess leukemia cases in workers in pliofilm operations in Ohio in the 1940's-50's by Infante and co-workers (20). More recently, Maltoni and Scarnato (21) have reported the formation of solid tumors in rats fed benzene. It is expected that these data will lead to further calls for the reduction of benzene exposure in the workplace.

In addition to the attention given marrow aplasia and leukemia, there has been intensive effort to determine whether benzene is mutagenic and/or teratogenic. Mutagenicity of benzene will be discussed by Tice et al. (this volume), but some comment on its potential for teratogenicity is appropriate. There have been several studies in which teratogenic responses to benzene exposure has been examined. Although there have been no reproducible observations of benzene-associated malformations, several laboratories have reported fetotoxic effects (22-25).

Most current work in the field of benzene toxicity and leukemogenesis is directed at a few goals: the identification of the toxic metabolite(s) of benzene, the determination of the mechanism by which benzene inhibits marrow function, the development of an animal model for benzene-induced leukemia, the determination of the effects of benzene on various in vivo and in vitro stem cell assays and the assessment of the effects of benzene on DNA synthesis, replication and repair which can lead to mutations or defects in cell proliferation.

For a number of years we have examined both benzene metabolism and how metabolic changes in benzene structure lead to toxicity. The aim of this paper is to review some of these studies and to suggest for further investigation areas which might lead to a better understanding of the etiology of benzene-induced hematopoietic disease.

RESULTS

Early in the course of our studies it became important to ask the question: Can changes in benzene metabolism influence benzene toxicity? The answer to this question required that we develop methods to evaluate both metabolism and toxicity (26). It was possible to measure overall benzene metabolism in animals by administering radioactive benzene, i.e. either ^{3}H- or ^{14}C-benzene, and assaying for radioactivity in collected urine. We were unable to detect unmetabolized benzene and concluded that the radioactivity in urine represented benzene metabolites.

The measurement of benzene toxicity presented a problem for us since the usual techniques for evaluating benzene toxicity involved prolonged exposure of the animals and eventual measurement of levels of the various circulating blood cell types. Since these events were remote in time from what we believed to be the primary metabolic events leading to bone marrow depression we developed an alternative approach which concentrated on measuring early events in benzene toxicity. This method as developed by Lee et al. (27) takes advantage of the fact that in the course of red blood cell production one of the most important synthetic events is the incorporation of iron into hemoglobin. Hemoglobin is synthesized in each of the erythroid precursors but most actively in the reticulocyte stage. While our experimental data suggested that exposure to benzene did not inhibit hemoglobin synthesis, we found that there was less uptake of radioactive iron (^{59}Fe) into red cells of treated animals. This observation suggested that erythrocyte proliferation and maturation were inhibited in benzene treated animals. We, therefore, elected to use the ^{59}Fe uptake method in conjunction with studies of metabolism in an attempt to understand the role of benzene metabolism in benzene toxicity.

Table 1 shows the extent of benzene metabolism in mice subcutaneously injected with benzene at two different doses (880 and 440 mg/kg) and examined 24 hrs after the single exposure. The ^{59}Fe uptake data shows that benzene reduced the production of red cells in a dose dependent fashion. As a test of this hypothesis we also exposed animals to toluene, a competitive inhibitor of benzene metabolism. Toluene did not affect red cell production when injected alone, but when simultaneously injected with benzene, it significantly reduced or completed ameliorated the benzene-induced inhibition of ^{59}Fe uptake, depending on the ratio of toluene to benzene used (Table 1).

Although it was clear from the in vitro studies that toluene was a competitive inhibitor of benzene metabolism, it was important to demonstrate that the co-administration of the two solvents in vivo did not alter benzene toxicity by causing a redistribution of benzene. Accordingly, we studied the concentrations of benzene in various organs of mice which had been injected with either benzene alone or with benzene plus toluene. Using the criterion that the exposure of an organ to benzene is reflected by the area under the curves described in Figure 1, we concluded that there was no difference in exposure to benzene in any of the organs regardless of the presence of toluene. It was clear that the organs with the highest fat content accumulated the most benzene.

Table 1. Effects of toluene on benzene metabolism and on
red cell ^{59}Fe Uptake in the Mouse

^{3}H-Benzene metabolism/24 hr

	Percent Administered Dose (mean ± SD)	Benzene Equivalents[a] (μmol)	Percent ^{59}Fe Utilization/24 hr (mean ± SD)
Benzene (880 mg/kg)	22.6±5.7 (17)[b]	72.6	4.9±3.4[c] (19)[b]
Benzene + toluene	9.9±3.9 (17)[b]	33.2	9.9±4.5[d] (17)[b]
Toluene (1720 mg/kg)			15.8±5.5 (19)[b]
Control			18.4±6.2 (22)[b]
Benzene (440 mg/kg)	35.8±1.1 (2)[e]	60.1	15.7±4.8[c] (21)[b]
Benzene + toluene	10.9±7.5 (2)[e]	18.3	22.0±5.9[d] (28)[b]
Toluene (1720 mg/kg)			23.3±4.6 (25)[b]
Control			24.2±4.7 (27)[b]

[a]Calculated as μmol equivalents of ^{3}H-benzene metaolized by a 30 g mouse

[b]Number of animals

[c]Significantly different by Students t test from both control and toluene groups, $p < 0.05$

[d]Significantly different by Students t test from group receiving benzene alone, $p < 0.05$

[e]Two groups of animals, three animals per group

Thus, the epidydimal fat contained the highest level of benzene and bone marrow, the target organ, also accumulated large amounts of benzene.

In the course of these studies we also measured the concentration of benzene metabolites in these organs (Fig 2). Two key observations were noted: (i) metabolite levels were considerably reduced in the presence of toluene, which supported the contention that toluene inhibited benzene metabolism rather than causing benzene redistribution, and (ii) extremely high levels of benzene metabolites were observed in bone marrow, exceeding those in blood by a factor of about ten.

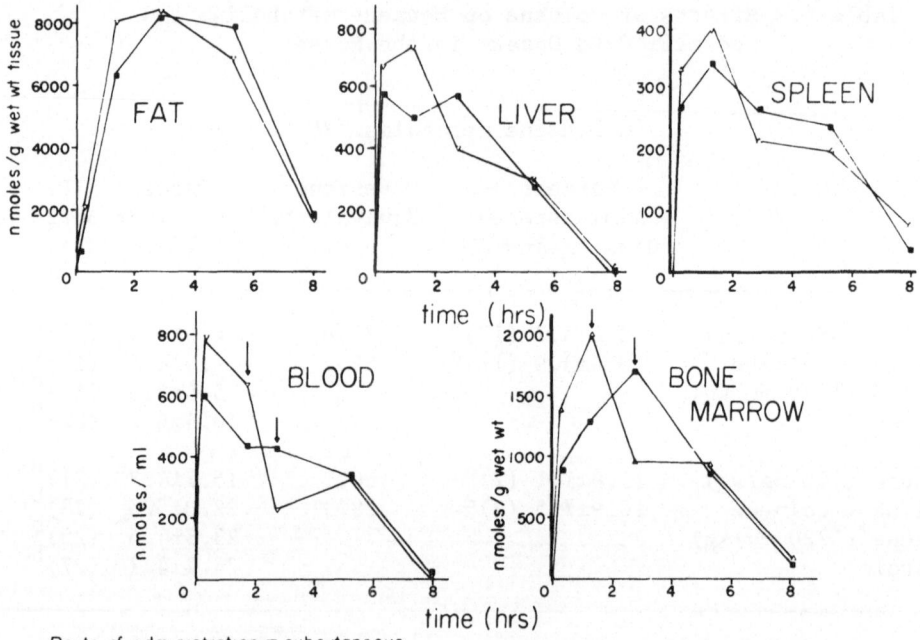

Route of administration — subcutaneous

Dose— Benzene 880 mg/kg (Δ) ; Toluene 1760 mg/kg + Benzene 880 mg/kg (■)-in oil
n = 6

Figure 1.   Recovery of ^{3}H-benzene in organs of mice given a single
subcutaneous dose (880 mg/kg) of ^{3}H-benzene alone (Δ) or with
toluene (1760 mg/kg) (■).   Groups of mice were sacrificed periodically during the 8 hrs following treatment and ^{3}H-benzene concentrations were determined in each organ.

By taking advantage of the sensitivity of the iron uptake
technique it was possible to measure signs of benzene toxicity
after a single exposure to benzene.   However, the more common
environmental situation is one of repeated exposures to benzene.
Therefore, we set up a series of studies to determine the effect
of repeated dosing of mice with benzene, measuring toxicity
by the iron uptake method. Benzene injected subcutaneously in mice
at 440 mg/kg/day for up to 20 days decreased iron uptake by at
most 60% (Fig 3).   When the dose was doubled to 880 mg/kg day,
given in a single injection, iron uptake was 80% inhibited after
10 days.   Complete inhibition was observed when mice were injected
with 400 mg/kg twice per day for ten days or with 880 mg/kg twice
per day for 6 days.

Figures 4 and 5 show the distribution of benzene and its
metabolites in various organs of the mouse during a study in
which the 440 mg/kg dose was injected twice per day.   Both

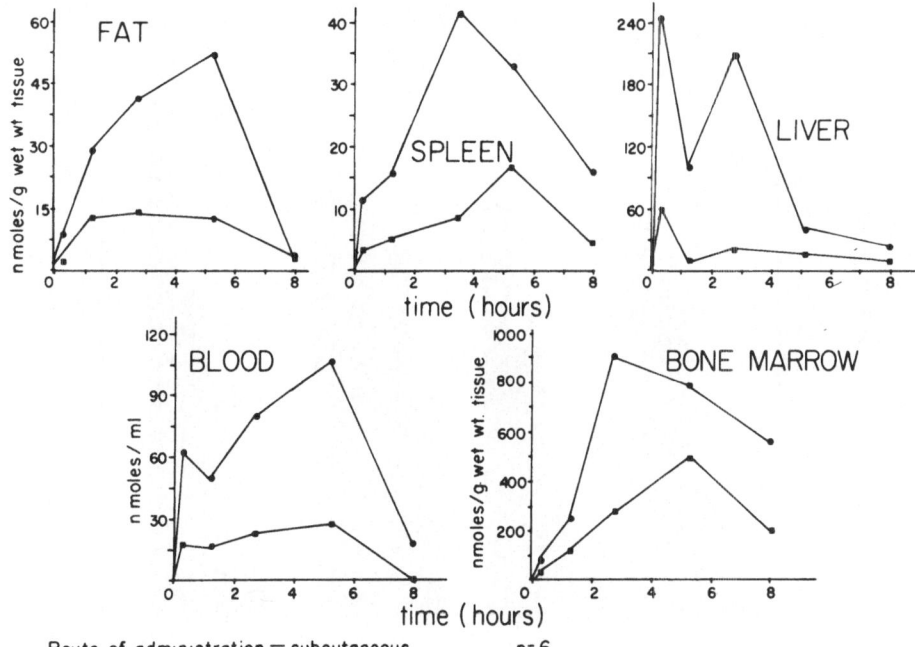

Figure 2. Recovery of ^{3}H-benzene metabolites in organs of mice. The experiment is described under Figure 1.

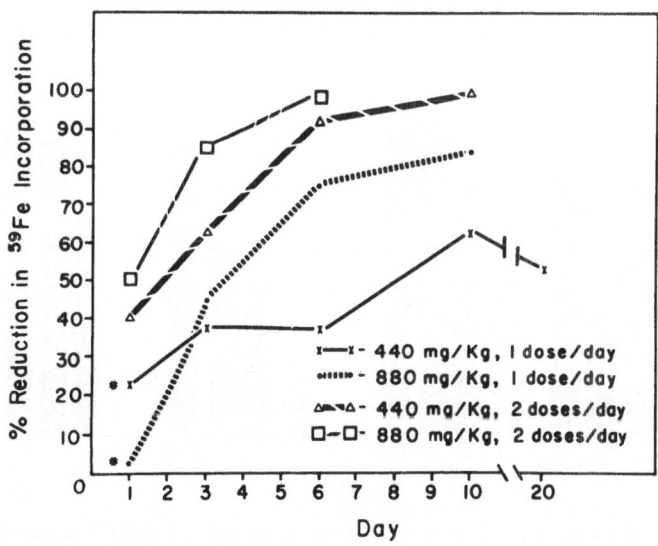

Figure 3. Reduction of ^{59}Fe incorporation into erythrocytes in mice given various doses of benzene.

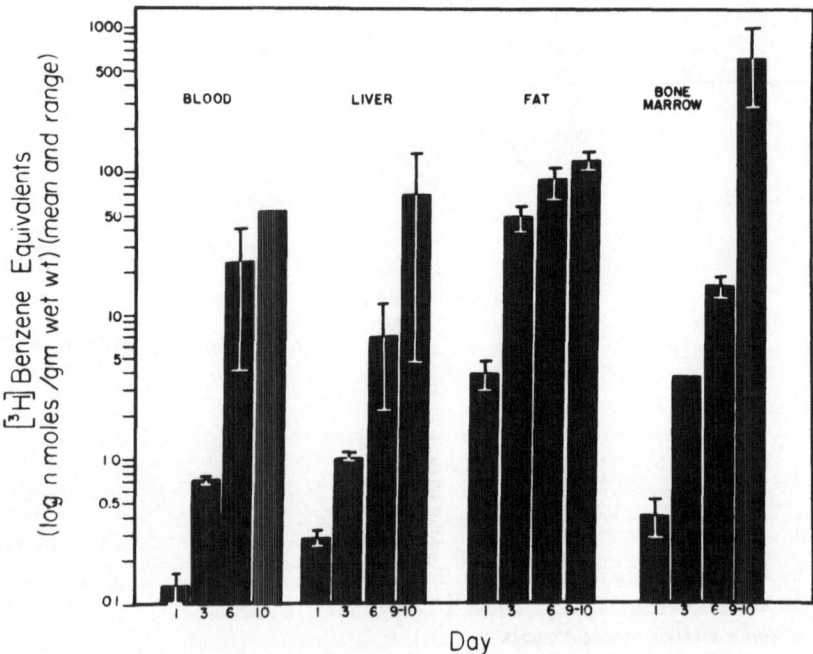

Figure 4.   Accumulation of ^{3}H-benzene in the organs of mice treated with ^{3}H-benzene (440 mg/kg) twice per day.

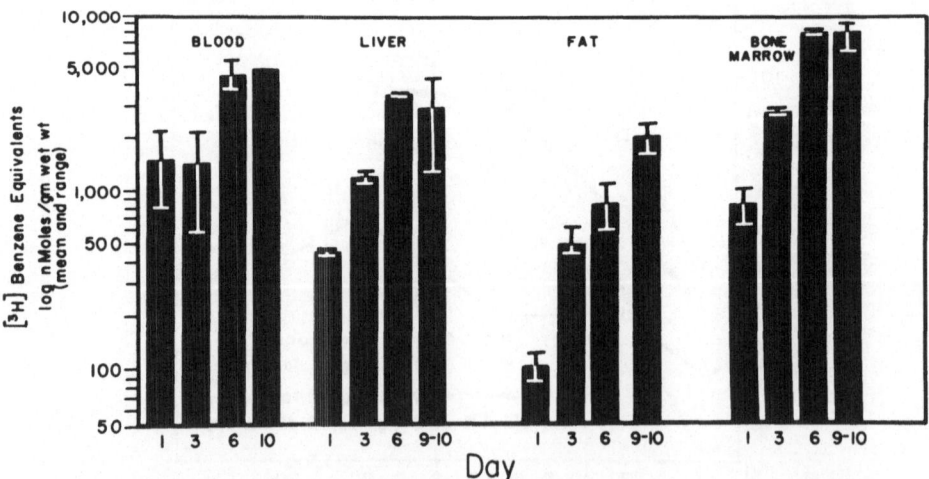

Figure 5.   Accumulation of ^{3}H-benzene metabolites in organs of mice treated with ^{3}H-benzene (440 mg/kg), twice per day.

benzene and its metabolites accumulated in body tissues during
the course of the study with the excessive accumulation in bone
marrow the most striking feature.

   During this time it was also observed that reactive metabolites of benzene covalently bound to cellular proteins (Table 2).
The binding was higher in liver than in marrow. Furthermore, as
the red color indicative of healthy marrow disappeared and was
replaced by fat and connective tissue, the associated loss of
marrow protein led to a decrease in measurable binding by day
9/10. While covalent binding to protein has been associated with
the cellular toxicity of many compounds, that connection has yet
to be made for benzene.

   The accumulation of benzene metabolites in bone marrow has
led to considerable speculation regarding the origin of those
metabolites. To determine whether they were formed in the marrow
or were transported from the liver to the marrow, a series of
studies were devised (28). Rats were subjected to either partial

Table 2.   Effect of duration of treatment with ^{3}H-benzene on
           irreversible binding of radioactivity to liver and
           bone marrow residues

Treatment	Liver[a] (nmoles ^{3}H-benzene equivalents) (gm dry wgt)	Bone Marrow[b]	Bone Marrow (pmoles in 2 mouse femurs)
1	185, 185[c]	7.4, 6.5	4.0, 6.2[c]
3	522, 596	34, 28	16, 13
6	716, 685	145, 247	80, 26
9-10	808, 954	61*	24, 40

[a]Each data point is taken from an independent pool of three livers

[b]Each data point for days 1, 3 and 6 represents a pool of 6 femoral
  marrow samples from 3 mice; data for days 9-10 represents 2 pools
  of 2 mice per group

[c]Replicate samples

*The second sample was not weighable but contained measurable
 radioactivity

Table 3.   Effect of partial hepatectomy on erythrocyte ^{59}Fe
           utilization in control and benzene treated rats

^{59}Fe Uptake In Erythrocytes (%)[a]

	Sham-Operated Rats	Partially Hepatectomized Rats
Control	31.6±13.3 (9)[b]	29.1±6.5 (10)
Benzene (2200 mg/kg)	15.4±6.7 (10)[c,d]	36.4±16.2 (10)[c,e]

[a]Based on administered dose

[b]Mean ± SD; the number of animals is given in parentheses

[c]Significantly different by Students t test from control sham-
operated rats (p<0.01)

[d]Significantly different by Students t test from benzene treated
partially hepatectomized rats (p<0.01)

[e]Not significantly different from control partially hepatectomized
rats

hepatectomy or sham operations, injected with benzene and iron
uptake was compared in treated versus control animals (Table 3).
Iron uptake was similar in control animals, whether sham-operated
or partially hepatectomized. Benzene, while reducing iron uptake
in the sham operated animals, was not hemotoxic in the partially
hepatectomized animals. We determined that the partially hepatectomized animals exhibited lower levels of benzene metabolites
in both excreted urine and in bone marrow as well as a lower level
of covalent binding in the marrow. These data support the concept
that one or more early stages in the metabolic activation of benzene
occurs in the liver.

The exact nature of the toxic metabolites of benzene is not
known. The various metabolites of benzene postulated by R.T.
Williams and his co-workers (29,30) in the late 1940's and early
1950's included phenol, hydroquinone, catechol, some trihydroxy-
lated derivatives, and some possible ring opening products. Much
of the interest in toxic metabolites of benzene has focused on
these compounds. In our laboratory, we have evaluated the urine
levels of some of these metabolites in a series of studies focusing
on differences in sensitivity to benzene between two strains of
mice, C57B1/6 and DBA/2.

Table 4. ^{3}H-Benzene metabolites in partially hepatectomized and sham operated rats[a]

	Sham Operated Rats	Partially Hepatectomized Rats
12 hr cumulative urinary metabolites (moles of ^{3}H-benzene equivalents)	203.7±102.7 (6)[b]	50.7 49.1±(2)[c]
Water soluble ^{3}H-metabolites in bone marrow (nmoles/g)	1737.7±82.8 (6)	763.5±23.9 (2)[c]
Covalently bound ^{3}H-metabolites in bone marrow (nmoles/g)	5.0±1.0 (6)	1.2±0.1 (2)[c]

[a]Benzene was administered at a dose of 2200 mg/kg sc., 8 hr after surgery. All animals were sacrificed 48 hours after surgery

[b]Mean±SD, the number of animals is given in parentheses

[c]Significantly different using Students t test from sham operated rats (p<0.05)

Table 5.   Effect of benzene on erythrocyte ^{59}Fe
utilization in C57BL/6 and DBA/2 mice

Strain	^{59}Fe Uptake In Erythrocytes (%)[a]		% Control
	Control	Treated[b]	
C57BL/6	57±7 (10)[c]	16±11 (10)[c]	28
DBA/2	33±11 (11)	0.2±0.04 (11)	0.6

[a]Based on administered ^{59}Fe dose, ^{59}Fe was given after the last
dose of benzene and uptake was evaluated 72 hr later

[b]Benzene was injected SC at a dose of 880 mg/kg, twice per day for
3 days

[c]Mean ± SD; the number of animals is given in parentheses

Table 5 shows the results of iron uptake studies in which
either C57B1/6 or DBA/2 mice were subcutaneously injected with
benzene, 880 mg/kg, twice per day for three days. Control values
for iron uptake were greater in C57B1/6 than in DBA/2 mice.
Benzene decreased iron uptake more in the DBA/2 mice than in the
C57B1/6 mice, indicating greater sensitivity in the former strain
to benzene toxicity. As shown in Table 6, the concentrations of
soluble and covalently bound metabolites in various organs,
and in bone marrow in particular, were higher in the more sensitive
DBA/2 animals. From a comparison of the hydrolyzed urinary
metabolites, we concluded that each strain excreted equal amounts
of catechol while the more sensitive DBA/2 mice excreted more
phenol and less hydroquinone (data not shown). At this point
it is not possible to relate these differences to greater production of phenol or to differential sequestration of hydroquinone
in the more sensitive strain.

DISCUSSION

The metabolic pathway for benzene metabolism as we currently
understand it is shown in Figure 6. In the first step, the microsomal mixed function oxidase, a benzene hydroxylase, converts
benzene to benzene oxide in the presence of oxygen and NADPH.
Benzene oxide can either be conjugated with glutathione to form a
permercapturic acid, hydrated by epoxide hydratase or non-enzymati-
cally rearranged to form phenol. Catechol is formed by the
oxidation of the glycol resulting from the action of the hydratase.
The mechanism of hydroquinone formation is not yet known. Tunek

Table 6. ³H-Benzene, and its water soluble and covalently bound metabolites in organs of C57BL/6 and DBA/2 mice given multiple doses of ³H-Benzene[a]

Organ	³H-Benzene[b]		Water Soluble Metabolites[b]		Covalent Binding[c]	
	C57BL/6	DBA/2	C57BL/6	DBA/2	C57BL/6	DBA/2
Bone marrow	---	---	2480±668[d]	4529±270	33±12[d]	52± 9
Liver	5±2	7±2	843±197[d]	1258±237	534±87	577±117
Kidney	8±1	6±1	807±135[d]	1134±165	539±70	650± 45
Blood	4±3	4±2	1051±127[d]	1389± 46	77± 9[d]	244± 42
Spleen	16±5	27±6	416± 44[d]	639±130	105±18[d]	146± 28
Lung	9±2	10±4	722±160[d]	1070±210	215±70	150± 13
Muscle	8±1	7±2	588± 36	771± 76	38± 3	35± 7

[a] ³H-Benzene was administered sc as 880 mg/kg (1 ml/kg), two doses per day for three days. Mice were sacrificed 16 hr after the last dose.

[b] Values represent nmole benzene equivalents/g wet wgt tissue (ml blood); M±SD for 3-4 values (10-12 values for liver).

[c] Values represent nmole benzene equivalents covalently bound/g dry tissue residue (M±SD) for 3-4 values.

[d] Significantly less than DBA/2 ($p < 0.05$).

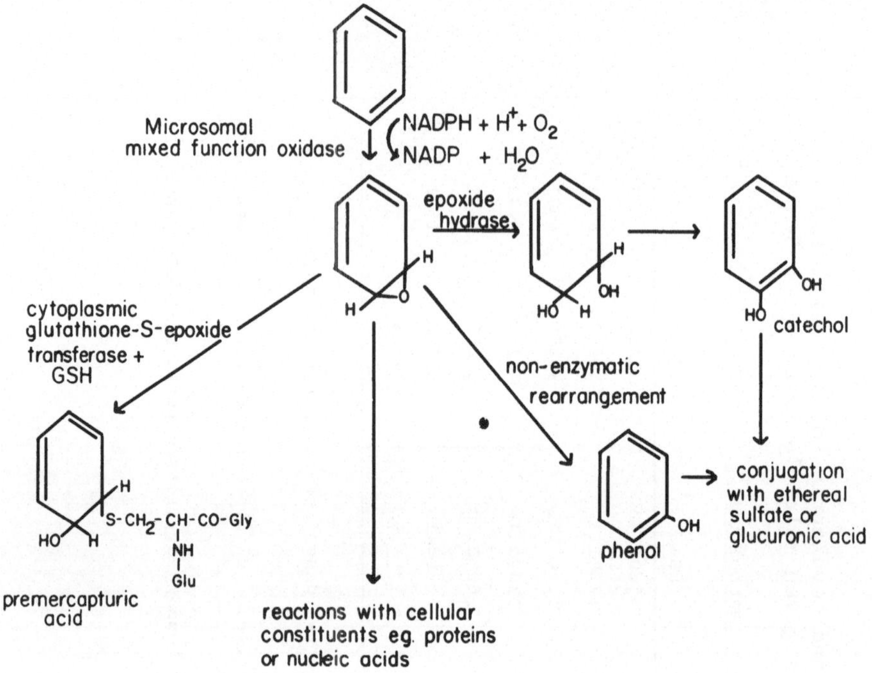

Figure 6.   Pathway of benzene metabolism

et al. (31) have suggested that although benzene oxide is a highly
unstable and reactive intermediate, it is unlikely to be the precursor to covalently bound benzenoid moieties. Their evidence
suggests that phenol is a more likely precursor to metabolites
capable of covalent binding. Hydroquinone may then be important
for covalent binding but catechol is not completely ruled out.
Specifically designating these compounds as the toxic metabolites
requires the assurance that the toxicity of benzene is expressed
by covalent binding. If other mechanisms are important a reevaluation of candidates for toxic metabolites will be required.

Based on our results, we can conclude that the sequence of
events occurring in benzene toxicity is the initial metabolic
activition of benzene in liver, followed by transfer of the meta-
bolite(s) to the marrow where toxicity results. Which specific
metabolite(s) is responsible, whether further metabolism is required,
and the molecular mechanism responsible for both the observed
toxicity and leukomgenic properties all remain unresolved.

ACKNOWLEDGEMENTS

The authors are indebted to Ms. Anne Strimpler and Mrs. Beatrice Engelsburg for their excellent technical assistance. This work was supported by NIH grant ES 00322.

REFERENCES

1. C. G. Santesson, Uber chronische Vergiftung mit Steinkohlen-theerbenzin; vier Todesfalle, Arch. Hyg. Berl. 31:336-376 (1897).
2. L. Selling, Benzol as a leucotoxin. Studies on the degenera-tion and regeneration of the blood and haematopoietic organs. Johns Hopkins Hosp. Reports 17:83-148 (1916).
3. M. C. Winternizt and A. D. Hirschfelder, Studies upon experimental pneumonia in rabbits: Parts I-III, J. Exptl. Med. 17:657-664 (1913).
4. A. D. Hirschfelder and M. C. Winternitz, Studies upon experimental pneumonia in rabbits: Part IV. Is there a parallelism between trypanocidal and pneumococcidal action of drugs? J. Exptl. Med. 17:666-678 (1913).
5. W. C. White, and A. M. Gammon, The influence of benzol inhalation on experimental pulmonary tuberculosis in rabbits. Trans. Assoc. Amer. Phys. 29:332-337 (1914).
6. J. P. Simonds and H. M. Jones, The effect of injections of benzol upon the production of antibodies. J. Med. Res. 33:197-209 (1915).
7. W. A. Camp and E. A. Baumgartner, Inflammatory reactions in rabbits with a severe leukopenia. J. Exptl. Med. 22:174-192 (1915).
8. L. Hektoen, The effects of benzene on the production of antibodies. J. Infec. Dis. 19:69-84 (1916).
9. A. Hamilton, The growing menace of benzene (benzol) poisoning in American industry. J. Amer. Med. Assoc. 78:627-630 (1922).
10. A. Hamilton, The lessening menace of benzol poisoning in American industry. J. Ind. Hyg. 10:227-233 (1928).
11. T. B. Mallory, E. A. Gall and W. J. Brickley, Chronic exposure to benzene (benzol). III. The pathologic results. J. Ind. Hyg. 21:355-393 (1939).
12. L. Greenburg, M. R. Mayers, L. Goldwater and A. R. Smith, J. Ind. Hyg. Toxicol. 21:395-420 (1939).
13. L. J. Goldwater, Disturbances in the blood following exposure to benzene. J. Lab. Clin. Med. 26:957-973 (1941).
14. C. Zenz, Benzene - attempts to establish a lower exposure standard in the United States. A review. Scand J. Work Environ. and Health 4:103-113 (1978).
15. R. Snyder and J. J. Kocsis, Current concepts of chronic exposure toxicity. Crit. Rev. Toxicol. 3:265-288 (1975).

16. E. Browning, Toxicity and Metabolism of Industrial Solvents,
    Chap. 1, Elsevier, Amsterdam, (1965).
17. E. C. Vigliani and G. Saita, Benzene and leukemia, New Eng.
    J. Med. 271:872-876 (1964).
18. E. C. Vigliani and A. Forni, Benzene, chromosome changes and
    leukemia. J. Occup. Med. 11:148-149 (1969).
19. M. Aksoy, S. Erden and S. Dincol, Leukemia in shoe workers
    exposed chronically to benzene. Blood 44:837-841 (1974).
20. P. F. Infante, J. K. Wagoner, R. A. Rinsky and R. J. Young,
    Leukemia in benzene workers. The Lancet July 9:76-78
    (1977).
21. C. Maltoni and C. Scarnato, First experimental demonstration
    of the carcinogenic effects of benzene. Med. Lavoro 70:
    352-357 (1979).
22. G. Watanabe and S. Yoshida, The teratogenic effect of benzene
    in pregnant mice. Acta Medica et Biologica17:285-291 (1970).
23. J. D. Green, B. K. J. Leong and S. Laskin, Inhaled benzene
    fetotoxicity in rats. Toxicol. Appl. Pharmacol. 46:9-18
    (1978).
24. P. S. Nawrot and R. E. Staples, Embryofetal toxicity and perato-
    genicity of benzene and toluene in the mouse. Teratology
    19:41A (1979).
25. F. J. Murray, J. A. John, L. W. Rampy, R. A. Knuna and B. A.
    Schwetz, Embryotoxicity of inhaled benzene in mice and
    rabbits. Amer. Ind. Hyg. Assoc. J. 40:993-998 (1979).
26. L. S. Andrews, E. W. Lee, C. M. Witmer, J. J. Kocsis and
    R. Snyder, Effects of toluene on the metabolism, disposition
    and hemopoietic toxicity of ^{3}H-benzene. Biochem. Pharmacol.
    26:293-300 (1977).
27. E. W. Lee, J. J. Kocsis and R. Synder, Benzene: Acute
    effect ^{59}Fe incorporation into circulating erythrocytes.
    Tox. Appl. Pharmacol. 27:431-436 (1974).
28. D. Sammett, E. W. Lee, J. J. Kocsis and R. Snyder, Partial
    hepatectomy reduces both metabolism and toxicity of
    benzene. J. Toxicol. Environ. Health 5:785-792 (1979).
29. D. V. Parke and R. T. Williams, Studies in detoxication.
    The metabolism of benzene containing (^{14}C) benzene. Biochem.
    J. 54:231-238 (1953).
30. D. V. Parke and R. T. Williams, Studies in Detoxication 54.
    The metabolism of benzene (a) The formation of phenyl-
    glucuronide and phenylsulfuric acid from ^{14}C-benzene; (b)
    The metabolism of ^{14}C phenol. Biochem. J. 55:337-340 (1953).
31. A. Tunek, K. L. Platt, M. Przybylski and F. Oesch, Multi-
    step metabolic activation of benzene. Effect of superoxide
    dismutase on covalent binding to microsomal macromolecules,
    and identification of glutathione conjugates using high
    pressure liquid chromatography and filed desorption mass
    spectrometry. Chem.-Biol. Interactions 3:1-17 (1980).

BENZENE METABOLITES: EVIDENCE FOR

AN EPIGENETIC MECHANISM OF TOXICITY

R. D. Irons and R. W. Pfeifer

Chemical Industry Institute of Toxicology
P.O. Box 12137
Research Triangle Park, NC 27709

Benzene is the most widely recognized chemical of industrial importance which is associated with toxicity to the blood and blood forming organs. Occupational exposure produces a variety of blood dyscrasias ranging from transient leukopenia or lymphocytopenia to aplastic anemia (1,2). In recent years interest has focused on the association of chronic benzene exposure with increased risk of leukemia (3-5) or lymphoma (6-8).

In our studies we have been impressed by the apparent sensitivity of the lymphocyte to benzene toxicity. Repeated exposure of experimental animals to high concentrations results in a lymphocytopenia accompanied by a decrease in bone marrow and spleen cellularity, which precedes any decrease in the number of other circulating cells (9-12). These changes coincide with alterations in cell cycle kinetics of bone marrow precursor cells and in chromosomal aberrations which suggest that benzene affects proliferating cells (9,13-15).

There is general agreement that benzene is not itself the principal molecular species responsible for bone marrow or lymphoid toxicity. The primary oxidation of benzene is effected through the cytochrome P-450-dependent monooxygenase system and, as is the case for the metabolic derivatization of many foreign compounds, oxidation results in the formation of biologically reactive intermediates (16,17) (Fig 1). Although benzene oxide, the immediate unstable product of benzene, has been proposed as a likely candidate for the proximate toxic species, Tunek et al. (18) have more recently

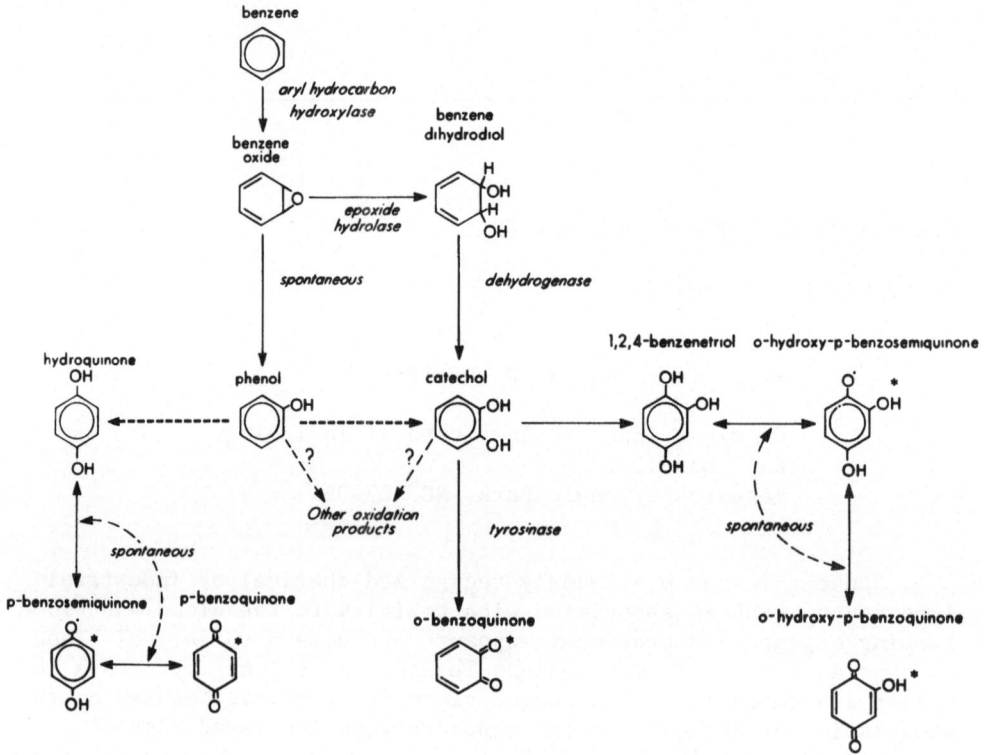

Figure 1.   Schematic diagram of the primary metabolism of benzene
showing established and putative pathways resulting in the production of potentially lymphotoxic metabolites.   Asterisks
denote putative or demonstrated alkylating activity toward intracellular nucleophiles.

demonstrated that phenol, and not benzene oxide, is predominantly
associated with the covalent binding of the parent compound to
microsomal protein.

The three principal hydroxy-metabolites of benzene are phenol,
catechol and hydroquinone. Under a variety of experimental conditions benzene toxicity to bone marrow and lymphoid organs is associated with the concentration of hydroquinone and catechol, but not
phenol, accumulating in target tissues (19-23). In vivo administration of catechol or hydroquinone also results in a reduction in
bone marrow and lymphoid cellularity and a depression in immune
function (24).

Proposals for mechanism of action of these compounds are forthcoming from studies of their effects on lymphocytes in culture.
We examined the ability of the hydroxy-metabolites of benzene to

affect lectin-induced blastogenesis in vitro. Phenol had no effect
over a wide concentration range whereas p-benzoquinone, hydroquinone,
1,2,4-benzenetriol and, to a lesser extent, catechol suppressed
mitogen response (21,25) (Table 1). This observation might, in
part, be explained by the fact that hydroquinone spontaneously
oxidizes under physiologic conditions to form p-benzoquinone (21,26).
Although catechol does not appreciably autoxidize, its oxidation
product, (1,2,4-benzenetriol) rapidly oxidizes, presumably to
form the analogous 2-hydroxy-p-benzoquinone.

Suppression of lectin-induced mitogen response by benzene
metabolites is attended by a number of intriguing observations.
First, the modulation of lymphocyte blastogenesis is biphasic:
short term pretreatment with low concentrations enhances mitogenesis
whereas exposure to higher concentrations suppresses mitogen response
(Table 2). p-Benzoquinone is a more potent suppressor of mitogen
response than hydroquinone. Second, the observed suppression in
blastogenesis occurs at exposure concentrations which do not
result in cell death or a reduction in energy production (25).
Another interesting finding is that blastogenesis is blocked by
these compounds concurrent with the failure of lymphocytes to
agglutinate (Table 2). The precise relationship between agglutination and those events which result in DNA synthesis is unknown.

Table 1.   Effect of benzene metabolites on
PHA-stimulated rat spleen cells[a]

Concentration (M)	Metabolite			
	Phenol	HQ	Triol	CAT
10^{-3}	1.38*	0.00*	0.00*	0.00*
10^{-4}	1.15	0.00*	0.00*	0.07*
10^{-5}	1.26*	0.03*	0.36*	0.75
10^{-6}	1.14	1.24*	1.24*	1.17
10^{-7}	1.14	1.28*	–	1.24

[a]Responses of Ficoll-purified cells pooled from four F-344 male
rats were assayed 48 hours after mitogen addition at an optimal
cell concentration (2 x 10^5 cells/200µL) and optimal dose of
PHA (5 µg/mL). Spleen cells were preincubated with metabolite
for 14 hours. Values are E/C ratios from raw mean dpm of
triplicate cultures. Asterisks denote statistical significance
at p = 0.05 level compared with control after ANOVA and comparison of means by Duncan's multiple range test.

Table 2.   Effect of hydroquinone and p-benzoquinone on PHA-
           stimulated rat spleen cells[a]

Concentration (x 10^{-7}M)	Hydroquinone		p-Benzoquinone	
	E/C Ratio	A.I.[b]	E/C Ratio	A.I.
4	2.07	++++	0.75	++++
6	1.90	++++	0.72	++++
8	1.56	++++	0.38	+++
10	1.36	++++	0.44	+++
20	0.32	+++	0.01	++
40	0.05	++	0.00	−
60	0.03	+	0.01	−
80	0.03	−	0.00	−
100	0.02	−	0.01	−

[a]Culture conditions and data representation same as in Table 1.

[b]Agglutination index (A.I.) indicates degrees of cell aggregation
and blast transformation after exposure to PHA as observed by
phase-contrast microscopy. Control cells (no metabolite) had
index of ++++; unstimulated cells had index of +.

Agglutination is thought by many to be a prerequisite for lectin-
induced blastogenesis and it certainly represents one of the earliest
events associated with cell division (27,28).

     Over the past decade a number of laboratories have extensively
studied the role of the cytoskeleton as a transducer both in the
communication of altered surface receptor conformation to the in-
terior of the cell and in the resulting modulation of cell response
to growth stimuli (29-32). These studies suggest that cell recognition and growth are cytoskeletal-dependent processes, and that
microtubule integrity is required for density-dependent lectin
stimulation of lymphocyte mitogenesis. Agents such as colchicine
or cytochalasin B, which are known to disrupt microtubules, micro-
filaments and their associated proteins, enhance blastogenesis
at low concentrations (33-35) but suppress blastogenesis at higher
concentrations (30-32,36,37). Berlin and Ukena (38) have reported
similar effects for microtubule disruptive agents on cell agglutination. The modulating effects of p-benzoquinone, hydroquinone and
1,2,4-benzene-triol on lymphocyte mitogenesis and agglutination
resemble those observed for microtubule disruptive agents.

   A number of sulfhydryl reagents inhibit microtubule assembly
in vitro, with those known to penetrate the cell (N-ethylmaleimide,
cytochalasin A) disrupting cytoskeletal-dependent functions involving
cell conformational change (39-41). These effects have been repeatedly observed at concentrations of sulfhydryl reagents which do
not impede energy production, deplete reduced glutathione levels
or alter lectin-binding to the cell (32, 38, 42-46). The sulfhydryl
reactivity of the quinones is well known (47, 48) and led us to
examine the possibility that quinone metabolites of benzene might
alter lymphocyte function through sulfhydryl-dependent disruption of
microtubule assembly. p-Benzoquinone, like N-ethylmaleimide or
cytochalasin A, reacts selectively with sulfhydryl groups through
a conjugate addition reaction, in which a highly polarized unsaturated
carbon-carbon bond is subject to attack by a reactive thiol acting
as a nucleophile (44, 47-50) (Fig 2). The semiquinone intermediate,
presumably formed in the autoxidation of hydroquinone to p-benzoquinone, would be expected to exhibit sulfhydryl reactivity but
not specificity.

   The sulfhydryl specificity of the suppression of lymphocyte
response by hydroquinone was confirmed in a series of experiments
which examined the effects of various prototype reagents on lectin-
induced lymphocyte blastogenesis and agglutination (Fig 3).
Hydroquinone suppressed both cell functions at approximately the
same concentrations as did a cell-penetrating sulfhydryl reagent,
N-ethylmaleimide (NEM). A poorly penetrating sulfhydryl reagent,
5,5'-dithiobis(2-nitrobenzoic acid) (DTNB), had no effect on either
blastogenesis or agglutination. Concentration-dependent protection
against these suppressive effects was achieved only in the presence

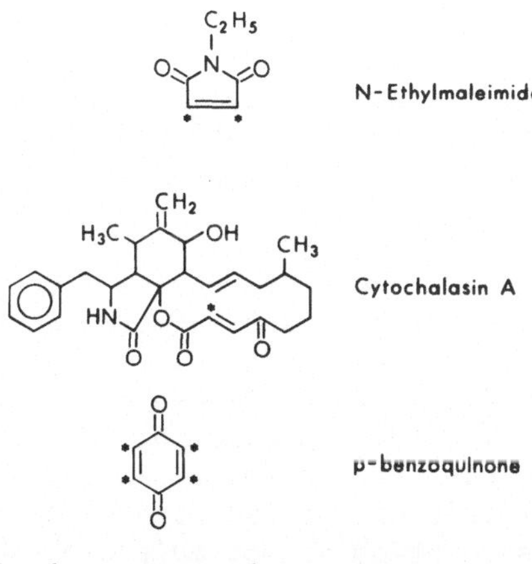

Figure 2. Compounds demonstrated to alkylate sulfhydryl groups
at physiologic pH via Michael addition. Asterisks denote carbons
subject to nucleophilic attack.

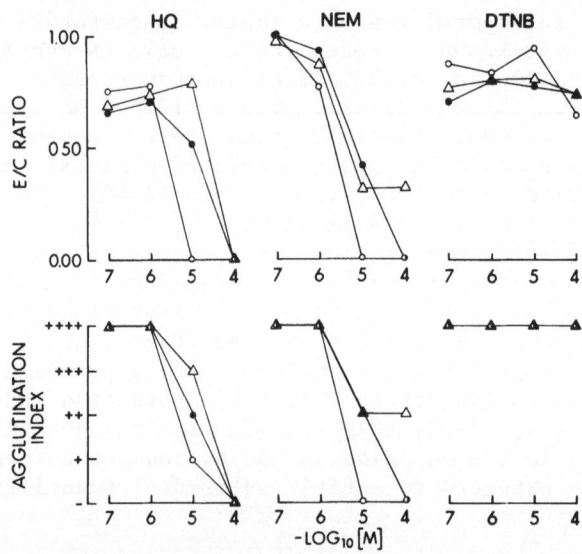

Figure 3.   Protection against HQ or NEM inhibition of PHA-
stimulated blastogenesis and agglutination in Ficoll-purified
rat lymphocytes by DTT.   Results expressed as E/C ratios of
the means of triplicate cultures.   S.D. of the mean did not vary
more than 10%.   Comparison of the effects of HQ and NEM, a cell
penetrating sulfhydryl agent; DTNB, a poorly penetrating sulfhydryl
reagent.   O, No DTT present during preincubation with HQ or
NEM;Δ,10^{-4} M DTT present during preincubation, ●, 10^{-5} M DTT
present during preincubation.

of sulfhydryl compounds such as cysteine or dithiothreitol.   Lysine,
serine or imidazole provided no protection against suppression at
physiologic pH.   We therefore concluded that hydroquinone was acting
at intracellular sulfhydryl sites critical for lectin-induced blastogenesis and agglutination, and that sublethal impairment of immune
function by hydroquinone could be mimicked by specific thiol alkylation by N-ethylmaleimide.

    We also examined the effects of benzene metabolites on assembly
of microtubules in vitro.   Phenol and catechol do not inhibit
tubulin polymerization; however, hydroquinone, p-benzoquinone and
1,2,4-benzenetriol inhibit microtubule assembly in a concentration-
dependent manner (Fig 4).   The linearity of a semilogarithmic plot
of inhibition versus concentration for p-benzoquinone and N-ethylmaleimide is characteristic of a first order reaction and suggests
a unimolecular process in which either molecule reacts directly with
tubulin.   Colchicine, which interacts directly with tubulin in a
site-specific manner, demonstrates linear inhibition kinetics.
Although minimally effective concentrations of p-benzoquinone and N-

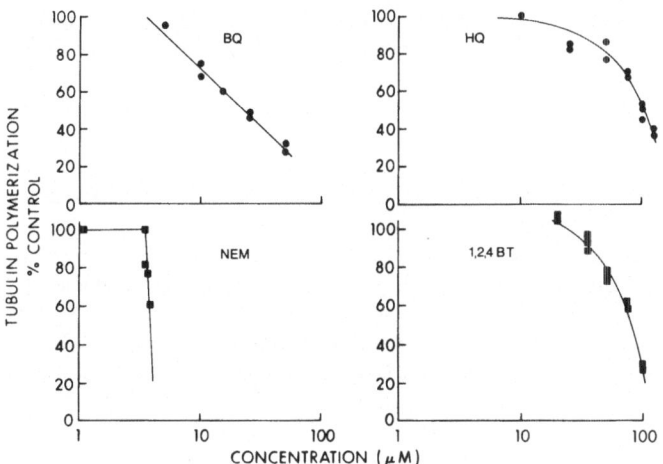

Figure 4.   Log concentration-inhibition curves for p-BQ, HQ,
NEM or 1,2,4-BT on purified tubulin polymerization in vitro.
Temperature-dependent polymerization was measured turbi-
dometrically at 350 nm after 10 min. incubation at 37°C.
Inhibition of polymerization was calculated as the percent
of control for incubation with different concentrations of
inhibitors.   Rat brain tubulin was isolated by the cycle
procedure as previously described (52).

ethylmaleimide are the same, differences in slope between the two
compounds suggest differences in affinity for some sulfhydryl sites.
These results are contrasted with the nonlinearity of the curves
exhibited by hydroquinone and 1,2,4-benzenetriol.   These data are not
linear on a log-log plot, reflecting complexity greater than second
order, which is consistent with a bimolecular process in which
hydroquinone or 1,2,4-benzenetriol must be oxidized to produce the
direct reacting molecular species.

     Although these findings suggest that p-benzoquinone is the
directly reacting species in the autoxidation scheme for hydroquinone,
a role for the semiquinone could not be dismissed on this basis
alone.   We therefore examined the effect of catechol on tubulin
polmerization in the presence of the enzyme, tyrosinase.   Catechol
alone had no effect on microtubule assembly but completely inhibited
tubulin polymerization in the presence of the enzyme.   Tyrosinase
converts catechol directly to the o-benzoquinone by way of a two-
electron transfer, forming no reactive intermediates in the reaction
(51).   The activation of catechol by tyrosinase therefore provides
confirmation of the role of quinones in the inhibition of tubulin
polymerization.

Characterization of the interaction between hydroquinone or p-benzoquinone with tubulin has revealed it to be highly specific. Hydroquinone binds covalently to the tubulin molecule with a concentration ratio of 1-2 moles hydroquinone per mole tubulin, sufficient to inhibit polymerization (52). In order to further assess the nature of the quinone reaction with tubulin we examined the effects of benzene metabolites on the decay of tubulin-colchicine binding activity (TCBA). Colchicine binds to the tubulin dimer at a specific site which does not directly involve sulfhydryl groups. Colchicine binding is technically reversible but extremely avid. The binding of colchicine to tubulin exhibits a first order decay over a period of hours in vitro, the biological significance of which is unknown, but it coincides with the irreversible loss of tubulin polymerization and the denaturation of the protein (53,54). TCBA decay is greatly attenuated by the addition of reagents, including dithiothreitol, glycerol and guanosine triphosphate (GTP), which stabilize free thiol groups on the protein (55-57). There are two different GTP binding sites on the tubulin dimer with hydrolysis of GTP occurring at one site during polymerization. GTP binding to tubulin is sulfhydryl dependent. Purified tubulin contains up to 12 titratable sulfhydryl groups, two of which demonstrate much greater reactivity with sulfhydryl reagents, such as N-ethylmaleimide. These two sulfhydryl groups have also been shown to be required for GTP binding to tubulin (57).

Quinones seemed to be acting at this site on the tubulin molecule. However, before TCBA decay could be used as a measure of specificity of interaction with GTP-tubulin sites, it was necessary to exclude the possibility that hydroquinone interfered directly with the reversible binding of colchicine to tubulin. A typical Scatchard analysis revealed that hydroquinone has no effect on the reversible binding of colchicine to tubulin, thus validating our proposed use of TCBA decay (Fig 5).

Hydroquinone greatly accelerates TCBA decay (Fig 6) in a process that is protected in a concentration-dependent manner by the addition of GTP (Fig 7). Similar findings were obtained with p-benzoquinone, 1,2,4-benzenetriol and N-ethylmaleimide. Further indication of the sulfhydryl specificity of the reaction was obtained from a series of experiments analogous to our studies in whole cells. Dithiothreitol, but not lysine or serine, affords complete protection against the effects of p-benzoquinone or N-ethylmaleimide on TCBA decay (Table 3). Reflecting the prerequisite for autoxidation in the activation of hydroquinone and 1,2,4-benzenetriol, incubation under nominally anaerobic conditions provides significant protection against acceleration of TCBA decay by these compounds but not for N-ethylmaleimide or p-benzoquinone which react directly with tubulin (Table 4).

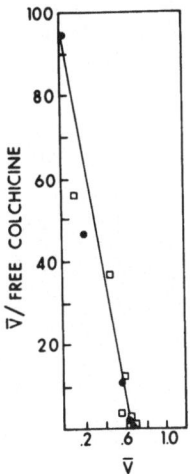

Figure 5.   Scatchard plot of colchicine binding to purified
rat tubulin.   The binding of [^{3}H] colchicine to tubulin (1 mg/mL)
120 min after incubation at 37°C alone (●) or in the presence
of 2 x 10^{-4} M hydroquinone (□) was examined for various
concentrations of colchicine.   The molar ratio of hydroquinone
to tubulin was 23:1.   Reproduced with permission from reference 52.

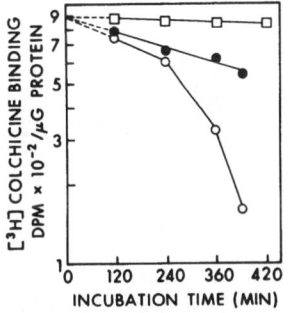

Figure 6.   Effect of hydroquinone and GTP on TCBA decay.
Purified tubulin contains 2 x 10^{-6} M [^{3}H] colchicine (0.1 μCi),
was incubated in the presence of 2.5 mM GTP (□), absence
of GTP (●) or the absence of GTP with 2 x 10^{-4} M hydroquinone
(0).   Reproduced with permission from reference 52.

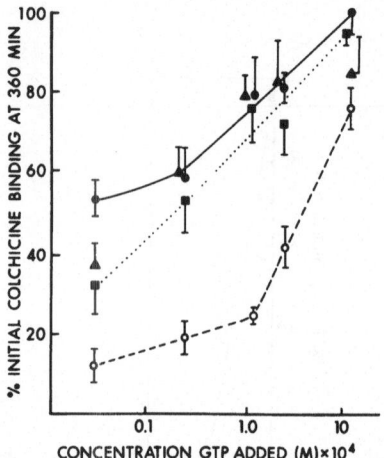

Figure 7.   Concentration-response curves for TCBA decay at
varying concentrations of added GTP.   Purified tubulin
preparations, containing 2 x 10^{-6} M [^{3}H] colchicine (0.1 μCi),
were incubated at 37°C for 360 min in the presence of varying
concentrations of GTP (abscissa); with or without (●) the
addition of 2 x 10^{-4} M hydroquinone (O), catechol (■) or
phenol (▲).   Values represent mean ± SEM for 3 (treated) and 4
(control) experiments.   Reproduced with permission from
Arch. Toxicol. 45:297–305, 1980.

In summary, benzene is toxic to bone marrow and lymphoid organs
in vivo with toxicity correlating with the concentration of hydroquinone and catechol accumulating in these tissues.   Hydroquinone
alters lectin-induced lymphocyte blastogenesis in a biphasic manner
similar to that observed for known microtubule disruptive agents.
Hydroquinone also interferes with microtubule assembly in vitro at
the same concentrations.   Both processes are sulfhydryl dependent:
the first involves intracellular sulfhydryl groups critical for
mitogenesis; the second involves interaction with sulfhydryl groups
critical for tubulin integrity.   These findings have been shown to
be consistent with a scheme in which the proximate toxic species,
p-benzoquinone, reacts with tubulin in a reaction involving a highly
polarized unsaturated carbon-carbon bond which is subject to attack
by particularly nucleophilic sulfhydryl groups.   The relatively
weak but specific sulfhydryl alkylating ability of the p-quinones
relative to other reactive intermediates such as o-quinones, may
account for their potential to interfere with cytoskeletal-dependent
processes in the absence of cell death.

Table 3.   Effect of various compounds on the acceleration of
            TCBA decay* by p-benzoquinone[+] or N-ethylmaleimide

	Untreated	DTT	Lysine	Serine
Control	100± 7.9†	119.2± 2.5	107.2± 7.0	91.6± 6.5
p-BQ	54.2± 2.6°	92.6± 9.5	50.4± 2.0°	53.4± 1.4°
NEM	63.8± 3.0°	114.3± 2.7	---	---

*   Purified tubulin preparations containing [3H] colchicine (0.1
       μCi; 2 x 10⁻⁶ M) were incubated at 37°C for 240 min in the
       presence of GTP (2.5 x 10⁻⁵M). Free and tubulin-bound
       colchicine were separated by Sephadex G-25 column chromato-
       graphy.

+   Concentrations of p-benzoquinone, dithiothreitol, N-ethylmalei-
       mide, serine and lysine were all 1 x 10⁻⁴M.

†   Values are expressed as percent of control colchicine binding
       at 240 min and represent the mean ± S.E.M. for 3 separate experi-
       ments calculated from the original data.

°   Significantly different from control at the p = 0.05 level using
       Duncan's multiple range test.

Table 4.   Effect of anaerobic conditions on the acceleration
           of TCBA decay by hydroquinone, 1,2,4-benzenetriol,
           p-benzoquinone or N-ethylmaleimide

| Treatment | Incubation* | |
	Aerobic	Anaerobic
Control	$100.0 \pm 4.7^+$	87.5 ± 4.4
Hydroquinone	$51.4 \pm 1.6^{\dagger \circ}$	96.9 ± 5.4
1,2,4-Benzenetriol	$36.6 \pm 4.8^{\dagger \circ}$	69.8 ± 6.7
p-Benzoquinone	$56.6 \pm 8.8^{\dagger}$	$42.4 \pm 0.7^{\dagger}$
N-ethylmaleimide	$42.2 \pm 5.0^{\dagger}$	$54.5 \pm 2.6^{\dagger}$

*   Aerobic conditions were the same as described in Table 3.
      Anaerobic incubations were conducted using degassed solutions
      saturated with N_2.

+   Values are expressed as % of control and represent the mean $\pm$
      S.E.M. for 3 separate experiments as described in Table 3.

†   Significantly different from control.

°   Significantly different from anaerobic incubation.

REFERENCES

1. L. J. Goldwater, Disturbances in the blood following exposure to benzol. J. Lab. Clin. Med. 26:957-973 (1941).

2. S. Laskin and B. D. Goldstein, Benzene toxicity, a critical review. J. Toxicol. & Environ. Health Suppl. 2:1-148 (1977).

3. M. Aksoy, K. Dincol, S. Erdem and G. Dincol, Acute leukemia due to chronic exposure to benzene. Am. J. Med. 52:160-166 (1972).

4. M. Aksoy, S. Erdem and G. Dincol, Types of leukemia in chronic benzene poisoning. A study in thirty-four patients. Acta Haematol. 55:65-72 (1976).

5. E. C. Vigliani and A. Forni, Benzene and leukemia. Environ. Research 11:122-127 (1976).

6. J. Bousser, R. Neydé and A. Fabre, Un case d'hémopathie benzolique trés retardeé à type de lymphosarcomatose. Bull. Soc. Medicale des Hopitaux de Paris 3:1000-1004 (1947).

7. I. R. Tabershaw and W. C. Cooper, A mortality study of petro- leum refinery workers. American Petroleum Institute. Medical Research Report #EA 7402. Sept. 15 (1974).

8. M. Aksoy, S. Erdem, K. Dincol, T. Hepyüksel and G. Dincol, Chronic exposure to benzene as a possible contributory etiologic factor in Hodgkin's Disease. Blut 28:293-298 (1974).

9. R. D. Irons, H. d'A. Heck, B. J. Moore and K. A. Muirhead, Effects of short-term benzene administration on rat bone marrow cell cycle kinetics in the rat. Toxicol. & Appl. Pharmacol. 51:399-409 (1979).

10. R. D. Irons, Benzene-induced myelotoxicity: Application of flow cytofluorometry for the evaluation of early proliferative change in bone marrow. Environmental Health Perspect. In press (1981).

11. R. D. Irons and B. J. Moore, Effect of short term benzene administration on circulating lymphocyte subpopulations in the rabbit: evidence of a selective B-lymphocyte sensitivity. Res. Comm. Chem. Path. Pharmacol. 27:147-155 (1980).

12. D. Wierda, W. F. Greenlee and R. D. Irons, Immunotoxicity of benzene metabolites in C57BL/6 mice. The Pharmacologist 22:260 Abstract (1980).

13. K. A. Muirhead, R. D. Irons, R. Bruns and P. K. Horan, A rabbit bone marrow model system for the evaluation of cytotoxicity: Characterization of normal bone marrow cell cycle parameters by flow cytometry. J. Histochem. Cytochem. 28:526-532 (1980).

14. S. Moeschlin and B. Speck, Experimental studies on the mechan- ism of action of benzene on the bone marrow (radioauto- graphic studies using ^{3}H-thymidine). Acta Haematol. 38:104-111 (1967).

15. R. Tice, D. L. Costa and R. T. Drew, Cytogenetic effects of inhaled benzene in murine bone marrow: Induction of sister chromatid exchanges, chromosomal aberrations, and cellular proliferation inhibition in DBA/2 mice. Proc. Natl. Acad. Sci. USA. 77:2148-2152 (1980).

16. L. M. Gonasun, C. Witmer, J. J. Kocsis and R. Snyder, Benzene metabolism in mouse liver microsomes. Toxicol. & Applied Pharmacol. 26:398-406 (1973).

17. R. Snyder and J. L. Kocsis, Current concepts of chronic benzene toxicity. CRC Critical Reviews in Toxicology, June, 265-288 (1975).

18. A. Tunek, K. L. Platt, P. Bentley, and F. Oesch, Microsomal metabolism of benzene to species irreversibly binding to microsomal protein and effects of modifications of this metabolism, Molec. Pharmacol. 14:920-929 (1978).

19. D. E. Rickert, T. S. Baker, J. S. Bus, C. S. Barrow and R. D. Irons, Benzene disposition in the rat after exposure by inhalation. Toxicol. & Appl. Pharmacol. 48:417-423 (1979).

20. R. D. Irons, J. G. Dent, T. S. Baker and D. E. Rickert, Benzene is metabolized and covalently bound in bone marrow in situ. Chem.-Biol. Interact. 30:241-245 (1980).

21. R. D. Irons, W. F. Greenlee, D. Wierda and J. S. Bus, Relationship between benzene metabolism and toxicity: A proposed mechanism for the formation of reactive intermediates from polyphenol metabolites, in: Biological Reactive Intermediates II, R. Snyder, D. V. Parke, J. J. Kocsis and D. A. Jollow, eds., Plenum Press, in press, (1981).

22. W. F. Greenlee and R. D. Irons, Modulation of benzene-induced lymphocytopenia in the rat by 2,4,5,2',4',5'-hexachloro-biphenyl and 3,4,3',4'-tetrachlorobiphenyl. Chem.-Biol. Interact. In Press (1981).

23. W. F. Greenlee, E. A. Gross and R. D. Irons, A study of the disposition of ^{14}C-labelled phenol, catechol and hydroquinone in the rat using whole body autoradiography. Chem.-Biol. Interact. In Press (1981).

24. D. Wierda and R. D. Irons, Reduction of progenitor B-lymphocytes in mice after hydroquinone and catechol administration. Fed. Proc. Abstract (1981).

25. R. W. Pfeifer and R. D. Irons, Inhibition of PHA-stimulated mito-genesis by benzene metabolites: Protection with sulfhydryl compounds. Toxicol. & Appl. Pharmacol. Abstract. (1981).

26. W. F. Greenlee and J. S. Bus, A proposed mechanism for benzene toxicity: Formation of reactive intermediates from polyphenol metabolities of benzene. The Pharmacologist 22:229. Abstract. (1980).

27. H. J. Wedner and C. W. Parker, Lymphocyte activation. Prog. Allergy 20:195-300 (1976).

28. N. R. Ling and J. E. Kay, The mechanism of lymphocyte activa-
     tion - Metabolic changes during lymphocyte stimulation, in:
     "Lymphocyte Stimulation," N. R. Ling and J. E. Kay, eds.,
     Amsterdam, 253-355 (1975).
29. G. M. Edelman, Surface alterations and mitogenesis in lympho-
     cytes, in: "Control of Proliferation in Animal Cells," B.
     Clarkson and R. Baserga, eds., Cold Spring Harbor Labora-
     tories, Cold Spring Harbor, NY, 357-377 (1973).
30. G. R. Gunther, J. L. Wang and G. M. Edelman, Kinetics of
     colchicine inhibition of mitogenesis in individual lympho-
     cytes. Exp. Cell Res. 98:15-22. (1976).
31. J. L. Wang, G. R. Gunther and G. M. Edelman, Inhibition by
     colchicine of the mitogenic stimulation of lymphocytes prior
     to the S phase. J. Cell Biol. 66:128-144 (1975).
32. W. C. Greene, C. M. Parker and C. W. Parker, Colchicine-
     sensitive structures and lymphocyte activation. J. Immunol.
     117:1015-1022 (1976).
33. M. Yoshinaga, A. Yoshinaga and B. H. Waksman, Regulation of
     lymphocyte response in vitro: potentiation and inhibition of
     rat lymphocyte responses to antigen and mitogens by
     cytochalasin B. Proc. Natl. Acad. Sci. USA 69:3251-3255
     (1972).
34. M. Ono and M. Hozumi, Effect of cytochalasin B on lymphocyte
     stimulation induced by concanavalin A or periodate. Biochem.
     Biophys. Res. Comm. 53:342-349 (1973).
35. M. Suthanthiran, K. H. Stenzel, A. L. Rubin and A. Novogrodsky,
     Augmentation of proliferation and generation of specific
     cytotoxic cells in human mixed lymphocyte culture reactions
     by colchicine. Cell. Immunol. 50:379-391 (1980).
36. P. Sherline and G. R. Mundy, Role of the tubulin-microtubule
     system in lymphocyte activation. J. Cell Biol. 74:371-376
     (1977).
37. G. M. Edelman, Surface modulation in cell recognition and
     cell growth. Science 192:218-226 (1976).
38. R. D. Berlin and T. E. Ukena, Effect of colchicine and vinblas-
     tine on the agglutination of polymorphonuclear leucocytes by
     concanavalin A. Nature New Biology 238:120-122 (1972).
39. G. I. Giordano and M. A. Lichtman, The role of sulfhydryl
     groups in human neutrophil adhesion, movement and particle
     ingestion. J. Cell Physiol. 82:387-396 (1973).
40. M. Tsan, B. Newman and P. A. McIntyre, Surface sulphydryl
     groups and phagocytosis-associated oxidative metabolic
     changes in human polymorphonuclear leucocytes. Brit. J.
     Hemat. 33:189-204 (1976).
41. J. G. R. Elferink and J. C. Riemersa, Effects of sulfhydryl
     reagents on phagocytosis and exocytosis in rabbit polymor-
     phonuclear leukocytes. Chem.-Biol. Interact 30:139-149
     (1980).

42. D. D. Chaplin and H. J. Wedner, Inhibition of lectin-induced
     lymphocyte activation by diamide and other sulfhydryl
     reagents. Cell Immunol. 36:303-311 (1978).

43. L. Sachs, M. Inbar and M. Shinitzky, Mobility of lectin sites
     on the surface membrance and the control of cell growth and
     differentiation, in: "Control of Proliferation in Animal
     Cells," Cold Spring Harbor Laboratories, NY, 283-296 (1973).

44. D. Lagunoff and H. Wan, Inhibition of histamine release from
     rat mast cells by cytochalasin A and other sulfhydryl
     reagents. Biochem. Pharmacol. 28:1765-1769 (1979).

45. M. T. Mazur and J. R. Williamson, Macrophage deformability and
     phagocytosis. J. Cell Biol. 75:185-199 (1977).

46. N. Chakravarty and Z. Echetebu, Plasma membrane adenosine
     triphosphatases in rat peritoneal mast cells and
     macrophages - the relation of the mast cell enzyme to
     histamine release. Biochem. Pharmacol. 27:1561-1569 (1978).

47. A. Michael, Das chinon vom standpunkt des enteropiegesetzes
     und der partialvalenzhypothese. J. Prakt. Chem. 79:418-431
     (1909).

48. M. Schubert, The interaction of thiols and quinones. J. Amer.
     Chem. Soc. 69:712-713. (1947).

49. R. H. Himes and L. L. Houston, The action of cytochalasin A
     on the in vitro polymerization of brain tubulin and muscle
     G-actin. J. Supramol. Struct. 5:81-90 (1976).

50. D. Lagunoff, The reaction of cytochalasin A with sulfhydryl
     groups. Biochem. Biophys. Res. Commun. 73:727-732 (1976).

51. R. P. Mason, Free radical metabolites of foreign compounds
     and their toxicological significance, in: "Reviews in
     Biochem. Toxicol.," E. Hodgson, J. R. Bend and R. M. Philpot,
     eds., Elsevier, New York, NY, 151-200 (1979).

52. R. D. Irons and D. A. Neptun, Effects of the principal hydroxy-
     metabolites of benzene on microtubule polymerization. Arch.
     Toxicol. 45:297-305 (1980).

53. R. Kuriyama and H. Sakai, Role of tubulin-SH groups in polymeri-
     zation to microtubules. J. Biochem. 76:651-654 (1974).

54. L. Wilson, J. R. Bamburg, S. B. Mizel, L. M. Grisham and K. M.
     Creswell, Interaction of drugs with microtubule proteins.
     Fed. Proc. 33:158-166 (1974).

55. Y. Ikeda and M. Steiner, Sulfhydryls of platelet tubulin:
     Their role in polymerization and colchicine binding.
     Biochem. 17:3454-3459 (1978).

56. M. G. Mellon and L. I. Rebhun, Sulfhydryls and the in vitro
     polymerization of tubulin. J. Cell Biol. 70:226-238 (1976).

57. K. Mann, M. Giesel, H. Fasold and W. Hasse, Isolation of
     native microtubules from porcine brain and characterization
     of SH groups essential for polymerization at the GTP sites.
     FEBS Lett. 92:45-48 (1978).

CYTOGENETIC EFFECTS OF INHALED BENZENE IN MURINE BONE MARROW

R. R. Tice, T. F. Vogt and D. L. Costa

Medical Department
Brookhaven National Laboratory
Upton, NY   11973

INTRODUCTION

Exposure to benzene is both an occupational and an environmental hazard. Health concerns are based on the extensive use of benzene in industry and local commerce (1) and a long-standing association with aplastic anemia and leukemia in occupationally-exposed workers (2-4, see Goldstein, these proceedings). Although negative for genotoxic activity in a variety of short-term in vitro bioassays (5-7), benzene is capable of inducing tumors in animals (8,9). The discrepency between these and other in vitro and in vivo results strongly suggests that a metabolite(s), rather than the parent compound, is primarily responsible for benzene's genotoxic activity.

Both cellular toxicity and carcinogenic potency appear to be causally-related to an agent's clastogenic potential (10,11). Exposure to benzene causes chromosomal aberrations in animals (1-4,12) and is associated with an increased frequency of such clastogenic events in occupationally-exposed workers (1-4,13-15). Several investigators have extended these observations in animals to other cytogenetic manifestations of DNA damage, e.g. micronuclei (16,17) and sister chromatid exchanges (SCEs) (18). Furthermore, two presumptive metabolites of benzene, catechol and hydroquinone, induce SCEs and inhibit cellular proliferation in human cells in vitro (19).

Extrapolating these effects of benzene to an assessment of its impact on human health requires considerably more scientific research. Some of the needed research includes: (i) an understanding of the mechanism(s) by which the various manifestations of benzene toxicity occur, (ii) the acquisition of dose-response information by exposure routes pertinent to human experience, and (iii) an analysis of whether

such factors as age, sex and/or genetic constitution can modulate the
magnitude of the response.

Using cytogenetic methodologies based on the incorporation of
5-bromodeoxyuridine (BrdUrd) into DNA (18,20,21), we have examined
benzene's ability to induce chromosomal aberrations or SCEs and to
inhibit cellular proliferation in murine bone marrow. Analysis of
SCE frequency appears especially useful since SCEs are, in many instances, an extremely sensitive indicator of direct DNA damage and
mutagenic potential (22-25, see Galloway and Tice, these proceedings).
Inhibition of cellular proliferation may be a sensitive measure of
cellular toxicity. Consequently, the former endpoint may indicate
leukemogenic potential, while the latter endpoint, bone marrow depression leading to anemia.

We have exposed mice to benzene by inhalation, the normal route
for human exposure, and have attempted to assess: (i) benzene's a-
bility to induce different kinds of genotoxic damage, (ii) the importance of concentration, age, sex and genetic constitution in modulating benzene's genotoxic potency, and (iii) the relationship between liver-specific metabolism and extrahepatic metabolism in determining the observed cytogenetic effects.

MATERIALS AND METHODS

In all of these studies, mice were obtained as weanlings from
Jackson Laboratory, ME, and fed and housed under the conditions described in detail elsewhere (18). Inhalation exposures were conducted in an isolation chamber previously described (26). Benzene vapor
was generated by bubbling filtered compressed air through liquid benzene and diluting the vapor effluent appropriately. Benzene concentrations were measured at half-hour intervals by a gas chromatograph
(Packard, Model 417, equipped with a column of 10% silicone SE-30 on
Chromasorb W-HP) using an automatic sampling valve. Some animals were
injected intraperitoneally (IP) with benzene dissolved in corn oil
(Mazola).

With one exception, mice were infused with BrdUrd (Sigma; 50
mg/Kg per hour) as described in detail elsewhere (21) beginning one
hour after completion of the benzene exposure and extending for up
to 30 hours. Two hours prior to animal sacrifice, Colcemid (Gibco;
1 mg/Kg) was injected intravenously into each animal. Femoral bone
marrow was obtained by flushing the femurs with phosphate buffered
saline (pH 7.0), incubated in 0.075 M KCl and then fixed with 3:1
methanol:glacial acetic acid. Flame-dried slides were prepared and
differentially stained with Giemsa (Harleco) using a modification
(18,27) of the technique of Goto et al. (28). Metaphase spreads were
analyzed microscopically for chromosomal aberrations (in 50 first
generation metaphase cells), SCEs (in 25 second generation metaphase

cells) and for replicative history, i.e. the number of cell generations completed in the presence of BrdUrd (in 100 randomly selected
metaphase cells). Any additional treatments are described in the
text.

RESULTS AND DISCUSSION

Induction of Genotoxic Damage in Murine Bone Marrow: 3000 ppm Benzene

     In our first experiment, male and female DBA/2 mice (approximate-
ly 10 months of age) were exposed to 3000 ppm benzene for 4 hours (9
am- 1 pm) (18). A 4 hour exposure to 3000 ppm benzene induced a significant increase in SCEs in bone marrow cells of both sexes, did not
induce a significant increase in chromosomal aberrations in bone marrow cells of either sex, and inhibited cellular proliferation in bone
marrow of male but not female mice (Table 1). Furthermore, the proliferative inhibition observed in male mice was even greater one day
after the exposure period (Table 2). This observation suggests the
delayed formation of toxic metabolites and/or a delayed cellular reaction (18).

     To examine the effects of enhanced liver metabolism on these
cytogenetic manifestations of benzene's genotoxicity, mice were injected with phenobarbital prior to their exposure to benzene. Phenobarbital is a well-known inducer of hepatic mixed function oxidases
(29) and has been shown to alter benzene's toxicity in vivo (1-4).
While having no effect on control values, pretreatment with phenobarbital: (i) enhanced benzene's ability to induce SCEs in bone marrow
cells of female mice but not of male mice, (ii) enabled benzene to
induce chromosomal aberrations in bone marrow cells of both sexes,
but more dramatically in male mice than in female mice, and (iii) enhanced benzene's ability to inhibit cellular proliferation in bone
marrow of male mice but not of female mice (Table 1). The chromosomal
aberrations were all of the chromatid- and not of the chromosome-type.
There was also no increase in chromosomal rearrangements of any kind
(18).

     The difference in both the magnitude and the type of genotoxic
response observed in the bone marrow of benzene-exposed male and female mice (Table 3) are intriguing for several reasons. First, these
results demonstrate a sex difference in the sensitivity of DBA/2 mice
to benzene. Second, the ability of phenobarbital pretreatment to alter the magnitude of different genotoxic responses observed in the
bone marrow supports the involvement of metabolism in benzene's genotoxic activity. Finally, the spectrum of response in male versus female mice, with and without phenobarbital pretreatment, suggests the
involvement of different metabolites of benzene in different genotoxic
responses.

Table 1. Cytogenetic effects of inhaled benzene in murine bone marrow (3000 ppm–4 hr)

Exposure Group	Mice (10 mo old)	SCE Frequency[a]	% Abnormal Cells[b]	AGT (hr)[c]
Control	Male	4.6±0.2(6)	8.8±1.2(5)	14.2±1.3(6)
	Female	4.5±0.3(6)	10.8±2.2(5)	13.4±0.9(6)
Benzene (3000ppm–4hr)	Male	8.7±0.4(10)*	12.8±2.0(5)	16.6±0.4(7)*
	Female	8.2±0.7(10)*	12.0±1.3(5)	12.2±0.9(11)
Phenobarbital**	Male	4.1±0.2(7)	10.8±2.0(5)	11.8±0.4(7)
	Female	4.2±0.2(6)	9.6±1.2(5)	12.5±0.2(6)
Phenobarbital** + Benzene (3000ppm–4hr)	Male	8.3±0.4(6)*	42.4±3.6(5)*	20.9±2.2(11)*
	Female	12.7±1.0(7)*	25.6±1.7(5)*	12.1±0.4(7)

[a]Mean frequency of SCE/cell ± S.E. between (n) animals

[b]Mean % abnormal cells ± S.E. between (n) animals; values reflect the presence of both achromatic lesions and chromatid-type aberrations

[c]Average Generation Time = mean cell cycle duration ± S.E. between (n) animals, calculated by dividing the BrdUrd exposure period by the frequency of first generation metaphase cells plus twice the frequency of second generation metaphase cells plus three times the frequency of third generation metaphase cells

*Statistically different from the appropriate control by Students t test at p< 0.01

**Animals injected with sodium phenobarbital (50 mg/Kg) ip, twice daily for 3 days prior to exposure to benzene

Table 2. Average generation time in bone marrow cells
during the first and second day after benzene
exposure (3000 ppm-4 hr)

Time Period(hr)[a]	Exposure Group	Mice (10 mo old)	AGT_b (hr)[b]
(i) 1-31	Control	Male	11.9±0.6(5)
		Female	10.9±0.4(5)
	Benzene	Male	17.2±1.3(6)*
		Female	11.1±0.6(6)
(ii) 21-51	Control	Male	12.2±0.8(5)
		Female	11.0±0.5(5)
	Benzene	Male	29.8±1.2(6)*
		Female	11.4±0.5(5)

[a]BrdUrd infusion period (30hr) beginning 1 hr (i) or 21 hr (ii)
after termination of benzene exposure

[b]Average Generation Time (see Table 1)

*Statistically different from the appropriate control by Students t
test at $p < 0.01$

Effect of Partial Hepatectomy on Benzene-Induced SCEs

We decided to further examine the importance of metabolism in
determining benzene's genotoxic activity by examining the impact of
a partial hepatectomy on benzene's ability to induce SCEs and to inhibit bone marrow proliferation in male DBA/2 mice. The theoretical
basis for this experiment was derived from studies by Snyder and his
co-workers (30, see Snyder, these proceedings) showing that the removal of most of the liver in rats resulted in the decreased ability
of benzene to interfer with red blood cell maturation.

We removed ∿60% of the liver in male DBA/2 mice (10-14 wks of
age), allowed recovery for seventeen hours and then exposed the animals (along with the appropriate controls) to 3000 ppm benzene for
four hours (9 am-1 pm) (31).

The results (Table 4) were surprising for three reasons. First,
instead of the expected doubling in SCE values for benzene-exposed
animals without partial hepatectomy, we observed a six-fold increase

Table 3. Effect of benzene inhalation on various cytogenetic end-
          points in bone marrow of DBA/2 mice, (10 mo old) with
          and without phenobarbitol pretreatment

Cytogenetic End Points[a]

Sex	Pheno-Na	SCE	Chromosomal Abberations	Inhibition of Cellular Proliferation
Male	no	+	−	−
	yes	+	++	++
Female	no	+	−	−
	yes	++	+	−

[a]See Table 1 for additional information

  − = no significant increase
  + = significant increase at p<0.01
 ++ = another significant increase at p<0.01

Table 4. Effect of partial hepatectomy on benzene (3000 ppm-4 hr)-
          induced SCE

Exposure Group	Partial Hepatectomy[a]	SCE Frequency[b]
Control	−	4.2±0.3(6)
	+	6.1±0.6(6)
Exposed	−	26.2±2.5(4)
	+	25.6±1.7(6)

[a]∿60% of the liver removed 17 hr prior to benzene exposure; male
DBA/2 (10 −14 weeks of age)

[b]Mean frequency of SCE/cell±S.E. between (n) animals

in SCE frequency. Second, bone marrow cell proliferation remained normal in these benzene exposed animals instead of exhibiting the expected inhibition. Finally, partial hepatectomy did not reduce benzene's ability to induce SCEs in bone marrow cells.

The inability of a partial hepatectomy to reduce benzene's genotoxic ability appears to conflict with the results of Snyder et al. (30, see Snyder, these proceedings). In the absence of additional experimental data, the difference in our results may be explained by experimental differences such as animal species (rat vs mouse), route of exposure (subcutaneous vs inhalation), recovery time after partial hepatectomy (7 hours vs 17 hours), dose of benzene (880 mg/Kg vs 12000 ppm-hr) or by the endpoint examined (iron-59 incorporation vs SCEs).

The dose of benzene to which the animals were exposed for four hours (3000 ppm) induced a maximal yield of SCEs and mixed function oxidase P450 levels did not increase in the remaining portion of the liver during the seventeen hour recovery time (data not presented). Thus, the most reasonable explanation for our respective results lies the difference in the biological endpoint under examination. Snyder's study (30, see Snyder, these proceedings) examined the incorporation of iron-59 into maturing red blood cells, a process occurring in a nonproliferating, differentiating cell population and induced by a mechanism probably unrelated to DNA damage. SCE formation, on the other hand, represents events occurring in proliferating cells in response to DNA damage (22-25). In view of our experimental results (cited earlier) suggesting possible metabolite specificity for DNA damage versus inhibition of cellular proliferation, our respective results may be both correct. The results of Snyder and his colleagues may indicate a liver-specific metabolite(s) responsible for delayed erythrocyte maturation while our results may indicate a genotoxic metabolite(s) formed directly from benzene in the bone marrow. Alternatively, and contrary to in vitro data on benzene (19), benzene, itself, may be responsible for the bone marrow-induced SCEs.

Effect of Co-Administered Toluene on Benzene-Induced SCEs

To determine if unmetabolized benzene was responsible for the formation of SCEs in the bone marrow of exposed animals, we next examined the ability of toluene, a well-known competitive inhibitor of benzene metabolism in vitro and in vivo (32,33, see Snyder, these proceedings), to affect the yield of benzene-induced SCEs.

DBA/2 male mice (10-14 weeks of age) were injected IP with toluene (32.4 mmoles/Kg) immediately prior to a four hour exposure to 3000 ppm benzene (31). Toluene did not induce SCEs or inhibit cell proliferation. Although the yield of benzene-induced SCEs was depressed by ∿40% in toluene pretreated animals, many of these doubly

exposed animals exhibited extreme respiratory depression. We thought
it likely that the depressed breathing rate may have contributed to
the reduction in SCEs and decided to circumvent this possible confounding effect by co-administering benzene and toluene by IP injection.

Benzene was dissolved in corn oil and injected IP into DBA/2
male mice (10-14 weeks of age) one hour prior to the beginning of
the BrdUrd infusion, a time corresponding to the end of the inhalation exposure. Using a range of benzene concentrations inducing low
to maximal yields of SCEs, we assessed the ability of equi-molar concentrations of co-administered toluene to affect the induction of
SCEs by benzene (Table 5). At the lowest concentration of benzene
(2.25 mmoles/Kg), toluene had no effect on benzene-induced SCE yields.
However, with increasing concentrations of both agents, toluene inhibited benzene's genotoxic activity to an increasing extent. Maximal inhibition of benzene' ability to induce SCEs ($\sim$90%) occurred
at concentrations of benzene which, in the absence of toluene, induced maximal yields of SCEs ($\sim$26 SCE/cell @ 22.5 mmoles/Kg). Presumably, at low levels of benzene and toluene, the enzymes responsible for benzene metabolism are in excess of the total number of benzene and toluene molecules present. Only at agent concentrations
exceeding the total number of enzyme binding sites would toluene inhibition of benzene metabolism be expected to occur. Bone marrow
cell proliferation and yields of chromosomal aberrations remained at
control values throughout this experiment (data not shown).

These results demonstrate an absolute requirement for the metabolic activation of benzene prior to the production of SCEs in bone
marrow cells of exposed animals.

Benzene Concentration-SCE Response: Modulating Effect of Sex and Genetic Constitution

Since 3000 ppm benzene induced approximately a six-fold increase
in SCEs in bone marrow cells of male DBA/2 mice when 10 to 14 weeks
of age, we next examined the ability of benzene to induce SCEs at
much lower ambient concentrations. We also examined whether sex and
genetic constitution could modulate benzene's genotoxic activity (34).
In this series of experiments, male and female mice (10-14 weeks of
age) from two isogenic strains--DBA/2 and C57B1/6--were exposed for
four hours to concentrations of benzene ranging from 28 to 5000 ppm.
These two mouse strains were chosen because they differ at the aryl-
hydrocarbon hydroxylase (AHH) inducible locus, a locus associated
with carcinogenic sensitivity in mice (35,36) and, perhaps, in man
(37).

The lowest concentration of benzene that induced a significant
increase in SCEs in male and female mice of both strains was 28 ppm

Table 5.   Effect of co-administered toluene on benzene (2.25-22.5 mmoles/Kg)-induced SCEs in bone marrow cells of DBA/2 male mice (10-14 weeks of age)

Exposure Group	Benzene mmoles/Kg[a]	SCE Frequency[b]	+ Toluene mmoles/Kg[a]	SCE Frequency[b]	% Decrease
Control		6.2±0.2(4)			
Corn oil		5.9±0.4(6)			
Toluene			18.50	6.2±0.3(4)	
Toluene			32.40	5.2±0.1(4)	
Benzene	2.25	10.9±0.2(4)	2.25	11.7±0.6(4)	0.0
	5.63	17.1±0.6(6)	5.63	11.2±0.9(5)	-50.1
	11.25	21.4±0.6(6)	11.25	9.6±1.1(4)	-76.0
	22.50	26.4±2.1(5)	22.50	8.1±1.0(5)	-89.2

[a]Dissolved in corn oil and administered IP (0.1 mL/5 gm body weight) one hour before initiation of BrdUrd infusion

[b]Mean frequency of SCE per cell ± standard error of the mean between (n) animals

(Table 6). Concentrations of benzene lower than 28 ppm were not tested. However, differential sex and strain sensitivities to ambient
benzene were readily apparent over the concentration range examined.
Benzene induced more SCEs in male mice than in female mice and more
SCEs in DBA/2 mice than in C57Bl/6 mice. SCE frequency increased
linearly to 3000 ppm benzene in the DBA/2 mice and to 2000 ppm in
the C57Bl/6 mice. Higher concentrations, i.e. >3000 ppm for DBA/2
mice and >2000 ppm for C57Bl/6 mice, resulted in a less than maximal
frequency of SCEs. Because the SCE frequency saturates at high doses
of benzene when the agent is injected IP, we believe that the most
plausibible explanation for this phenomenon is that the high doses
of benzene affected the central nervous system of the animals which
resulted in a depressed breathing rate and a decrease in the uptake
of benzene via the lungs.

In early epidemiolical surveys an increased susceptibility to
the adverse health activity of benzene was ascribed to women (1-4).
However, no current survey supports a sex-related differential sensitivity in humans for either benzene-induced anemia or leukemia
(1-4). This may not be surprising considering the extent of the
difference in SCE levels observed between the two sexes and the confounding effect of genetic constitution on SCE yields that we have
observed in mice. The ability of benzene at concentrations as low
as 28 ppm to induce a significant increase in SCEs in mice, regardless of the sex or genetic constitution of the animal, is especially
interesting. OSHA currently considers 10 ppm benzene to be an occupationally safe exposure level for an eight hour working day (Time
Weighted Average = 10 ppm with a 25 ppm ceiling) (see Goldstein,
these proceedings). These levels are within the same order of magnitude as the lowest exposure level used in our experiments, an exposure level which elicited a positive genotoxic response in mice.
Although these positive effects in mice cannot be directly extrapolated to human exposure conditions and to human physiology and the
relevance of SCE induction to cancer remains unknown, the data does
suggest that more research is needed to evaluate the current "safe"
levels of benzene.

Age-Response to Inhaled Benzene

In evaluating the data from these various studies, one obvious
and very significant observation is the age-related differential
sensitivity of mice to the genotoxic effects of benzene (Table 7).
"Aged" male mice, approximately 10 months of age, exhibited only a
doubling in SCE frequency while "young" male mice, 3 months of age,
exhibited almost a six-fold increase in SCEs. Furthermore, while
exposure to benzene inhibited bone marrow cellular proliferation in
the older male mice, inhibition which was enhanced one day after the
exposure (Table 2), no inhibition was observed in the younger male
mice examined under the same conditions (data not shown, 38).

Table 6. Effect of inhalation of benzene on SCE frequencies in bone marrow
cells of mice

Conc. Benzene (ppm/ppm–hr)	DBA/2 SCE Frequency[a]		C57B1/6 SCE Frequency[a]	
	Male	Female	Male	Female
Control	4.3±0.3(6)	4.6±0.4(4)	4.1±0.5(5)	4.7±0.4(4)
28/112	7.4±0.5(4)	6.6±0.5(4)	6.2±0.4(5)	6.2±0.5(3)
489/1960	12.7±0.9(4)	9.5±0.6(4)	8.2±0.4(4)	7.5±0.5(3)
1018/4072	16.2±1.4(5)	9.6±1.0(4)	9.8±1.0(5)	7.5±0.5(4)
2082/9541	22.2±1.4(6)	11.2±0.2(3)	16.6±1.8(3)	9.3±0.2(3)
3044/12182	26.2±2.5(4)	11.4±0.5(5)	11.2±1.4(4)	9.0±0.9(5)
5276/21104	17.1±0.9(4)	10.6±0.8(4)	9.3±1.0(4)	8.7±0.7(4)

[a]Mean frequency of SCE per cell ± standard error of the mean between (n) animals

*Statistically different from the appropriate control by Students t test at $p < 0.01$

Table 7.   Effect of age on benzene-induced bone marrow
           damage in DBA/2 male mice

	Control		Benzene*	
Age (mo)	SCE Frequency[a]	AGT (hr)[b]	SCE Frequency[a]	AGT (hr)[b]
3	4.3±0.3(4)	12.2±0.3(4)	26.2±2.5(4)[c]	13.5±0.5(4)
10	4.6±0.2(6)	14.2±0.3(6)	8.5+0.6(7)[c]	16.6±0.4(7)[c]

*3000 ppm for 4 hours

[a] Mean frequency of SCE/cell ± standard error of the mean between (n) animals

[b] Average Generation Time = mean cell cycle duration ± S.E. between (n) animals; see Table 1 for additional information

[c] Statistically different by Student's t test at $p < 0.01$

      The biological basis for these age differences is not known.
The differences may result from alterations in metabolism, distribution of metabolites or parent compound, target cell sensitivity or
to levels of DNA repair.   There is no appropriate epidemiological
information for age-related occupational exposure of humans to benzene suggesting an age-related differential sensitivity.   However,
the extensive genetic heterogeneity among humans would make the
identification of age-related effects difficult to discern.

Route of Exposure

      In one additional analysis, SCE data obtained from the two routes
of exposure to benzene--inhaled and IP injection--were compared (Table
8).   SCE frequency increased in a linear fashion as a function of benzene concentration for both routes of exposure.   For one manifestation
of genotoxic damage--SCEs--and under our experimental conditions, an
ambient exposure of 120 ppm benzene for four hours equaled the effect
of an IP injection of 1.0 mmoles/Kg.   Experiments to determine the relative potency of benzene by these two routes of exposure are in progress.   However, based on the known pharmacokinetic data for benzene
metabolism in vivo (33,39), we have tentatively concluded that a
greater bone marrow genotoxic effect results from inhaled benzene
than when injected IP (40).

Table 8. Effect of benzene exposure on bone marrow damage
          in male DBA/2 mice (10-14 weeks of age)

Inhalation*		IP Injection**	
Conc. (ppm)	SCE Frequency[a]	Conc. mmoles/Kg	SCE Frequency[a]
Control	4.3±0.3(6)	Control	6.1±0.3(6)
28	7.4±0.5(4)	.22	9.4±0.8(4)
217	10.3±0.9(4)	2.25	10.9±0.2(4)
489	12.7±0.9(4)	5.63	17.0±0.6(6)
1018	16.2±1.4(5)	11.25	21.4±0.6(6)
2082	22.2±1.4(6)	22.50	26.4±2.1(5)
3044	26.2±2.5(4)	33.75	36.0±3.3(4)
5276	17.1±0.9(4)	45.00	36.4±2.1(4)

*3000 ppm for 4 hours.

**Single IP injection of benzene dissolved in corn oil, total
  volume 0.1 mL/5 gm

[a]Mean frequency of SCE/per cell ± S.E. between (n) animals

CONCLUSION

     The results of our cytogenetic investigation into the genotoxic
potential of benzene in murine bone marrow can be summarized as follows:

     (i) The different responses to benzene observed in male and in
female DBA/2 mice (with and without phenobarbital pretreatment), suggest that different metabolites of benzene are responsible for the
induced chromosomal aberrations and SCEs and for the inhibition of
cellular proliferation in bone marrow tissue in older mice.

     (ii) The results from the partial hepatectomy and toluene stu-
dies indicate that while metabolism of benzene is necessary before
genotoxic damage can occur in bone marrow cells, the liver may not
be the main site of production for the SCE-inducing metabolite.

(iii) A four-hour exposure to ambient concentrations of benzene as low as 28 ppm induces a significant increase in bone marrow geno-toxic damage as measured by SCE formation.

(iv) The modification of the magnitude of the benzene-induced SCE frequency in murine bone marrow cells by age, sex and genetic constitution emphasizes the importance of considering these parameters when attempting to extrapolate animal data to human health risks.

(v) The ability to directly correlate two different routes of benznene exposure for genotoxic damage should permit a better under-standing of the pharmacokinetic distribution of benzene and its meta-bolites in animals and a better evaluation of benzene data for human health risks.

(vi) Our results suggest that the pharmacokinetic, metabolic and molecular mechanisms involved in benzene's ability to cause anemia and leukemia are extremely complex. If different metabolites of ben-zene are responsible for the induction of chromosomal aberrations, SCEs and cellular proliferation inhibition, then a good animal model for anemia may not be a good animal model for leukemia and vice-versa. Considering benzene's occupational and environmental importance, re-search in this area is not only of considerable interest but also of economic and societal importance. Furthermore, an understanding of the biological processes involved in benzene's genotoxic activity may lead to a better understanding of the human health hazards as-sociated with the many other genotoxic agents to which man is exposed.

ACKNOWLEDGEMENT

This research was conducted at Brookhaven National Laboratory under contract with the U.S. Department of Energy, the U.S. Environ-mental Protection Agency and the National Institutes of Health. Ac-cordingly, the U.S. Government retains a nonexclusive, royalty-free licence to publish or reproduce the published form of this contri-bution or allow others to do so for U.S. Government purposes. We gratefully acknowledge the editorial assistance of Drs. K. Schaich and R. D. Benz.

REFERENCES

1. T. J. Haley, Evaluation of the health effects of benzene, Clin. Toxicol. 11:531-548 (1977).
2. R. Snyder and J. J. Kocsis, Current concepts of chronic benzene toxicity, CRC Crit. Rev. Toxicol. 3:265-288 (1975).
3. E. C. Vigliani and A. Forni, Benzene and leukemia, Environ. Res. 11:122-127 (1976).

4. B. J. Dean, Genetic toxicology of benzene, toluene, xylenes, and phenols, Mutat. Res. 47:75-97 (1978).

5. P. Gerner-Smidt and U. Friedrich, The mutagenic effect of benzene, toluene, and xylene studied by the SCE technique, Mutat. Res. 58:313-316 (1978).

6. H. Lebowitz, D. Brusick, D. Matheson, M. Reed, S. Goode, and G. Roy, The genetic activity of benzene in various short-term in vitro and in vivo assays for mutagenicity, Seventh Annual Meeting of the Environmental Mutagen Society, San Francisco, Abstract (1976).

7. V. Ray, Are benzene effects limited to the chromosomal level?, in, Banbury Report 1, "Assessing Chemical Mutagens: The Risks to Humans", V. K. McElheny, ed., Cold Spring Harbor Laboratory, Cold Spring Harbor, N.Y., 201-206 (1979).

8. C. A. Snyder, B. D. Goldstein, A. R. Sellakumar, I. Bromberg, S. Laskin, and R. E. Albert, The inhalation toxicology of benzene: incidence of hematopoietic neoplasms and hemato-toxicity in AKR/J and C57Bl/6J mice, Toxicol. Appl. Pharmacol. 54:323-331 (1980).

9. C. Maltoni and C. Scarnato, First experimental demonstration of the carcinogenic effects of benzene: Long-term bioassays on Sprague-Dawley rats by oral administration, Med. Lavoro 5:352-357 (1979).

10. W. C. Dewey, H. H. Miller, and D. B. Leeper, Chromosomal aberrations and mortality of x-irradiated mammalian cells: emphasis on repair, Proc. Natl. Acad. Sci. USA 68:667-671 (1971).

11. T. Tsutsu, M. Umeda, H. Maizumi, and M. Saito, Comparison of mutagenicity, inducibility of single-strand breaks and chromosome aberrations in cultured mouse cells by potent mutagens, Gann 68:609-617 (1977).

12. P. Philip and M. K. Jensen, Benzene induced chromosome abnormalities in rat bone marrow cells, Acta Pathol. Microbiol. Scand. 78(A):489-490 (1970).

13. A. Forni, E. Pacifico, and A. Limonta, Chromosome studies in workers exposed to benzene or toluene or both, Arch. Environ. Health 22:373-378 (1971).

14. G. Hartwich and G. Schwanitz, Chromosomenuntersuchungen nach chronischer Benzol-exposition, Dtsh. Med. Wschr. 97:45-49 (1972).

15. I. M. Tough, P. G. Smith, W. M. Court-Brown, and D. G. Harnden, Chromosome studies on workers exposed to atmospheric benzene: The possible influence of age, Europ. J. Cancer 6:49-55 (1970).

16. M. Hite, M. Pecharo, I. Smith, and S. Thornton, The effect of benzene in the micronucleus test, Mutat. Res. 77:149-155 (1980).

17. J. Meyne and M. S. Legator, Sex-related differences in cyto-genetic effects of benzene in the bone marrow of Swiss mice, Environ. Mutagenesis 2:43-50 (1980).

18. R. R. Tice, D. L. Costa, and R. T. Drew, Cytogenetic effects of
     inhaled benzene in murine bone marrow: Induction of sister
     chromatid exchanges, chromosomal aberrations and cellular
     proliferation inhibition in DBA/2 mice, Proc. Natl. Acad.
     Sci. USA 77:2148-2152 (1980).
19. K. Morimoto and S. Wolff, Increase of sister chromatid exchanges
     and perturbations of cell division kinetics in human lympho-
     cytes by benzene metabolites, Cancer Res. 40:1189-1193
     (1980).
20. S. A. Latt, Microfluorimetric detection of deoxyribonucleic
     acid replication in human metaphase chromosomes, Proc.
     Natl. Acad. Sci. USA 70:3395-3399 (1973).
21. E. L. Schneider, R. R. Tice, and D. Kram, Bromodeoxyuridine
     differential chromatid staining techniques: A new approach
     to examining sister chromatid exchange and cell replication
     kinetics, in, "Methods in Cell Biology", D. M. Prescott, ed.,
     Academic Press, New York, NY 20:379-409 (1978).
22. H. Kato and H. Shimada, Sister chromatid exchanges induced by
     Mitomycin C: A new method of detecting DNA damage at the
     chromosomal level, Mutat. Res. 28:459-464 (1975).
23. P. Perry and H. J. Evans, Cytological detection of mutagen-
     carcinogen exposure by sister chromatid exchanges, Nature
     258:121-125 (1975).
24. T. Nakanishi and E. L. Schneider, In vivo sister chromatid
     exchange: A sensitive measure of DNA damage, Mutat. Res.
     60:329-337 (1979).
25. A. V. Carrano, L. H. Thompson, P. A. Lindhl, and J. L. Minkler,
     Sister chromatid exchange as an indicator of mutagenesis,
     Nature 271:551-553 (1978).
26. S. Laskin, M. Kuschner, and R. T. Drew, Studies in pulmonary
     carcinogenesis, in, "Inhalation Carcinogenesis", M. G. Hanna,
     P. Nettesheim, and J. R. Gilbert, eds., U.S.A.E.C. Symposium
     Series 18, Springfield, CONF-691001, CFSTI, NBS, U. S. Dept.
     of Commerce, 321-350 (1970).
27. R. R. Tice, M. A. Bender, J. L. Ivett, and R. T. Drew, Cyto-
     genetic effects of inhaled ozone, Mutat. Res. 58:293-304
     (1978).
28. K. Goto, T. Akamatsa, H. Shimazu, and T. Sugiyama, Simple
     differential Giemsa staining of sister chromatids after
     treatment with photosensitive dyes and exposure to light
     and the mechanism of staining, Chromosoma 53:223-230 (1975).
29. A. H. Conney, R. Welch, R. Kuntzman, R. Chang, M. Jacobson,
     M. Finster, and J. A. Wolff, Effect of environmental chemi-
     cals on the metabolism of drugs, carcinogens, and normal
     body constituents in man, Ann. N. Y. Acad. Sci. 179:155-172
     (1971).
30. D. Sammett, E. W. Lee, J. J. Kocsis, and R. Snyder, Partial
     hepatectomy reduces both metabolism and toxicity of benzene,
     J. Toxicol. Environ. Health 5:785-792 (1979).

31. R. R. Tice, T. F. Vogt, and J. L. Ivett, Cytogenetic effects of benzene in murine bone marrow. II. The effect of prior partial hepatectomy or co-administered toluene, submitted for publication.

32. L. S. Andrews, E. W. Lee, C. M. Witmer, J. J. Kocsis, and R. Snyder, Effects of toluene on the metabolism, disposition, and hemopoietic toxicity of [^{3}H] benzene, Biochem. Pharmacol. 26:293-300 (1977).

33. A. Sato and T. Nakajima, Dose-dependent metabolic interaction between benzene and toluene in vivo and in vitro, Toxicol. Appl. Pharmacol. 48:249-256 (1979).

34. R. R. Tice, T. F. Vogt, and D. L. Costa, Cytogenetic effects of benzene in murine bone marrow. III. Modulating effect of sex and genetic constitution, submitted for publication.

35. D. W. Nebert, S. A. Atlas, T. M. Guenthner, and R. E. Kouri, The Ah locus: Genetic regulation of the enzymes which metabolize polycyclic hydrocarbons and the risk for cancer, in, "Polycyclic Hydrocarbons and Cancer", 2nd ed., H. V. Gelboin and P. O. P. T'so, eds., Academic Press, New York, NY, 346-389 (1978).

36. D. W. Nebert and N. M. Jensen, The Ah locus: Genetic regulation of the metabolism of carcinogens, drugs, and other environmental chemicals by cytochrome P-450 mediated monooxygenases, CRC Critical Rev. Biochemistry 6:401-437 (1979).

37. B. Paigen, H. L. Gurtoo, J. Minowada, E. Ward, L. Houten, K. Paigen, A. Reilly, and R. Vincent, Genetics of aryl hydrocarbon hydroxylase in the human population and its relationship to lung cancer, in, "Polycyclic Hydrocarbons and Cancer", 2nd ed., H. V. Gelboin and P. O. P. T'so, eds., Academic Press, New York, NY, 391-406 (1978).

38. R. R. Tice, unpublished data.

39. D. E. Rickert, T. S. Baker, J. S. Bus, C. S. Barrow, and R. D. Irons, Benzene disposition in the rat after exposure by inhalation, Toxicol. Appl. Pharmacol. 49:417-423 (1979).

40. D. L. Costa and R. R. Tice, unpublished data.

DISCUSSION

Q. Ma (Western Illinois University): I wasn't sure whether the 3,000 ppm benzene exposure was in one application into an enclosed chamber or as a continuous flow?

A. Tice (BNL): It was a continuous flow measured throughout the four-hour period and represents an integrated dose.

Q. Sivak (Arthur D. Little): Would you put a couple of thousand ppm of benzene in context with work-place and environmental concentrations.

A. Tice (BNL): Historically, typical exposure levels were in excess of a thousand ppm in certain factories in Italy and in the U.S.
In a Science article published a few years ago, benzene levels were
measured in a garage during a normal paint stripping operation and
the levels were on the order of a couple of hundred parts per million. The current OSHA standards set occupational exposure levels
at 10 ppm for an eight-hour day. We observed a significant increase
in SCEs after a four-hour exposure to 28 parts per million in the
10 to 14 week-old mice, an exposure level not greatly different from
the current OSHA "safe" limits.

Q. Drew (BNL): Along these same lines the Hematology Group at
Brookhaven has been looking at stem cells viability in bone marrow
and we see a reduction in the number of stem cells after nine daily
exposures to 25 ppm.

Q. Snyder (Thomas Jefferson): Ray, one question about the relationship between the benzene by IP and by inhalation. In recent years,
one concept in the area of metabolic activation and induction of
reactive intermediates is that toxicity seems to occur when you
exceed detoxication mechanisms. This would suggest that the kinetics
of the various pathways change as the dose increases. I wonder if
at the lower end of your dose response curves the kinetics aren't
quite different than what they are at the higher end?

A. Tice (BNL): I also feel sure that the spectrum and ratios of
various metabolites of benzene alters as a function of dose. From
the shape of the dose response curve and for one end point--SCEs--
I get a much greater induction of SCEs at the low end of the concentration scale than I would expect from the linear regression analysis. I might interpret that to mean that the responsible reactive
intermediate occurs more frequently at low doses of benzene than at
higher doses. At higher concentrations of benzene the metabolite(s)
capable of inducing chromosomal aberrations or of inhibiting cell
proliferation kinetics may be more prevalent.

Q. Wieland (BNL): Ray, do you have any hypothesis on what may
account for the difference between sexes or between young and
old animals? For example, do you think it might be a manifestation
of a change in DNA repair capacity?

A. Tice (BNL): DNA repair capacity is said to change as a function
of age, although to what extent is still questionable. I expect
that the differences are due more to changes in metabolism. There's
a fair amount of data which shows that as animal age, the capacity
to metabolize alters. I don't have a good answer for the difference
between sexes. Hopefully, we will be able to examine the sex-depen-
dent response by neonatal imprinting. I also don't know of any
good benzene data in humans which suggests that there is a sex-
related difference in metabolism or, for that matter, an age-related

difference. Perhaps, like with a lot of other things, it just has not been looked at closely enough. Bob, do you have any comment.

Q. Snyder (Thomas Jefferson): The only interesting note in that regard is that in the early benzene literature there is a good deal of emphasis on the greater susceptiblity of women than men. However, there is no good evidence now to indicate any difference of suscepti- bility between males and females.

Q. Maltoni (Italy): There are experimental data suggesting that female rats are more susceptible than males to benzene's tumorigenic effects. Zymbal Gland carcinoma, which is the most common tumor in benzene-treated animals occurs only in females and not in males.

BENZENE LEUKEMOGENESIS

Bernard D. Goldstein and Carroll A. Snyder*

Department of Environmental & Community Medicine
College of Medicine & Dentistry of New Jersey
Rutgers Medical School
Piscataway, NJ   08854

The regulation of benzene has become a highly controversial subject. Attempts to alter the occupational standard for benzene from 10 ppm to a 1 ppm Time-Weighted Average (TWA) by OSHA have resulted in major litigation which, despite a recent Supreme Court decision, is still not complete. In this paper we plan: (a) to review the evidence justifying the assignment of a causal relationship between benzene and acute myelogenous leukemia in man, (b) to describe the uncertainties concerning the quantitative aspects of this relationship, and (c) to suggest research approaches necessary to resolve this uncertainty, thereby providing an answer to the controversy concerning appropriate regulation of benzene. In addition we will summarize some of our own efforts involving the inhalation toxicology of benzene, including brief descriptions of studies which we believe are pertinent to the question of benzene leukemogenesis.

There no longer should be any reasonable doubt that benzene causes leukemia in man (1). The evidence comes from three different approaches: accumulated individual case reports; epidemiological studies including estimations of relative risk; and basic bio-medical knowledge which supports the inherent plausibility of this causal relationship. Before discussing benzene leukemogenesis it would be useful to review the non-leukemic hematological effects of benzene. The chronic effects of benzene are primarily related to bone marrow toxicity (1,2). This is observed in animals and in man, and when severe is manifested as a marked decrease in bone

*Dept. of Environmental Medicine, New York University Medical Center. New York, NY   10016

marrow precursors and in circulating blood cells. This is usually
termed aplastic anemia, which itself carries a high mortality rate.
Lesser degrees of bone marrow toxicity are observed, ranging from
mild decreases of only one of the formed elements in the blood,
i.e. red blood cells, white blood cells, platelets, to pancytopenia
in which there is a combined anemia, leukopenia, and thrombocyto-
penia. In laboratory animals the extent of these effects are clearly
proportional to dose. In workplaces with significant benzene exposure, a wide range of hematological toxicity has often been observed,
perhaps reflecting differences in point exposure, but also presumably
reflecting host variation. Factors such as age, sex, familial
predisposition, and body fat content have been suggested to play
a role in host variation (3-7). In addition to quantitative decreases, benzene also produces qualitative abnormalities in circulating blood cells. Such abnormalities include macrocytic red blood
cells, shortening of the red cell life span, alterations in porphyrin
pathway compounds in red cells and urine, abnormal morphology and
function of granulocytes including a decrease in phagocytic ability,
a lowered leukocyte alkaline phosphatase activity, and changes in
platelet morphology and function (8-17). In our own studies we have
observed that prolonged red cell glycerol hemolysis time occurs
within a few weeks of exposure of mice to 100 ppm or 300 ppm benzene,
six hours daily, 5 days weekly (18). This observation is also
consistent with an alteration in red cell membrances as a manifestation of benzene hematotoxicity.

One can consider that the ultimate qualitative change for a
cell is development of the autonomy that characterizes cancer.
Current concepts of carcinogenesis stress the central role of the
chromosome. Demonstration of chromosomal damage suggests a high
likelihood, although not incontrovertible proof, that an agent is
capable of causing cancer. There is ample evidence that benzene
exposure results in chromosomal abnormalities in animals and man
(see review by Wolman (19)). The importance of such genotoxic
effects is of course of major interest to thie symposium and will
be discussed by Dr. Tice and his colleagues.

Studies of workers exposed to benzene include that of Tough,
et al., who in 1965 observed significant chromosomal damage in
cultured lymphocytes (20). In a subsequent study of workers from
three separate factories along with the appropriate control individuals a significant increase in unstable chromosomal aberrations was
observed in exposed workers from one factory with ambient benzene
concentrations ranging from 25 to 150 ppm. In the other factory
with similar benzene exposure levels, both exposed and control
groups exhibited similar elevated levels of unstable chromosomal
aberrations while in the third factory in which benzene levels were
approximately 12 ppm neither exposed or controlled individuals
exhibited increased levels of chromosomal damage (21). Relatively
extensive studies on benzene exposed workers have been reported

from Italy by Forni et al. (22,23). Among their studies was the observation of a statistically significant increase in chromosomal aberrations in a group of rotogravure workers exposed to benzene alone or to benzene plus toluene. The benzene levels ranged well over 100 ppm. The frequency of chromosomal aberrations in workers exposed to toluene alone did not differ from an age and sex matched control group.

The persistence of chromosomal abnormalities in individuals who have recovered from benzene-induced pancytopenia (22,23) is particularly of interest in view of the many reports of acute myelogenous leukemia many years after cessation of exposure to benzene. Among the more recent studies is a highly controversial report by Picciano who noted an increase in chromosomal aberration rates in the peripheral lymphocytes of 52 workers employed in chemical factory when compared to 44 persons observed for pre-employment examinations (24). The workers reportedly were exposed to benzene levels of less that 10 ppm, i.e. less than the current U.S. standard.

Hematologists have long known that patients with aplastic anemia, from whatever cause, run a higher risk of developing acute myelogenous leukemia than does the general population (25). Thus, in addition to the cytogenetic changes, the known relationship between aplastic anemia and acute myelogenous leukemia has led to the general notion of a benzene causality for the latter disorder. Moreover, many case reports have described individuals with aplastic anemia clearly associated with significant benzene exposure who have been followed through a pre-leukemia phase into frank acute leukemia or erythroleukemia (1), an otherwise rare variant of acute myeloblastic leukemia. This information argues against any for-tuitous occurrence of leukemia in individuals with benzene-induced aplastic leukemia and further suggests some degree of specificity of benzene on hematopoietic precursor cells.

Obviously, carefully controlled epidemiological studies are superior to individual case studies for the assessment of relative risk in man. Unfortunately, a misconception has arisen, particularly among regulatory agencies, that the only evidence at all appropriate for consideration in deciding questions of cancer causality for man is epidemiological studies or the direct determination of carcino-genesis in animals. This simply is not true. Reasonable scientists will not restrict themselves to any narrow source of information in coming to decisions. Rather, all available information obtained by any valid scientific discipline should be considered.

The epidemiological evidence supporting a causal relationship between benzene and acute myeloblastic leukemia is particularly notable for the variety of different approaches which have been used. These approaches include: classic cohort studies evaluating

the mortality rates of a well-defined work force: case-control
approaches in which leukemic patients are questioned concerning
their possible exposure to benzene; and studies in which the identification of a cluster of leukemia cases in a benzene exposed work
group has led to estimation of the increased risk for the exposed
cohort. Among the latter approaches has been that of Vigliani and
Saita (26) who estimated a twenty-fold increase in risk for acute
leukemia among shoe workers exposed to benzene in two Italian cities.
More recently Aksoy and his colleagues in Turkey have described
an epidemic of benzene hematotoxicity temporally related to the
introduction of a benzene-containing glue in the leather industry
in Istanbul (27-30). In their report of 26 cases with acute leukemia
or pre-leukemia observed from 1967 to September 1973 (more cases
have since been reported) they calculated a yearly incidence among
the workers at risk of 13 per 100,000. This risk was significantly
greater than the number of 6 per 100,000 usually expected in Western
countries. Although there has been some controversy associated
with these findings, the calculated risk appears more than likely
to be an underestimate of the relative risk of acute leukemia in that
population since age adjustment of the data was not performed. It
should be noted that levels of benzene measured in the Turkish and
in the Italian workplaces from which these studies are derived were
often well over 100 ppm.

More classical epidemiological studies evaluating the mortality
rates of well-defined occupational exposure groups have also associated benzene exposure with acute leukemia. Such approaches
have the advantage of closely defining the population at risk, but
do suffer problems related to death ascertainment. An extensive
series of studies in the rubber industry by the Occupational Health
Study Group of the University of North Carolina report finding an
increase in deaths due to cancer of the lymphatic and hematopoietic
system at four tire manufacturing plants (31-33). Studies in the
rubber industry by Monson and Nakano also note an increase in leukemia
as compared to other neoplasms (34-35). Evaluation of the cause of
death of 3637 members of the American Chemical Society noted
that the most highly significant increase in tumors occurred in the
lymphatic and hematopoietic system (36). Similarly, in Great Britain,
the highest level of leukemia deaths occurred in the occupational
group described as professional, technical workers and artists (37).

Undoubtedly the most controversial of the epidemiological
studies relating benzene to acute myeloblastic leukemia is that of
Infante et al. (38). In 1977 these NIOSH investigators reported
7 deaths due to leukemia in a cohort of white male rubber industry
workers who died between January 1950 and June 1975 and who were
employed at any time during the period 1940-49. The expected number
of leukemia deaths was 1.38 based on the general population, and 1.48
based upon a control group obtained from white males employed elsewhere in the same state. This increase in leukemia deaths is clearly

statistically significant and provides further evidence of a causal association between benzene exposure and leukemia. Inasmuch as it was believed that benzene levels at these two factories had been within the allowable standard, the findings of Infante et al. played a role in the decision by OSHA to reduce the allowable benzene standards from 10 ppm to 1 ppm (TWA). However, subsequent testimony at the OSHA hearing suggested that, in fact, benzene levels had been well above the allowable standard in some areas of these factories.

More recently Ott et al have evaluated the mortality of 594 individuals occupationally exposed to benzene (39). Three cases of leukemia were noted as compared to 0.8 expected, a finding with a p value of .047. Relatively extensive monitoring data is presented in this study. The data from this study has been used, along with the data from the studies of Infante et al. (38) and of Aksoy (27-30), by the EPA Cancer Assessment Group for their estimation of the leukemogenic risk of benzene. While open to criticism, this is probably the best estimation now available. A major point is their use of a linear model extrapolating back to zero benzene levels. As discussed below, in the absence of any clear evidence distinguishing between this model and a threshold model, it is appropriately prudent to assume that no threshold exists and that one molecule of benzene can result in leukemogenic alteration in hematopoietic stem cell.

Recently, Greene et al. (40) studied cancer mortality among printing plant workers and noted excess deaths from leukemia in those who may have had exposure to benzene. Negative studies have also been reported. A reasonably extensive approach by Thorpe found no increase in leukemia mortality rates in European employees of a major petroleum company. Although the numbers are small, coke oven workers, who are occupationally exposed to benzene as well as other potenital carcinogens, have not been noted to have an increase in leukemia risk (41).

We have been involved in a series of studies evaluating the inhalation toxicology of benzene (42-47). These were begun under the direction of Professor Sidney Laskin and have been performed at the New York University Medical Center Department of Environmental Medicine. Our procedures have been described previously in detail (42-47). Briefly, randomly distributed control and test male animals were exposed in identical 1.6 M^3 stainless steel incubation chambers. Exposure was generally for six hours daily, five days weekly, lifetime. Completed studies included evaluation of Sprague-Dawley rats and AKR mice at both 100 and 300 ppm benzene, and CD1 and C57B1 mice at 300 ppm. Of note in all the studies was the particular sensitivity of lymphocytes to benzene. Lymphocytopenia was consistently the earliest and most dramatic change in the circulating formed elements. Granulocytopenia was not observed. In fact, there was a tendency toward granulocytosis in mice, particularly in the CD1 strain which exhibited myeloproliferative disorders. A mild to

moderate fall in red blood cell count was observed in the mice but
not in the Sprague-Dawley rats. All of the mouse strains were more
sensitive to the cytopenic effects of benzene than were the Sprague-
Dawley rats as evidenced both by the hematological parameters and
by survival (Table 1). Additional findings included a statistically
significant increase in cytogenetic abnormalities in the bone marrow
cells of AKR mice exposed to 300 ppm benzene. However, the differences from the control group were not dramatic and were observed only
after other clearcut evidence of hematotoxicity was present. Pathological findings in the more seriously affected animals included
bone marrow aplasia and the presence of hemosiderin granules in the
spleen. This may reflect some degree of hemolysis particularly as
there was a tendency toward an increased reticulocyte count in AKR
mice exposed to 300 ppm benzene.

AKR mice spontaneously develop a lymphoid leukemia with an
incidence of about 90%. This was not observed to occur earlier or
more frequently in the benzene-exposed group. In fact, at 300 ppm
evidence of lymphoma was observed in only 2% of the exposed animals
as compared to 91% of the controls. This presumably represents
the lymphopenic effects of benzene (45).

In contrast to the spontaneous disorder occuring in AKR mice,
C57Bl mice are known to be susceptible to an inducible lymphoma.
Exposure of C57Bl mice to 300 ppm benzene was found to produce a
statistically significant increase in the incidence of this thymic
lymphoma along with a shorter expression period (45). This is the
first statistically significant evidence of a hematological neoplasm
in animals exposed to benzene. Tumors of Zymbal glands have recently
been reported by Maltoni (48).

We also observed myelogenous leukemia in these studies. Lifetime exposure of 40 CD-1 mice to 300 ppm benzene 6 hours daily, 5
days weekly, resulted in one case of acute myelogenous leukemia,
one case of chronic myelogenous leukemia, and a third animal with
a myeloproliferative syndrome. The chronic myelogenous leukemia
was observed in a mouse exposed for 27 weeks. The mouse developed
a white blood count of $160,000/mm^3$, well over 90% of which were
granulocytes with a marked shift to the left. At autopsy the bone
marrow showed increased prominence of myelopoietic elements with
a shift to the left and there was extensive myelopoietic infiltration of the periosteal area, connective tissue and muscle. Increased
numbers of immature myelopoietic cells were observed in spleen and
liver. Just prior to death in the 29th week of exposure the CD-1
mouse with acute myelogenous leukemia was noted to have 30% blast
cells in a peripheral blood smear. The total white blood count
was normal (aleukemic leukemia). Leukemic infiltration was observed
in bone marrow, liver and spleen. The morphology of this blast
cell leukemia and the pathological findings are distinctly different

Table 1. Summary of findings in rats and mice chronically exposed to benzene*

Strain	Benzene (ppm)	Median Survival (Weeks) Test	Median Survival (Weeks) Control	Significant Findings
Sprague Dawley Rats	100	69	72	Mild lymphocytopenia; one case of chronic myelogenous leukemia.
	300	51	65	Moderate lymphocytopenia.
AKR mice	100	31	33	Moderate lymphocytopenia, mild anemia, no increase in lymphatic leukemia.
	300	11	39	Severe lymphocytopenia, aplastic anemia, neutrophilia, no increase in lymphatic leukemia.
C57BL mice	300	41	75	Moderate lymphocytopenia, moderate anemia, neutrophilia, development of thymic lymphomas.
CD-1 mice	300	23	42	Moderate lymphocytopenia, moderate anemia, neutrophilia; one case of acute myelogenous leukemia; one case of chronic myelogenous leukemia.

*Exposures were 6 hours per day; 5 days per week for lifetime.

from those observed in the spontaneous stem cell leukemia previously
reported in this strain. The CD-1 mouse with an apparent myeloproliferative disorder had a variable degree of leukocytosis in the four
weeks prior to death with counts ranging above $100,000/mm^3$. Just
prior to death the count decreased to $15,000/mm^3$ with 93% granulocytes and a marked shift to the left. At autopsy there was a
marked bone marrow myeloid hyperplasia. Spleen size was slightly
increased and no site of infection was observed.

In addition to the findings in CD-1 mice, a case of chronic
myelogenous leukemia was observed in a Sprague-Dawley rat exposed to
100 ppm benzene. After 35 weeks of exposure the white blood count
began to increase to a count of over $600,000/mm^3$ 14 weeks before
death. The peripheral blood smear revealed more that 90% granulocytes
with a marked shift to the left. At autopsy, the bone marrow was
packed with myeloid elements and there was leukemic infiltration of
muscle, liver, kidney, lymph nodes, and a grossly enlarged spleen.

In this study, the incidence of myeloproliferative disorders
observed in the benzene exposed animals is not statistically different from that observed in the control animals. However, the results
are suggestive since the respective types of leukemia observed have
not been reported previously in normal animals of these strains.
Percy and Jonas (49) did not observe a single instance of acute or
chronic myelogenous leukemia in a study of 1000 consecutive necropsies
of CD-1 mice, nor has myelogenous leukemia been observed in untreated
Sprague-Dawley rats (50). Our own experience with Sprague-Dawley
rats used in lifetime inhalation carcinogenesis studies reveals no
incidence of myelogenous leukemia in a total of 2000 completely
autopsied control animals during the period 1966 to 1979.

In interpreting this low incidence of myelogenous leukemia
certain points must be kept in mind concerning how benzene differs
from the usual chemical compound subjected to toxicological appraisal
in a carcinogenesis assay. First, we already know it is leukemogenic
in man. Therefore, these findings are not to be considered in the
usual sense of deciding whether or not a compound poses a potential
carcinogenic risk to man. Secondly, benzene is not a new compound
which has been subjected to limited animal testing. Rather, despite
over six decades of relatively extensive animal studies, myelogenous
leukemia has never before been reported. It is safe to say that if
it was all that easy to observe, we would not have been the first
to see it. Thirdly, benzene is an inherent component of our chemical
age. It cannot be banned. Epidemiological studies are unlikely
to provide an answer to the current major controversy concerning
whether or not a threshold exists for leukemogenesis in man. Such
an answer may conceivably come out of an animal model. We believe
these findings are the first indication that an animal model of
benzene-induced myeloid leukemia can be developed.

The question of the shape of the benzene leukemogenic dose-response curve is not only an interesting problem in toxicology but one that has important ramifications in terms of the appropriate regulation of benzene in the workplace and general environment. If, in fact, the dose-response curve extrapolates linearly back to zero then clearly the lower the dose the less the likelihood of developing leukemia. In this case, the level of the standard would be based upon practical considerations keeping in mind economic and social costs which are beyond the expertise of the natural scientist. However, if leukemia due to benzene should only arise as a consequence of severe aplastic anemia then obviously regulations designed to prevent aplastic anemia would also prevent benzene-induced leukemia. If this is the case, then the present standard of 10 ppm for occupational exposure is reasonable and there appears to be no need to go to a more stringent standard.

Discussion of the impact of our scientific uncertainties concerning benzene on the regulation of this major industrial chemical inevitably brings up the question of whether our priorities in dealing with these environmental agents are misplaced. In the case of benzene considerable funds have been spent on the regulatory process and only a relatively small amount on the scientific efforts which could resolve this problem. It is not unreasonable to estimate that a scientific answer to the question of the shape of the benzene dose-response curve for leukemia could be obtained for perhaps one-tenth of one percent of the cost of legal fees spent since 1977 on the issue of benzene regulation.

REFERENCES

1. B. D. Goldstein, Hematotoxicity in man, in, "Benzene Toxicity, A Critical Evaluation", S. Laskin and B. Golstein, eds., Toxicol. Environ. Health, Suppl. 2:69-105 (1977).
2. R. Snyder and J. J. Kocsis, Current concepts of chronic benzene toxicity, CRC Critical Rev. Toxicol. 265-288 (1975).
3. M. Aksoy, S. Erdem, G. Eradogan, and G. Dincol, Acute leukaemia in two generations following chronic exposure to benzene, Hum. Hered. 24:70-74 (1974).
4. T. A. Doskin, Effect of age on the reaction to a combination of hydrocarbons, Hygiene and Sanitation 36:379-384 (1971).
5. L. Greenburg, M. R. Mayers, L. Goldwater, and A. R. Smith, Benzene poisoning in rotogravure printing, J. Indust. Hyg. Toxicol. 21:395-420 (1939).
6. T. Ito, Study on the sex difference in benzene poisoning Report 1. On the obstacles in benzene workers, Showa Igakukai Zasshi 22:268-272 (1963).
7. T. B. Mallory, E. A. Gall, and W. J. Brickley, Chronic exposure to benzene (benzol). III. The pathologic results, J. Indust. Hyg. Toxicol. 21:355-377 (1939).

8. M. Aksoy, K. Dincol, T. Akgun, S. Erdem, and G. Dincol,
    Haematological effects of chronic benzene poisoning in 217
    workers, Brit. J. Indust. Med. 28:296-302 (1971).

9. L. J. Goldwater, Disturbances in the blood following exposure
    to benzol, J. Lab. Clin. Med. 26:957-973 (1941).

10. H. Kahn and V. Muzyka, The chronic effect of benzene on prophyrin
    metabolism, Work-Environm. Health 10:140-143 (1973).

11. H. A. Kahn and V. I. Muzyka, The effect of benzene on the
    -aminolevulic acid and porphyrin content in the cerebral
    cortex and in the blood, Indust. Hyg. and Profession
    Assoc. Disorders 3:59-60 (1970).

12. P. Shubik, U. Saffioti, W. Lijinsky, G. Pietra, H. Rappaport,
    B. Toth, C. R. Raha, L. Tomatis, R. Feldman, and H. Ramahi,
    Studies on the toxicity of petroleum waxes, Toxicol. Appl.
    Pharmacol. 4:1-62 (1962).

13. L. A. Erf and C. P. Rhoads, The hematological effects of benzene
    (benzol) poisoning, J. Indust. Hyg. Toxicol. 20:421-435 (1939).

14. T. A. Koslova and A. P. Volkova, Blood picture and phagocytic
    activity of leucocytes in workers having contact with
    benzene, Gig Sanit. 25:29-34 (1960).

15. R. Girard, M. L. Mallein, J. Bertholon, P. Coeur, and F. Tolot,
    Leucocyte alkaline phosphatase and benzene exposure, Med.
    Lavoro 61:502-508 (1970).

16. G. Pollini and R. Colombi, Changes in the osmotic resistance
    of the leucocytes in persons exposed to benzene, Lavora
    Umano 16:177-184 (1964).

17. A. Monterverde, C. Grazioli, and E. Fumagalli, The effect of
    fibrinogen and platelets on the thromboelastogram in benzene
    poisoning, Medici del Lavora 54:95-102 (1967).

18. B. D. Goldstein, M. G. Rozen, and C. A. Snyder, Prolonged red
    cell glycerol hemolysis in mice inhaling benzene, Soc. of
    Toxicol. Conf., March, 1980.

19. S. R. Wolman, Cytologic and cytogenetic effects of benzene,
    Toxicol. Environ. Health Supplement 2:63 (1977).

20. I. M. Tough and W. M. Court-Brown, Chromosome aberrations and
    exposure to ambient benzene, Lancet 1:684 (1965).

21. I. M. Tough, P. G. Smith, W. M. Court-Brown, and D. G. Harnden,
    Chromosome studies on workers exposed to atmospheric benzene,
    Europ. J. Cancer 6:49-55 (1970).

22. A. M. Forni, A. Cappellini, E. Pacifico, and E. C. Vigliani,
    Chromosome changes and their evolution in subjects with
    past exposure to benzene, Arch. Environ. Health 23:385-391
    (1971).

23. A. Forni, D. Pacifico, and A. Limonta, Chromosome studies in
    workers exposed to benzene or toluene or both, Arch.
    Environ. Health 22:373-378 (1971).

24. D. Picciano, Cytogenetic study of workers exposed to benzene,
    Environ. Res. 19:33-37 (1979).

25. W. Damashek, What do aplastic anemia, PNH, and hypoplastic
    leukemia have in common?, Blood 30:251 (1967).

26. E. C. Vigliani and G. Saita, Benzene and leukemia, New England J. Med. 271:872-876 (1964).

27. M. Aksoy, K. Dincol, T. Akgun, S. Erdem, and G. Dincol, Haematological effects of chronic benzene poisoning in 217 workers, Brit. J. Indust. Med. 28:296-302 (1971).

28. M. Aksoy and S. Erdem, Follow-up study on the mortality and the development of leukemia in 44 pancytopenic patients with chronic exposure to benzene, Blood 52:285 (1978).

29. M. Aksoy, K. Dincol, S. Erdem, T. Akgun, and G. Dincol, Details of blood changes in 32 patients with pancytopenia associated with long-term exposure to benzene, Brit. J. Indust. Med. 29:56-64 (1972).

30. M. Aksoy, S. Erdem, and G. Dincol, Leukemia in shoe-workers exposed chronically to benzene, Blood 44:837-841 (1974).

31. D. Andjelkovich, J. Taulbee, M. Symons, and T. Williams, Mortality experience of a cohort of rubber workers, J. Occup. Med. 19:397-405 (1977).

32. A. J. McMichael, D. A. Andjelkovic, and H. A. Tyroler, Cancer mortality among rubber workers: an epidemiological study, Annals N. Y. Acad. Sci. 271:125-137 (1976).

33. A. J. McMichael, R. Spirtas, J. F. Gamble, and P. M. Tousey, Mortality among rubber workers: Relationship to specific jobs, J. Occup. Med. 18:178-185 (1976).

34. R. R. Monson and K. K. Nakano, Mortality among rubber workers. I. White male union employees in Akron, Ohio, Am. J. Epidemiology 103:297-303 (1976).

35. R. R. Monson and K. K. Nakano, Mortality among rubber workers, II. Other employees, Am. J. Epidemiology 103:297-303 (1976).

36. F. P. Li, J. L. Fraumeni, N. Mantel, and R. W. Miller, Cancer mortality among chemists, J. Nat. Cancer Inst. 43:1159-1164 (1969).

37. A. M. Adelstein, Occupational mortality:Cancer, Ann. Occup. Hyg. 15:53-57 (1972).

38. P. F. Infante, R. A. Rinsky, J. K. Wagoner, and R. J. Young, Leukemia in benzene workers, Lancet 2:76-78 (1977).

39. M. G. Ott, et al., Mortality among individuals occupationally exposed to benzene, Arch. Environ. Health 33:3-   (1978).

40. M. H. Greene, R. V. Hoover, R. L. Eck, and J. F. Fraumeni, Jr., Cancer mortality among printing plant workers, Environ. Res. 20:66-73 (1979).

41. C. K. Redmond, B. R. Strobino, and R. H. Cypress, Cancer experience among coke by-product workers, Annals N.Y. Acad. Sci. 271:102-115 (1976).

42. C. A. Snyder, S. Laskin, and B. D. Goldstein, An extractive method for determination of benzene in blood by gas chromatography, Am. Indust. Hyg. Assoc. J. 36:868-   (1975).

43. C. A. Snyder, M. N. Erlichman, B. D. Goldstein, and S. Laskin, An extraction method for determination of benzene in tissue by gas chromatography, Am. Indust. Hyg. Assoc. J. 38:272-276 (1977).

44. C. A. Snyder, B. D. Goldstein, A. Sellakumar, S. R. Wolman,
     I. Bromberg, M. N. Erlichman, and S. Laskin, Hematoxicity
     of inhaled benzene to Sprague-Dawley rats and AKR mice at
     300 RPM, J. Toxicol. Environ. Health 4:605-618 (1978).
45. C. A. Snyder, B. D. Goldstein, A. R. Sellakumar, I. Bromberg,
     S. Laskin, and R. E. Albert, The inhalation toxicology of
     benzene: incidence of hematopoietic neoplasms and hemato-
     toxicity in AKR/J and C57BL/6J mice, Toxicol. Appl.
     Pharmacol. 54:323-331 (1980).
46. C. A. Snyder, M. N. Erlichman, S. Laskin, B. D. Goldstein,
     and R. E. Albert, Pharmakokinetics of repetitive benzene
     exposures at 3000 ppm and 100 ppm in AKR mice and Sprague-
     Dawley rats, Toxicol. Appl. Pharmacol., 57:164-171, 1981.
47. J. P. Green, C. A. Snyder, J. Lo Bue, B. D. Goldstein, and
     R. E. Albert, Acute and chronic dose-response-effects of
     inhaled benzene on multi-potential hematopoietic stem
     (CFV/S) and granulocyte/macrophage progenitor (GM/CFV/C)
     cells in CD/1 mice, Toxicol. Appl. Pharmacol., in press.
48. C. Maltoni and C. Scarnato, First experimental demonstration
     of the carcinogenic effects of benzene: Long-term bioassays
     on Sprague-Dawley rats by oral administration, Med. Lav.
     70:352-357 (1979).
49. D. H. Percy and A. M. Jonas, Incidence of spontaneous tumors
     in CD-1 HaM/ICR mice, J. Natl. Cancer Inst. 46:1045-1046
     (1971).
50. W. F. MacKenzie and F. M. Garner, Comparison of neoplasms in
     six sources of rats, J. Natl. Cancer Inst. 50:1243-1257
     (1973).

DISCUSSION

Q. Infante (U.S. Dept. of Labor): I would like to make one comment
about benzene exposure levels. Obviously no one has the answer about
what level of benzene induces leukemia in humans and whether you need
aplastic anemia to develop leukemia. In one of the epidemiological
studies that has not been mentioned here, Ott and others at Dow
observed a significant excess of leukemias in workers exposed to
benzene levels below a time-weighted averages of 9 ppm. This study
demonstrates that you don't need high exposure levels of benzene
to get leukemia.

A. Goldstein (Rutgers): We have tremendous problems with this kind
of study. Your industrial hygienist are measuring the levels of
benzene in this factory and overall the levels are low. But, how
do you know that the unexpected cases of leukemia are not in workers
who are always making mistakes, turning the valve the wrong way,
or whatever. It would not take too many such individuals before
you would have a situation where leukemias occur at what is supposed
to be a low-exposure situation.

Now, does this mean we should have more stringent standards in order to prevent this situation or that more care should be taken with the standards we already have.

Q. Hatch (Northrop): Could someone comment on the solid tumors produced by ingestion of benzene?

A. Maltoni(Italy): We have found in our experiment with benzene on rats an increased incidence of Zymbal gland carcinoma, which is, to me, the most important finding. The incidence of this tumor in our colony of animals is around one to two per thousand. We have found 8 Zymbal gland carcinoma among 30 females exposed to 250 mg/Kg body weight, five times weekly for one year. When we exposed rats to 550 mg/Kg the tumor incidence decreased from 8 to 2 per 30 animals. For all cancers other than those of the hemo-lympholenticular type, we have also found a doubling of mammary carcinoma plus several cases of tumors rare in our colony--such as hepatoma, two skin carcinomas, and one angiosarcoma. Each cancer, in itself, dosen't mean very much, but taken together, they may be significant. The most striking result is that when you pool all malignancies in the two treated groups versus the control, we have found a straight dose-response relationship. I would also like to say that we are continuing studies with 500 mg/Kg benzene and that we have ongoing a long-term experiment by inhalation of 300 ppm.

Q. Infante (U.S. Dept. of Labor): Excuse me, I just want to make one comment on Dr. Goldsteins' scenario about a few people turning valves or whatever. That may well be the case and it may not be the case. You may, in fact, have hypersusceptilibity of response, because we know that in individuals working alongside each other, some have developed aplastic anemia and some have not. What you said could possibly be the case but the point is, we just don't know.

A. Goldstein (Rutgers): You're absolutely right. I agree with you 100%. We don't know.

GENETIC EFFECTS OF ETHYLENE DIBROMIDE IN DROSOPHILA MELANOGASTER

P. Kale and J. W. Baum

Safety and Environmental Protection Division
Brookhaven National Laboratory
Upton, NY   11973   (U.S.A.)

INTRODUCTION

Drosophila, an organism known to be one of the best available
systems for mutation detection, can also be used for detecting
very low concentrations of airborne mutagens. 1,2-Dibromoethane
also called ethylene dibromide (EDB), is a chemical widely used
in industry, agriculture, and homes. EDB was selected as a test
chemical in these studies since its mutagenic properties had
previously been demonstrated in Drosophila (1) and in Tradescantia
(2). In Tradescantia, extensive data on the dose versus mutation
relation were available and these data were considered useful in
evaluating the comparative sensitivity of the two systems for
detecting airborne mutagens.

MATERIALS AND METHODS

Exposures were performed in a specially designed dynamic
exposure facility developed by others for studies on Tradescantia
(2). EDB was vaporized by bubbling with purified air. The air
containing known concentrations of EDB was circulated through the
exposure chamber. Air employed in both the control and exposure
chamber was purified by heating it over copper oxide at 670°C and
then filtering it through charcoal. Vernier-operated needle valves
were used to control the amount of gaseous EDB entering the main
air stream in the exposure chamber. Flow rate for the EDB air
mixture entering the exposure chamber was 2 L/min. Air pressure
in the chamber was atmospheric. Temperature and humidity in the
exposure and the control chambers were monitored continuously and
held at 24 ± 1°C and about 60 percent relative humidity. EDB

concentration was measured at both inlet and outlet of the chambers using a Varian 3700 gas chromatograph with a flame ionization detector. Readings were taken every two hr except at night during long exposures. The average concentration was then computed. Acute treatments, corresponding to some of the chronic treatments, were made by increasing chemical concentrations approximately 30 times with concommitant decreases in exposure periods (3).

Immediately after treatment, the treated males were mated individually to 6 Basc virgins for one day after which they were separated and again mated individually to a second set of six virgins. These matings continued for 12 days to obtain 12 daily broods in all experiments except one where daily brooding was continued for 19 days. Inseminated females were allowed to lay eggs for six days and then discarded. F_1 progeny from each male were pair-mated separately in each brood. The pair matings were scored from days 12-16 after culturing. Suspected lethals were confirmed by testing for one more generation. Most cultures were maintained at 24 ± 1°C and at 60-65 percent relative humidity.

The F_1 pair matings for each brood were cultured and scored separately for each individual male. Poisson statistics was used to identify the clusters of mutations either of common origin or due to possible prediposition of some treated males to the chemical. A Chi-square test for goodness of fit to the Poisson distribution was made for each brood. These values (not given here) did not indicate clustering in any of the broods in any of the experiments reported herein.

RESULTS

The summed data are given in Table 1 for spermatozoa, spermatids, and spermatocytes. Mutation induction by acute and chronic treatments at the four lower exposures were compaired quantitatively using a Chi-square test. At the three lower exposures, 2 ppm-hr through 18 ppm-hr, the differences in mutation induction by the two types of treatments is not significant for any cell type except for spermatids exposed to 2.10 ppm-hr. At 30 ppm-hr, however, acute exposure produced a significantly higher number of mutations in spermatozoa, spermatids, and spermatocytes. Figures 1, 2 and 3 show plots of percent mutations vs. exposure for the three tested germ cell stages at the lower four exposure levels. The dose response appears to be linear for spermatids and spermatocytes. However, it is not adequately defined for chronic exposures to spermatozoa due to the small response and it has a strong upward curvature (greater effect at higher doses) for acute exposures. Weighted regression statistics for these data are given in Table 2.

Table 1. Induced six-linked recessive lethal mutations in successive daily broods from D. melanogaster males treated with acute and chronic concentrations of ethylene dibromide

DBE conc. ppm	Exposure Time hr	Exposure* ppm-hr	Exposure Type	Spermatozoa Broods 1 to 2			Spermatids Broods 3 to 6			Spermatocytes Broods 7 to 12		
2	100	200	Chronic	L 15	T 2547	% 0.59	2	780	0.25	–	–	–
2	75	150	Chronic	L 15	T 2736	% 0.55	29	974	3.0	–	–	–
1	60	60	Chronic	L 13	T 4863	% 0.2	41	1138	3.6	51	1638	3.1
1	30	30	Chronic	L 6	T 4290	% 0.14	53	5189	1.0	63	3414	1.8
31.1	1.2	31	Acute	L 12	T 2300	% 0.52	78	3347	2.3	111	3593	3.1
0.25	72	18	Chronic	L 5	T 2632	% 0.19	49	3788	1.3	22	1564	1.4
0.40	2.6	18	Acute	L 1	T 2090	% 0.04	42	4317	1.0	72	3684	2.0

Table 1 (continued)

DBE conc. ppm	Exposure Time hr	Exposure* ppm-hr	Exposure Type		Spermatozoa Broods 1 to 2	Spermatids Broods 3 to 6	Spermatocytes Broods 7 to 12
0.21	36	8	Chronic	L	3	14	14
				T	2439	3589	1439
				%	0.12	0.40	0.99
0.50	1.5	6.8	Acute	L	0	15	37
				T	2314	4746	6304
				%	0.0	0.31	0.14
0.21	11	2.3	Chronic	L	0	10	5
				T	982	2956	3693
				%	0.0	0.34	0.14
0.70	0.58	2.1	Acute	L	1	3	10
				T	2357	4770	6914
				%	0.04	0.06	0.14

Total Controls: 33 lethals per 49,467 chromosomes tested: 0.067%
*Due to the exponential approach to equilibrium in the exposure chambers the integrated exposure is not equal to the product of column one and two for short exposure time.

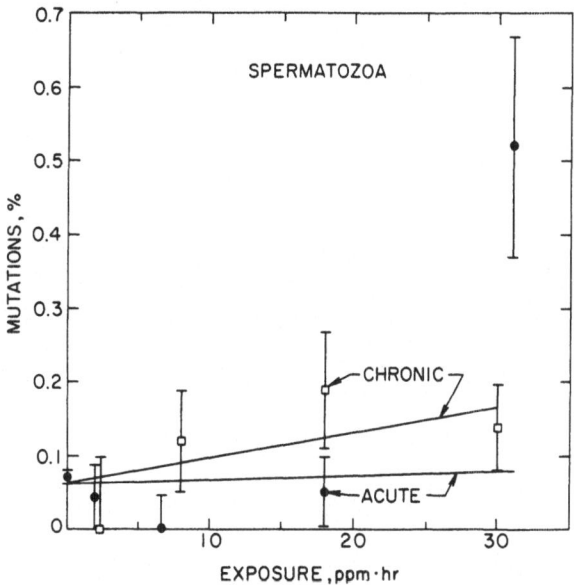

Figure 1. Dose response for recessive lethal mutations in Drosophila melanogaster spermatozoa (broods 1 and 2) exposed to gaseous 1,2-dibromoethane.

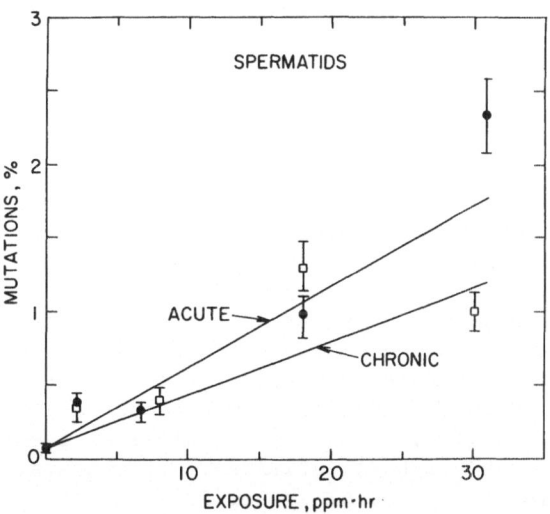

Figure 2. Dose response for recessive lethal mutations in Drosophila melanogaster spermatids (broods 1 and 2) exposed to gaseous 1,2-dibromoethane.

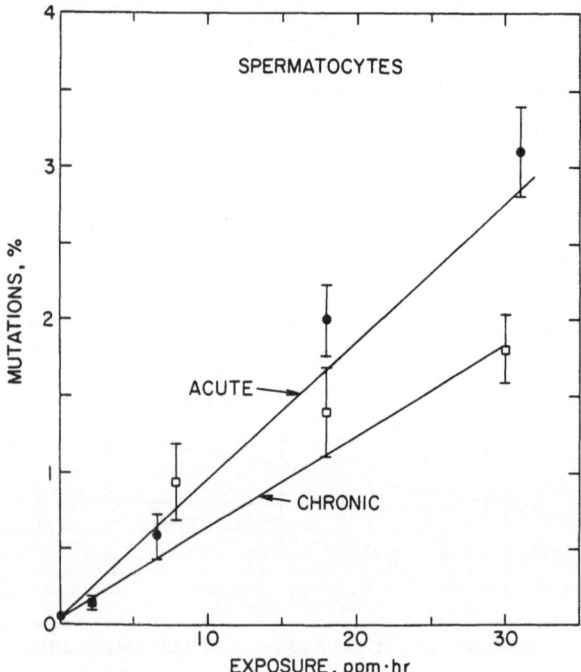

Figure 3. Dose response for recessive lethal mutations in Drosophila
melanogaster spermatocytes (broods 1 and 2) exposed to gaseous 1,2-
dibromoethane.

DISCUSSION

     Our experiments indicate that EDB exposures as low as 0.21 ppm
are detectable (4). This demonstrates the high sensitivity of
Drosophila for detection of gaseous mutagens. At present, low
concentrations of gaseous mutagens in ambient air are being studied
utilizing a sensitive clone of Tradescantia. This clone has been
exposed to gaseous EDB in the same facility used in the above
studies at an exposure of 18 ppm-hr. In Tradescantia this exposure
induced twice the spontaneous mutation rate. In Drosophila, the 18
ppm-hr exposure induced mutation rates 1-2 times the spontaeous
rate in sperm but about 17 times the spontaeous rate in spermatids
and in spermatocytes (10-12 times if the slope of the least-squares
regression fit is used as best estimate). It appears, therefore,
that Drosophila is also a useful system for monitoring polluted
air for its mutagenicity.

     Important factors which must be considered in mutagen testing
and the wider problem of quantitative assessment of genetic damage
are the shape of the dose-response curve and the changes in response

Table 2. Weighted linear regression fits to sex-linked recessive lethal mutations induced by chronic and acute exposures to gaseous 1,2-dibromoethane from 2 to 31 ppm-hr

	Spermatozoa		Spermatids		Spermatocytes		Total	
	acute	chronic	acute	chronic	acute	chronic	acute	chronic
Intercept %	0.060 ±0.011	0.063 ±0.003	0.065 ±0.012	0.070 ±0.012	0.059 ±0.012	0.066 ±0.012	0.050 ±0.012	0.072 ±0.012
Slope %ppm-hr	0.0006 ±0.002	0.0031 ±0.0017	0.057 ±0.0053	0.039 ±0.0041	0.092 ±0.0069	0.061 ±0.0066	0.060 ±0.0035	0.034 ±0.0025
Coefficient of determination	0.08	0.69	0.96	0.93	0.97	0.98	0.90	0.97

as a function of dose rate.   This concern has been very well demonstrated in Neurospora.   In an adenine-requiring strain of Neurospora,
(diepoxybutane) (DEB) – induced reversions increased linearly after
protracted treatment with low concentrations, increased very steeply
after short treatment with a high concentration and an intermediate
response was observed after fractionated treatment with high concentrations (5,6).   Our data compare favorably with these results.
The slopes obtained from curves for acute treatments are significantly higher for spermatids and spermatocytes than those obtained
from chronic exposures.   According to Auerbach (7), the dose rate
effect observed in Neurospora is due to a repair mechanism which
is inhibited by DEB, but able to recover during treatment at low
dose rate.   A similar interpretation has also been given by Kilby
(6) who found conditions which could convert the linear dose-response
relation obtained with DEB at low dose rates into the upward bending
relationship characteristic at higher dose rates.   Thus, these
dose-rate effects appear to result from the inhibition of a repair
process in metabolically active cells.

    In Drosophila, spermatids and spermatocytes are metabolically
active, but spermatozoa are not.   In radiation mutagenesis these
metabolically inactive cells do not show a dose-rate effect.   With
EDB we have observed a greater number of mutations induced by
high concentration acute treatments in all three cell types, including spermatozoa where a dose-rate effect was not expected.
However, at lower exposure levels, there was no significant difference in mutation production in any of three cell types by acute
compared to chronic exposures.   We, therefore, feel that the
greater number of mutations at higher acute doses is not due to
inhibition of a DNA repair process in the target cells but, rather,
the result of greater systemic deactivation or elimination of the
mutagen at low dose rates.

    For other, higher animals, few data are available on gene
mutation induction comparing chronic and acute exposure.   Using
dominant lethal and cytogenetic endpoints, Sram (7) reported a
dose-rate effect for dominant lethal induction but not for cytogenetic endpoints in mice chronically and acutely exposed to sodium
arsenite.

    Lack of a dose-rate effect, especially at lower exposures, has
an important bearing on risk estimation for chemical mutagens.   In
the absence of information on dose-rate effects by chemicals,
Committee 17 (8), appointed by the Council of the Environmental
Mutagen Society, relied on radiation dose-rate effect information
in assuming that similar dose-rate effects might play a role in
chemical mutagenesis.   Committee 17 states that "In making comparisons between chemical and radiation mutagenesis, we have frequently
corrected for a radiation dose-rate effect: since acute doses are
commonly administered in testing procedures, whereas chronic or

low-dose irradiation produces only about one-third as many mutations
as dose acute irradiation, we have assumed that the same holds for
environmental chemical mutagenesis" (see 8, page 61). Our results
and those of Sram (7) on gene mutations demonstrating the lack of
a dose-rate effect at lower doses, suggest that the Committee's
assumption of the one-third as many mutations for chronic versus
acute exposures may be inappropriate and could underestimate the
mutation frequency at low dose rates. For example at less than
18 ppm hr for EDB, we found no detectable difference between chronic
and acute response even though the responses were large (about 10
times the spontaeous rates).

Thus, the factor of one-third should only be used when it is
clear that the acute chemical exposure is in the "high" dose
region. To be sure of this, comparisons between chronic and acute
exposures are necessary. However, it would be better to employ
the actual factor deduced from such comparisons rather than using
a factor inferred from radiation studies which may result from
important mechanistic differences in mutation induction. Since
metabolic factors could work either as inactivating mechanisms
(as suggested for EDB) or as activating mechanisms, the dose-rate
factor (chronic/acute) may even be greater than one for some substances. Therefore, since exposure to environmental chemicals
are often at low chronic levels, chronic testing protocols for
purposes of quantitative risk evaluations should be included.

CONCLUSIONS

Using EDB we have been able to demonstrate that Drosophila
can detect concentrations as low as 0.2 ppm in air when exposed
for a relatively short period of 11 hours. In this bioassay system
the exposure period can be extended up to 700 hours. At low exposure levels, mutagenic response was found to be proportional to
the integrated exposure in ppm hr. Also it is possible to detect
airborne mutagens in the parts per billion range using proper germ
cell stages in this system. Drosophila may be extremely useful as
a biological monitor to detect airborne genotoxic agents in polluted
areas. Thus, the system complements Tradescantia (9), another
sensitive system being used for this purpose.

ACKNOWLEDGEMENTS

We highly appreicate the laboratory assistance of Ann Porteous,
Gloria Emanovsky, and Leora Ben-Ami during the course of this work.
EDB exposures were made by Paul Klotz. Alan Keuhner provided
computer software for doing the weighted linear regression analyses.
Research carried out under Contract EY-76-C-02-0016 with the U.S.
Department of Energy.

REFERENCES

1. E. Vogel and J. L. R. Chandler, Mutagenicity testing of cycla-
    mate and some pesticides in Drosophila melanogaster,
    Experientia 30:621-623 (1974).
2. A. H. Sparrow, L. A. Schairer, and R. Villalobos-Peitrini,
    Comparison of somatic mutation rates induced in Tradescantia
    by chemical and physical mutagens, Mutat. Res. 26:265-276
    (1974).
3. P. G. Kale and J. W. Baum, Sensitivity of Drosophila
    melanogaster to low concentrations of gaseous mutagens. III.
    Dose-rate effects, Environ. Mutatgenesis, in press (1981).
4. P. G. Kale and J. W. Baum, Sensitivity of Drosophila
    melanogaster to low concentrations of gaseous mutagens. II.
    Chronic exposures, Mutat. Res. 68:59-69 (1979).
5. C. Auerbach and D. Ramsy, Analysis of a case of mutagen specifi-
    city in Neurospora crassa. III. Fractionated treatment
    with Diepoxybutane (DEB), Mol. Gen. Genet. 109:285-291 (1970).
6. B. J. Kilbey, The manipulation of mutation-induced kinetics in
    Neurospora crassa, Molec. Gen. Genet. 123:73-76 (1973).
7. R. J. Sram, Relationship between acute and chronic exposures
    in mutagenicity studies in mice, Mutat. Res. 41:25-42 (1976).
8. Committee 17, Environmental mutagenic hazards, Science 187:
    503-514 (1975).
9. L. A. Schairer, J. van't Hof, C. G. Hayes, R. M. Burton, and
    F. J. de Serres, Exploratory monitoring of air pollutants
    for mutagenicity activity with the Tradescantia stamen hair
    system, Environ. Health Perspect. 27:51-60 (1978).

MUTAGENIC AND ONCOGENIC EFFECTS OF CHLOROMETHANES, CHLOROETHANES

AND HALOGENATED ANALOGUES OF VINYL CHLORIDE*

Peter F. Infante and Theodora A. Tsongas

Office of Carcinogen Identification and Classification
Health Standards Program,
Occupational Safety and Health Administration
Department of Labor
200 Constitution Avenue, N. W.
Washington, D. C. 20210 (U.S.A.)

INTRODUCTION

In the early 1970's, information bearing on the mutagenicity and carcinogenicity if vinyl chloride (VC) in experimental test systems and in humans came to the attention of the public health community. As a result, emphasis was placed on the evaluation of potential geno-toxic and carcinogenic effects of a number of halogenated hydrocarbons structurally related to VC. This response was of particular concern as many of these substances are in high volume production and have a large number of individuals exposed in the occupational setting. As a result of pollution from industrial effluents, many of these sub-stances have been identified in community water supplies. Some of these substances also have been or still are contained in consumer products.

Because of the magnitude of potential adverse health effects from exposure to such chemicals and the potential toll on individuals in terms of physical, psychological and economic hardships, government, industry and labor must focus on the identification and control of those substances which pose potential hazards to human health.

*The views expressed do not necessarily represent those of the Occu-
 pational Safety and Health Administration.

The purpose of this presentation is to briefly indicate some of the mutagenic and carcinogenic effects of chloromethanes, chloroethanes, some chlorinated olefins and their brominated analogues in experimental test systems. An exhaustive literature search was not conducted for mutagenic effcts, and negative carcinogenesis bioassays are not presented unless they are part of a study where positive results were observed. However, all available epidemiologic study results of workers with potential exposure to these substances were reviewed.

DISCUSSION

Table 1 indicates data for 18 halogenated hydrocarbons by annual U.S. production, estimated number of workers exposed and OSHA permissible exposure levels (PEL) for exposure expressed as 8-hour time-weighted-averages (TWA) in parts per million (ppm). Annual U.S. production in millions of pounds is based on information for the year 1979 as obtained from the International Trade Commission. The estimated numbers of workers exposed are based on the National Occupational Health Survey (NOHS) Hazard Summary Report, Interim Data for 08/06/80, compiled by the National Institute for Occupational Safety and Health (NIOSH).

As indicated in Table 1, there is no limit to the level of occupational exposure for four of these substances. Available information indicates most of these substances are in high volume production, and that estimated numbers of workers potentially exposed range from 900 for 1,1-dichloroethane to more than 2 million for 1,1,1-trichloroethane.

Table 2 indicates the major uses and the major occupations with potential for exposure to the substances shown in Table 1.

Mutagenicity of Chloromethanes

The evidence presently available on the mutagenicity of the chloromethanes is summarized in Table 3. The systems that have been tested, the results of those tests, and literature citations are presented for each substance. The literature cited is not a complete listing but is intended to be representative of the data available on each substance. The table does not include information on the effects that chloromethanes may have on the mutagenic potential of other substances.

When assayed in a dessicator system, chloromethane is highly mutagenic in S. typhimurium test strain TA1535, both with and without rat liver homogenate (1), and in strain TA100 with and without S-9 mix (2,3). These data indicate that chloromethane induces base pair substitutions in DNA of these tester strains. However, negative

Table 1.   Estimates of U.S. annual production, number of workers
           exposed and OSHA standards for 18 halogenated hydrocarbons

Substance	Annual U.S. Production in Millions of Pounds (1979)	Estimated No. of Workers Exposed	OSHA PEL (TWA in ppm)
Chloromethanes:			
Chloromethane	463	40,500	100
Dichloromethane	633	2,175,500	500
Chloroform	356	215,000	50
Carbon Tetrachloride	714	1,380,000	10
Chloroethanes:			
Chloroethane	582	142,000	1000
1,1-dichloroethane	NUS	900	100
1,2-dichloroethane (EDC)	11,791	33,000– 2,000,000	50
1,1,1-trichloroethane	716	2,610,000	350
1,1,2-trichloroethane	37	72,000	10
1,1,1,2-tetrachloroethane	NUS	ND	None
1,1,2,2-tetrachloroethane	ND	7,000	5
Pentachloroethane	NUS	ND	None
Hexachloroethane	NUS	1,500	1
Chlorinated Olefins:			
1,1-dichloroethene (VDC)	200	6,500– 58,000	None
1,1,2-trichloroethene (TCE)	301	282,000	100
Tetrachloroethylene (PCE)	773	500,000	100
Brominated Analogues:			
Bromoethene (VB)	5*	360– 26,000	None
1,2-dibromoethane (EDB)	330	9,000– 660,000	20

* = Undisclosed; ND = No data available; NUS = No known U.S.
Production; TWA = 8-hour time weighted average; PEL = Permissible
exposure level; ppm = Parts per million

Table 2.   Major industrial uses and potential occupations
with exposure to 18 carcinogenic substances

Substances	Major Uses	Major Occupational Exposure Industries
Chloromethane	Catalyst carrier in low-temperature polymerization, tetramethyl lead, silicones; refrigerant; medicines; fluid for thermometric and thermostatic equipment; methylating agent; extractant in high pressure aerosols; herbicides.	Aerosol packagers; drug makers; flavor extractors; low temperature solvent workers; methylation workers; methyl cellulose makers, polystyrene foam makers; refrigeration workers; rubber makers; vapor pressure thermometer makers.
Dichloromethane	Paint removers; special photographic film; fumigant; degreasing solvent; solvent mixtures for cellulose esters and ethers; textile and leather coatings; refrigeration; local anesthetic; pharmaceutical and food extraction; plastic processing; spotting agent; dewaxing; organic synthesis; propellant for aerosols; blowing agent in foams.	Aerosol packagers; anesthetic makers; bitumen makers; degreasers; fat extractors; flavoring makers; leather finish workers; oil processors; paint remover makers; resin makers; solvent workers; stain removers.
Chloroform	Fluorocarbon refrigerants and propellants; fluorocarbon plastics; dyes and drugs; general solvent; analytical chemistry; fumigant; insecticides.	Chemists; drug makers; fluorocarbon makers; lacquer workers; polish makers; silk synthesizers; solvent workers.
Carbon tetrachloride	Refrigerants and propellants, especially chlorofluorohydrocarbons; metal degreasing; agricultural fumigant; chlorinating organic compounds; production of semiconductors.	Dry cleaners; degreasers; fat processors; firemen; fluorocarbon makers; grain fumigators; ink makers; insecticide makers;

Table 2. (continued)

Substance	Major Uses	Major Occupational Exposure Industries
Carbon tetrachloride (cont.)		lacquer makers; metal cleaners; propellant makers; refrigerant makers; rubber makers; solvent workers; wax makers.
Chloroethane	Ethylating agent in the manufacture of tetraethyl lead, dyes, drugs and ethyl cellulose; refrigerant; local anesthetic.	Anesthetists; dentists; drug makers; ethylation workers; fat and oil processors; nurses; perfume makers; phosphorus and sulfur processors; refrigeration workers; resin makers; tetraethyl lead makers; wax makers; physicians.
1,1-dichloroethane	Medicines; extraction solvent; fumigant; in stone, clay and glass products.	Custodians; clerical workers; electricians; assemblers; agricultural and biological technicians.
1,2-dichloroethane (EDC)	Chemical manufacturing; fumigant; fuel additive; organic solvent; manufacture of chlorinated chemicals; manufacture of tetraethyl lead; paint and varnish removers.	Adhesive makers; bakelite processors; camphor workers; drycleaners; exterminators; gasoline blenders; insecticide makers; metal degreasers; ore upgraders; solvent workers; textile cleaners; vinyl chloride makers.

Table 2. (continued)

Substance	Major Uses	Major Occupational Exposure Industries
1,1,1-trichloro-ethane	Solvent for cleaning precision instruments; aerosol propellant; metal degreasing; pesticide.	Degreasers; dry-cleaners; machinery cleaners; metal degreasers; propellant makers; stain removers.
1,1,2-trichloro-ethane	Solvent for fats, oils, waxes, resins; other products; organic synthesis; manufacture of 1,1-dichloroethylene.	Organic chemical synthesizers; solvent makers.
1,1,1,2-tetra-chloroethane	Solvent, general manufacturing.	No data available.
1,1,2,2-tetra-chloroethane	Solvent, cleansing and degreasing metals; paint removers, varnishes, lacquers; photographic film; resins and waxes; extraction of oils and fats; alcohol denaturant; organic synthesis; insecticides; weed killer; fumigant; manufacture of tetrachloroethylene, 1,1-dichloroethylene, artificial silk, leather and pearls.	Biologists; dry-cleaners; fat processors; fumigators; gasket makers; herbicide workers; insecticide workers; lacquer workers; metal cleaners; mineralogists; oil processors; paint makers; phosphorus processors; resin makers; soil treaters; solvent workers; varnish makers; waxers.
Pentachloro-ethane	Solvent for oil and grease in metal cleaning; separation of coal from impurities by density difference.	No data available
Hexachloroethane	Organic synthesis; retarding agent in fermentation; camphor substitute in nitrocellulose; rubber accelerator; pyrotechnics	Cleaners and maintenance workers; millwrights; machine operators; plumbers and pipefitters;

Table 2. (continued)

Substance	Major Uses	Major Occupational Exposure Industries
Hexachloroethane (cont.)	and smoke devices; solvent; explosives; medicine.	electricians; paper product makers; lumber and wood product makers.
1,1-dichloro-ethene (VDC)	Manufacture of copolymers and modacrylic fibers.	Chemical workers; plastics workers.
1,1,2-trichloro-ethene (TCE)	Metal degreasing; organic solvent.	Chemical workers; metal workers; textile processors.
Tetrachloro-ethylene (PCE)	Dry cleaning solvent; vapor-degreasing solvent; drying agent for metals and other solids; vermifuge; heat-transfer medium; manufacture of fluorocarbons.	Degreasers; cellulose ester processors; drug makers; drycleaners; electroplaters; ether processors; fumigant workers; gum processors; metal degreasers; printers; rubber workers; soap workers; solvent workers; tar processors; vacuum tube makers; wax makers; wool scourers.
Bromoethene (VB)	Flame retardant, textiles, chemical manufacturing.	Chemical workers; textile workers.
1,2-dibromoethane (EDB)	Fumigant; fuel additive; organic solvent.	Chemical workers; petroleum workers; exterminators and fumigators.

Table 3. Evidence for mutagenicity of chloromethanes

Substance	Test System		Results (+/-)	References
Chloromethane – CH$_3$Cl				
	Salmonella typhimurium	TA1535, TA100	+	1,2,3
	Salmonella typhimurium	TA98, TA1537	-	4
	Tradescantia paludosa		+	5
Dichloromethane (Methylene Chloride) – CH$_2$Cl$_2$				
	Salmonella typhimurium	TA98, TA100	+	2,3,6,7
	Bacillus subtilis		-	7
	Saccharomyces cerevisiae	D7	+	8
	Saccharomyces cerevisiae	D3	-	3
	Drosophila melanogaster		-	9
Chloroform – CHCl$_3$				
	Salmonella typhimurium	TA1535, TA1537, TA1538, TA98, TA100	-	2,3,10
	Saccharomyces cerevisiae	D7	+	8
	Escherichia coli	K-12	-	10
Carbon tetrachloride – CCl$_4$				
	Salmonella typhimurium	TA1535, TA1537, TA1538, TA1950, TA100, TA98, G46	-	2,3,10,11
	Saccharomyces cerevisiae	D7	+	8
	Escherichia coli	K-12	-	10

results were obtained with S. typhimurium tester strains TA98 and
TA1537, which would detect frame shift mutations (4). Chloromethane
was also found to produce chromosomal aberrations at pollen tube
mitosis in Tradescantia paludosa (5).

Methylene chloride was found to produce base pair substitutions
in S. typhimurium TA100 (2,3,6,7), frame shifts in TA98 (6,7), and
mitotic gene conversion and recombination in S. cerevisiae strain
D7 (8). Negative results were obtained in mutagenicity tests with
S. cerevisiae strain D3 (3), B. subtilis (7), and D. melanogaster (9).

Chloroform has been found to have a mutagenic effect in assays
with S. cerevisiae strain D7 (8), but has shown negative results in
other systems tested, including S. typhimurium strains TA1535, TA1537,
TA1538, TA98, TA100 and in E. coli K-12 (2,3,10). Similarily, carbon
tetrachloride has shown no mutagenic effect in S. typhimurium strains
TA1535, TA1537, TA1538, TA98, TA100, or in strains G46 and TA1950
(2,3,10,11). Assays using E. coli K-12 also showed negative results
(10). Positive evidence of mutagenicity has been found only in
S. cerevisiae D7 (8). Salmonella and Saccharomyces assays of chloroform and carbon tetrachloride have been conducted with and without
metabolic activation and in desiccators. Strains G46 and TA1950 have
also been tested in host mediated assays (11). It has been suggested
by Simmon et al., that the lack of mutagenic activity of many of the
alkyl halides in the standard Salmonella/microsome assay is due to
the volatility of these compounds, and that when tested in desiccators they have been shown to be mutagenic (3). However, these
investigators (3) were unable to demonstrate that chloroform is a
mutagen in Salmonella after extensive testing in suspension and in
desiccators.

Experimental Oncogenicity of Chloromethanes

All of the chloromethanes appear to be carcinogenic in experimental animals. These study results are summarized in Table 4. A
draft interim report of a carcinogenesis bioassay on chloromethane
performed by the Chemical Industry Institute of Toxicology (CIIT)
was forwarded to the U.S. Environmental Protection Agency (EPA) in
October, 1980 (12). At that time, a total of 21 male B6C3F1 mice
exposed by inhalation to 1000 ppm chloromethane, 6 hours/day, 5 days/
week for up to 21 months had been diagnosed with neoplasms of the
kidney (adenomas and renal adenocarcinomas). Kidney tumors were not
observed in female mice, nor in control mice of either sex. Since
published data for male B6C3F1 mice indicated only three renal tubular
adenomas/carcinomas in 2,543 controls, the report concluded that
the oncogenic response was compound related. No such tumors were
reported in animals exposed to 50 or 225 ppm chloromethane; however,
histopathologic examination of tissue was not completed at the time
of the report. Thus, final interpretation of the study results cannot be made from the October 1980 report.

Table 4. Oncogenic response in experimental animals exposed to chloromethanes

Data Source	Substance	Route	Species	Site	Sex	Reference
CIIT*	methyl chloride	inhal	mouse	kidney	M	12
			rat	--		
Dow*	methylene chloride	inhal	rat	salivary gland	M	13
			hamster	--		
NCI/NTP	methylene chloride	gav	mouse	(due 1981)		
			rat			
NCI/NTP	chloroform	gav	mouse	liver	M F	14
			rat	kidney	M	
				thyroid	F	
NCI/NTP	carbon tetrachloride	gav	mouse	liver	M F	14
				adrenal gland	M F	
			rat	liver	M F	

*Preliminary data

--Negative results

The oncogenicity of methylene chloride is being studied by the
Toxicology Research Laboratory of the Dow Chemical Company (13).
Male and female rats and hamsters were exposed by inhalation to 0,
500, 1500 or 3500 ppm methylene chloride for 6 hours/day, 5 days/week
for 2 years. Preliminary results indicate an oncogenic response in
the male rats. Of 124 rats in each exposure group (0, 500, 1500,
or 3500 ppm), salivary gland sarcomas were observed in 1, 0, 5 and
11 animals, respectively. According to the report, this type of
tumor is found historically in 0-2 percent of control animals.
Therefore, the report concluded "there was an apparent association
between the increased incidence of sarcomas in the salivary gland
of male rats and prolonged exposure to 1500 or 3500 ppm methylene
chloride." In this study, one must question the selection of the
hamster rather than the mouse as the second species for testing since
"there are compelling reasons for rejecting the hamster for chronic
toxicity/oncogenicity testing. Compared to the rat and mouse only
a meager data base is available on spontaneous tumor incidence.
Moreover, it is known that hamsters have several chronic degenerative
diseases that result in poor longevity, making the species unsuitable
by EPA's criteria" (12). Furthermore, as indicated in Table 4, previous carcinogenesis bioassay testing results for the group of chloromethanes would suggest that the mouse is the most sensitive species.
Final interpretation of the Dow study (13), however, will have to
await an evaluation of the final report.

Methylene chloride also has been re-tested by the NCI/NTP Bioassay Program. The results are expected in the latter part of 1981.
Interpretation of the results from an earlier NCI bioassay was not
possible because of the high mortality experienced in the study
groups.

Chloroform has been reported to induce liver cancer in the
mouse, kidney tumors in the male rat and thyroid tumors in the female rat (14).

Carbon tetrachloride administered by oral gavage as a positive
control has induced hepatocellular carcinoma, adrenal adenoma and
pheochromocytoma in both male and female mice and liver tumors in
male and female rats (14).

Mutagenicity of Chloroethanes

A summary of the evidence for the mutagenicity of the chloroethanes is presented in Table 5 by test systems, results and literature citations.

1,1,-Dichloroethane (3), 1,1,2-trichloroethane (3,21), 1,1,1,2-
tetrachloroethane (3,16) and hexachloroethane (22) were found not to
be mutagenic when assayed with S. typhimurium test strains TA1535,
TA1537, TA1538, TA98 and TA100. Hexachloroethane also showed no
evidence of mutagenicity when tested with S. cerevisiae strain D4
(22).

Table 5. Evidence for mutagenicity of chloroethanes

Substance	Test System	Results(+/-)	References
Chloroethane		ND*	15,16
1,1-dichloroethane	Salmonella typhimurium TA1535, TA100, TA1537, TA1538, TA98	-	3
1,2-dichloroethane	Salmonella typhimurium TA98, TA100, TA1530, TA1535	+	7,17,18,19
	Salmonella typhimurium TA1538	-	17,18,19
	Escherichia coli pol A$^+$/pol A$^-$	+	17,18,19
	Bacillus subtilis	-	7
	Drosophila melanogaster	+	20
1,1,1-trichloroethane	Salmonella typhimurium TA100	+	3
1,1,2-trichloroethane	Salmonella typhimurium TA98, TA100, TA1535, TA1537, TA1538	-	3,21
1,1,1,2-tetrachloroethane	Salmonella typhimurium TA98, TA100, TA1535, TA1537, TA1538	-	3,16
1,1,2,2-tetrachloroethane	Salmonella typhimurium TA1530, TA1535	+	17,18
	Salmonella typhimurium TA1538	-	17,18
	Escherichia coli pol A$^+$/pol A$^-$	+	17,18
	Saccharomyces cerevisiae D7	+	8
Pentachlorethane		ND*	15
Hexachloroethane	Salmonella typhimurium TA1535, TA1537, TA1538, TA98, TA100	-	22
	Saccharomyces cerevisiae D4	-	22

*ND = No data available

1,2-dichloroethane has been tested for mutagenicity in a number of systems and has been found to be mutagenic in S. typhimurium test strains TA1530, TA1535, TA100 and TA98, and in DNA polymerase-deficient E. coli (pol A$_1^-$) and in Drosophila melanogaster (7,17,18,19,20). Negative results were obtained when 1,2-dichloroethane was tested with S. typhimurium TA1538 (17,18,19), and with B. subtilis (7).

1,1,1-trichloroethane was said to be weakly mutagenic when assayed in a desiccator with S. typhimurium TA100 (3).

1,1,2,2-tetrachloroethane was found to be mutagenic when tested in assays using S. typhimurium TA1530 and TA1535, E. coli (pol A$_1^-$) (17,18), and S. cerevisiae D7 (8), but was not found to be mutagenic for S. typhimurium TA1538 (17,18).

No information was available on the mutagenicity of chloroethane or pentachloroethane (15,16).

Experimental Oncogenicity of Chloroethanes

A summary of study results for the chloroethanes is shown in Table 6. All of the substances for which study results were available appear to demonstrate an oncogenic response in the mouse (14,23, 24,25). With the exception of 1,1,1,2-tetrachloroethane, for which liver adenomas only were observed in female mice, hepatocellular carcinomas were induced by all of these compounds and usually in both sexes of mice. A positive response was observed in the rat only with 1,2-dichloroethane (EDC). The latter substance also induced tumors at multiple sites in both the mouse and the rat. The NTP/NCI Bioassay Program is currently completing the retesting of 1,1,1-trichloroethane; interpretation of the earlier bioassay results was not possible because of poor survival.

Experimental Oncogenicity of Chlorinated Analogues of Vinyl Chloride

A summary of data indicating the carcinogenic response of 1,1-dichloroethene (VDC), 1,1,2-trichloroethene (TCE) and tetrachloroethylene (PCE) is shown in Table 7. VDC has been reported to induce adenocarcinoma of the kidney in the male mouse (26) and mammary carcinoma in the female rat (27). Lee et al. (28) have reported angiosarcoma of the liver in the mouse and angiosarcoma of the lymph node and subcutaneous tissue in the rat exposed to VDC.

Hepatocellular carcinoma has been induced in mice exposed to TCE by inhalation or oral gavage (29,30). Hepatocellular carcinoma also has been induced in mice exposed to PCE (31). TCE appears to be mutagenic in several systems, while there has been less consistent positive response for VDC and very little evidence for the mutagenicity of PCE (see review by Fabricant and Chalmers (32).

Table 6. Oncogenic response in experimental animals exposed to chloroethanes from NCI/NTP bioassay program

Substance	Route	Species	Site	Sex	Reference
1,1-dichloroethane	gav	mouse	liver	M	14
1,2-dichloroethane	gav	mouse	lung	M F	23
			liver	M	
			mammary gland	F	
			uterus	F	
		rat	hemangiosarcoma(s)	M	
			forestomach	M	
			mammary gland	F	
			subcutaneous	M	
1,1,2-trichloroethane	gav	mouse	liver	M F	14
			adrenal gland	F	
		rat	--		
1,1,1-trichloroethane	gav	mouse	(due 1981)		
		rat			
1,1,1,2-tetrachloroethane	gav	mouse	liver	F	*
		rat	--		
1,1,2,2-tetrachloroethane	gav	mouse	liver	M F	24
		rat	--		
Pentachloroethane	gav	mouse	liver	M F	*
		rat	--		
Hexachloroethane	gav	mouse	liver	M F	25
		rat	--		

*Preliminary data, NCI/NTP unpublished results   -- Negative results   s = spleen

Table 7. Oncogenic response in experimental animals exposed to VDC, TCE, and PCE

Data Source (reference)	Substance	Route	Species	Site	Sex
Maltoni, 1977 (26)	1,1-dichloroethene (VDC)	inhal	mouse	kidney	M
Maltoni et al., 1977 (27)	1,1-dichloroethene (VDC)	inhal	rat	mammary gland	F
Lee et al., 1977 (28)	1,1-dichloroethene (VDC)	inhal	mouse rat	angiosarcoma (L) angiosarcoma (*)	M M
NCI/NTP (29)	1,1,2-trichloroethene (TCE)	gav	mouse	liver	M F
MCA (30)	1,1,2-trichloroethene (TCE)	inhal	mouse	liver	M F
NCI/NTP (31)	tetrachloroethylene (PCE)	gav	mouse	liver	M F

L = liver

* = lymph node and subcutaneous tissue

Experimental Oncogenicity of Bromoethene (VB) and 1,2-Dibromoethane (EDB)

As shown in Table 8, three studies have demonstrated the carcinogenicity of 1,2-dibromoethane. This oncogenic response has been demonstrated in multiple organs by either oral or inhalation exposure (33,34). The site of entry of the substance has been among those sites demonstrating a carcinogenic response, i.e., oral gavage has resulted in the induction of forestomach cancer, while inhalation has resulted in nasal cancer. Lung cancer has been observed following both routes of administration.

In the NIOSH study (34), animals were exposed by inhalation to 20 ppm EDB and were also given a diet containing 0.05% disulfiram (tetraethyl thiuram disulfide), an inhibitor of aldehyde dehydrogenase used in the management of alcoholism. In addition to the cancer sites indicated in Table 8, recent evaluation of the nasal cavity of the rats exposed in the NIOSH study has resulted in the observation of nasal cancer in about 30% of the exposed animals. The mutagenicity of EDB has been reviewed in detail by Fabricant and Chalmers (32).

Inhalation exposure of bromoethene (vinyl bromide) to rats also demonstrates the induction of liver angiosarcoma and hepatocellular carcinoma, as well as zymbal gland tumors. A contrast of data for the incidence of liver angiosarcoma among Sprague-Dawley rats exposed to either vinyl bromide (35) or vinyl chloride (36) is shown in Table 9. A much greater carcinogenic response is demonstrated for vinyl bromide as compared to vinyl chloride. The incidence of liver angiosarcoma is lower for animals exposed to 1250 ppm vinyl bromide as contrasted with the incidence from exposure at levels of 50 and 250 ppm because of high mortality in the highest exposure group. Clearly, the brominated compounds EDB and VB demonstrate a greater carcinogenic response in animals than their chlorinated analogues.

Epidemiologic Evidence of Carcinogenicity

A summary of the epidemiologic study results for workers exposed to these substances is shown in Table 10. Observed and expected deaths are shown for all cancers and for selected site-specific cancer deaths in cases where such analyses were conducted. For the historical prospective cohort mortality studies, the power to detect a 50 percent excess in cancer mortality at the 95 percent confidence level also was calculated.

Axelson et al. (37) reported on the mortality experience of 518 male workers exposed to TCE in the 1950's and 1960's, and followed through 1975. Among the subcohort who had achieved at least 10 years since first exposure, 9 deaths from all cancers were observed compared to 9.5 expected. The number of deaths from cancer was too small to

Table 8. Oncogenic response in experimental animals exposed to EDB and VB

Data Source (reference)	Substance	Route	Species	Site	Sex
NCI/NTP (33)	1,2-dibromoethane (EDB)	gav	mouse	forestomach	M F
				lung	M F
			rat	forestomach	M F
				spleen (A)	M
				liver (H)	F
NCI/NTP (unpublished)	1,2-dibromoethane (EDB)	inhal	mouse	lung	M F
				mammary gland	F
			rat	nasal	M F
				mammary gland	F
				spleen (A)	M
NIOSH (34)	1,2-dibromoethane (EDB)	inhal + oral disulfiram	rat	liver (H,A)	M F
				spleen (A)	M F
				mesentary/ omentum (A)	M F
				kidney	M F
EPL (35)	bromoethene (VB)	inhal	rat	liver (H,A)	M F
				zymbal gland	M F

H = hepatocellular

A = angiosarcoma

Table 9. Incidence of hepatic angiosarcoma induced by inhalation
          exposure to vinyl bromide and vinyl chloride in Sprague-
          Dawley rats, sexes combined

Vinyl Bromide*

Exposure Level (ppm)	No. of Animals	No. with liver Angiosarcoma	%
0	228	1	0.3
10	240	17	7
50	240	86	36
250	240	122	51
1250	240	84	35
2500	No Animals Exposed		

* from reference 35

Vinyl Chloride**

Exposure Level (ppm)	No. of Animals	No. of liver Angiosarcoma	%
0	120	0	0
10	120	0	0
50	59	1	2
250	59	4	7
1250	No Animals Exposed		
2500	59	13	22

** from reference 36 (BT1, BT2 & BT15)

Table 10. Observed and expected deaths from studies of workers exposed to selected halogenated hydrocarbons

Source	Substance	Total Cancer Deaths			Site Specific Cancer Deaths			
		OBS.	EXP.	POWER*	Site	OBS.	EXP.	RR.
Axelson et al., (1978) ≥10 years latency	TCE	9	9.5	0.40				N/A
Ott et al., (1976) ~15 years latency	VDC	1	0.5	0.09	Lung	1	0.2	5.0
Ott et al., (1980) ~15 years latency	EDB	6	4.3	0.24				N/A
Friedlander et al., (1978) ≥20 years latency	CH_2Cl_2	7	12.5	0.48				N/A
Blair et al., (1979)	TCE PCE CCl_4 Benzene	87	67.9**		Lung	17	10.0	1.7**
					Liver	4	1.7	2.4
					Leukemia	5	2.2	2.3
					Cervix Uteri	10	4.8	2.1**
					Skin	3	0.7	4.3**
Blair (1980)	TCE PCE	244	223.7		Lung	62	58.7	1.1
					Esophagus	10	5.4	1.9
					Liver	10	6.1	1.6
Katz and Jowett (1981)	TCE PCE CCl_4	141	147.4		Lung	10	10.2	1.0
					Kidney	7	2.7	2.6
					Bladder	5	2.6	1.9
					Genitals	4	0.8	5.0
					Skin	4	1.9	2.1
					Lymphosarcoma	6	3.4	1.8

* = In this case the power = ability to detect a 50% increase significant at the 0.05 level.

** = $p < 0.05$ N/A = Not analyzed in study   RR = Relative Risk

analyze by site-specific risk. On the basis of experimental bioassay
and known toxicity of TCE in humans, one might expect an excess of
liver cancer. None was observed. The authors stated that "the cancer risk to man from trichloroethylene can by no means be ruled out
from this study, particularly with regard to uncommon malignancies
such as liver cancer."

Ott et al. (38) reported on the mortality experience of 138 employees exposed to VDC at any time between 1942 and 1968, and followed through 1973. Mortality among the exposed group was compared to
the expected based on age and calendar time period adjusted U.S. white
male mortality rates. Only five deaths were observed compared to 7.5
expected. These data are not shown in Table 10. For individuals with
15 or more years of latency, one cancer death was observed, compared
to 0.5 expected. Clearly, the sample size is too small to allow any
meaningful conclusion.

Ott et al. (39) reported on the mortality experience of 161 men
exposed to EDB in two production units. For individuals employed in
production unit two, seven total malignancies were observed versus
2.8 expected, p<0.05. Two of these individuals, who died from lung
cancer, had been exposed to arsenicals for 1.5 and 20 months, respectivelly, whereas they had been exposed to EDB for 102 and 111 months,
respectivelly. In production unit one, there were two deaths from
cancer observed versus 3.6 expected. These data are not shown in
Table 10. The authors concluded that the findings of the investigation neither rule out nor establish EDB to be a human carcinogen.
Although of limited sample size, the observations in production unit
two raise suspicion of a potential excess cancer risk among workers
exposed to EDB.

Mortality patterns among 330 dry cleaning and laundry workers
potentially exposed to CCl4, TCE, PCE, and petroleum solvents were
studied by Blair et al. (40) and analyzed by the proportionate mortality method. Age, race, and sex specific proportional mortality
for persons in the U.S. between 1957 and 1977 served as the basis
for calculating expected deaths. A significant excess of cancer
mortality was observed, with 87 cancer deaths observed and 67.9
expected. Malignancies of the respiratory system, skin, and uterine
cervix were significantly elevated. Because these workers were exposed to multiple substances known to be carcinogenic, it is difficult
to attribute this excess to one substance.

Blair (41) recently reported the results of a proportionate mortality study of workers in the metal polishing and plating industry.
These workers were potentially exposed to various metals, corrosive
and caustic alkaline solutions, and solvents such as TCE and PCE.
Mortality patterns among white male metal platers who died between
1951 and 1969 were compared with the expected number based on cause-
specific proportionate mortality for U.S. white males. Excess cancer

s of the esophagus and liver were present in the study population. Excess cancers of other organs, such as lung and digestive system, were not statistically significant. As with the dry cleaner population, exposures to more than one substance known to induce cancer experimentally do not allow the determination of any specific etiologic factor.

In a study of female occupational mortality in Wisconsin, Katz and Jowett (42) examined the causes of death for 671 white females whose death certificates indicated they had been employed as laundry and dry cleaning workers. The deaths occurred between 1963 and 1977. They found significant excesses in the proportion of deaths due to cancers of the kidney, genitals (unspecified), bladder, uterine cervix, skin, and lymphosarcoma among these women as compared to the proportions observed among 65,559 deceased women employed in other occupations and in low wage occupations. The occupation of housewife was eliminated from the analyses. The authors did not specify how the expected values were generated to calculate PMRs, but it is assumed that they were based upon Wisconsin rather than U.S. deaths for the years 1963-1977. Data were not age-adjusted. No estimate of length of employment was possible with this type of study. Because the occupational code included both laundry and dry cleaners, it is not certain how many of the women were actually exposed to solvents. The use of several different solvents in the last 50 years in laundry and dry cleaning, including perchloroethylene, Stoddard solvent, trichloroethylene, carbon tetrachloride, and fluorocarbons, makes it difficult to evaluate the relationship between exposure and cause of death. Furthermore, lack of age adjustment renders any interpretation of the study results difficult.

Friedlander et al. (43) examined the mortality experience of active, disabled and retired employees who were chronically exposed to dichloromethane using historical prospective cohort and proportionate mortality analyses. The cohort of 751 male hourly workers included all those who were employed in 1964 in the methylene chloride area. Of this group, 252 men were stated to have had a minimum of 20 years work in the methylene chloride exposure area as of July 1, 1964.

Proportionate mortality ratios based on available death certificates from deaths that occurred from 1956 through 1976 showed no excess of deaths due to malignant neoplasms over the number expected. Expected numbers of deaths due to specific causes were calculated from deaths among other hourly employees at Kodak Park. Standardized mortality ratios calculated for the 252 members of the methylene chloride cohort with _20 years latency showed no significant increase in risk of death due to cancer. Three separate sets of expected values were based on rates in control populations from Kodak Park, New York State, and the United States. However, the power of this study to detect an excess of total cancer is only 0.48. Hence, there was less than a 50% chance of detecting an excess in total cancer mortality, if in fact one was present. The power to detect an excess of site-specific cancer would be much lower.

CONCLUSIONS

Epidemiologic studies available for review do not allow the assessment of carcinogenic risk among humans exposed to these substances because the number of workers available for study who had achieved an adequate latency period was too small. Some studies lacked specificity in that they could not differentiate the potential for any effect to be related to a particular chemical exposure. Some of the studies also lacked sensitivity by not being able to distinguish those exposed from those not exposed to chlorinated solvents. Information presented further indicates that most of the carcinogenic substances studied are in high-volume production with large estimated numbers of workers exposed. Retention of personnel records containing information necessary for epidemiologic study of health hazards is not a requirement in the U.S. and this adds to the problem of insufficient sample size.

On the basis of these observations, it is apparent that, in general, carcinogenic risk of a specific chemical substance to humans must be estimated through the conduct of experimental studies. Epidemiologic studies should be conducted when adequately sized cohorts of workers who have achieved sufficient latency are available. Insensitive epidemiologic studies should not be used as a basis for delay in public health decisions when adequate experimental data indicate cancer induction.

REFERENCES

1. A. W. Andrews, E. S. Zawistowski, and C. R. Valentine, A comparison of the mutagenic properties of vinyl chloride and methyl chloride, Mutat. Res. 40: 273-276 (1976).
2. V. F. Simmon, Structural correlations of carcinogenic and mutagenic alkyl halides, in: "Structural Correlates of Carcinogenesis and Mutagenesis -- A Guide to Testing Priorities", Dept. of Health, Education and Welfare, Washington, D.C., DHEW Publ. No. FDA 78-1046, 163-171 (1978).
3. V. F. Simmon, K. Kauhanen, and R. G. Tardiff, Mutagenic activity of chemicals identified in drinking water, in: "Progress in Genetic Toxicology", D. Scott, B. A. Bridges, and F. H. Sobels, eds., Elsevier/North Holland Inc., New York, N.Y., 249-258 (1977).
4. Dupont, Chloromethane mutagenicity test, Haskell Laboratory 1977, TSCA Sect. 8(d) Submission 8 DHQ-1078-0204, Office of

Toxic Substances, U.S. Environmental Protection Agency, Washington, D.C. (1978).

5. H. H. Smith, and T.A. Lofty, Comparative effects of certain chemicals on Tradescantia chromosomes as observed at pollen tube mitosis, Amer. J. Bot. 41:589-593 (1954).

6. W. M. F. Jongen, G. M. Alink, and J. H. Koeman, Mutagenic effect of dichloromethane on Salmonella typhimurium, Mutat. Res. 56:245-248 (1978).

7. T. Kanada, and M. Uyeta, Mutagenicity screening of organic solvents in microbial systems, Mutat. Res. 54:215 (1978).

8. D. F. Callen, R. C. Wolf, and R. M. Philpot, Cytochrome p-450 mediated genetic activity and cytotoxicity of seven halogenated aliphatic hydrocarbons in Saccharomyces cerevisiae, Mutat. Res. 77:55-63 (1980).

9. L. M. Filippova, O. A. Panjshin, and R. G. Kostianovsky, Chemical mutagens. IV. Genetic activity of geminal systems, Genetika 8:134-148 (1967).

10. H. Uehleke, H. Greim, M. Krämer, and T. Werner, Covalent binding of haloalkanes to liver constituents, but absence of mutagenicity on bacteria in a metabolizing test system, Mutat. Res. 38:114 (1976).

11. R. Braun, and J. Schöneich, The influence of ethanol and carbon tetrachloride on the mutagenic effectivity of cyclophosphamide in the host-mediated assay with Salmonella typhimurium, Mutat. Res. 31:191-194 (1975).

12. J. E. Gibson, CIIT's comments before the U.S. Environmental Protection Agency, proposed test rule for chloromethane, Chemical Industry Institute of Toxicology, Research Triangle Park, N.C. October 28 (1980).

13. D. J. Roberts, Dow Chemical Company report to Chemical Information Division, Office of Toxic Substances, U.S. EPA, November 14 (1980) (EPA Document Control No.: FYI-OTS-1180-0097.).

14. E. K. Weisburger, Carcinogenicity studies on halogenated hydrocarbons. Envir. Health Perspect. 21:7-16 (1977).

15. U.S. Environmental Protection Agency, Ambient Water Quality Criteria for Chlorinated Ethanes, EPA 440/5-80-029, Washington, D.C. October (1980).

16. L. Fishbein, Potential halogenated industrial carcinogenic and mutagenic chemicals. II. Halogenated saturated hydrocarbons, Sci. Total Environ. 11:163-195 (1979).

17. H. Brem, A. B. Stein, and H. S. Rosenkranz, The mutagenicity and DNA-modifying effect of haloalkanes. Cancer Res. 34:2576-2579 (1974).

18. H. S. Rosenkranz, Mutagenicity of halogenated alkanes and their derivatives, Environ. Health Perspect. 21:79-84 (1977).

19. J. McCann, V. Simmon, D. Streitwieser, and B. N. Ames, Mutagenicity of chloroacetaldehyde, a possible metabolic product of 1,2-dichloroethane (ethylene dichloride), chloroethanol (ethylene chlorohydrin), vinyl chloride, and cyclophosphamide, Proc. Natl. Acad. Sci. U.S.A. 72:3190-3193 (1975).

20. P. O. Nylander, H. Olofsson, B. Rasmuson, and H. Svahlin, Muta-
     genic effects of petrol in Drosophila melanogaster. I. Effects
     of benzene and 1,2-dichloroethane, Mutat. Res. 57:163-167
     (1978).

21. U. Rannug, A. Sundvall, and C. Ramel, The mutagenic effect of
     1,2-dichloroethane on Salmonella typhimurium. I. Activation
     through conjugation with glutathione in vitro. Chem. Biol.
     Interact. 20:1-16 (1978).

22. M. H. Weeks, R. A. Angerhofer, R. Bishop, J. Thomasino, and
     C. R. Pope, The toxicity of hexachloroethane in laboratory
     animals, Am. Ind. Hyg. Assoc. J. 40: 187-199 (1979).

23. National Cancer Institute, Carcinogenesis bioassay of 1,2-di-
     chloroethane, Technical Report series no. 55, DHEW publica-
     tion no. (NIH) 78-1361, Government Printing Office, Washing-
     ton, D.C. (1978).

24. National Cancer Institute, Carcinogenesis bioassay of 1,1,2,2-
     tetrachloroethane, Technical report series no. 27, DHEW pub-
     lication no. (NIH) 78-827, Government Printing Office,
     Washington, D.C. (1978).

25. National Cancer Institute, Carcinogenesis bioassay of hexachloro-
     ethane, Technical report series no. 68, DHEW publication no.
     (NIH) 78-1318, Government Printing Office, Washington, D.C.
     (1978).

26. C. Maltoni, Recent findings on the carcinogenicity of chlorin-
     ated olefins, Envir. Health Perspec. 21:1-5 (1977).

27. C. Maltoni, G. Cotti, L. Morisi, and P. Chieco, Carcinogenicity
     bioassays of vinylidene chloride, Med. Lav. 68:241-262 (1977).

28. C. C. Lee, J. C. Bhandari, J. M. Winston, W. B. House, R. L.
     Dixon, and J. S. Woods, Carcinogenicity of vinyl chloride and
     vinylidene chloride, J. Toxic. Envir. Health 4:15-30 (1978).

29. National Cancer Institute, Carcinogenesis bioassay of trichloro-
     ethylene, Technical report series no. 2, DHEW publication no.
     (NIH) 76-802, Government Printing Office, Washington, D.C.
     (1976).

30. Manufacturing Chemists Association, Final report of audit find-
     ings, administered trichloroethylene (TCE) chronic inhalation
     study at Industrial Bio-Test Laboratory, Inc., Decatur, Ill.
     (1978).

31. National Cancer Institute, Carcinogenesis bioassay of tetra-
     chloroethylene, Technical report series no. 13, DHEW publica-
     tion no. (NIH) 77-813, Government Printing Office, Washington,
     D.C. (1977).

32. J. D. Fabricant, and J. H. Chalmers, Evidence of the mutagen-
     icity of ethylene dichloride and structurally related com-
     pounds, in: "Ethylene Dichloride: A Potential Health Risk?",
     B. Ames, P. Infante, and R. Reitz, eds., Banbury Report 5,
     Cold Spring Harbor laboratory, Cold Spring Harbor, N.Y.,
     309-329 (1980).

33. National Cancer Institute, Carcinogenesis bioassay of 1,2-dibromo-
     ethane. Technical report series no. 86, DHEW publication no.

(NIH) 78-1336, Government Printing Office, Washington, D.C. (1978).

34. H. B. Plotnick, W. W. Wiegel, D. E. Richards, K. L. Cheever, and C. Kommineni, Dietary disulfiram enhancement of the toxicity of ethylene dibromide, in: "Ethylene Dichloride: A Potential Health Risk?", B. Ames, P. Infante, and R. Reitz, eds., Banbury Report 5, Cold Spring Harbor Laboratory, Cold Spring Harbor, N.Y., 279-286 (1980).

35. W. M. Busey, Oncogenic potential of vinyl bromide during chronic inhalation exposure - rats, Experimental Pathology Laboratories, Inc., Herndon, V.A., unpublished report (1979).

36. C. Maltoni, Vinyl chloride carcinogenicity: An experimental model for carcinogenicity studies, in: "Origins of Human Cancer", H. H. Hiattt, J. D. Watson, and J. A. Winsten, eds., Cold Spring Harbor Laboratory, Cold Spring Harbor, N.Y., 119-146 (1977).

37. D. Axelson, K. Anderson, C. Hogstedt, B. Holmberg, G. Molina, and A. de Verdier, A cohort study on trichloroethylene exposure and cancer mortality, J. Occup. Med. 20:194-196 (1978).

38. M. G. Ott, W. A Fishbeck, J. C. Townsend, E. J. Schneider, A health study of employees exposed to vinylidene chloride, J. Occup. Med. 18:735-738 (1976).

39. M. G. Ott, H. C. Scharnweber, and R. R. Langner, Mortality experience of 161 employees exposed to ethylene dibromide in two production units, Br. J. Ind. Med. 37:163-168 (1980).

40. A. Blair, P. Decoufle, and D. Grauman, Causes of death among laundry and dry cleaning workers, Am. J. Publ. Health 69: 508-511 (1979).

41. A. Blair, Mortality among workers in the metal polishing and plating industry, 1951-1969, J. Occup. Med. 22:158-162 (1980).

42. R. M. Katz, and D. Jowett, Female laundry and dry cleaning workers in Wisconsin: A mortality analysis, Am. J. Public Health 71:305-307 (1981).

DISCUSSION

Q. Sivak (A. D. Little): I don't want to get into the polemic of what increased incidences of liver tumors in rodents mean but I think that it may be a rather shallow basis on which to extrapolate expectations for liver tumors in humans. My second comment is with respect to the recent findings that low molecular weight haloalkanes are extremely prevelant in a lot of people's drinking water. It means we're dealing with a very pervasive and widespread competing risk and that the level of low molecular weight haloalkanes have to be taken into consideration when doing epidemiology studies.

A. Infante (U.S. Dept of Labor): I think we could get into a long discussion about liver cancers in animals. We recently had three months of discussion for our carcinogen policy in the Department of

Labor and these issues were discussed. Experts from around the world
concluded that this information should be accepted as demonstrating
induction of cancer. With regard to your second point, I'm not quite
clear on what it is you're saying.

Q. Sivak (A. D. Little): Well, one must look at the total exposure
of an individual before trying to implicate either the workplace or
the ambient environment. I think that it may be extremely difficult
to determine whether it's a workplace exposure or an ambient water-
drinking exposure or both that might be contributing to what looks
to be only a very modest level of increased carcinogenic risk.

A. Infante (U.S. Dept. of Labor): This is exactly why we study
industrial populations where you have much higher levels of exposure.
You're suggesting that there may be a contribution from background
and I think this is a good point. Perhaps individuals who are
designing carcinogenesis bioassays need to take this into consideration rather than just exposing the animals to one substance at a
time in a fairly clean environment. Humans are exposed to a lot of
different chemicals at the same time. However we tend to look
at the risk for one substance at a time and never put them all together. I agree with you that more work should be done in this area.

Q. White (BNL): I'd like to know what has been the rationale for
deciding that a compound like vinyl chloride should be controlled
with a permissible exposure limit and a similar compound like bis-
chloromethyl ether should have a zero exposure level.

A. Infante (U.S. Dept. of Labor): I can't speak for bischloro-
methyl ether because I wasn't involved in those studies at the
time the standard was promulgated, but I don't think it was promulgated at a zero exposure level. I would assume that the industry
itself, realizing that the induction of lung cancers in very young
workers was detrimental, decided not to use it. This seems very wise,
because, obviously, industry is getting along very well without this
compound. Your question is then why do we have a one part per million exposure limit for vinyl chloride. Our exposure standards in
the occupational setting are based on more than just health effects.
They're also based on feasibility; feasibility from the standpoint
of--can you measure the substance at the level that you're going to
recommend. If you can't measure it that low, then you have a standard you can't enforce. The second aspect of feasibility is the cost
to the industry. We don't ban substances in the occupational setting.
We set standards for them. We are not allowed to set standards that
would adversely impact on the economy of industry more than what they
could absorb. All these factors are taken into consideration. So,
in fact, when you see Dr. Maltoni's data, where he has induction of
cancer from low levels of vinyl chloride exposure and it's close to
our current standard, you say, "Well, what are we going to do about
it?" Probably something should be done, but you have to look at the

situation based on the economic impact on the industry.   Furthermore, there are so many substances for which we have totally inappropriate standards vis a vi carcinogenicity that part of our
priority-setting scheme is to look at the level of exposure and
how many workers are exposed.   There may be problems besides vinyl
chloride that we should address first; not that vinyl chloride
should be ignored.

Q. Maltoni (Italy):   If I can add a comment to Dr. Infante's
presentation, I would say that the time is coming in which you
cannot think any more in terms of a unique exposure, but rather
of a multitude of concurrent exposures which may interact synergistically in regard to cancer induction.

VINYL CHLORIDE: A MODEL CARCINOGEN FOR RISK ASSESSMENT

Cesare Maltoni, Giuseppe Lefemine, Adriano Ciliberti,
Guiliano Cotti and Donata Carretti

Institute of Oncology and Tumour Center
Bologna, Italy

INTRODUCTION

In 1971, an extensive, integrated program of long term
studies on carcinogenicity was initiated at the Institute of
Oncology and Tumor Center of Bologna. The main objectives of
the program were: to test the carcinogenic potency of several
major airborne industrial compounds; to obtain information
on the site and type of tumors induced by the compounds; to
assess the level of risk in quantitative terms through dose
response data; to determine the relative potency of the different
compounds; to evaluate the effects of route of administration
on tumor type and incidence; and to determine the role of different biological factors such as species, strain, sex and age on
tumor yields.

Since 1971, several compounds have been tested by one or more
routes, in one or more species and strains and at different
concentrations. These agents include: vinyl chloride,
vinylidene chloride, ethylene dichloride, trichloroethylene,
methylene chloride, acrylonitrile, propylene, benzene, styrene,
trichlorofluormethane, dichlorodifluormethane and chlorodifluoro-
methane. Vinyl chloride (VC) has been the industrial agent of
greatest interest since it represents a successful attempt to
assess risk to man by experiments in vinyl chloride carcinogenicity in animals.

MATERIALS AND METHODS

The animals are breeds (Sprague-Dawley rats, Wistar rats, Swiss mice, golden hamsters) which have been routinely employed in our laboratory for many years. The animals are weighed every two weeks during treatment and at least every eight weeks after the end of the treatment. Every two weeks, the animals are examined for any gross changes. All animals are kept alive until spontaneous death. A full autopsy is performed on each animal. Specimens taken for routine histology include: brain, Zymbal glands, interscapular brown fat, salivary glands, tongue, lungs, liver, kidneys, adrenals, spleen, pancreas, stomach, intestine, uterus, gonads and any other organ with an observable pathological lesion. Sections from prepared specimens are routinely stained with Hematoxylin-Eosin. All slides are screened by a junior pathologist and then reviewed by a senior pathologist.

Our inhalation facilities permit the simultaneous exposure of 6,000 laboratory rodents (rats, mice and hamsters). The inhalation chambers are built of stainless steel and glass. An automatic gas chromatography system is used to monitor the exposure level in the inhalation experiments. For ingestion studies, the test compound is administered by gavage, mixed with food or dissolved in drinking water. All test compounds and carrier material are routinely examined to determine that they meet purity standards. Extra-virgin olive oil is used as the carrier in gavage studies. Each treatment is performed by the same individual; this is particularly important in gavage treatments, since the animals become accustomed to the same operator.

All data are submitted to statistical analysis. However, we must stress that although statistical analysis provides an extremely important tool for interpreting the meaning of the results of long-term bioassays, there may be nonstatistically significant differences between exposed and control groups which could still be meaningful from an oncological point of view. This is particularly true in the case of tumors which occur only rarely in our animal colony. Therefore, the data are evaluated both from statistical and from biological perspectives.

RESULTS

Vinyl chloride has been tested on animals of different species, strain, sex and age. The choice of the animals was made with the intention of having an integrated system of complementary biological models which could express a wide range of neoplastic reponses. VC was administered by four different routes (inhalation, ingestion, intraperitoneal and subcutaneous). The monomer was administered at different concentrations (fifteen by inhalation, six by ingestion, three by intraperitoneal injections and one by subcutaneous injection), for various periods of time, by continuous or intermittent treatment.

Eighteen experiments were initiated, seventeen experiments have ended and have been evaluated and one experiment is still ongoing. A total of 6,680 animals were used in these studies.

The ability of VC to induce tumors was determined on the basis of one or more of the following parameters: (a) a sharply enhanced tumor incidence; (b) rare or exceptional tumor occurrence in the animal colony; (c) a dose-response relationship for tumor formation; and (d) an association with precarcinogenic lesions.

The following conclusions can be drawn from the data produced in the seventeen VC experiments. Exposure to VC caused tumors in all animal systems. The distribution of different tumor types in different sites, e.g. Zymbal gland carcinoma, liver angiosarcoma, nephroblastoma, neuroblastoma, mammary gland carcinoma and fore-stomach papilloma, etc., clearly demonstrated that VC is a multipotential carcinogen (Table 1). Some of the same types of tumors were observed in all of the animal species studied, e.g. liver angiosarcomas, while other tumors, e.g. nephroblastoma, were observed in only one animal system. VC induced a dose-response increase in tumors when given by inhalation or ingestion (Table 2). The duration and schedule for the VC inhalation treatment greatly affected the neoplastic response (Table 3). Qualitatively and quantitatively, the neoplastic response to inhaled VC is greatly modulated by the species, e.g., for liver angiosarcoma--Swiss mice > rats > Golden hamsters (Table 4); the strain, e.g. for Zymbal gland carcinoma--Sprague-Dawley > Wistar (Table 5); the age, e.g. for liver angiosarcoma--newborn > 11 week old rats (Table 6); and the sex, e.g. for liver angiosarcoma--female rats > male rats (Table 2 for example). Newborn animals appeared to be extremely responsive to the carcinogenic ability of VC by easily developing liver tumors; not only angiosarcomas but also hepatocarcinomas. Exposure of pregnant animals to VC was also carcinogenic in embryos.

The results of the experiments on Sprague-Dawley rats initiated to study the effects of different doses of VC, when given by inhalation and by ingestion, were submitted to statistical analysis by the Fisher exact probability test ($p < 0.05$). The concentration of VC inducing a significant number of excess cancer cases in these experiments are given in Table 7.

In our opinion, the following results, although not statistically significant according to the Fisher exact probability test should be specifically noted as biologically of importance.

Table 1. Vinyl chloride: tumors on experimental rodents presently correlated to the monomers.

Species	Angiosarcomas of liver	Tumors of brain	Tumors of lung	Lymphomas and leukemias	Hepatomas	Angiosarcomas angiomas of other sites
Rat	+	+			+	+
Mouse	+		+			+
Hamster	+			(+)		(+)

Species	Nephrobastomas	Sebaceous cutaneous carcinomas	Other epithelial tumors	Mammary Carcinomas	Forestomach papillmoas and acanthomas	Melanomas
Rat	+	+	(+)	+	+	
Mouse			(+)	+	(+)	
Hamster			(+)		+	(+)

+   statistically significant increase
(+) meaningful increase

Table 2A. Incidence (%) of Zymbal gland carcinomas
in Sprague-Dawley rats after exposure to
vinyl chloride: inhalation vs ingestion

Inhalation*

Concentration (ppm)	% Animals with Tumors		
	Male	Female	Total
30,000	56.6	60.0	58.3
10,000	33.3	20.0	26.7
6,000	10.3	13.3	11.9
2,500	3.3	3.3	3.3
500	10.0	3.3	6.7
250	-	-	-
200	5.0	1.7	3.3
150	-	6.7	3.4
100	-	1.7	0.8
50	2.3	2.8	2.5
25	5.0	1.7	3.3
10	1.7	1.7	1.7
5	-	1.7	0.8
1	1.7	-	0.8
0	0.9	0.8	0.9

*4 hr/d, 5 d/wk, 52 wk; 2025 animals total, 60-360 animals per group

Ingestion**

Concentration (mg/Kg b.w.)	% Animals with Tumors		
	Male	Female	Total
50.00	2.5	-	1.2
16.65	2.5	2.5	2.5
3.33	-	-	-
1.0	2.7	4.0	3.3
0.3	-	-	-
0.03	-	-	-
0	-	1.7	0.9

**Gavage: 5 times/wk, 52 (or 59) wk; 920 animals total, 80 or 150
animals per group.

Table 2B. Incidence (%) of liver angiosarcomas in
           Sprague-Dawley rats after exposure to
           vinyl chloride: inhalation vs ingestion

Inhalation*

Concentration (ppm)	% Animals with Tumors		
	Male	Female	Total
30,000	16.6	43.3	30.0
10,000	10.0	13.3	11.7
6,000	10.3	33.3	22.0
2,500	20.0	23.3	21.7
500	-	20.0	10.0
250	3.4	6.7	5.1
200	11.7	8.3	10.0
150	1.7	8.3	5.0
100	-	1.7	0.8
50	1.1	7.2	4.2
25	1.7	6.7	4.2
10	-	1.7	0.8
5	-	-	-
1	-	-	-
0	-	-	-

*4 hr/d, 5 d/wk, 52 wk; 2025 animals total, 60-360 animals per group.

Ingestion**

Concentration (mg/Kg b.w.)	% Animals with Tumors		
	Male	Female	Total
50.00	20.0	22.5	21.2
16.65	10.0	15.0	12.5
3.33	-	-	-
1.0	1.3	2.7	2.0
0.3	-	1.4	0.7
0.03	-	-	-
0	-	-	-

**Gavage; 5 times/wk, 52 (or 59) wk; 920 animals total,; 80 or 150
  animals per group.

Table 3. Incidence (%) of liver angiosarcomas and Zymbal gland carcinomas in Sprague-Dawley rats after vinyl chloride inhalation: effect of exposure regimen

Conc. (ppm)	Schedule*	% Animals with Tumors** liver angiosarcomas M	F	T	Zymbal gl. ca. M	F	T
10,000	I	10.0	13.3	11.7	33.3	20.0	26.7
10,000	II	-	-	-	17.8	13.3	15.5
10,000	III	1.7	-	0.8	13.5	1.7	7.6
10,000	IV	1.7	-	0.8	8.5	6.7	7.6
10,000	V	-	1.7	0.8	3.3	10.2	6.7
6,000	I	10.3	33.3	22.0	10.3	13.3	11.9
6,000	II	-	3.3	1.7	20.0	10.0	15.0
6,000	III	-	-	-	10.0	5.0	7.5
6,000	IV	3.4	1.7	2.5	8.5	-	4.2
6,000	V	-	1.7	0.8	10.0	5.0	7.5

*4 hr/d, 5 d/wk, 52 wk (I); 60 animals per group.

4 hr/d, 5 d/wk, 17 wk (II); 60 animals per group.

4 hr/d, 5 d/wk, 5 wk (III); 120 animals per group.

1 hr/d, 4 d/wk, 25 wk (IV); 120 animals per group.

4 hr/d, 1 d/wk, 25 wk (V); 120 animals per group.

**M=male; F=female; T=total.

Table 4. Incidence (%) of liver angiosarcomas after chronic
vinyl chloride inhalation: effect of species

Concentration (ppm)	% Male Animals with Tumors			
	Sprague-Dawley Rats[a]	Wistar Rats[b]	Swiss Mice[c]	Golden Hamsters[d]
10,000	10.0	29.6	3.8	–
6,000	10.3	11.5	6.7	3.3
2,500	20.0	12.0	20.7	–
500	–	10.0	20.0	6.7
250	3.4	3.7	30.0	–
50	–	–	3.3	–
0	–	–	–	–

[a] 4 hr/d, 5 d/wk, 52 wk; 60 animals per group.

[b] 4 hr/d, 5 d/wk, 52 wk; 30–40 animals per group.

[c] 4 hr/d, 5 d/wk, 30 wk; 60–150 animals per group.

[d] 4 hr/d, 5 d/wk, 30 wk; 30–62 animals per group.

Table 5. Incidence (%) of Zymbal gland carcinoma after chronic vinyl chloride inhalation: effect of strain

Concentration (ppm)	% Male Animals with Tumors	
	Sprague-Dawley Rats[a]	Wistar Rats[b]
10,000	33.3	7.4
6,000	10.3	7.7
2,500	3.3	-
500	10.0	-
250	-	-
50	-	-
0	-	-

[a] 4 hr/d, 5 d/wk, 52 wk; 60 animals per group.

[b] 4 hr/d, 5 d/wk, 52 wk; 30-40 animals per group.

Table 6. Incidence (%) of liver angiosarcomas in Sprague-Dawley rats after exposure to vinyl chloride by inhalation: effect of age

Concentration[a] (ppm)	% Animals with Tumors*					
	Newborn rats[b,c]			11 weeks old rats[b,d]		
	M	F	T	M	F	T
10,000	25.0	45.0	34.1	1.7	-	0.8
6,000	27.8	50.0	40.5	-	-	-

* M=male; F=female; T=total

[a] 4 hr/d, 5 d/wk, 5 wk
[b] Age at start of exposure
[c] 43-45 animals per group
[d] 126 animals per group

Table 7A. Vinyl chloride: concentrations where total cancer
           bearing Sprague-Dawley rats were significantly
           in excess*

Sex	Inhalation[a] (ppm)	Ingestion[b] mg/Kg b.w.
Male	30,000; 10,000; 6,000; 2,500; 500; 250; 200; 50	50
Female	30,000; 10,000; 6,000; 2,500; 500; 200; 150; 50	50

*Fisher exact probability test ($p \leq 0.05$)

[a] 4 hr/d, 5 d/wk, 52 wk; 2025 animals, 60-360 animals per group
[b] Gavage; 5 times/wk, 52 wk; 920 animals, 80 or 150 animals per group

Extra-hepatic angiosarcomas of different sites: These tumors
are observed at a very low incidence in untreated Sprague-
Dawley rats of our colony. Our data strongly suggest a VC
dependency in the appearance of these tumors in these rats. This
relationship is supported by the excessive incidence of extra-hepatic
vascular tumors in mice treated with VC.

Hepatomas: A few cases of hepatomas have been observed in
VC-treated groups of animals, when exposed as adults. This tumor
is exceptionally rare in our Sprague-Dawley rat colony, and none
have been observed in the control animal groups of the seventeen
experiments. Moreover, a high incidence of hepatomas was observed in Sprague-Dawley rats following neonatal exposure to a
high concentration of VC for a short period.

Considering their rare occurrence, or total absence, in our
animal colony, particular attention should also be given to
the appearance of the following tumors, even when not statistically
significant in animals exposed to low doses.

Zymbal gland carcinomas: An excess of these tumors appears
to be present in Sprague-Dawley at airborne concentrations down
to 50 and 25 ppm.

Table 7B. Vinyl chloride: tumors significantly in excess*

Tumor	Sprague-Dawley Rats	Inhalation[a] (ppm)	Exposure	Ingestion[b] (mg/Kg b.w.)
Zymbal gland carcinoma	Male	30,000;10,000		
	Female	30,000;10,000		
Liver angiosarcoma	Male	30,000; 2,500; 200		50
	Female	30,000; 6,000; 2,500; 500; 200; 150; 50		50; 16.65
Nephroblastoma	Male	2,500; 200, 150, 100		
	Female	500; 250		
Neuroblastoma	Female	10,000		
Mammary gland adenocarcinoma	Female	150; 50; 25, 10; 5		
Forestomach papilloma	Male	30,000		
	Female	30,000		

*Fisher exact probability test (p<0.05)
[a]4 hr/d, 5 d/wk, 52 wk; 2025 animals, 60-360 animals per group.
[b]Gavage; 5 times/wk, 52 wk; 920 animals, 80 or 150 animals per group.

Liver anigosarcomas: This tumor is extremely rare in our animal colony (4 cases in several thousand untreated animals). Therefore, one must consider their appearance as important even when not statistically different. Their appearance at airborne concentrations below 50 ppm (5 liver angiosarcomas out of 120 animals at 25 ppm, and 1 liver angiosarcoma out of 120 animals at 10 ppm), and at ingested doses of 1 mg/Kg b.w. (3 liver angiosarcomas out of 150 animals), and at 0.3 mg/Kg b.w. (1 liver angiosarcoma out of 150 animals) is especially striking.

Nephroblastomas: The onset of a few of these tumors, observed after inhalation treatment at doses below 100 ppm and in groups treated by ingestion with 50 and 16.65 mg/Kg b.w. is not accidental in our opinion, given the extreme rarity of these tumors in rats.

Neuroblastomas: To date, this tumor has never been observed by us in control or otherwise treated Sprague-Dawley rats used in our laboratory. Therefore, we consider their occurrence as dependent on VC treatment even at inhaled exposure levels below 10,000 ppm, e.g. 6,000 and 2,500 ppm.

The ability of low concentrations of VC to induce tumors may be better evaluated by considering all tumor types (Table 8). Below 10 ppm and 0.3 mg/Kg b.w., no increase in the specifically

Table 8. Induction of tumors at low concentrations of vinyl chloride in Sprague-Dawley rats*

---

at 25 ppm   : over 120 animals; 5 liver angiosarcomas, 4 Zymbal gland carcinomas and 1 nephroblastoma.

at 10 ppm   : over 120 animals; 1 liver angiosarcoma, 2 extra-hepatic angiosarcomas and 2 Zymbal gland carcinomas.

at 1 mg/Kg   : over 150 animals; 3 liver angiosarcomas, 1 extra-hepatic angiosarcoma, 1 hepatoma and 5 Zymbal gland carcinomas.

at 0.3 mg/Kg   : over 150 animals; 1 liver angiosarcoma and 1 hepatoma.

---

*not significantly different by the Fisher exact probability test ($p \leq 0.05$) but biologically meaningful based on the rare occurrence of the specified tumors in the animal colony.

VC related tumors were observed in the experiments on Sprague-Dawley rats.

CONCLUSION

As outlined historically in Table 9, these experimental studies led to the discovery that VC is carcinogenic and to an approach for quantifying the risk. As a direct consequence, of our investigation considerable effort has been made to control occupational exposure to this industrial carcinogen. Long-term carcinogenicity bioassays on VC were a crucial step in the field of environmental and occupational carcinogenesis which, in turn, are among the most important areas of public health.

The VC project and the other bioassays on industrial compounds of the program being carried out in our Institution have demonstrated that the experimental approach is a unique tool for:

1. identifing environmental and occupational carcinogens;

2. predicting target organs and type of tumors;

3. assessing the risk in quantitative terms in relation to dose;

4. providing a highly standardized model for the classification of compounds by potency; and

5. improving our scientific knowledge aimed at selecting the most suitable experimental protocols for extrapolating animal data to humans.

Table 9. The history of Vinyl Chloride carcinogenicity

1961	: VC was found to produce liver enlargement and microscopic hepatic degenerative changes (1).
1970	: Zymbal gland carcinomas were reported in rats exposed to 30,000 ppm of VC, by inhalation (2).
1970	: An increase in atypias in respiratory cells was observed among workers heavily exposed to VC (3).
1971 (July)	: A vast project of long-term carcinogenicity bioassays on VC was started in Bentivoglio, near Bologna, Italy (the BT Project)

1972   : Zymbal gland carcinomas, nephroblastomas and
(August)   liver angiosarcomas were observed in rats
                exposed to VC by inhalation (Maltoni, BT Project)

1973   : The first data of the BT Project were released
(April)   to the scientific community: the oncogenic effect
                was observed down to 250 ppm (3).

1973   : Splenomegalic liver disease was found among
                polyvinyl chloride production workers (4).

1973   : For the first time a case of liver angiosarcoma
(December)   in a worker of polyvinyl chloride production was
                correlated to VC exposure (5).

1974   : On the basis of the BT Project data indicating
(February)   a carcinogenic effect at 250 ppm, OSHA proposed
                a TLV of 50 ppm (Maltoni: testimony at the I OSHA
                Hearing on VC, 1974).

1974   : The BT project data showed that VC is a multi-
(February)   potential carcinogen, producing a variety of tumors,
                in different animal species (Maltoni: testimony at
                the I OSHA Hearing on VC, 1974).

1974   : The BT project data indicated a carcinogenic effect
                at 50 ppm. OSHA proposed new stricter rules
                (Maltoni: communication at the II OSHA Hearing on
                VC, 1974; 6).

1974   : Early epidemiological observation (paralleling the
                experimental information) indicated an increase in
                tumors other than liver angiosarcomas (of brain,
                lung, liver, hemolymphoreticular tissues) among
                workers of VC-PVC industries (7).

1974   : BT project data showed that VC had carcinogenic
                effects in rats also when given by ingestion (8).

1976   : In rats of the BT project exposed to VC by inhala-
                tion, angiosarcomas were observed down to the level
                of 25 ppm, and Zymbal gland carcinomas down to the
                level of 10 ppm (9). (These data were unaccount-
                ably omitted in IARC Monograph 19, February 1979).

1979   : Presentation of the final results of the effects
                of long-term exposure to a range of 14 doses by
                inhalation (from 30,000 to 1 ppm) and of 6 doses by
                ingestion (from 50 mg to 0.03 mg/Kg b.w.) on
                Sprague-Dawley Rats, at the XXe Reunion du Club de
                Cancerogenese Chimique, Fondation Curie, Paris" (10).

REFERENCES

1. T. R. Torkelson, F. Oyen and V. K. Rowe, The toxicity of vinyl chloride as determined by repeated exposure of laboratory animals. Am. Ind. Hyg. Assoc. J. 22:354 (1961).

2. P. L. Viola, A. Bigotti and A. Caputo, Oncogenic response of rat skin, lungs and bones to vinyl chloride. Cancer Res. 31:516 (1971).

3. C. Maltoni, Occupational carcinogenesis. II International Symposium on cancer detection and prevention, Bologna 1973, in: Advances in Tumor Prevention, Detection and Characterisation. Excerpta Medica (Amsterdam) 2:19 (1974).

4. H. K. Marsteller, W. K. Selbach, R. Muller, S. Juhe, C. E. Lange, H. G. Rohner and G. Veltman, Chronic toxic liver damage in workers of PVC producing plants. Deut. Med. Wochshcr 98:2311 (1973).

5. J. L. Creech and M. N. Johnson, Angiosarcomas of liver in the manufacture of polyvinyl chloride. J. Occupational Med. 16:150 (1974).

6. C. Maltoni and G. Lefemine, Carcinogenicity bioassays of vinyl chloride: Current results, in: Toxicity of Vinyl Chloride-Polyvinyl Chloride. New York Academy of Science, New York, 195 (1975).

7. J. K. Wagoner, Statement before the Subcommittee on the Environment of the U.S. Senate Commerce Committee (1974).

8. C. Maltoni, A. Ciliberti, L. Gianni and P. Chieco, Insorgenza di angiosarcomi in ratti in seguito a somministrazione per via orale di cloruro di vinile. Gli Ospedali della Vita 2, fasc. 1:65 (1975).

9. C. Maltoni, Vinyl chloride carcinogenicity: an experimental model for carcinogenesis studies, in: Origins of Human Cancer. Cold Spring Harbor Laboratory, 119-146 (1977).

10. C. Maltoni, G. Lefemine, A. Ciliberti, G. Cotti and D. Carretti Vinyl chloride carcinogenicity bioassays (BT project) as an experimental model for risk identification and assessment in environmental and occupational carcinogenesis, in: Epidemiologie animale et epidemiologie humaine: le cas du chlorure de vinyle monomere. Publications Essentielles (Paris) 15-112, (1980).

DISCUSSION

Q. Infante (U.S. Dept. of Labor): Given the same species, how much difference in carcinogenic response do you see by the various routes of administration?

A. Maltoni (Italy): I would say that the ratio of certain cancer types is very much changed by ingestion.

Q. Infante (U.S. Dept. of Labor): Rather than incidence, how about in terms of site-specific cancer induction?

A. Maltoni (Italy): I would say that by inhalation for vinyl chloride we get a broader spectrum of tumors.

INHALATION ANESTHETICS

Vincent F. Simmon

Genex Corporation
12300 Washington Avenue
Rockville, MD   20852 (U.S.A.)

INTRODUCTION

A number of epidemiological studies have suggested that adverse
health and reproductive effects are associated with exposure
of operating room personnel to inhalation anesthetics (1-5). Most
of the modern volatile anesthetics are alkyl halides, a class of
compounds that has come under scrutiny because of the demonstrated
in vivo carcinogenic and in vitro mutagenic activities of some of
its members. Alkyl halides having one or both of these activities
include, but are not limited to, methyl iodide, chloroform, bis
(α-chloroethyl) ether, bis (chloromethyl) ether, chloromethyl methyl
ether, vinyl chloride and trichloroethylene.

With the exception of chloroform, all of these carcinogens are
mutagenic in the Salmonella typhimurium assay developed by Ames (6).
Anesthetics are structurally similar to many known carcinogens and
mutagens. Therefore, we undertook a series of experiments to assess
the mutagenicity of these commonly used anesthetics and some of
their purported metabolites in S. typhimurium 7-16 . Some of
these compounds also were evaluated for their ability to induce
mutations in Escherichia coli and to damage DNA in a differential
toxicity assay.

MATERIALS AND METHODS

Due to the volatility of these compounds, variations of the
standard Ames protocol were employed; bacteria were exposed in
suspension and/or in a desiccator. All assays were conducted both

in the absence and in the presence of a metabolic activation system
prepared from the livers of young adult male rats pre-treated with
Aroclor 1254 (6). Exposure concentrations were chosen to bracket
anesthetic concentrations and included doses high enough to
cause significant lethality to the microorganisms used in the assay.

RESULTS

Halothane

Halothane was not mutagenic towards strains TA98 or TA100 when
tested in suspension using stationary phase cells in buffer, in
suspension using logarithmically growing cells, or in desiccator
experiments (7-13).

Halothane Metabolites

Trifluoroacetic acid, a major urinary metabolite of halothane,
was tested for mutagenic activity both in agar incorporation assays
and in spot tests. No evidence of mutagenic activity was observed
(7).

The volatile metabolites CF_2CBrCl, CF_2CHCl, and CF_3CHCl_2 were
tested with strains TA98 and TA100 is desiccators, in suspension
using stationary cells in buffer, and in suspension using logarithmically growing cells (13). The metabolite CF_2CBrCl increased the
number of revertants per 10^8 survivors of strain TA100 by approximately five fold when tested in suspension using logarithmically
growing cells. A slight (40% above controls), dose-related increase in revertants of strain TA100 was observed in desiccator
experiments with CF_2CHCl.

Five non-volatile, conjugated metabolites, R-SCFCHCR,
$R-SCF_2CH_2Cl$, R-SCFCHBr, $R-SCF_2CH_2Br$ and $R-SCF_2CHBrCl$
($R=CHC_3ONHC(COOH)HCH_2-$) were tested in S. tryphimurium by the
standard Ames agar incorporation protocol using strains TA1535,
TA1537, TA1538, TA98 and TA100, with and without metabolic activation
preparation (14). None were found to be mutagenic. However, the
three conjugates R-SCFCHCl, R-SCFCHBr and $R-SCF_2CHBrCl$ were more
toxic to a Bacillus subtilis rec$^-$ strain than to a rec$^+$ strain in a
differential toxicity assay (14). Four of the conjugates
($R-SCF_2CHCl$, $R-SCF_2CH_2Cl$, $R-SCF_2CH_2Br$ and R-SCFCHCl) were subsequently
tested for mutagenic activity in suspension assays with logarithmically growing TA1537 and TA100 (15). All four were found to
increase the number of revertants per 10^6 surviving cells with strain
Ta100; no effect was observed with strain TA1537. Appropriate
reconstruction experiments were conducted to demonstrate that the
presumed mutagenic effect was not artifactual.

Enflurane, Methoxyflurane and Isoflurane

These three halogenated ether anesthetics were tested for mutagenic activity in desiccators and with stationary phase cells in buffer and were not found to be mutagenic (9).

Fluroxene

Fluroxene was the first halogenated ether anesthetic and it enjoyed use from 1954 until 1975 when its production was discontinued, presumably due to its flammability. Fluroxene was found to be mutagenic when activated by a liver homogenate activation system prepared from the livers of Aroclor pretreated rats, mice or hamsters (8). Mutagenicity was observed with S. typhimurium strains TA1535 and TA100 and with Escherichia coli WP2 (uvrA), but not with S. typhimurium strains TA1537 or TA98. The mutagenic effect was observed only with stationary phase cells in suspension assays; assays in desiccators were negative. No assays with logarithmically growing cells were conducted. Assays with uninduced rodent liver preparations and with human liver preparations also gave nonmutagenic responses. Trifluoroethanol, a major rodent metabolite of fluroxene, was not mutagenic in any of the assays.

Trichloroethylene

Trichloroethylene, a purported weak carcinogen, has had limited use in Europe as an anesthetic. It is a weak mutagen in S. typhimurium TA100 with metabolic activation when assayed in desiccators or in liquid suspension with stationary phase cells (11,17).

Divinyl Ether

This anesthetic is flammable and is not in use in the United States. It was shown to be mutagenic towards TA100, both in the absence and presence of an exogenous metabolic activation system (more so with activation) when assayed in desiccators, and only with metabolic activation when assayed in suspension (16).

Nitrous Oxide and Cyclopropane

These two anesthetics were not mutagenic when assayed in desiccators or in stationary phase suspension assays with S. typhimurium TA1535 or TA100 (16).

Assays of Urine

It has been reported that mutagens can be detected in the urine of anesthesiologists (18). We have examined urine from patients anesthetized with halothane, enflurane, methoxyflurane or isoflurane

(7-9), urine from rats (both uninduced and phenobarbital induced) exposed to fluroxene, and urine from medical students and anesthesiologists in both scavenged and unscavenged operating rooms (8,19). In the two latter cases, the urine was tested both before and after concentration using XAD-2 resin columns. In no case did we observe a mutagenic effect associated with exposure to these anesthetics, and it is our contention that the previous report (18) is in error or describes an artifact.

CONCLUSIONS

The Salmonella/microsome assay has been used extensively to detect mutagenic chemicals that may also be carcinogens. The vast majority of data indicating a high correlation between Salmonella mutagens and carcinogens is based on agar incorporation of the test chemicals. This procedure was not applicable to testing of most volatile chemicals so modifications were necessary. Using this modified approach, we found none of the inhalation anesthetics commonly used in the United States to be mutagenic. Some known (and several presumptive) metabolites of these anesthetics were found to be weakly mutagenic under special assay conditions (using suspension assays with logarithmically growing cells). The sensitivity of this procedure, compared to the standard procedure, is not known; therefore, it is difficult to draw conclusions regarding the potential mutagenic/carcinogenic importance of these metabolites. However, the results argue against the use of anesthetic systems that recycle exhaled air (and metabolites) to the patient. In recent years, hospitals in the United States have adopted procedures to eliminate or reduce the exposure of operating room personnel to anesthetics. It will remain for epidemiologists to determine if exposure to anesthetics using current practices represents a hazard to patients and operating room personnel.

ACKNOWLEDGEMENT

This work is the result of a collaboration between the laboratory directed by the author at S.R.I. International and the co-authors named in references 7-17.

REFERENCES

1. E. N. Cohen, J. W. Belville and B. W. Brown, Anesthesia, pregnancy and miscarriage: A study of operating room nurses and anesthetists, Anesthesiology 35:343-347 (1971).
2. T. H. Corbett, R. G. Cornell, K. Liedling and J. L. Endres, Incidence of cancer among Michigan nurse anesthetists, Anesthesiology 38:260-263 (1973).

3. R. Doll and R. Peto, Mortality among doctors in different occupations, Br. Med. J. 1:1433–1436 (1977).

4. R. P. Knill-Jones, B. J. Newman and A. A. Spencer, Anaesthetic practice and pregnancy: Controlled survey of male anesthetists in the United Kingdom, Lancet 2:807–809 (1975).

5. P. O. Pharoah, E. Alberman and P. Doyle, Outcome of pregnancy among women in anesthetic practice, Lancet 1:34–36 (1977).

6. B. N. Ames, J. McCann, E. Yamasaki, Methods for detecting carcinogens and mutagens with Salmonella/microsome mutagenicity test, Mutat. Res. 31:347–364 (1975).

7. J. M. Baden, M. Brinkenhoff, R. S. Wharton, B. Hitt, V. F. Simmon and R. I. Mazze, Mutagenicity of volatile anesthetics: halothane, Anesthesiology 45:311–318 (1976).

8. J. M. Baden, M. J. Kelley, R. S. Wharton, B. A. Hitt, R. I. Mazze, and V. F. Simmon, Mutagenicity of fluroxene, Anesthesiology 45:695–701 (1976).

9. J. M. Baden, M. J. Kelley, R. S. Wharton, B. A. Hitt, V. F. Simmon, and R. I. Mazze, Mutagenicity of halogenated ether anesthetics, Anesthesiology 45:346–350 (1977).

10. J. M. Baden, M. J. Kelley, V. F. Simmon, S. A. Rice and R. I. Mazze, Fluroxene mutagenicity, Mutation Res. 58:183–191 (1978).

11. J. M. Baden, M. J. Kelley, R. I. Mazze and V. F. Simmon, Mutagenicity of inhalation anesthetics; Nitrous oxide, cyclopropane, trichloroethylene and divinyl ether, Br. J. Anaesth. 51:417–421 (1979).

12. J. M. Baden, V. F. Simmon, Mutagenic effects of inhalational anesthetics, Mutation Research 75:169–189 (1980).

13. H. N. Edmunds, J. M. Baden and V. F. Simmon, Mutagenicity studies with volatile metabolites of halothane, Anesthesiology 424–429 (1979).

14. K. Sachdev, E. N. Cohen, V. F. Simmon, Genotoxic and mutagenic assays of halothane metabolites in Bacillus subtilis and Salmonella typhimurium, Anesthesiology 53: 31–39 (1980).

15. K. Sachdev, M. V. Peirce, V. F. Simmon, Mutagenicity of halothane metabolites to log-phase cultures of Salmonella typhimurium (submitted for publication).

16. J. M. Baden, M. Kelley, R. I. Mazze and V. F. Simmon, Mutagenicity of inhalation anaesthetics: trichloroethylene, divinyl ether, nitrous oxide and cyclopropane, Br. J. Anaesth. 51: 417–422 (1979).

17. V. F. Simmon, K. Kauhanen and R. G. Tardiff, Mutagenic activity of chemicals identified in drinking water, Progr. Genet. Toxicol. (Proc. Symp.) 2:249–258 (1977).

18. E. C. McCoy, R. Hankel, H. S. Rosenkranz, J. G. Giuffrida and D. V. Bizzari, Detection of mutagenic activity in the urines of anesthesiologists: a preliminary report, Environ. Health Perspectives 21:221–223 (1977).

19. J. M. Baden, M. J. Kelley, A. M. Chung, and K. E. Mortelmans, Lack of mutagens in urine of operating room personnel, Anesthesiology 53:195–198 (1980).

DISCUSSION

Q. Lockard (Kentucky): Would you please comment on your dose
rates being expressed as percents?

A. Simmon (Genex): In all of these studies we bracketed the
normal anesthetic exposure levels and we also tried to use an
agent concentration which was clearly toxic to the microorganisms.
If tested in liquid suspension, the agent was added volume per
volume; if used in a dessicator, the agent was added as a known
liquid volume or the % airborne concentration in a sealed
dessicator was calculated and subsequently verified by gas
chromotography.

Q. Borg (BNL): Has the original finding of increased rate of
miscarriages in operating room personnel been verified?

A. Simmon (Genex): It is commonly believed that such an effect
does occur among operating room personnel. There are some 12-15
studies directed at this question and as far as I know, no
study has proved that an increased rate of miscarriages does not
occur.

Q. Borg (BNL): A comment, since you have used the present
profile of Ames tester strains, which are said to be quite
insensitive to the mutagenic activity of most haloalkanes. Ames
has recently developed another series of Salmonella strains with
different base sequences, which appears to be very sensitive to
oxidants and haloorganics. It may be worthwhile to re-examine
these anesthetics and their metabolites using these new strains
when they become available.

A. Simmon (Genex): I think the statement that the current Ames
tester strains are insensitive to halogenated compounds, or
haloalkanes, is not entirely correct. Using the appropriate
technology, for example a dessicator, mutagenic responses in
these strains can be detected at parts per billion for some of
these compounds.

Q. Tice (BNL): The question I have is really applicable to any
of the airborne agents we've been talking about during this
symposium. Do you have any indication as to whether the constant
exposure of operating room personnel to low levels of these
anesthetics would alter their sensitivity to some carcinogenic
agents? Perhaps we should be just as much concerned about
secondary sensitization effects as we are about the primary
mutagenic and carcinogenic ones?

A.   Simmon (Genex): That problem is very important. In experiments
where animals have been exposed to anesthetics, their liver
metabolism has been altered with assayed in vitro. There is just
no way at the present to conclude whether that altered metabolism
will make someone more or less sensitive to carcinogenic induction
by other compounds.

MUTAGENIC AND CARCINOGENIC EFFECTS OF FORMALDEHYDE

Craig J. Boreiko, David B. Couch and James A. Swenberg

Chemical Industry Institute of Toxicology
P. O. Box 12137
Research Triangle Park, NC   27709   (U.S.A.)

INTRODUCTION

Formaldehyde is an important commodity chemical with widespread and diverse industrial applications (1,2). Formaldehyde production in the United States has been steadily increasing for a number of years, and was close to 3 million tons in 1978 (1). World-wide formaldehyde production approached 12 million tons (1). Approximately 50% of the formaldehyde produced in the United States is utilized in urea-formaldehyde and phenol-formaldehyde resin production (1,2). These resins are used in turn for the manufacture of particle board, plywood and home-insulation foams. Formaldehyde resins are also used as molding compounds for the manufacture of dinnerware, appliances and telephones. Other uses of formaldehyde include the production of herbicides, pharmaceuticals and permanent-press clothing. A significant portion of the U.S. work force is thus involved in the manufacture and utilization of formaldehyde or formaldehyde based products (3).

In addition to potential occupational exposure to formaldehyde, people may be exposed to formaldehyde from public sources. The incomplete combustion of hydrocarbons yields measureable concentrations of formaldehyde, and exhaust from vehicles and incinerators may contain significant concentrations of formaldehyde (4,5). Formaldehyde concentrations as high as 0.16 ppm, with a daily average of 0.06 ppm, have been measured in photochemical smog in Los Angeles (6). Cigarette smoke may contain up to 40 ppm of formaldehyde (7). Between 0.4 and 8.1 ppm of formaldehyde was detected in 38 of 68 homes insulated with urea-formaldehyde foam insulation (8). The formaldehyde levels in the remaining homes were below the limits of detection (0.4 ppm) for the

analytical methods used. Formaldehyde concentrations in the
air within mobile homes constructed of particle board has
reportedly ranged from 0.03 to 2.5 ppm; approximately two-thirds
of the homes contained formaldehyde concentrations less than
0.5 ppm (9). Since the formaldehyde measurements within these
dwellings were determined in response to consumer complaints in both
of these studies (8,9), they may not be representative of the
levels which exist in all homes of similar construction.

The current OSHA threshold limit value for formaldehyde
specifies an 8 hr Time-Weighted Average (TWA) of 3 ppm (3).
Although this standard is based on the irritant properties of
formaldehyde, airborne formaldehyde concentrations between 1.5 and
3.0 ppm have caused at least moderate discomfort in 20% of
exposed human subjects in clinical studies (5). The adverse
effects reported in these studies, and in the workplace, include
eye and upper respiratory tract irritation, headaches, gastointes-
tinal distress and allergic dermatitis (1,4,5,8). The severity
of an adverse response, and the formaldehyde concentration needed
to elicit the response, vary greatly among individuals. Unusually
sensitive persons can experience some discomfort at formaldehyde
concentrations of 0.05 ppm or lower (4,5). Other individuals
with occupational exposure to formaldehyde report the development
of tolerance to irritant effects during the course of a workday
(3).

Formaldehyde-Induced Mutagenicity

The ability of formaldehyde to induce genetic alterations in
biological organisms has been under study for several decades (10).
Formaldehyde is known to exert weak mutagenic effects in fungi,
yeast, Drosophila and some bacteria. However, a number of in
vitro test systems, including the standard Ames Salmonella assay,
have found formaldehyde to lack mutagenic activity (4,11,12).
In more recent studies, formaldehyde has been found to be genotoxic
to some types of cultured mammalian cells (4,12). Combined,
these studies suggest that the genotoxic effects of formaldehyde
are complex and varied.

In an effort to define some of the variables operative in
formaldehyde mutagenesis, we undertook a series of experiments
which examined the effects of formaldehyde on Salmonella typhimurium.
At the start of this study, we considered the possiblity that
formaldehyde's failure to exert genotoxic effects in the standard
Ames assay might be a function of the molecular mechanism of
formaldehyde mutagenesis. If formaldehyde were not capable of
inducing one of the specific base changes required for reversion
in the Ames tester strains, its mutagenic potential would remain
undetected. Since forward mutation assays partially circumvent this
theoretical limitation, we examined the ability of formaldehyde

to produce forward mutations to 8-azaguanine resistance in S. typhimurium.

The ability of formaldehyde to induce mutations to 8-azaguanine resistance was studied in S. typhimurium strain TM677. In accordance with the procedures of Skopek et al. (13), logarith- mically growing cultures of TM677 were incubated at 37°C for two hours in the presence of dilute formalin prepared in phosphate- buffered saline. After treatment, approximately 10^6 cells per dish were plated in the presence of 8-azaguanine for the enumera- tion of resistant bacteria, and $\geq$ 200 cells were plated per dish in the absence of the selective agent to assess cytotoxicity.

Formaldehyde was determined to be a weak mutagen for strain TM677. A dose dependent increase in mutant frequency, up to 7- times background, was observed at concentrations of formaldehyde between 2 and 8 μg/mL (Table 1). The survival of treated cells relative to solvent controls indicates that formaldehyde is also quite cytotoxic to these cells. The inclusion of an S9 mixture during treatment did not influence the results.

Table 1. Mutagenicity and cytotoxicity of
formaldehyde to S. Typhimurium TM677

Concentration[a] (μg/mL)	Percent Survival[b,c]	Mutant Fraction[c] (x 10^5)
0	100	5.7 ± 3.2
2	92 ± 8	7.6 ± 4.1
4	59 ± 12	12.6 ± 5.0
6	27 ± 7	22.7 ± 5.6
8	17 ± 7	36.6 ± 22.2

[a]Formaldehyde was administered as dilute formalin.

[b]Survival is expressed relative to solvent controls.

[c]Mean ± standard deviation (n = 5).

Since the observed increases in mutant frequency were small
and were paralleled by significant cytotoxicity, reconstruction
experiments were performed to insure that the observed increase
in mutants was not actually due to selective cytotoxicity which
favored the survival of spontaneous mutants over wild-type cells.
Studies (data not shown) were conducted by subjecting defined
mixtures of wild-type and mutant cells to the mutagenesis protocol
and then analyzing the relative proportions of the mutant and
wild-type phenotype among the survivors. No differential
cytotoxicity for the two cell types by formaldehyde was observed.

We have since determined that the ability of formaldehyde to
induce mutations in a forward mutation assay, but not in the
standard Ames test, probably does not reflect upon the molecular
mechanism of formaldehyde mutagenesis. Rather, technical
differences between the two assays may account for the opposing
results. The standard Ames test does not include a preincubation
step such as is used in the forward mutation assay. Our initial
results indicate that if a preincubation step is incorporated
into the Ames test, formaldehyde is detected as a weak mutagen.
Similarly, we have found that omission of the preincubation step
from the forward mutation assay results in a loss of formaldehyde
mutagenesis. A plausible explanation for these observations
might be found in the studies of Dr. Barber, presented elsewhere
in this volume. The 45°C temperatures maintained at the start of
the standard Ames assay may cause rapid evaporation of volatile
test compounds during treatment. This could cause false-negative
results in the study of chemicals such as formaldehyde which
possess low boiling points.

Cell Transformation Studies

Observations that formaldehyde possesses mutagenic properties
suggest that formaldehyde may also be potentially carcinogenic.
In an effort to define this carcinogenic potential, we have studied
the effects of formladehyde in the C3H/10T½ cell transformation
system. Cell transformation systems employ cultures of non-tumori-
genic cells which can be converted to a tumorigenic state following
exposure to a carcinogen. The process of oncogenesis in cell culture
often appears to mimic certain aspects of in vivo carcinogenesis,
suggesting that cell transformation systems may provide a biologically relevant in vitro means for detecting potential chemical
carcinogens.

The C3H/10T½ cell transformation system utilizes a cell line
of mouse embryo fibroblasts (C3H/10T½ Cl 8) which displays strong
density-dependent controls upon cell division (14,15). The
technical details of the transformation system have changed little
since its inception (15) and have been recently reviewed (16).

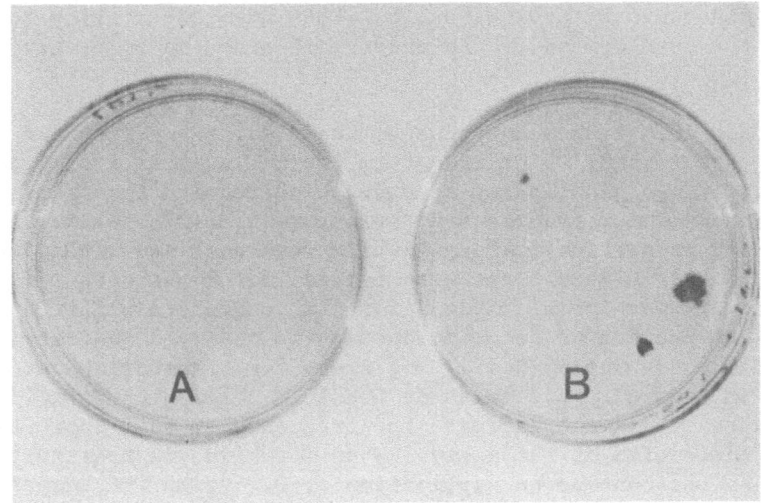

Figure 1.   Select fixed and stained dishes from transformation
studies.   A.   An untreated dish of C3H/10T½ Cl 8 cells held at
confluence 5 wk.   Note the evenly staining monolayer of cells.
B.   A carcinogen treated dish of C3H/10T½ Cl 8 cells.   The
darkly staining areas are foci of transformed cells.

Briefly, if 2000 C3H/10T½ cells are inoculated into a cell
culture disn and provided regularly with fresh medium over a six
week incubation, they will cease cell division after forming an
evenly staining confluent monolayer across the bottom of the dish
(Fig la).   If, however, growing C3H/10T½ cultures are treated
with a carcinogen, some of the treated cells may be oncogenically
transformed and lose their density-dependent controls over cell
division.   These transformed cells continue to proliferate after
a confluent monolayer has formed and produce multilayered foci
of transformed cells.   These foci are easily detected by their
deep-staining pattern in dishes stained with Giemsa (Fig lb).
Cells isolated from transformed foci are frequently tumorigenic
when injected into syngenic hosts.

The transformation of C3H/10T½ cells can, in certain instances,
be observed to proceed through the discrete stages of initiation
and promotion (16) in a fashion quite similar to multiple-stage
carcinogenesis in vivo (17).   When C3H/10T½ cells are treated with
subthreshold (initiating) doses of chemical carcinogens (16,18)
or radiation (16,19), very few transformed foci are produced.
Rather, treatment with an initiating agent produces cells with a
·latent tranformed potential.   If C3H/10T½ cultures containing
initiated cells are then continuously exposed to a tumor promoter,
such as 12-0-tetradecanoyl-phorbol-13-acetate (TPA), transformed

foci will develop. Tumor promoters are generally considered not
to be carcinogenic by themselves, but they have the ability to
enchance the development of the oncongenic state in initiated
cells.

The C3H/10T½ system can thus be utilized as a screen for
complete carcinogens, which transform the cells after a single
treatment, and for initiating agents, which require the subsequent
application of tumor promoters to transform C3H/10T½ cells. The
transforming properties of formaldehyde were examined within this
framework. C3H/10T½ cultures were treated for 24 hr with
commercial 37% w/w formaldehyde diluted in phosphate buffered
saline. The pattern of focus production we observed indicated
that formaldehyde was an initiating agent for transformation. As
is shown in Figure 2, treatment of C3H/10T½ cells with formaldehyde
alone generally did not increase the frequency of foci above the
limits of detection of the assay. When formaldehyde treatment was
followed by the continuous application of 0.1 μg/mL TPA, numerous
transformed foci were produced in a dose-dependent fashion. This
finding is significant for several reasons. First, it provides a
scientific basis for postulating that formaldehyde may posssess
carcinogenic activity in vivo. Second, it suggests that studies on
in vivo formaldehyde toxicity should give consideration to multi-
stage mechanisms being operative in the expression of carcinogenic
properties.

Formaldehyde Carcinogenicity

In spite of formaldehyde's known properties as a mutagen and
general irritant, there have been few reports on the effects of
chronic exposure in animals. In order to facilitate assessment
of formaldehyde's toxicological effects, CIIT undertook a study
to investigate the effects of long-term inhalation exposure of
rats and mice to formaldehyde. Interim reports on this study
have recently been published (22,23), detailing the experimental
protocol and findings through sacrifice at the 18 month. The
results of this study through the sacrifice at the 24 month, as
well as some more recent findings on the effects of formaldehyde
in vivo are discussed below.

The CIIT inhalation study was conducted at the Battelle
Laboratories in Columbus, Ohio. Fisher-344 rats and B6C3F1 mice
were exposed to 0, 2, 6 or 15 ppm of formaldehyde vapor for 6
hr per day, 5 days per week for two years. For both species,
240 animals were exposed (120 animals per sex) at each dose level.

During the first year of the study, mortality in rats was
less than 5%. A sharp increase in mortality was noted among rats
exposed to 15 ppm of formaldehyde after 15 months. This increase
in mortality was primarily due to the occurrence of tumors in the

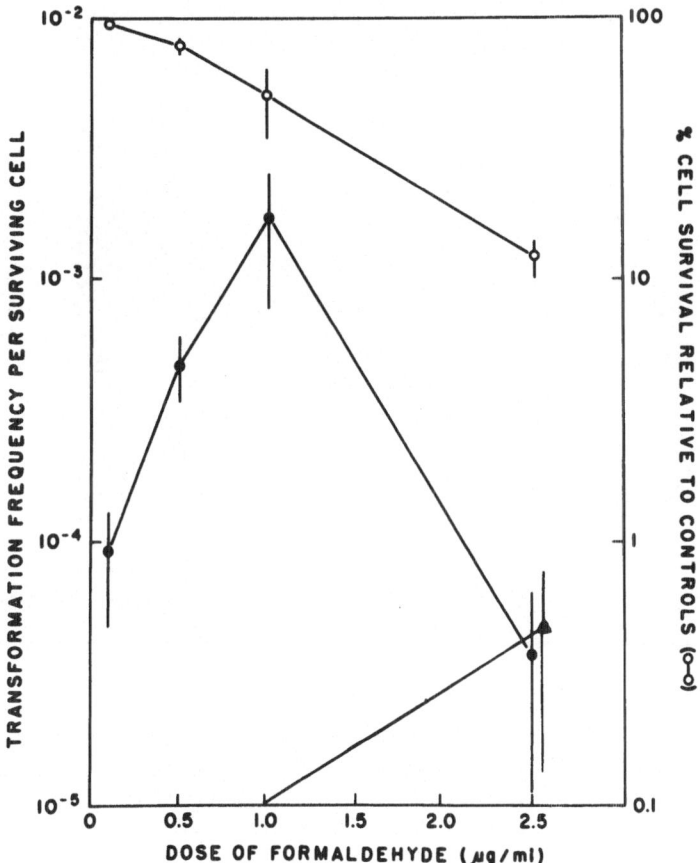

Figure 2.   Frequency of C3H/10T½ transformation following formaldehyde
treatment.   Exposure of cells to 0.1, 0.5, or 1.0 µg/mL of formaldehyde did not result in detectable levels of transformation.   A low
level of transformation was observed at 2.5 µg/mL (▲).   When formaldehyde treatment was followed by the application of tumor promoters,
numerous transformed foci were produced with a frequency that
generally increased with formaldehyde dose (●).   The cytotoxicity
of formaldehyde is also shown (O).   The procedural details for
these studies are reported elsewhere (20,21).

nasal turbinates (Fig 3).   Many of the tumors observed were large
osteolytic neoplasms, first detected as a localized swelling over
the nasal bones (Fig 3).   The majority of such tumors were squamous
cell carcinomas.   Microscopic examination of nasal turbinates from
other animals which had been sacrificed, or had died during the
course of the study, revealed the presence of additional neoplasms

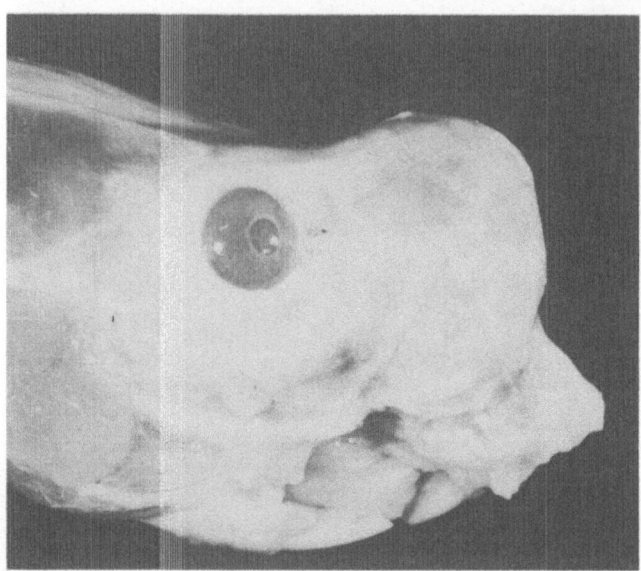

Figure 3.   Gross photography of a squamous cell carcinoma that
has grown through the nasal and maxilla bones of a rat exposed
to 15 ppm formaldehyde.

(Fig 4).   Tumor incidence data through the sacrifice at 24 months
(24) is shown in Table 2.   In exposed rats, no nasal carcinomas
were observed at 0 or 2 ppm formaldehyde.   Three tumors were observed at 6 ppm, and 95 tumors at 15 ppm, indicative of a very
sharp dose response.   In exposed mice, two nasal tumors were
observed at 15 ppm.   No nasal carcinomas were observed at lower dose
levels.

     Having established that formaldehyde can be carcinogenic for
rats and mice, research in our laboratories has been directed to
defining the mechanisms which produce this effect.   Constraints
of space will limit this discussion to a brief summary of a few
of these studies.   One of the first studies concerned the selectivity of the carcinogenic effects of formaldehyde for the nasal
turbinates.   Preliminary studies indicate this selectivity may be
imparted by the extreme water solubility and chemical reactivity
of formaldehyde and the obligatory nose breathing character of
rodents.   Autoradiographic studies were performed on tissue
sections taken from rats exposed to ^{14}C-formaldehyde (25).
These investigations revealed that inhaled formaldehyde was concentrated in the anterior portion of the nose (Fig 5).   Formaldehyde
binding was greatly diminished in the olfactory portions of the
nose and was limited to the naso-pharynx (Fig 6).   Thus, formaldehyde deposition strongly correlates with tumor location.

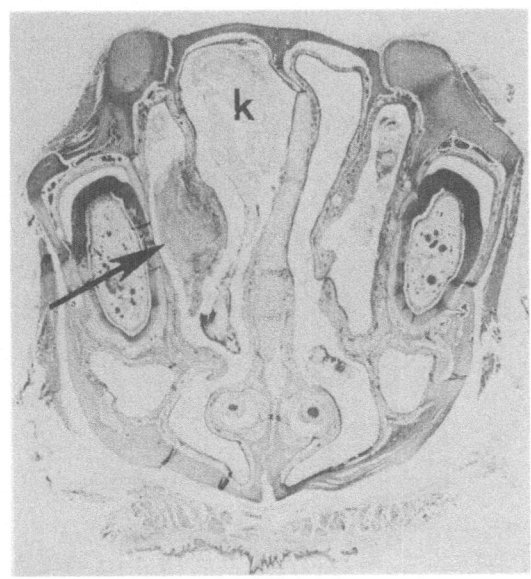

Figure 4. Macrophotograhy of an early squamous cell carcinoma (arrow) arising in the nasoturbinate of a rat exposed to 15 ppm formaldehyde. Note the excessive keratin (k) in the adjacent nasal cavity.

Table 2.   Incidence of nasal tumors following inhalation exposure to formaldehyde

Formaldehyde Concentration	Number of Tumor-bearing Animals Number of Animals at Risk	
	Mouse	Rat
0	0/220	0/201
2 ppm	0/220	0/205
6 ppm	0/220	3/204
15 ppm	2/85	95/220

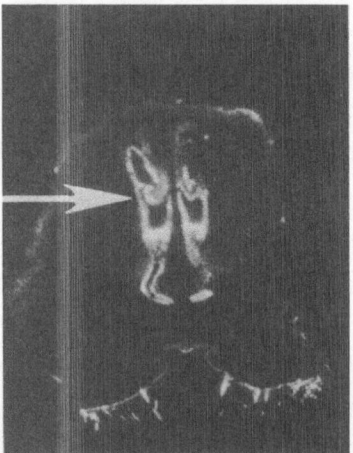

Figure 5. Autoradiograph of the anterior portion of a rat's nose following a 6 hr exposure to 15 ppm of ^{14}C-formaldehyde. Note the heavy deposition of radioactivity (arrow).

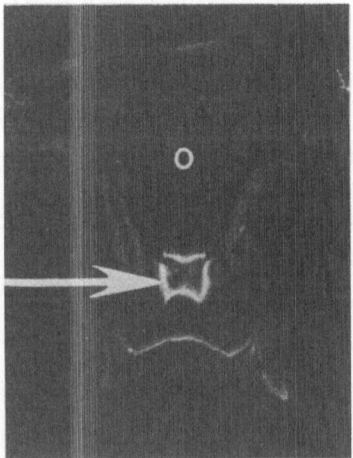

Figure 6. Autoradiograph of the posterior nasal cavity of the rat in Figure 5. Note the absence of radioactivity in the olfactory region (0) and its presence in the naso-pharynx (arrow).

A possible explanation has also been found for the species differences in formaldehyde carcinogenicity. Studies by Barrow et al. (26) have quantitated the minute volumes of both rats and mice during acute (10 minute) exposures to formaldehyde. Morphometric analyses of the nasal cavities of rats and mice have also been conducted (25). These data can be used to calculate the theoretical dose of formaldehyde being administered per cm^2 of nasal tissue during an inhalation exposure. Such calculations indicate that a mouse will receive approximately one half of the formaldehyde dose (per cm^2 of tissue at risk) received by a rat during a 10 minute exposure to 15 ppm of formaldehyde vapor (26). This calculation correlates well with the tumor incidence data which found the incidence in mice at 15 ppm to be equivalent to the incidence in rats at 6 ppm.

Very little is known about the actual mechanisms of formaldehyde carcinogenesis, although much of the available data suggests that the induction of neoplasms by formaldehyde may be a multi-stage process. For example, a time- and concentration-dependent progression of histological abnormalties were noted in rat turbinates during the chronic study (27). Exposed rats first developed squamous metaplasia, followed in order by squamous hyperplasia, carcinoma in situ and invasive carcinoma. A multi-step mechanism for formaldehyde carcinogenesis is further supported by the finding that formaldehyde vapor induces extensive cell turnover in the nasoturbinate target cells (25), a cell type normally characterized by low rates of cell replication. A high rate of cell turnover could help stage the tissue for oncogenic insult, may be responsible for the fixation of oncogenic lesions in DNA or may have a promoting influence on the development of malignancies. The cell transformation studies discussed earlier support this latter possibility.

Other factors that may be involved in formaldehyde's multi-stage carcinogenic effects have been discussed elsewhere (24,25). Speculation concerning the mechanism of formaldehyde induced neoplasia is extremely important in the risk assessment process. Statistical models used to calculate the carcinogenic risk formaldehyde poses to rats provide dramatically different results when one compares a single-hit linear extrapolation model to models which assume a multi-stage mechanism (24). Multi-stage models predict a 1:100,000 risk of nasal carcinomas in rats at concentrations between 0.98 and 2.3 ppm, values close to current occupational exposure limits. A linear extrapolation model predicts that this low incidence cannot be achieved until formaldehyde concentrations are reduced to 0.005 ppm, a level below that which is commonly found in ambient air. Epidemiology studies reported to date have been limited in scale, but have not found human occupational exposures to formaldehyde to be associated

with an increased incidence of neoplasia (28). The wide disparity
between the safe levels of formaldehyde predicted by these different
statistical models make it apparent that decisions concerning the
regulation of formaldehyde exposure can only be made after the
mechanisms operative in formaldehyde carcinogenesis have been
established.

REFERENCES

1. H. R. Gerberich, A. L. Stautzenberger and A. C. Hopkins,
    Formaldehyde, in: Kirk-Othmer Encyclopedia of Chemical
    Technology, 3rd ed., John Wiley and Sons, New York
    11:231-250 (1980).
2. J. F. Walker, Formaldehyde, 3rd ed., Robert E. Krieger
    Publishing Co., Huntington, New York (1975).
3. National Institute for Occupational Safety and Health,
    CDC/PHS/HEW, Criteria for a recommended standard:
    Occupational exposure to formaldehyde. DHEW (NIOSH)
    Pub. No. 77-126 (1976).
4. J. F. Kitchens, R. E. Casner, G. S. Edwards, W. E. Harward,
    III and B. J. Macri, Investigation of selected
    potential environmental contaminants: Formaldehyde.
    Final technical report. Atlantic Research Corp., Alexan-
    dria, VA. Report No. ARC 49-5681. EPA/560/2-76-009 (1976).
5. Formaldehyde - An assessment of its health effects. A report
    prepared for the Consumer Product Safety Commission by the
    National Academy of Sciences, Washington, D.C. (1980).
6. A. P. Altshuller, and S. P. McPherson, Spectrophotometric
    analysis of aldehydes in the Los Angeles atmosphere,
    J. Air. Pollut. Control Assoc. 13:109-111 (1963).
7. C. J. Kensler, and S. P. Battista, Components of cigarette
    smoke with cilary-depressant activity; their selective
    removal by filters containing activated charcoal granules.
    New Engl.J. Med. 269:1161-1166 (1963).
8. A. V. Sardinas, R. S. Most, M. A. Giulietti, and P. Honchar,
    Health effects associated with urea-formaldehyde foam
    insulation in Connecticut, J. Environ. Health 41:270-272
    (1979).
9. P. A. Breysse, Formaldehyde in mobile and conventional
    homes, Environ. Health & Safety News 25, 1 through 6 (1977).
10. C. Auerbach, M. Moutschen-Dahmen and J. Moutschen, Genetic
    and cytogenetical effects of formaldehyde and related
    compounds, Mutat. Res. 39:317-362 (1977).
11. A. Koops, and B. Butterworth, In vitro microbial and mutagenicity
    studies of formaldehyde. E. I. DuPont de Nemours &
    Company, Inc., Haskell Laboratory for Toxicology and
    Industrial Medicine, Wilmington, DE (1976).

12. D. Brusick, Genetic and transforming activity of formaldehyde, in: Proceeding of the Third CIIT Conference on Toxicology: Formaldehyde Toxicity. Chemical Industry Institute of Toxicology, Research Triangle Park, North Carolina, November 20-21, 1980. In press (1980).

13. T. R. Skopek, H. L. Liber, J. J. Krolewski, and W. G. Thilly, Quantitative forward mutation assay in Salmonella typhimurium using 8-azaguanine resistance as a genetic marker. Proc. Natl. Acad. Sci. (USA) 75:410-414 (1978).

14. C. A. Reznikoff, D. W. Brankow, C. Heidelberger, Establishment and characterization of a cloned line of C3H mouse embryo cells sensitive to post-confluence inhibition of cell division. Cancer Res. 33:3231-3238 (1973).

15. C. A. Reznikoff, J. S. Bertram, D. W. Brankow, C. Heidelberger, Quantitative and qualitative studies of chemical transformation of a clone of C3H mouse embryo cells sensitive to post-confluence inhibition of cell division. Cancer Res. 33:3239-3249 (1973).

16. S. Mondal, The C3H/10T½ Cl 8 mouse embryo cell line: Its use for the study of carcinogenesis and tumor promotion in cell culture, in: Advances in Environmental Toxicology. Vol. 1. Mammalian Cell Transformation by Chemical Carcinogens, Mishra N., Dunkel V., Mehlman, M., eds. Senate Press, Princeton Junction, New Jersey 181-211 (1980).

17. J. D. Scribner and R. Suss, Tumor initiation and promotion. Int. Rev. Exp. Pathol. 18:137-198 (1978).

18. S. Mondal, D. W. Brankow, and C. Heidelberger, Two-stage chemical oncogenesis in cultures of C3H/10T½ cells. Cancer Res. 36:2254-2260 (1976).

19. S. Mondal and C. Heidelberger, Transformation of C3H/10T½ Cl 8 mouse embryo fibroblasts by ultraviolet radiation and a phorbol ester, Nature 260:710-711 (1976).

20. C. J. Boreiko and D. L. Ragan, Formaldehyde effects in the C3H/10T½ cell transformation assay, in: Proceedings of the Third CIIT Conference on Toxicology: Formaldehyde Toxicicity Chemical Industry Institute of Toxicology, Research Triangle Park, North Carolina, November 20-21, 1980. In press (1980).

21. D. Ragan and C. J. Boreiko, Initiation of C3H/10T½ Cell Transformation by formaldehyde. Cancer Letters (in press) (1981).

22. J. A. Swenberg, W. D. Kerns, R. I. Mitchell, E. J. Gralla, and J. L. Pavkov, Induction of squamous cell carcinomas of the rat nasal cavity by inhalation exposure to formaldehyde vapor, Cancer Res. 40:3398-3402 (1980).

23. J. A. Swenberg, W. Kerns, K. Pavkov, R. Mitchell and E. J. Gralla, Carcinogenicity of formaldehyde vapor: interim findings in a long-term bioassay of rats and mice, in: Mechanisms of Toxicity and Hazard Evaluation, Proceedings of the Second International Congress on

Toxicology held in Brussels, Belgium, July 6-11, 1980.
B. Holmstedt, R. Lauwerys, M. Mercier and M. Roberfroid,
eds., 283-286 (1980).

24. J. E. Gibson, Risk assessment using a combination of testing
     and research results, in: Proceedings of the Third CIIT
     Conference on Toxicology: Formaldehyde Toxicity.
     Chemical Industry Institute of Toxicology, Research
     Triangle Park, North Carolina, November 20-21, 1980.
     In press (1980).

25. J. A. Swenberg, E. A. Gross, J. Martin, J. A. Popp, Mechanisms
     of formaldehyde toxicity, in: Proceedings of the Third CIIT
     Conference on Toxicology Formaldehyde Toxicity. Chemical
     Industry Institute of Toxicology, Research Triangle Park,
     North Carolina, November 20-21, 1980. In press (1980).

26. C. S. Barrow, W. H. Steinhagen and J. C. F. Chang,
     Formaldehyde sensory irritation, in: Proceedings of the
     Third CIIT Conference on Toxicology: Formaldehyde
     Toxicity. Chemical Industry Institute of Toxicology,
     Research Triangle Park, North Carolina, November 20-21,
     1980. In press (1980).

27. W. D. Kerns, D. J. Donofrio and K. L. Pavkov, The chronic
     effects of formaldehyde vapor inhalation in rats and
     mice. A preliminary report, in: Proceedings of the Third
     CIIT Conference on Toxicology: Formaldehyde Toxicity.
     Chemical Industry Institute of Toxicology, Research
     Triangle Park, North Carolina, November 20-21, 1980.
     In press (1980).

28. R. J. Levine, Respiratory disease among West Virginia
     morticians, in: Proceedings of the Third CIIT Conference
     on Toxicology: Formaldehyde Toxicity. Chemical Industry
     Institute of Toxicology, Research Triangle Park, North
     Carolina, November 20-21, 1980. In press (1980).

DISCUSSION

Q. Sivak (A.D. Little): About six months ago, at the request of
my contracting officer, the National Cancer Institute, we looked at
the effect of formaldehyde in the Balb/c-3T3 Transformation System,
a close relative of 10T½ System. We found that a three day
exposure to doses between 2 and 20 µg/mL resulted in a direct dose-
related induction of neoplastic transformation. We didn't test
whether formaldehyde was a promoter which can be done in that
system, but it was quite clear that formaldehyde, either given
directly into the medium, or when medium which was exposed was
applied to the cells, tranformations occurred.

A. Boreiko (CIIT): Right, this result has also been obtained at
ICI using the BHK_{21} Transformation System. This could just be a

characteristic of 10T½ cells. We don't want to generalize to other
cell types or to other species at this point.

Q. Hatch (Northrop): I'd like to make one comment for the complete-
ness of discussion. We've tested formaldehyde for its ability to
enhance viral transformation in hamster embryo cells exposed to
the gas phase in closed treatment chambers and the results were
positive. So this is another species to be added to the list.

Q. Leiber (PPG): While formaldehyde has fairly strong genotoxic
activity, is it not also possible that it may have a promotional
activity. Could this possibility be tested by using subthreshold
doses of initiators and then applying formaldehyde in a promotional
scheme?

A. Boreiko (CIIT): Yes, that is a good possibility. Formaldehyde
may also be a strong tumor promoter, principally on the basis of
the rather extreme cell turnover and hyperplasia that you see in
the turbinates following exposure. To my knowledge, formaldehyde
has not been tested in skin studies. We do have studies ongoing
in the 10T½ System to test the promotional capacity of formaldehyde.

Q. Galloway (Litton Bionetics): Have you any information about
sex differences in toxicity with formaldehyde? I don't know all
the details, but we have had some differences in toxicity with
inhalation of formaldehyde between sexes and also between pregnant
and nonpregnant animals. I was wondering if you knew any more
about it?

A. Boreiko (CIIT): To my knowledge, there is no sex difference.

Q. Hatch (Northrop): Craig, some of your colleagues have speculated that the type of damage that formaldehyde induces might be
very efficiently repaired. Would you care to comment on that?

A. Boreiko (CIIT): I couldn't comment on that.

THE MICROBIAL MUTAGENICITY OF NITROARENES

Robert Mermelstein, Herbert S. Rosenkranz and
Elena C. McCoy

Joseph C. Wilson Center for Technology
Xerox Corporation, Rochester, NY 14644 (U.S.A.)
and
Center for the Environmental Health Sciences
Case Western Reserve Univerity
Cleveland, OH 44106 (U.S.A.)

INTRODUCTION

There are several reasons which warrant a detailed and system-
atic study of the mutagenicity of aromatic nitro compounds in
bacteria.

1. Our interest in nitroarenes was first aroused during an
investigation into the source of the mutagenic activity detected
in the extracts of certain xerographic copies (1,2). The initial
observation that extracts of several photocopies were found to be
mutagenic in Salmonella tester strain TA98 was made by Lofroth and
coworkers (3). Through considerable detective work, we were able to
trace the mutagenic activity from the extracts of selected copies
to selected toners and then, to one specific type of carbon black
used as the colorant in the toners. Prolonged Soxhlet extraction
of the carbon black, followed by fractionation of the extract,
resulted in a neutral polar fraction, which contained less than
3% of the mass, yet more than 85% of the mutagenicity of the extract.
The two principal compounds, responsible for most of the observed
mutagenicity, were identified as 1,6- and 1,8-dinitropyrene (4).

The various nitro derivatives of pyrene attainable through
direct nitration were synthesized and evaluated in the Salmonella
mutagenicity assay. The activity of all six nitropyrenes toward
strain TA98 appeared to be unusually high. The mutagenic activity
of 1,8-dinitropyrene appeared to be about 27 times higher than

that of 3-amino-1-methyl-5H-pyrido (4,3,-b) indole (5), heretofore
considered the most mutagenic substance, and about 200 times
higher than that of the well known benzo(a)pyrene-7b,8a-diol-9a,10a-
epoxide in strain TA98 (6). Further, the mutagenicity of 1,8-dini-
tropyrene was not significantly diminished when tested in the nitroreductase-deficient tester strain TA98NR (2). These observations
encouraged us to look more closely at aromatic nitro compounds in
general.

   Prior to leaving this issue, it is important to point out that
Cabot Corporation, the manufacturer of the above carbon black, modified the production conditions for this item to result in a substantial reduction in the level of nitropyrenes without affecting the
other important chemical and physical properties of the carbon
black. The modified production conditions were implemented in
early 1980 and the resultant carbon black was supplied to all
previously known manufacturers of xerographic toners. Copies
made with toners using the modified carbon black do not yield
mutagenic extracts, verifying that the problem had been solved.

   2. Nitroarenes are ubiquitous in the environment as a result
of natural and man-made activities. Nitroaromatics are widely used
as intermediates in the synthesis of various organic compounds,
explosives and in pharmaceuticals. Polycyclic aromatic hydrocarbons
(PAH) are formed in various combustion processes (7). Their facile
reaction to form nitrated derivatives under acid catalysed conditions
has been documented (8-10). Some of the more likely sources of
nitroaromatics in the environment may include incinerators, coal
fired power plants, polluted air, and exhausts from internal
combustion engines (10-21).

   3. The biological activity of nitroaromatics is presumably
due to their reduction to hydroxylamines by either bacterial or
mammalian enzymes (22-26). It has been previously noted that
bacterial nitroreductases are especially abundant and active and
this may result in the formation of large amounts of the biologically active intermediates (27-30).

   4. Prior studies with nitrofurans (31-35) and their congeners
indicated that mutagenicity in Salmonella required the presence
of plasmid pkM101 which codes for error-prone DNA repair enzymes
(36-39). Similar error-prone repair enzymes have been implicated
in carcinogenesis (40-41). Therefore, a study of the requirement
for the error-prone mechanism for the expression of mutagenicity
of the various nitroarenes, first in bacterial, then in mammalian
systems, appeared to be warranted.

   5. The ready availability of molecules varying in size, shape
and number of functional groups permits the detailed study of the
mechanism responsible for the biological effects of these chemicals.

6.   Assuming that emission of nitrated PAH into the environment
is inherent in various combustion processes, it might be possible to
influence the degree of nitration, through modification of the
combustion process.   The availability of PAH molecules nitrated to
different degrees (e.g., mono- to tetra-substituted) and a determination of their mutagenicity might provide some indication whether
optimization of the combustion process to minimize the emission of
highly mutagenic species is a plausible approach.

SALMONELLA

Molecular Size

Nitronaphthalenes.   The mutagenicity of a series of nitro-
substituted naphthalenes, and their derivatives, toward various
strains of Salmonella are shown in Table 1.   These compounds show
only modest levels of mutagenicity as will be evident in a comparison with other nitroarenes.   1-Nitronaphthalene displayed activity
only in the plasmid bearing strains TA98 and TA100.   The increased
activity of 1-hydroxylaminonaphthalene in the same tester-strains
is consistent with the hypothesis that it may be an intermediate
in the bacterial reduction of the corresponding nitro compound.
The mutagenicity of the 1-hydroxylaminonaphthalene is undiminished,
and actually slightly increased, when assayed in the nitro-reductase
deficient strains (TA100NR, TA98NR).   This may indicate that the
bacterial nitroreductase is capable of reducing the hydroxylamine
to the free amine (which requires activation by microsomal enzymes
to display mutagenicity), and this process is blocked in the NR
strains.   Moreover, it should be noted that unlike 1-nitronaphtha-
lene, the 1-hydroxylamino analog did not require the presence of
the plasmid to express its base-substitution activity, i.e. it
was active in strain TA1535.

2-Nitronaphthalene was not dependent upon the plasmid for
mutagenic activity.   The postulated penultimate metabolite of 2-
nitronaphthalene, 2-hydroxylaminonaphthalene, has only been tested
in strain TA1535.   However, in that strain its mutagenic activity,
as expected, exceeded that of its 2-nitro precursor.   Although the
2-hydroxylamino analog was more mutagenic than the isomeric 1-
hydroxylaminoaphthalene, the two nitronaphthalenes exhibited almost
equal activites (in strain TA100) suggesting that the bacterial
nitroreductases do not discriminate (or are rate-limiting) in the
reduction of these two chemicals.   Such reduction is obligatory as
evidenced by the finding that mutagenic activity was not significant
when nitroreductase-deficient tester strains were used.   This lack
of substrate discrimination may be unique to bacteria.   In animal
studies 2-nitronaphthalene but not the 1-nitro isomer, has been
shown to induce cancer (42,43) while both 1-hydroxylamino- as well
as 2-hydroxylaminonaphthalene are carcinogens.

Table 1. Mutagenicity of nitronaphthalenes

Compound	Revertants/Nanomole							
	TA1535	TA1535NR	TA100	TA100NR	TA1537	TA1537NR	TA98	TA98NR
1-Nitronaphthalene	•	•	0.96	•	•	•	0.05	•
1-Hydroxylamino naphthalene	2.9*	•	10.6	11.8	•	•	1.9	2.9
1-Nitroso-2-naphthol	•	•	0.14	•	•	•	0.23	•
2-Nitronaphthalene	1.4	0.005	1.3	0.11	0.02	•	0.20	0.01
2-Hydroxylamino- naphthalene	8.7*							
2-Nitroso-1-naphthol	•	•	0.09	•	•	•	•	
1,3-Dinitro- naphthalene	0.15	•	7.28	•	0.76	•	0.89	•
1,5-Dinitro- naphthalene	•	•	4.70	•	0.22	•	3.30	•
1,8-Dinitro- naphthalene	•	•	•	•	•	•	•	•
2,4-Dinitro-1- naphthol	•	•	0.35	•	0.09	•	0.66	0.13
1,3,6,8-Tetranitro- naphthalene	•	•	0.30	•	3.4	3.4	0.2	0.6

*Data taken from H.S. Rosenkranz and L. A. Poirier (28).

The reduction of the nitro groups to the corresponding
hydroxylamines is assumed to proceed through the nitroso intermediates. If this is the case, the nitrosonaphthalenes should
be more mutagenic than the corresponding nitronaphthalenes, and less
than the hydroxylamines. The results obtained show that the two
nitrosonaphthalenes tested are very weak mutagens, casting doubt on
the suggested pathway.

The mutagenic activity and specificity of the four dinitro-
naphthalenes examined were strongly dependent on positional effects.
Those compounds which showed appreciable mutagenic activity were
mainly base-substitution mutagens, although, unlike the mononitro-
naphthalenes, they also induced mutations of the frameshift type.
It is noteworthy that 1,8-dinitronaphthalene was devoid of mutagenic activity. Also, the activity of the 1,3,6,8-tetranitrona-
phthalene was lower in strains TA100 and TA98 than the corresponding
dinitro compound, an observation that will recur when the tri- and
tetra-cyclic series are considered.

Nitroquinolines. Two of the three isomeric nitroquinolines
tested, 5-nitro- and 6-nitroquinoline, possessed low mutagenic
activity mainly of the base-substitution type (Table 2). Interestingly, the base-substitution activity of only the 5-nitro isomer
required the plasmid (TA100) for expression. This is reminiscent
of the situation with the 1- and 2-nitronaphthalenes (Table 1)
which might be regarded as analogs of 5- and 6-nitroquinolines,
respectively. 8-Nitroquinoline (an analog of 1,8-dinitronaphtha-
lene) was devoid of direct-acting mutagenicity. The mutagenicity
of the 5-nitro- and 6-nitro isomers is due to the reduction of
the nitro moiety, because the activity was substantially diminished
in nitroreductase-deficient bacteria. In view of the demonstrated
carcinogenicity of 8-nitroquinoline (44), it was of interest to
determine whether this lack of mutagenicity for Salmonella
typhimurium reflected an inability of the bacteria to reduce this
chemical to the corresponding hydroxylamine, the penultimate metabolite. Preliminary results indicate (45) that a hydroxylamine
obtained from 8-nitroquinoline by chemical reduction is mutagenic
for Salmonella.

Table 2. Mutagenicity of nitroquinolines

Compound	TA1535	TA100	TA100NR	TA98	TA98NR
		Revertants/Nanomoles			
5-Nitroquinoline	0.01	0.20	0.04	0.11	0.02
6-Nitroquinoline	0.18	0.09	0.00	0.05	0.00
8-Nitroquinoline	0.00	0.00	0.00	0.00	0.00

Nitrofluorenes and Nitrofluorenones. The five nitrofluorenes
listed in Table 3 show activity in strains TA1538, TA98 and TA1537
with minimal or no activity in TA1535 indicating that they possess
frameshift type activity. The observed activity in strain TA100
presumably reflects a loss of mutagenic specificity as a result of
the introduction of the pkM101 plasmid into strain TA1535 (32). The
observed mutagenic activities in strains TA1538 and TA98 increase
with the number of nitro groups in the molecule; reaching a maximum
with trinitrofluorenone and is decreased for the tetranitrofluorenone.
Oxidation at the nine-position of the molecule results in increased
mutagenicity in strains TA1538 and TA98 but slight decrease in TA
1537. This latter strain is an indicator strain for frameshift
mutagens causing mutations as a result of intercalation between DNA
base pairs without adduct formation (45).

The behavior of 2-nitrofluorene and the corresponding hydroxylamine is quite similar to 1-nitronaphthalene. The parent nitro
compound is active in strain TA98 and its mutagenicity is substantially reduced in the nitroreductase deficient strain TA98NR. Wang
and coworkers (14) have shown that 2-hydroxylaminofluorene is about
eight times as active as the corresponding nitro compound in strain
TA98 and the mutagenicity of the hydroxylamine is unchanged in strain
TA98NR. These results are consistent with the hypothesis that the
2-hydroxylaminofluorene is an intermediate in the bacterial reduction
of the 2-nitrofluorene.

The results summarized in Table 3 are consistent with a
combination of two mechanisms, primarily frameshift mutations due
to adduct formation but accompanied by some small, yet finite,
amount of intercalation. The primary basis of the mutagenic action
of these chemicals appears to be a reduction of the nitro function
to the hydroxylamines, which, following esterification to form an
electrophile, are capable of reacting with the guanine moiety of
DNA (22-27) to form the corresponding adducts. There is evidence
to suggest that the resultant bulky adducts are capable of displacing
the base to which they are now covalently linked and swivel into the
plane of the helix (25) resulting in a frameshift mutation. Consistent with this mechanism are the following observations:

(a) All of the nitrated fluorene compounds have higher
activity in strains TA1538 and TA98 than in TA1537. Simple intercalation causes only a low level of activity in TA1537 (46-48).

(b) The substantial decrease in activity in the nitroreductase-
deficient strain TA98NR supports the hypothesis that reduction of
the nitro moiety is required for the expression of the mutagenic
activity.

(c) The substantial decrease in activity with strain TA1978,
the uvrB+ analog of TA1538, supports the hypothesis that the

Table 3. Mutagenicity of nitrofluorenes and nitrofluorenones

Compounds	TA1535	TA100	TA1537	Revertants/Nanomole TA1537NR	TA1977	TA1538	TA1978	TA98	TA98NR
2-Nitrofluorene	0.01	6.1	0.04	-	-	6.5	0.07	13.9	0.8
2-Nitrofluorene*								71.9	5
2-Hydroxyl aminofluorene*								592	586
2,7-Dinitro-fluorene	-	5.9	290	305	0.2	346	0.60	471	34
2,7-Dinitro-9-fluorenone	-	457	195	245	2.8	1702	48	1459	184
2,4,7-Trinitro-9-fluorenone	1.9	159	572	555	7.3	2860	81	2125	722
2,4,5,7-Tetranitro-9-fluorenone	5.0	287	186	213	1.0	2190	25	860	160

*C.Y. Wang and coworkers (14b).

formation of a covalent adduct between DNA and the "activated"
fluorenes is responsible for the observed mutagenicity (47-49).

(d) 2,7-Diaminofluorene (not shown) exhibits no activity
in TA1537 and only minimal activity for strain TA98. If intercalation were the prime mechanism for the mutagenicity of nitrofluorenes and nitrofluorenones, amino analogs should behave in a
similar manner.

The evidence in support of a small part of the reactions proceeding through a simple intercalation mechanism is as follows:

(a) All five of the nitrated fluorenes/fluorenones cause a
small but measureable increase in the helix-to-coil transition
temperature of isolated DNA. The increase ranges from 0.2 °C for
2-nitrofluorene to 4.4 °C for 2,4,7-trinitro-9-fluorenone (50).

(b) The listed compounds show a modest yet finite activity
in strain TA1977. The residual activity in this uvrB+ analog of
TA1537 ranges from 0.06% for 2,7-dinitrofluorene to 1.3% for 2,4,7-
trinitro-9-fluorenone. This is a measure of intercalative activity
(46-50).

(c) An alternate estimate of the extent of intercalation is
obtained from the ratios of activities in strains TA1978/TA1538.
This is based on the assumption that in TA1978 no alternate DNA
repair pathways involving fluorene adducts are operative.

(d) The unchanged mutagenic activity in strains TA1537 and
TA1537NR indicates that reduction of the nitro function is not
required for the expression of intercalation.

(e) The comparable level of mutagenicity in strain TA1537 of
2,7-dinitro-9-fluorene and 2,7-dinitrofluorenone suggests that the
state of oxidation at the nine position is not an important factor
for intercalation.

Nitropyrenes. At first sight, the pattern of mutagenic activity
for the six nitropyrenes listed in Table 4 appears similar to that
of the nitrofluorenes. The potent mutagenic activity in strains TA
1537, TA1538 and TA98 is indicative of frameshift activity. The
absence of activity in TA1535 and E. coli WP2uvrA (not shown) excludes base-pair substitution mutations. The activity detected in
TA100 presumably reflects a decreased mutagenic specificity of this
tester strain due to the presence of plasmid pKM101 (32).

The absence of any significant activity in strains TA1977 and
TA1978, the uvrB+ analogs of TA1537 and TA1538, indicates that the
mutagenic activity is due to some type of adduct formation with
DNA (47,49).

Table 4. Mutagenicity of nitropyrenes

Compound	Revertants/Nanomole										
	TA1535	TA100	TA1537	TA1977	TA1538	TA1978	TA98	TA98NR	TA98*
1-Nitropyrene	•	63	67	•	85	•	484	35	453
1,3-Dinitro-pyrene	•	8360	13,400	•	15,600	•	28,600	4,900	144,760
1,6-Dinitro-pyrene	•	•	33,000	303	12,100	•	36,350	37,850	183,570
1,8-Dinitro-pyrene	•	•	11,800	•	9,950	•	72,900	75,500	254,000
1,3,6-Trinitro-pyrene	•	5750	20,100	37	15,650	7	31,400	28,330	40,700
1,3,6,8-Tetra-nitropyrene	•	750	3,300	6	1,350	3	7,700	5,200	15,600

*Overnight resting – rather than exponentially growing bacteria used.

There is little, if any, evidence for intercalation of the
nitropyrenes between DNA base-pairs. Direct intercalators, e.g.
9-aminoacridine, quinacrine, proflavin, etc., show only modest
activity in strain TA 537 and not the high values listed in Table
4 (46,48,51). Further, intercalators are active only in strain
TA]537, while the nitropyrenes are active in TA1538 and its
derivatives as well (52). Physical-chemical studies on mixtures
of DNA and nitropyrenes provided no evidence for intercalation.
There is an absence of spectral shifts, no increase in the temperature of the thermal helix-to-coil transition and no effect on the
sedimentation rate of closed circular DNA (52). Finally, we have
already cited above the absence of activity in strains TA1977/TA1978.

A more detailed examination of the results in Table 4, particularly the responses in the less frequently used tester strains
indicates several important differences between the nitrofluorenes
and nitropyrenes. In strains TA98 and TA98NR, the successive addition of nitro groups results in increased activity which reaches a
maximum with 1,8-dinitropyrene and then declines for 1,3,6-trinitro-
and 1,3,6,8-tetranitropyrene. In strain TA1538, the highest values
are observed with 1,3-dinitro- and 1,3,6-trinitropyrene, followed
by 1,6-dinitro-, 1,8-dinitro- and 1,3,6,8-tetranitro- and 1-nitro-
pyrene. In strain TA1537, the highest value was noted for 1,6-
dinitropyrene, followed by 1,3,6-trinitropyrene, 1,3- and 1,8-
dinitropyrenes and then the tetra and mononitropyrenes. The reason
for this behavior, especially the lack of parallel behavior between
strain TA1538 and TA98, is not clear but presumably reflects the
activity of plasmid pKM101 present in TA98, i.e., an effect of
error-prone DNA repair enzymes on the nitropyrene-modified DNA.

Perhaps more significant is the observation that the mutagenic
activity of three of the six compounds was not decreased when tested
in strain TA98NR. This nitroreductase-deficient derivative of TA98
has been shown to lack the enzyme which activates nitrofurans,
nitronaphthalenes and nitrofluorenes to mutagens. A priori this
would suggest that the test chemicals are direct-acting mutagens
and do not require the usual reductive conversion to the corresponding hydroxylamines.

Since the possibility of intercalation has already been excluded
(see above), and a direct in vitro reaction between unmodified nitropyrenes and DNA has been ruled out as well, i.e. no susceptibility
to S nuclease; no decrease in the temperature of the thermal
helix-to-coil transition, an alternate hypothesis would suggest
that nitropyrenes are converted to hydroxylamines which upon
esterification may act as electrophiles. However, such a reduction
process must be carried out by enzymes other than the "classical"
nitroreductase. Indeed, bacterial strains lacking such nitroreductases were isolated recently. Such strains, while fully responsive

to nitrofurans, niridazole, and 4-nitroquinoline-1-oxide, are not optimally mutagenized by nitropyrenes. A more complete characterization of these tester strains and their use in the detection of nitroarenes in environmental mixtures will be presented elsewhere (53,54).

The microbial mutagenicity data presented herein were obtained with the standard Salmonella assay developed by Ames and his associates (55). That assay has been criticized, on occasion, because it is a reverse mutational system, from histidine-dependence to histidine-independence. By necessity the genetic events scored will be a function of the original forward mutation introduced into the tester strains, i.e. mutation to histidine-dependence. With respect to the tester strains TA1538/TA98, which respond maximally to the nitropyrenes, the genetic event scored occurs at a site enriched with respect to guanine and cytosine residues. Accordingly the tester strains will respond maximally to chemicals which preferentially form adducts with these base moieties. Because nitropyrenes are widely distributed in the environment, it is imperative that the possible risks posed by these chemicals be assessed. It would therefore be appropriate to determine whether the extraordinarily high mutagenicity of nitropyrenes as well as their mutagenic specificity is possibly a unique reflection of events scored only in the reverse mutational assay. Accordingly the Salmonella tester strains were used in a forward mutational assay (to 5-methyltryptophan-resistance). Preliminary results indicate that the mutagenic potency was retained in the forward assay. Thus while 1,8-dinitropyrene in strain TA1538 displayed an activity of 9,950 mutants per nanomole in the standard (reverse) assay (Table 4), it had an activity of 10,800 in the forward system (56). It remains to be established whether this pattern will be maintained with the other tester strains and the other nitropyrenes.

Except for the entries in the last column, all of the data in Table 4 were obtained with exponentially growing bacteria. Such a procedure was used as a result of a large collaborative study which indicated improved intra- and inter- laboratory precision when growing bacteria were used (57,58). In our study with nitropyrenes, the use of overnight (resting) cultures yielded more reproducible and substantially enhanced specific activities. It is plausible that this reflects the state of anaerobiosis of the bacteria. Overnight (resting) cultures having exhausted some of their nutrients, enter the fermentative state. Presumably, the intermediates formed from nitropyrenes, i.e., hydroxylamines, are oxygen-labile and are capable of reverting readily to the oxidized form. Thus, the reduced oxygen tension of fermenting bacteria would favor the formation of the hydroxylamines which would be reflected in increased mutagenicity.

It may be appropriate to note that 1,8-dinitropyrene exhibited in strain TA98, the highest mutagenic activity found to date in the literature, varying between 72,900 and 254,000 revertants per nanomole for growing and resting cells respectively. This activity should be compared to that of Trp-2 (3-amino-1-methyl-5H-pyrido [4,3, b] indole) in the presence of microsomal enzymes, 2760 revertants per nanomole, previously considered the most mutagenic chemical (5).

Effect of "Microsomes" on the Mutagenicity of Nitropyrenes

Addition of NADP and rodent microsomes, i.e. S9 mix (55) (which more precisely is a postmitochondrial supernatant), resulted in great decreases in the mutagenicity of nitropyrenes (Table 5). Generally, for microsomes derived from uninduced animals, those from mice were the most efficient, while those originating from rats were the least efficient in reducing the mutagenic activity. The reduction in activity was most striking when the S9 were derived from the livers of Arochlor 1254-induced animals. Optimal inactivation of the nitropyrenes required the presence of NADP. This ability of S9 to block the mutagenicity of nitropyrenes was abolished upon heating (100°C, 5 min). Preliminary studies indicate that this enzyme activity, when derived from rat livers, was sedimentable (100,000 x g, 1 hr) suggesting that it is microsome-associated.

The relevance of these findings to the estimation of the risk posed by these chemicals is self-evident. Presumably, human exposure to nitrated polycyclic aromatic hydrocarbons is via the skin or through inhalation. Accordingly these chemicals may undergo microsomal metabolism.

Microsomes have been shown to possess nitroreductase activity (59-70) although the involvement of soluble enzymes (cytosol) has been demonstrated as well (67-70). The inactivation of the nitropyrenes could, therefore, be due to the efficient reduction of these species to the corresponding amines. However, it is reasonable to assume that the resultant aminopyrenes would be converted by microsomal cytochrome P-450 dependent monoxygenases to the mutagenic hydroxylamines. Predicated on the ability of S9 to activate such aminopyrenes, microsomes derived from Arochlor 1254-induced animals should be especially active in this oxidation. The results summarized in Table 5 do not support such a mechanism. On the contrary, microsomes from Arochlor 1254-treated rodents are more efficient in reducing the mutagenicity of nitropyrenes than microsomes from untreated animals.

Alternatively, it is conceivable that nitropyrenes are oxidized by S9 to substances (quinones, diols, etc.) which are devoid of mutagenic activity. In this connection it should be recalled that, unlike benzo(a)pyrene, unsubstituted pyrene is not activated by S9 to a mutagen (56,71). This could be due either to the inertness of

Table 5. Effect of microsomal preparations on the mutagenicity of nitropyrenes

Pyrene	Mutagenicity of Control (Rev/μg)	Percent of Control*					
		M	MA	R	RA	H	HA
1-Nitro**	13,820	14	4	68	1.2	32	13
1,3-Dinitro	485,200	6	0.3	31	0.0	9	0.0
1,6-Dinitro	615,200	6	1.2	16	0.2	25	0.2
1,8-Dinitro	631,200	6	0.1	6	0.0	1.4	0.0
1,3,6-Trinitro	77,600	0.4	0.1	36	0.4	13	2.2
1,3,6,8-Tetranitro	10,820	0.0	0.0	15	1.2	7	0.7

Samples were tested in TA98; resting cultures were used.

*Abbreviations: M, R and H refer to the inclusion in the assay of hepatic microsomes derived from mice, rats and hamsters, respectively. MA, RA and HA indicate that the microsomes were obtained from Arochlor 1254-induced animals.

**The purity of the sample of 1-nitropyrene used was lower than usual, 97.5%.

The mutagenic activities were calculated from the initial slopes of the dose-response curves. The microsomal effects of each species were determined separately and are expressed as percent of their controls, i.e. mutagenic activity in the absence of S9. The mean values of the mutagenic activities of the individual determinations for each compound are listed in the table.

the pyrene molecule or to its metabolic conversion to products
which are also not mutagenic. Certainly these aspects of the microsomal metabolism of nitropyrenes deserve much more consideration.
The expected availability of radioactive nitropyrenes will be of
great value in this respect.

The Role of Bacterial Nitroreductases

As mentioned previously, bacteria deficient in the "classical"
nitroreductase do not respond to the mutagenic action of nitrofurans,
niridazole, nitronaphthalenes and nitrofluorenes. On the other hand,
these microorganisms are fully sensitive to the mutagenic action of
other nitro-containing chemicals such as 4-nitroquinoline-1-oxide,
nitroacridine and nitropyrenes (2,46,72). This has led to the
suggestion that bacteria possess a family of nitroreductases differing in substrate specificities. The strains deficient in the
"classical" nitroreductase were isolated as nitrofuran- and nirida-
zole-resistant. Similarly, to obtain strains lacking a nitroreductase specific for nitropyrenes, clones resistant to the inhibitory
action of the various nitropyrenes were selected. Upon verification that they responded normally to a panel of non-nitrated mutagens, their responses to a series of nitro-containing chemicals
were evaluated. The behavior of a strain selected with 1,8-dinitro-
pyrene (TA98/1,8DNP$_6$) is illustrated in Table 6. As expected,
the mutagenicity of the 1,8-isomer was greatly decreased in this
strain. However, the responses of the other nitropyrenes were
also affected, although to differing extents. While the 1,3-
isomer exhibited only 1.9% residual activity, the 1,6-compound
retained 25% of its activity. On the other hand the activity of
the trinitro- and tetranitro-analogs were only slightly affected.
Since the last three named chemicals are also not recognized by
the "classical" nitroreductase, it might therefore be hypothesized
that they owe their mutagenic activity to yet another nitroreductase.
Finally, the TA98/1,8 DNP$_6$ is fully sensitive to the mutagenic
action of niridazole, which is a substrate for the "classical"
enzyme.

The above findings suggesting the presence of a family of
bacterial nitroreductases, raise questions regarding their function
and also whether this phenomenon is unique to bacteria. Certainly
the results obtained thus far indicate the necessity of undertaking
further studies to ascertain (a) the total number of nitroreductases,
(b) the location and function of the genes regulating their synthesis,
(c) their physical chemical characteristics and (d) their substrate
specificity. Before embarking upon such an extensive investigation
it was deemed advisable to eliminate the possibility that the bacteria
possessed a single nitroreductase but that its substrate specificity
was altered as the result of a mutationally generated amino acid
sequence. If the bacteria contained only one enzyme, then it can be
predicted that double-mutants, i.e., resistant to the mutagenicity

Table 6. Mutagenic response of a <u>Salmonella</u> TA98 derivative
          lacking a nitroreductase specific for nitrated
          polycyclic aromatic hydrocarbons

Chemical	Revertants per μg		
	TA98	TA98/1,8-DNP	
1-Nitropyrene	3,115	578	(18)
1,3-Dinitropyrene	496,800	9,440	(1.9)
1,6-Dinitropyrene	630,000	160,000	(25)
1,8-Dinitropyrene	870,000	20,120	(2.3)
1,3,6-Trinitropyrene	60,000	38,050	(63)
1,3,6,8-Tetranitropyrene	10,820	9,760	(90)
Niridazole	1,820	1,820	(100)
4-Nitroquinoline-1-oxide	687	578	(84)
2,7-Dinitro-9-fluorenone	1,669	157	(9.4)

The values in parenthesis refer to the percent reduction in specific
activity in strain TA98/1,8-DNP$_6$ compared to TA98.

of nitrofuran-type chemicals as well as to nitropyrenes, should not
exist. However, starting with either TA98/1,8 DNP$_6$ or with TA98/NR,
double mutants could be readily isolated (Table 7). All of them
exhibited the expected pheonotype and, moreover, they retained their
response to the mutagenic action of 4-nitroquinoline-1-oxide thereby
confirming that this chemical requires a third enzyme.

If one could construct a strain entirely lacking a specific
nitroreductase, even an inactive one, the determination of the
activity of a second nitroreductase will resolve the question as to
whether the two activities result from a mutated site or from the
presence of two different gene products. <u>Salmonella</u> strain TA98
deficient in the "classical" nitroreductase were obtained as the
result of frameshift mutations. Such mutants should be devoid of
recognizable gene products due to premature chain termination.
Indeed, as expected (Table 8) such strains show a further decrease
in their response to niridazole, yet their response to 1,8-dinitro-
pyrene and 4-nitroquinoline-1-oxide remains unaltered. These
results suggest the existence of at least three different nitro-

Table 7. Properties of derivatives of <u>Salmonella</u> TA98
deficient in two nitroreductases

Strain	Percent Residual Activity		
	1,8-DNP	Niridazole	NQO
TA98	100	100	100
TA98/NR	104	4	100
TA98/1,8-DNP$_6$	2	100	84
TA98/NR/1,8-DNP$_1$	1.3	0	100
TA98/NR/1,8-DNP$_2$	1.4	0	100
TA98/NR/1,8-DNP$_5$	3.9	0	100
TA98/1,8-DNP$_6$/NR$_1$	2	0	100
TA98/1,8-DNP$_6$/NR$_8$	2	0	100
TA98/1,8-DNP$_6$/NR$_{11}$	2	0	100
TA98/1,8-DNP$_6$/NR$_{12}$	2	0	100

Abbreviations: 1,8-DNP, 1,8-dinitropyrene; NQO, 4-nitroquinoline-1-oxide.

Strains designated TA98/NR/DNP$_6$ were TA98/NR strains rendered
resistant to 1,8-DNP; similarly TA98/1,8-DNP$_6$/NR denotes TA98
1,8-DNP strains made resistant to niridazole.

All strains retained plasmid pKM101 and the deep-rough character.
Moreover they all responded normally to non-nitrated mutagens.

reductases and encourage us to undertake the more extensive
studies mentioned above.

Mutagenic Intermediates of Nitroarenes

As has been mentioned on several occasions, it is assumed that
the mutagenicity of nitroarenes derives from their conversion to
hydroxylamines which, following esterification to electrophiles,
react with the base moieties of DNA. It is assumed that this reaction occurs intracellularly and is catalyzed by the nitroreductases

Table 8. Properties of <u>Salmonella</u> <u>typhimurium</u> TA98
          strains deficient in classical nitroreductase
          as the result of frameshift mutations

	Percent Residual Activity		
	Niridazole	1,8-DNP	NQO
TA98	100	100	100
TA98NR	4.4	104	100
TA98NIR301	0.19	106	100
TA98NIR302	0.19	100	100
TA98NIR303	0.16	100	100
TA98NIR304	0.22	100	100

Abbreviations: 1,8-DNP, 1,8-dinitropyrene; NQO, 4-nitroquinoline-
1-oxide.

Strain TA98NR is deficient in "classical" nitroreductase selected
as niridazole resistant. In strains TA98NIR301-TA98NIR304 this
deficiency was introduced as the result of frameshift mutations.

discussed above. The mutagenicity of nitroarenes in bacteria
deficient in nitroreductases is then ascribed to the block in the
conversion to the proximate hydroxylamines. It could be expected
that supplying bacteria with preformed hydroxylamines will result
in (a) an increased mutagenic potency when compared to the parent
nitro compounds and (b) a bypassing of the blockage in the nitroreductase-deficient bacteria. Indeed, the studies with nitro- and
hydroxylaminonaphthalenes (Table 1) have confirmed this prediction.
It must be recalled, however, that hydroxylamines are notoriously
oxygen-labile. (This is the possible reason for the increased
mutagenicity of nitropyrenes in resting cells (see above) and the
increased mutagenicity of nitroarenes when the initial exposure
of the bacteria is anaerobic (28)). The inactivating effect of
S9 preparations (see above) on the mutagenicity of nitropyrenes
suggests that if it is due to a nitroreductase activity, that
reduction proceeds all the way to the amine stage possibly because
the hydroxylamine intermediate is enzyme-bound and therefore not
free to enter the cell and react with the cellular DNA (72). In
this respect S9 preparations would behave differently towards these
higher nitroarenes than they do for simpler chemicals, as the
ability of S9 to convert 2-nitronaphthalene, 2-nitrofluorene and
niridazole to substances mutagenic in <u>Salmonella</u> tester strains
deficient in the "classical" nitroreductase has demonstrated (28).
On the other hand, S9 preparations were consistently unable to
activate 1,8-dinitropyrene to a preparation mutagenic for the nitroreductase-deficient TA98/1,8-DNP$_6$ strain (56).

In view of the above hypotheses and experimental findings, a procedure was desired to generate these possibly unstable hydroxylamines chemically in the assay system. The standard procedure (73,74) for the reduction of aromatic nitro groups to the corresponding hydroxylamines involves treatment with zinc dust and ammonium chloride. This procedure was found to be satisfactory for our purposes and compatible with the Salmonella assay system. The results of representative experiments with 2-nitrofluorene and 1,8-dinitropyrene are presented in Table 9. As predicted, the mutagenicity of the chemicals in strain TA98 is increased following reduction, which is consistent with the production of a proximate metabolite and more strikingly, the nitroreductase-deficiency of strains TA98NR and TA98/1,8DNP$_6$ respectively, is by-passed. This procedure has now been applied to a number of other chemicals with similar results (45). The production of hydroxylamines has been followed spectrophotometrically and by the generation of colored complexes with FeCl$_3$. Chemical characterizations are planned.

LACK OF RECOMBINOGENIC EFFECTS OF NITROPYRENES IN YEAST

In view of the potent mutagenicity of nitropyrenes for Salmonella, we sought to determine their genetic activity in a higher system. The recombinogenic activity in Saccharomyces cerevisiae D4

Table 9. Chemical reduction of 1,8-dinitropyrene to a mutagen

	TA98	Revertants per μg TA98/1,8-DNP$_6$	TA98NR
2,Nitrofluorene[a]	5.0		0.9
2-Nitrofluorene + Zn/NH$_4$Cl	5.9		6.4
1,8-Dinitropyrene[b]	327,500	7,428	
1,8-Dinitropyrene[b]+ Zn/NH$_4$Cl	450,800	205,600	

Bacteria were incubated for 45 min at 37°C in the presence or absence of zinc dust and NH$_4$Cl whereupon soft agar was added and the standard Salmonella assay procedure was carried out. The presence of Zn and NH$_4$Cl alone did not result in mutagenic activity.

[a]Experiment done at 33.3 mg of 2-nitrofluorene per reaction mixture, and the activity is expressed as revertants per microgram.

[b]Activity calculated from the linear portion of the dose response curve.

was selected for the study. The methods of exposure used were that described by Zimmermann (75) and recent modification thereof (76). In both assays, as shown in Tables 10 and 11, all of the nitropyrenes, with the exception of 1,3,6-trinitropyrene, were uniformly devoid of genetic activity. Trinitropyrene occasionally gave borderline positive results. On the other hand, 4-nitroquinoline-1-oxide demonstrated consistently potent recombinogenic activity.

Table 10. Gene conversion induced by nitropyrenes in Saccharomyes cerevisiae D4*

Chemical	µg/ml	Conversion Frequency x 10^6		
		Exp I	Exp II	Exp III
Solvent	-	0.17	0.17	0.20
Pyrene	500	0.20	0.14	
	1000	0.19	0.11	
1-Nitropyrene	50	0.17		
	500	0.20	0.24	
1,3-Dinitropyrene	50	0.20		
	250	0.22		
	500	0.22	0.16	
1,6-Dinitropyrene	12.5	0.17		
	125	0.20		
	500	0.24	0.19	
1,8-Dinitropyrene	12.5	0.21		
	125	0.24		
	500	0.24	0.14	
1,3,6-Trinitropyrene	17	0.15		
	50	0.14		
	100			0.41
	500	0.18	0.28	0.33
	1000	0.19	0.24	0.28
4-Nitroquinoline -1-oxide	0.01	6.94	5.67	3.27
Ethyl methanesulfonate	0.01	3.40	1.89	1.84

*Determined by the procedure of Zimmerman (75)

Table 11.   Gene conversion induced by nitropyrenes in
Saccharomyces cervisiae D4*

Chemical	µg/plate	Convertants per Plate		
		Exp 1	Exp II	Exp III
None	0	33	50	49
Pyrene	500	36	41	
	750	27	42	
1-Nitropyrene	100	26	56	54
	333	37	56	
	1000	33	46	
1,3-Dinitropyrene	100	44	48	43
	500	46		47
	1000	44		41
1,6-Dinitropyrene	100	43	37	41
	500	45		53
	1000	33		62
1,8-Dinitropyrene	100	45	48	
	500	45		
	1000	44		
1,3,6-Trinitropyrene	100	72	111	135
	333	63	126	102
	500			111
	1000	104	60	147
1,3,6,8-Tetranitro-pyrene	100	34	56	51
	333	38	49	
	1000	37		
4-Nitroquinoline--1-oxide	0.3	609	1275	736
Ethyl methanesulfonate	2.4	1072	1703	1347

*Plate incorporation procedure.

Recently, it was demonstrated that the genetic activity of nitrophenylenediamines was detected in Saccharomyces cerevisiae if the time period of pre-incubation was prolonged (77). However, a series of experiments in which incubation was prolonged to 72 hours did not result in a demonstration of the recombinogenic activity of nitropyrenes.

There are a number of possible explanations for the apparent lack of activity of nitropyrenes in yeast. The likelihood that nitropyrenes are incapable of penetrating into yeast cells can be excluded, since in the modified test procedure the slightly larger benzo(a)pyrene moiety was capable of entering the cells and resulted in activity. Alternatively, it is possible that yeast, even while possessing a nitroreductase capable of activating 4-nitroquinoline-1-oxide to a mutagen, lacks the enzyme able to reduce nitropyrenes. It has already been demonstrated (see above) that in bacteria the two enzyme activities are separate.

SUMMARY

1. The nitroarenes examined are mutagenic in Salmonella. Generally mutagenicity increases with the number of aromatic rings and nitro substituents.

2. Nitronaphthalenes and nitroquinolines exhibit base-pair substitution type activity.

3. Nitrofluorenes and nitropyrenes give rise to frameshift type mutagenic activity.

4. Preliminary results indicate that in strain TA1538 nitropyrenes are active in both the standard reverse and forward mutational assay.

5. 1,8-Dinitropyrene, the most mutagenic of the nitroarenes examined in strain TA98 is about 30 times more active than Trp-2, hitherto the most mutagenic substance examined.

6. The mutagenic activity of nitropyrenes is dramatically diminished in the presence of microsomal enzymes derived from rat, mouse and hamster. This activity is most striking when the livers of Arochlor 1254 induced animals are used.

7. The presence of a series of bacterial nitroreductases in Salmonella was demonstrated.

8. Evidence was presented that hydroxylamines are key intermediates in the mutagenicity of nitroarenes.

9. With the possible exception of 1,3,6-trinitropyrene,
    nitropyrenes do not exhibit recombinogenic effects in
    yeast.

The data presented herein demonstrate that nitroarenes are a
unique class of microbial mutagens. In view of their wide distribution in the environment, the basis of their activity and their
effects on mammalian cells deserves elucidation.

ACKNOWLEDGEMENT

The authors sincerely appreciate the leadership and stellar
support of Mr. Horace Becker. We have benefitted from numerous
discussions with our colleagues, Dennis Kiriazides and Dr. Zev
Leifer. We gratefully acknowledge the superb technical assistance
of our colleagues, Mrs. Monika Anders, Ms. Marsha Butler, Ms. Lynn
Petrullo and Dr. George E. Karpinsky, as well as express our thanks
to Jeraldine Allen for preparation of the manuscript.

REFERENCES

1. R. Mermelstein, H. S. Rosenkranz, E. C. McCoy, M. Butler,
        D. K. Kiriazides, and D. R. Sanders, Nitropyrenes: Isola-
        tion, identification, and reduction in mutagenic impurities
        in a certain carbon black in xerographic toners, Environ.
        Mutagenesis 2:285 (1980).
2. H. S. Rosenkranz, E. C. McCoy, D. R. Sanders, M. Butler,
        D. K. Kiriazides, and R. Mermelstein, Nitropyrenes: Isola-
        tion, identification, and reduction of mutagenic impurities
        in a carbon black and toners, Science 209:1039-1043 (1980).
3. G. Lofroth, E. Hefner, I. Alfheim, and M. Moller, Mutagenic
        activity in photocopies, Science 209:1037-1039 (1980).
4. W. H. H. Gunther, S. Kaplan, J. VanDusen, F. Elder, D. Evans,
        and R. Mermelstein, manuscript in preparation.
5. T. Sugimura and M. Nagao, Mutagenic factors in cooked foods,
        CRC Crit. Rev. Tox., 189-209 (1979).
6. T. Kuroki, Comparative mutagenicity of diol epoxides of benzo-
        (a)pyrene and benz(a)anthracene in V79 Chinese hamster
        cells and Salmonella typhimurium, in: "Polycyclic Hydro-
        carbons and Cancer" Vol. 2 , H. V. Gelboin and P. O. P.
        Ts'o, eds., Academic Press, New York, NY, 123-124 (1978).
7. B. D. Crittenden and R. Long, in: "Carcinogenesis - a Comprehen-
        sive Survey", R. I. Freudenthal and P. Jones, eds., Raven
        Press, New York, NY, Vol. 1, 209 (1977).
8. P. B. D. de la Mare and J. H. Ridd, in: "Aromatic Substitution,
        Nitration and Halogenation", Academic Press, New York, NY,
        187 (1969).

9. J. N. Pitts, Jr., K. A. Van Cauwenberghe, D. Grosjean, J. P.
     Schmid, D. R. Fitz, W. L. Belser, Jr., G. B. Knudson, and
     P. M. Hynds, Atmospheric reactions of polycyclic aromatic
     hydrocarbons: facile formation of mutagenic nitro deriva-
     tives, Science 202:515-519 (1978).
10. J. N. Pitts, Jr., Photochemical and biological implications
     of the atmospheric reactions of amines and benzo(a)pyrene,
     Phil. Trans. Roy. Soc. London A 290:551-576 (1979).
11. J. Jager, Detection and characterization of nitro derivatives
     of some polycyclic aromatic hydrocarbons by fluorescence
     quenching after thin-layer chromatography: Application to
     air pollution analysis, J. Chromatog. 152:575-578 (1978).
12. Y. Y. Wang, S. M. Rappaport, R. F. Sawyer, R. E. Talcott, and
     E. T. Wei, Direct-acting mutagens in automobile exhaust,
     Canc. Lett. 5:39-47 (1978).
13. H. Tokiwa, R. Nakagawa, K. Morita, and . Kamachi, Analysis of
     mutagenic nitro compounds in environmental samples (in
     Japanese), Environ. Mutagen. Soc. of Japan, Abstracts,
     p. 18, (1979).
14a. C. M. King, C. Y. Wang, and P. O. Warner, Evidence for the
     presence of nitro aromatics in airborne particulates,
     Proc. Amer. Assoc. Cancer Res., 21: 83 (1980).
14b. C. Y. Wang, M. S. Lee, C. M. King, and P. O. Warner, Evidence
     for nitroaromatics as direct-acting mutagens of airborne
     particulates, Chemosphere 9:83-87 (1980).
15. H. E. Kubitscheck and D. M. Williams, Mutagenicity of fly ash
     from fluidized-bed combuster during start-up and steady
     state operating conditions, Mutat. Res. 77:287-291 (1980).
16. G. Lofroth, Comparison of the mutagenic activity from diesel
     and gasoline powered motor vehicles to carbon particulate
     matter, in: "Second Symposium on Application of Complex
     Environmental Mixtures", Plenum Press, New York, NY in
     press (1981).
17. T. C. Pedersen and J.-Siak, Characterization of direct-acting
     mutagens in diesel exhaust particulates by thin-layer chroma-
     tography, Ann. Meeting Amer. Chem. Soc. Div Environ.
     Chem., Abstract (1980).
18. D. Schuetzle, T. J. Prater, T. Riley, A. Durisin, and
     I. Salmeen, Analysis of nitrated derivatives of PAH and
     determination of their contribution to Ames assay mutageni-
     city for diesel particulate extracts, Fifth Intern. Symp.
     Polynuclear Aromatic Hydrocarbons, Columbus, Ohio, Abstract,
     (1980).
19. E. T. Wei, Y. Y. Wang, and S. M. Rappaport, Diesel emissions
     and the Ames test: A commentary, J. Air Poll. Control.
     Assoc. 30:267-271 (1980).
20. D. Schuetzle, F. S.-C. Lee, T. J. Prater, and S. B. Tejada,
     The identification of polynuclear aromatic hydrocarbon
     derivatives in mutagenic fractions of diesel particulate
     extracts, Intern. J. Environ. Analyt. Chem. 9:1-52 (1981).

21. National Academy of Sciences, U.S., Health effects of exposure
     to diesel exhaust. The Report of the Health Effects Panel
     of the Diesel Impact Study Committee, National Research
     Council - National Academy of Sciences, Washington, D.C.,
     (1981).

22. H. Bartsch, G. Malaveille, H. F. Stich, E.C. Miller, and
     J. A. Miller, Comparative electrophilicity, mutagenicity,
     DNA repair induction activity and carcinogenicity of some
     N- and O-acyl derivatives of N-hydroxy-2-amino-fluorene,
     Cancer Res. 37:1461-1467 (1977).

23. E. Kriek, On the interaction of N-2-fluorenylhydroxylamine with
     nucleic acids in vitro , Biochem. Biophys. Res. Commun. 20:
     793-799 (1965).

24. E. Kriek, J. A. Miller, U. Juhl, and E. C. Miller, 8-(N-2-
     fluorenyl-acetamido) guanosine, an arylamidation reaction
     product of guanosine and the carcinogin N-acetoxy-N-2-fluor-
     enylacetamide in neutral solution, Biochem. 6:177-182 (1967).

25. A. F. Levine, L. M. Fink, I. B. Weinstein, and D. Grunberger,
     Effect of N-2-acetylfluorene modification on the conformation
     of nucleic acids, Cancer Res. 34:319-327 (1974).

26. E. C. Miller, Nucleic acid guanine: Reaction with the carcino-
     gen N-acetoxy-2-acetylaminofluorene, Science 153:1125-1127
     (1966).

27. J. A. Miller, Carcinogenesis by chemicals: An overview, Cancer
     Res. 30:559-576 (1970).

28. H. S. Rosenkranz and L. A. Poirier, An evaluation of the
     mutagenicity and DNA modifying activity in microbial systems
     of carcinogens and non-carcinogens, J. Natl. Cancer Inst.
     62:873-892 (1979).

29. J. McCann and B. N. Ames, The Salmonella/microsome mutagenicity
     test: Predictive value for animal carcinogenicity, in:
     "Origins of Human Cancer", Book C, H. H. Hiatt, J. D. Watson,
     and J. A. Winstein, eds., Cold Spring Harbor Laboratory,
     Cold Spring Harbor, NY, 1431-1450 (1977).

30. B. N. Ames, Identifying environmental chemicals causing muta-
     tions and cancer, Science 204:587-593 (1979).

31. D. R. McCalla, D. Voutsinos, and P. L. Olive, Mutagen screening
     with bacteria: Niridazole and nitrofurans, Mutat. Res. 31:
     31-37 (1975).

32. J. McCann, N. E. Spingarn, J. Kobori, and B. N. Ames, Detection
     of carcinogens as mutagens: Bacterial tester strains with
     R factor plasmids, Proc. Natl. Acad. Sci. (U.S.A.) 72:979-983
     (1975).

33. Y. Tazima, T. Kada, and A. Murakami, Mutagenicity of nitrofuran
     derivatives including furylfuramide, a food preservative,
     Mutat. Res. 32:55-80 (1975).

34. C. Y. Wang, K. Muraoka, and G. T. Bryan, Mutagenicity of nitro-
     furans, nitrothiophenes, nitropyrroles, nitroimidazole,
     aminothiophenes, and aminothiazoles in Salmonella
     typhimurium, Cancer Res. 35:3611-3617 (1975).

35. T. Yahagi, M. Nagao, K. Hara, T. Matsushima, T. Sugimura, and G. T. Bryan, Relationships between the carcinogenic and mutagenic or DNA-modifying effects of nitrofuran derivatives, including 2-(2-Furyl)-3-(5-nitro-2-furyl)acrylamide, a food additive, Cancer Res. 34:2266-2273 (1974).

36. A. Goze and R. Devoret, Repair promoted by plasmid pKM101 is different from SOS repair, Mutat. Res. 61:163-179 (1979).

37. C. Monti-Bragadin, S. Venturini, and P. A. Todd, Interaction between two error-prone DNA repair systems in Escherichia coli, FEMS Microbiol. Lett. 2:125-128 (1977).

38. G. C. Walker, Isolation and characterization of mutants of the plasmid pKM101 deficient in their ability to enhance mutagenesis and repair, J. Bacteriol. 133:1203-1211 (1978).

39. P. A. Todd, C. Monti-Bragadin, and B. W. Glickman, MMS mutagenesis in strains of Escherichia coli carrying the R46 mutagenic enhancing plasmid: phenotypic analysis of Arg+ revertants, Mutat. Res. 62:227-237 (1979).

40. P. Moreau and R. Devoret, Potential carcinogens tested by induction and mutagenesis of prophage in Escherichia coli K12, in: "Origins of Human Cancer", Book C, H. H. Hiatt, J. D. Watson, and J. A. Winstein, eds., Cold Spring Harbor Laboratory, Cold Spring Harbor, NY, 1451-1472 (1977).

41a. M. Radman, G. Villani, S. Boiteux, M. Defais, P. Caillet-Fauquet, and S. Spadari, On the mechanism and genetic control of mutagenesis induced by carcinogenic mutagens, in: "Origins of Human Cancer", Book B, H. H. Hiatt, J. D. Watson, and J. A. Winstein, eds., Cold Spring Harbor Laboratory, Cold Spring Harbor, NY, 903-922 (1977).

41b. M. Radman, Inducible pathways in deoxyribonucleic acid repair, mutagenesis, and carcinogenesis, Biochem. Soc. Trans. 5: 1194-1199 (1977).

42. L. Tomatis, C. Agthe, H. Bartsch, J. Huff, R. Montesano, R. Saracci, E. Walker, and J. Wilbourn, Evaluation of the carcinogenicity of chemicals: A review of the monograph program of the International Agency for Research on Cancer (1971-1977), Cancer Res. 38:877-885 (1978).

43. R. A. Griesemer and C. Cueto, Jr., Toward a classification scheme for degrees of experimental evidence for the carcinogenicity of chemicals for animals, in: "Molecular and Cellular Aspects of Carcinogen Screening Tests" IARC Scientific Publication No. 27, R. Montesano, H. Bartsch, and L. Tomatis, eds., International Agency for Research on Cancer, Lyon, 259-281 (1980).

44. M. Takahashi, T. Shirai, S. Fukushima, K. Hosoda, S. Yoshida, and N. Ito, Carcinogenicity of 8-nitroquinoline in Sprague-Dawley rats, Cancer Lett. 4:265-270 (1978).

45. G. E. Karpinsky and H. S. Rosenkranz, The anaerobe-mediated mutagenicity of 2-nitrofluorene and 2-aminofluorene for Salmonella typhimurium, Environ. Mutag. 2:353-358 (1980).

46. H. S. Rosenkranz and R. Mermelstein, The Salmonella mutagenicity
    and the E. coli Pol A+/Pol A⁻ repair assays: Evaluation of
    relevance to carcinogenesis, in: "The Predictive Value of
    in vitro Short-Term Screening Tests in the Evaluation of
    Carcinogenicity", G. M. Williams, R. Kroes, H. W. Waaijers,
    and K. W. van de Poll, eds., Elsevier/North Holland Biomed-
    ical Press, 5-26 (1980).

47. B. N. Ames, F. D. Lee, and W. F. Durston, An improved bacterial
    test system for the detection and classification of mutagens
    and carcinogens, Proc. Natl. Acad. Sci. (U.S.A.) 70:782-786
    (1973).

48. W. T. Speck and H. S. Rosenkranz, Proflavin: An unusual mutagen,
    Mutat. Res. 77:37-43 (1980).

49. B. N. Ames, A bacterial system for detecting mutagens and
    carcinogens, in: "Mutagenic Effects of Environmental
    Contaminants", H. E. Sutton and M. I. Harris, eds., Academic
    Press, New York, NY, 57-66 (1972).

50. E. C. McCoy, E. J. Rosenkranz, H. S. Rosenkranz, and R.
    Mermelstein, Nitrated fluorene derivatives are potent frame-
    shift mutagens, Environ. Mutagenesis, in press (1982).

51. E. C. McCoy, E. J. Rosenkranz, L. A. Petrullo, and H. S.
    Rosenkranz, Frameshift mutations: Relative roles of simple
    intercalation and of adduct formation, Environ. Mutagensis,
    in press (1982).

52. R. Mermelstein, D. K. Kiriazides, M. Butler, E. C. McCoy, and
    H. S. Rosenkranz, The extraordinary mutagenicity of nitro-
    pyrenes in bacteria, Mutat. Res., 89:187-196 (1981).

53. E. C. McCoy, Z. Leifer, H. S. Rosenkranz, and R. Mermelstein,
    Characterization of nitroreductase-deficient Salmonella
    tester strains useful in the detection of nitroarenes in
    environmental mixtures, Environ. Mutagenesis, in press (1981).

54. H. S. Rosenkranz, E. C McCoy, R. Mermelstein, and W. T. Speck,
    A cautionary note on the use of nitroreductase-deficient
    strains of Salmonella typhimurium for the detection of
    nitroarenes as mutagens in complex mixtures including diesel
    exhausts, Mutat. Res. 91:103-105 (1981).

55. B. N. Ames, J. McCann, and E. Yamasaki, Methods for detecting
    carcinogens and mutagens with the Salmonella/mammalian-micro-
    some mutagenicity test, Mutat. Res. 31:347-364 (1975).

56. E. C. McCoy, et al., unpublished results.

57. C. Cheli, D. DeFrancesco, L. A. Petrullo, E.C. McCoy, and
    H. S. Rosenkranz, The Salmonella mutagenicity assay: Repro-
    ducibility, Mutat. Res. 74:145-150 (1980).

58. H. S. Rosenkranz, G. Karpinsky, and E. C. McCoy, Microbial
    Assays: Evaluation and application to the elucidation of the
    etiology of cancer, in: "Short-term Mutagenicity Test Systems
    for Detecting Carcinogens", K. Norpoth and R. C. Garner,
    eds., Springer-Verlag, New York, NY, 19-57 (1980).

59. H. S. Rosenkranz and W. T. Speck, Mutagenicity of metronidazole: Activation by mammalian liver microsomes, Biochem. Biophys. Res. Commun. 66:520-525 (1975).

60. H. S. Rosenkranz and W. T. Speck, Activation of nitrofurantoin to a mutagen by rat liver nitroreductase, Biochem. Pharmacol. 25:1555-1556 (1976).

61. L. A. Poirier and J. H. Weisburger, Enzymic reduction of carcinogenic aromatic nitro compounds by rat and mouse liver fractions, Biochem. Pharmacol. 23:661-669 (1974).

62. M. B. Aufrere, B. A. Hoener, and M. Vore, Reductive metabolism of nitrofurantoin in the rat, Drug Metab. and Disp. 6:403-411 (1978).

63. J. L. Blumer, R. F. Novak, S. V. Lucas, J. M. Simpson, and L. T. Webster, Jr., Aerobic metabolism of niridazole by rat liver microsomes, Mol. Pharm. 16:1019-1030 (1979).

64. J. Kratka, G. Goerz, W. Vizethum, and R. Strobel, Dinitrochlorobenzene: Influence on the cytochrome P-450 system and mutagenic effects, Arch. Dermatol. Res. 266:315-318 (1979).

65. P. L. Olive, Correlation between metabolic reduction rates and electron affinity of nitroheterocytes, Cancer Res. 39:4512-4515 (1979).

66. F. J. Peterson, R. P. Mason, J. Hovsepian, and J. L. Holtzman, Oxygen-sensitive and -insensitive nitroreduction by Escherichia coli and rat hepatic microsomes, J. Biol. Chem. 254:4009-4014 (1979).

67. C. Y. Wang, B. C. Behrens, M. Ichikawa, and G. T. Bryan, Nitroreduction of 5-nitrofuran oxidase and reduced nicotinamide adenine dinucleotide phosphate cytochrome c reductase, Biochem. Pharmacol. 23:3395-3404 (1974).

68. C. Y. Wang, C. W. Chiu, and G. T. Bryan, Nitroreduction of carcinogenic 5-nitrothiophenes by rat tissues, Biochem. Pharm. 24, 1563-1568 (1975).

69. M. R. Boyd, A. W. Stiko, and H. A. Sasame, Metabolic activation of nitrofurantoin - possible implications for carcinogenesis, Biochem. Pharmacol. 28:601-606 (1979).

70. N. Harada and T. Omura, Participation of cytochrome P-450 in the reduction of nitro compounds by rat liver microsomes, J. Biochem. 87:1539-1554 (1980).

71. T. Kawachi, T. Yahagi, T. Kada, Y. Tazima, M. Ishidate, M. Sasaki, and T. Sugiyama, Cooperative Program on Short-Term Assays for Carcinogenicity in Japan, in: "Molecular and Cellular Aspects of Carcinogen Screening Tests", R. Montesano H. Bartsch, and L. Tomatis, eds., IARC Scientific Publication No. 27, International Agency for Research on Cancer, Lyon, 323-330 (1980).

72. E. C. McCoy, L. A. Petrullo, H. S. Rosenkranz, and R. Mermelstein, 4-Nitroquinoline-1-oxide: Factors determining its mutagenicity in bacteria. Mutat. Res., 89: 151-159 (1981).

73. O. Kamm, Organic Synthesis, Vol. 1, John Wiley, New York, 445–447 (1964).

74. W. J. Hickinbottom, Reactions of Organic Compounds, Longmans, Green, and Co., London (1950).

75. F. K. Zimmermann, Procedures used in the induction of mitotic recombination and mutation in the yeast Saccharomyces cerevisiae, Mutat. Res. 31:71–86 (1975).

76. D. Brusick, Principles of Genetic Toxicology, Plenum Press, New York, 199–202 (1980).

77. R. W. Mayer and C. J. Goin, Induction of mitotic recombination by certain hair–dye chemicals in Saccharomyces cerevisiae, Mutat. Res. 78:243–252 (1980).

MUTAGENIC EFFECTS IN HUMAN AND

MOUSE CELLS BY A NITROPYRENE

C. F. Arlett, J. Cole, B. C. Broughton, J. Lowe and
B. A. Bridges

M.R.C. Cell Mutation Unit, and
Sussex Centre for Medical Research
University of Sussex
Falmer, Brighton BN1 9QG, England

INTRODUCTION

The demonstration of a mutagenic activity in photocopy
processes (1) and its identification and removal as an impurity
in carbon black from some photocopying toners (2) is an
excellent example of scientific detective work. The activity
resides in nitropyrenes which are present in carbon blacks
manufactured under some conditions. Studies with various
Salmonella strains (2) showed that nitropyrenes were particularly potent frame-shift mutagens and that they may either
be direct acting or require activation by an enzymatic pathway
not involving the classical nitroreductase. The use of a
standard S9 microsome fraction substantially reduced the
mutagenic potency of a set of nitropyrenes.

The possibility that nitropyrenes may act directly or be
activated by a novel enzymatic pathway in bacteria, necessitated
that their mutagenic activity be examined in mammalian cells in
vitro. Accordingly, we set out to test for mutagenicity with
1,8-dinitropyrene (DNP) in mouse lymphoma L5178Y cells and in
various human fibroblast cell strains. The L5178Y cells are
particularly useful since a number of mutagenic end points can
be examined independently with some degree of specificity (3-5).

Human fibroblast cell strains established from individuals
with genetic diseases characterized by defective DNA repair
systems can also be used to demonstrate specificity using a
lethal (6) or mutagenic response (7).

397

Since DNP may be regarded as a novel mutagen, its genotoxic activity was compared with 4-nitroquinoline-1-oxide (4NQO), another nitro-compound. 4NQO is a powerful mutagen for both microbial and mammalian systems and is activated by both soluble and micro-somal reductases (8).

MATERIALS AND METHODS

Routine culture, survival and mutation experiments with the mouse lymphoma cell line L5178Y have been described in detail elsewhere (4,5,9,10). These cells were grown in suspension culture. Colony formation was measured in soft agar to assess survival and mutation induction.

Human skin fibroblast cell lines were obtained from clinically normal or xeroderma pigmentosum (XP) (11) individuals (Table 1). The routine culture and design of survival and mutation experiments have been described elsewhere (4,12,13).

The 1,8-dinitropyrene and 4NQO were kindly supplied by Dr. Robert Mermelstein of the Xerox Corporation. The DNP was purified by high pressure liquid chromatography. Initially we experienced considerable difficulties in obtaining a practical solvent but ultimately selected dimethyl sulfoxide (DMSO). Maximum solubility achieved at 37°C was 500 µg/mL. The 4NQO was also dissolved in DMSO.

RESULTS

Cell killing in L5178Y cells

The response of the mouse lymphoma cells to the lethal effects of DNP and 4NQO is summarized in Table 2. Over the range of con-centrations and treatment times tested with DNP there was no evidence of any consistent dose-related cytotoxicity. However, 4NQO was clearly toxic; the concentrations required to achieve cell killing being at the ng level.

Cell killing in human fibroblasts

Three human cell strains were tested for their response to DNP. We were unable to find any evidence for cell lethality after either 2 or 22 hr treatments at concentrations up to 10 µg/mL (Table 3). The 22 hr treatment times at 10 µg/mL represent the limit beyond which clear cell death due to the solvent DMSO became apparent. In one experiment (data not shown), 5 µg/mL treatments were applied to cell strains 1BR and XP2BI for up

Table. 1. Characteristics of the human cell strains used in this study

Designation	Phenotype of Individual	Age at Biopsy	Cellular Characteristics	Source	Reference
1BR	Normal	23	Normal	University of Sussex Brighton, U.K.	14
GM730	Normal	45	Normal	Camden Culture Collection, U.S.A.	15
XP4LO	xeroderma pigmentosum	10	defective in excision repair, complementation Group A	Dr. P. Hall-Smith Brighton, U.K.	14
XP2BI	xeroderma pigmentosum	16	defective in excision repair, complementation Group G	Dr. M. Taylor, University of Birmingham U.K.	16

Table 2. Toxicity of 1,8-dinitropyrene (DNP) and
4-nitroquinoline-1-oxide (4NQO) to mouse
lymphoma (L5178Y) cells▲

(i) <u>1,8-dinitropyrene</u> Solvent control (mean of 10) 92.8 $\pm$ 3.9
$\pm$ S.E.

Length of Treat. (hr)	Exper. #	Concentration of DNP (µg/mL)					
		0.1	0.25	0.5	0.625	1.25	2.5
2	1				60.5	86.5	86.5
	2					111.5	97.5
24	1				83.5	87.5	92.0
	2					94.7	65.0
	3					100*	100*
48	1	123.7	99.8	95.5	70.3	97.2	66.6
	2	100*	100*		70.3	94.3	106.0
	3					65.0	58.7
	4					78.5	
72	1					119.8	45.7

(ii) 4-nitroquinoline-1-oxide Solvent control (mean of 7) 85.1 $\pm$ 3.5%
$\pm$ S.E.

Length of Treat. (hr)	Exper. #	Concentration of DNP (µg/mL)										
		9.5	19	24	38	45	37	76	95	150	190	380
2	1		62		61			44	15	17	10	
	2	94	80				51		35	19	14	1.7
	3			60		20			9			
	4			51		42			28			
	5								27			
Mean 2		94	71	56	61	31	51	44	23	18	12	1.7
24		61	34			31						

▲Exponentially growing cells were treated in suspension culture
in sealed foil-covered plastic containers for 2 hr (initial cell
density 5 x 10^5 cells/mL), 24 hr (initial cell density 1 x 10^5
cells/mL), 48 hr (initial cell density 4 x 10^4 cells/mL) or 72 hr
(initial density 1 x 10^4 cells/mL). Both compounds to be tested
were dissolved in DMSO (maximum final concentration 0.5% v/v),
and control cultures were solvent treated. At the end of the
treatment period the cultures were centrifuged, washed twice and
resuspended in fresh medium, to be (a) cloned immediately in soft
agar medium or (b) diluted in suspension culture to be grown for
7-8 days to monitor growth rate. Survival data presented represents
the absolute cloning efficiency (%) in soft agar immediately
after treatment, except those marked (*) for which survival was
measured by back extrapolation from growth rate in suspension
culture. In these experiments the growth rate of the treated
cultures could not be distinguished from the control cultures.

Table 3. The lethal response of human cell strains to
          1,8-dinitropyrene and 4-nitroquinoline-1-oxide

(i) 1,8-dinitropyrene

Length of Treat. (hr)	# of Exper. 1	Concentration of DNP (μg/mL)			Cell Strains
		1	5	10	
2	1	122*	–	99	IBR
22	3	100	111	100	
2	2	121	–	118	XP2BI
22	3	121	123	71	
2)		95	–	90	XP4LO
)	1				
18)		125	–	85	

(ii) 4-nitroquinoline-1-oxide

No. of Experiments	Concentration of 4NQO (ng/mL for 2 hr)								Cell Strains
	9.5	19	38	57	95	190	570	950	
5					52	36	2.4	0.5	1BR
2		23	4.4	0.6	0.1				XP2BI
3					8.8	0.7	0.04	0.01	GM730

* Expressed as a percent of the control untreated platings. The
  cloning efficiency of 1BR = 6-88 percent, of XP2BI = 15-100 percent, of XP4LO = 16 percent, of GM730 = 58-98 percent during
  these experiments.

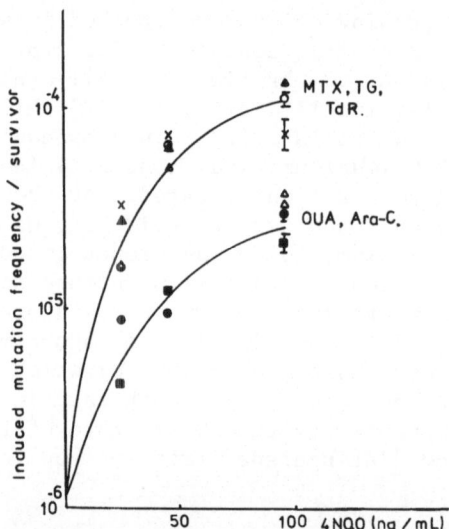

Figure 1. Mutation frequency in mouse lymphoma L5178Y cells at five
loci induced by 4-nitroquinoline-1-oxide (4NQO) (2hr treatment)
Key: ● = ouabain (OUA); O = 6-thioguanine (TG); X = methotrexate
(MTX); ■ = 1-β-D arabinofuranosyl cytosine (Ara-C); ▲ = 4mM, △
= 2mM thymidine (TdR). Each point represents the mean induced
mutation frequency per survivor of 2-4 experiments, except in TdR
selective medium, where individual experiment results have been
plotted. Error bars indicate S.E. of the mean. Protocol: $2-4 \times 10^7$
cells at 5×10^5 cells/mL were treated with 4NQO for 2 hr. After
removal of the mutagen, the cells were washed twice, resuspended in
fresh medium and counted using a haemocytometer. Samples were diluted
to 2 x 10 /mL in soft agar medium and plated to determine survival
immediately after treatment. A minimum of ∿1×10^7 estimated
viable cells were diluted in suspension culture medium to be
counted and subcultured every day for 7-8 days (the expression
period) for the determination of: (a) the growth rate in suspension
culture after treatment; (b) cloning efficiency in non-selective
soft agar medium at 24 hr intervals after treatment; (c) mutation
frequency in the five selective media: Oua (1×10^{-3} M); TG (1.8 x
10^{-4} M): araC (10^{-6} M); MTX (5×10^{-8} M) and TdR (2 or 4×10^{-3} M).
In every experiment the mutation frequency in each selective
medium was tested at several expression times. The induced
mutation frequency (M.F.) plotted was calculated by the method
of Arlett and Harcourt (23)

     I.M.F. = max treated M.F./survivor –

              max control M.F./survivor in the same experiment
Maximum treated mutation frequencies generally occurred at
48-96 hr expression time for OUA, MTX, araC, TdR and 144-192 hr
for TG.

to seven days and no cell killing was observed. As with the mouse
lymphoma cells, DNP was clearly non-toxic to human cells.
Exposure to 4NQO resulted in substantial cell lethality with
a significant difference in sensitivity between XP and normal
cell strains. However, it should also be noted that the two
normal cell strains, GM730 and 1BR, also exhibited a wide
difference in sensitivity (Table 3).

Mutation induction in L5178Y cells

The induction by 4NQO of mutants resistant to five distinct
selective agents is illustrated in Figure 1. A two-hour exposure
to 4NQO induced mutants in a dose dependent fashion for these
selective systems. The induced mutation frequencies for resistance
to methotrexate (MTX), 6 thioguanine (TG) and thymidine (Tdr) were
higher than for resistance to ouabain (Oua) or 1-β-D-arabinofuranosyl-
cytosine (Ara-C). At the highest concentration of 4NQO used (95
ng/mL), a four-fold difference in mutation frequency was observed.
A summary of the induction of mutants resistant to these same
selective agents by DNP (48 hr treatment time) is provided in
Figure 2. Again, a dose-dependent increase in resistant
mutants was observed for all selective systems with higher
frequencies of mutants resistant to TG and MTX than to Oua
or Ara-C at each dose level.

Mutation induction in human cells

We elected to concentrate on mutation induction in xeroderma
pigmentosum cells and the results of these studies are summarized
in Table 4. XP2BI cells were exposed to DNP in two separate
experiments; each study gave no indication of mutation induction
by this agent. We have found, however, that the control mutation
frequencies in these experiments were all rather high, suggesting
that the long exposures to high concentrations of DMSO (up to
2%) may have had some mutagenic effect. This effect may have
conceivably masked a small mutagenic effect of DNP. Further
work will be required to clarify this point. Exposure of XP2BI
cells to 4NQO was unequivocal in showing a powerful mutagenic
response. No 4NQO experiments have yet been performed with
normal cells.

DISCUSSION

1,8-Dinitropyrene was essentially non-lethal to both mouse
lymphoma L5178Y cells and normal or XP human cells. In contrast,
4NQO was extremely toxic to cells of both species. Among the human
cell strains, XP cells from complementation group G 16 were more
sensitive than normal cells to 4NQO as has been reported for other
XP cells 17 . Considerable differences between cell strains from

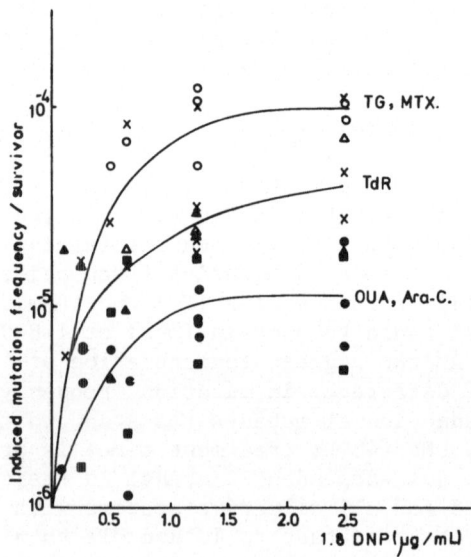

Figure 2.   Mutation frequency in mouse lymphoma L5178Y cells at five
loci induced by 1,8-dinitropyrene (DNP) (48 hr treatment)
Key: ● = ouabain (OUA);   ■ = 1-β-D arabinofuranosyl cytosine (Ara-C);
X = methotrexate (MTX);   O = 6-thioguanine (TG);   ▲ = 4mM, △ = 2mM
thymidine (TdR).   Points represent results from individual experiments Protocol.   DNP dissolved in DMSO was added to 100 mL of
exponentially growing cells in suspension culture, at 4 x 10⁴
cells/mL in 260 mL plastic Falcon flasks.   The bottles were sealed,
protected from light and incubated at 37°C for 48 hr.   The DMSO
concentration in DNP treated and solvent treated control cultures
was 0.5% v/v.   At the end of the treatment period, the mutagen was
removed and the cells plated for survival and M.F., as in the 4NQO
experiments.   In the control cultures, 4 x 10⁶ cells were present
at the start, and ∿5 x 10⁷ cells at the end of the treatment period.
DNP at all concentrations tested had little effect on the growth
rate during the treatment time, or the subsequent cloning efficiency
and growth rate during the expression time.   The maximum DNP
concentration possible, using DMSO as solvent (max. final
concentration 0.5% v/v) was 2.5 μg/mL.

Table 4.   Induction of 6-thioguanine (2.5 µg/mL)
resistant clones in XP2BI

(i)   1,8-dinitropyrene

		Concentration (µg/mL-22 hr)			
		0	1	5	10
Experiment 1	Survival	100	115[a]	123	72
	Mutation	17.9[b]	16.8	23.0	22.4
Experiment 2	Survival	100	126	160	80
	Mutation	14.3	16.0	23.6	16.4

(ii) 4-nitroquinoline-1-oxide

		Concentration (ng/mL-2 hr)			
		0	3.8	7.6	11.4
Experiment 1	Survival	100	60	25	36
	Mutation	3.0	37.2	44.7	52.9

[a]% survival determined at the end of the treatment period and
expressed relative to the control, untreated sample.   The
cloning efficiency in DNP experiment 1 was 69 percent, in
experiment 2 it was 47 percent.   For the 4NQO experiment it
was 100 percent.

[b]Mutation frequency per survivor x 10^{-6}.   Data were obtained
using the bulk culture bottle technique of Cox and Masson
(22) modified for our conditions (12, 13).

two normal individuals were also observed. It is not known whether
this variability is due to differences in DNA repair capacity,
a probable explanation for the differential sensitivity of XP
cells, or due to different levels of metabolic activation among
the normal cell strains.

DNP induction of mutants resistant to a set of five metabolic
poisons was unequivocal in L5178Y cells. Mutants resistant to MTX,
TG, Tdr, Oua, and AraC were induced in these cells in a dose-
dependent fashion after exposure to DNP for 48 hr. While we have
also obtained resistant mutants but at a much lower frequency after
24 hr treatments, 2 hr treatments were ineffective. Circumstantial
evidence from previous studies on mutagenesis with cultured
mammalian cells suggests that mutants resistant to TG may arise both
by a change in or a loss of genetic information 18 while Oua
resistance is the consequence of base pair substitution 19 .
Since ouabain resistance depends upon alterations in a vital
function the maintenance of the Na+/K+ balance of the cell, mediated
by a membrane bound ATPase 20 , loss of such an enzyme by deletion
or frameshift mutation would be lethal. DNP acts like a powerful
frameshift in Salmonella (2), and this result implies mechanistic differences between bacteria and mouse cells. In our
experiments, DNP is a novel mutagen by being essentially non-lethal
but mutagenic to eukaryotic cells.

With human cells, the first set of experiments attempted to
induce TG-resistant mutants in an XP cell strain by exposing these
cells to DNP for 22 hr. This cell strain was selected because, like
the mouse lymphoma cells, XP cells are lacking in the excision of
UV or "UV-like" damage 7 . The interpretation of the data for the
induction of TG resistant mutants is equivocal and, considering the
results obtained with the L5178Y cells, induction of such mutations
in XP cells may require treatment times much longer than 22 hr.
These experiments are in progress.

Definite mutation induction by 4NQO occurred in both L5178Y
and XP2BI cells and may potentially be used to quantitate the
effectiveness of DNP in mouse cells. A conventional technique
used to compare two or more mutagens where the mode of action may
be different or unknown is to compare the mutation frequency
against survival 5,21 . In this way, differences in chemistry
may possibly be negated. A display of this kind, comparing the
mutagenic ability of DNP and 4NQO in L5178Y cells for resistance
to the five selective agents, is shown in Figure 3. The absence
of any consistent dose-dependent lethality by DNP precludes the
use of such a comparison leaving a dilemma with respect to answering
the question of the mutagenic potency of DNP.

There are other interesting implications of these observations
on mouse lymphoma cells. Studies with the bacterial strains show
that the mutagenic effect of DNP is substantially reduced by the

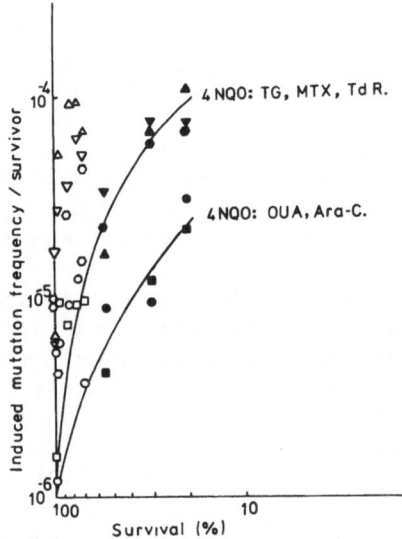

Figure 3.   Comparison of 4NQO and DNP induced mutation frequencies
at five loci in mouse lymphoma L5178Y cells.
Points show the mean induced mutation frequency per survivor,
plotted against survival immediately after the end of the treatment period. (1-4 experiments per point).   4NQO - 2 hr treatment.
DNP - 48 hr treatment.

DNP	Selective agent	4NQO
O	ouabain	●
Δ	6-thioguanine	▲
□	1-β-D arabinofuranosyl cytosine	■
▽	methotrexate	▼
⬡	thymidine (pooled results using 2 and 4 mM TdR)	⬢

Induced mutation frequency calculated by the method of Arlett and
Harcourt (23). Dose response curves for 4NQO have been drawn.

addition of a post-mitochondrial fraction of rat liver extract
(2). The metabolic capacity of L5178Y cells is largely unknown,
but it clearly allowed or promoted the mutagenic activity of DNP.
It should be remembered that the levels of DNP used in these L5178Y
mutation experiments were significantly greater than that used in
the bacterial studies. All the data presented are for 48 hr
exposures to DNP; convincing data for 24 hr treatment time were
difficult to obtain. One possibility currently under investigation
is that an inducible enzymatic activity may be present.

SUMMARY

DNP was mutagenic although non-toxic for mouse lymphoma
L5178Y cells when assayed for induced resistance to 6-thioguanine,
methotrexate, ouabain, excess thymidine and 1-β-D-arabinofuranosyl-
cytosine. 4-Nitro-quinoline-1-oxide (4NQO) was both toxic and
mutagenic with respect to these same five mutation assays. It is
not possible to scale the effectiveness of DNP against 4NQO because of the lack of toxicity of the former compound.

In mouse cells, DNP-induced ouabain resistant mutants are
believed to result from base pair substitutions, whereas in bacteria
only frameshift mutations have been observed. This differential
mutagenic response suggest that DNP's mechanism of action may
be different in mouse cells and bacteria. Long treatment times
(in excess of 24 hr) were required to detect DNP-induced mutants,
suggesting the possibility of an inducible activation system.

The pattern of cytotoxicity revealed in human cells is
similar to that in L5178Y cells. Xeroderma pigmetosum cells showed
more sensitivity to the toxic action of 4NQO than normal cells.
The induction of thioguanine-resistant mutants in XP cells by DNP
is equivocal. Knowledge obtained with the studies using L5178Y
cells suggests an increase in treatment times may be necessary
to detect induced mutants in these human cells.

ACKNOWLEDGEMENT

This work was supported, in part, by the Commission of the
European Communities Environmental Research Programme Contract
No. 182-79-1 Env. U.K.

REFERENCES

1. G. Lofroth, E. Hefner, I. Alfheim, and M. Møller, Mutagenic
       activity in photocopies. Science 209:1037-1039 (1980).
2. H. S. Rosenkrantz, E. C. McCoy D. R. Sanders, M. Butler,
       D. K. Kiriazides, and R. Mermelstein, Nitropyrenes:
       Isolation, identification and reduction of mutagenic
       impurities in carbon black and toners, Science 209:
       1039-1043 (1980).
3. C. F. Arlett, Mutagenicity in cultured mammalian cells.
       in: "Progress in Genetic Toxicology", D. Scott, B. A.
       Bridges and F. H. Sobels, eds. Elsevier Scientific Press,
       Amsterdam, 141-154 (1977).
4. C. F. Arlett, Genetic markers for mutagenesis studies in
       mammalian cells. in: "DNA Repair: A Laboratory Manual

of Research Procedures", P. C. Hanawalt and E. C. Friedberg, eds. Marcel Dekkar, New York, NY, 545-568 (1981).

5. J. Cole and C. F. Arlett, Methyl methanesulphonate mutagenesis in L5178Y mouse lymphoma cells, Mutat. Res. 50:111-120 (1978).

6. C. F. Arlett and S. A. Harcourt, Cell killing and mutagenesis in repair-defective human cells. in: "DNA Repair Mechanisms", P. C. Hanawalt, E. C. Friedberg and C. F. Fox, eds., Academic Press, New York, NY, 633-636 (1978).

7. C. F. Arlett and A. R. Lehmann, Human disorders showing increased sensitivity to the induction of genetic damage. Ann. Rev. Genet. 12:95-115 (1978).

8. M. Nagao and T. Sugimura, Molecular biology of the carcinogen 4-nitroquinoline-1-oxide. Adv. in Cancer Res. 23:131-169 (1975).

9. J. Cole and C. F. Arlett, Ethyl methanesulphonate mutagenesis with L5178Y mouse lymphoma cells. A comparison of ouabain, thioguanine and excess thymidine resistance. Mutat. Res. 34:507-526 (1976).

10. A. M. Rogers, R. Hill, A. R. Lehmann, C. F. Arlett and V. W. Burns, The induction and characterization of mouse lymphoma L5178Y cell lines resistant to 1-β-arabinofuranosyl-cytosine. Mutat. Res. 69:139-148 (1980).

11. K. H. Kraemer, Xeroderma pigmentosum. in: "Clinical Dermatology", D. J. Dennis, R. L. Dobson and J. McGuire, eds., Harper and Low, New York, NY, 1-33 (1981).

12. C. F. Arlett, Mutagenesis in repair-deficient human cell strains. in: "Progress in Environmental Mutagenesis", M. Alacevic, ed., Elsevier Scientific Press, Amsterdam, 161-174 (2980).

13. C. F. Arlett, S. A. Harcourt, A. R. Lehmann, S. Stevens, M. A. Ferguson-Smith and W. N. Morley, Studies on a new case of xeroderma pigmentosum (XP3BR) from complementation group G with cellular sensitivity to ionizing radiation. Carcinogenesis 1:745-751 (1980).

14. A. R. Lehmann, S. Kirk-Bell, C. F. Arlett, S. A. Harcourt, E. A. de Weerd-Kastelein, W. Keijzer and P. Hall Smith, Repair of ultraviolet light damage in a variety of human fibroblast cell strains. Cancer Res. 37:904-910 (1977).

15. C. F. Arlett, Survival and mutation in gamma-irradiated human cell strains from normal or cancer-prone individuals. in: "Radiation Research: Proceedings of the Sixth International Congress of Radiation Research, S. Okada, M. Imamura, T. Terashima and H. Yamagushi, eds., Japanese Association for Radiation Research, Tokyo, 596-602 (1979).

16. W. Keijzer, N. G. J. Jaspers, P. J. Abrahams, A. M. R. Taylor, C. F. Arlett, B. Zelle, H. Takebe, P. D. S. Kinmont and D. Bootsma, A seventh complementation group in exicision-deficient xeroderma pigmentosum, Mutat. Res. 62:183-190 (1979).

17. H. Takebe, J-H. Furuyama, Y. Miki and S. Kondo, High sensitivity of xeroderma pigmentosum cells to the carcinogen 4-nitro-quinoline-1-oxide. Mutat. Res. 15: 98-100 (1972).

18. R. Cox, Comparative mutagenesis in cultured mammalian cells, in: "Progress in Environmental Mutagenesis", M. Alacevic, ed., Elsevier Scientific Press, Amsterdam, 33-46 (1980).

19. R. M. Baker, Nature and use of ouabain-resistant mutants, in: "Banbury Report 2: Mammalian Cell Mutagenesis", A. W. Hsie, J. P. O'Neill and V. K. McElheny, eds., Cold Spring Harbor Laboratory, Cold Spring Harbor, NY, 237-247 (1979).

20. R. M. Baker, D. M. Brunette, R. Mankovitz, L. H. Thompson, G. F. Whitmore, L. Siminovitch and J. E. Till, Ouabain-resistant mutants of mouse and hamster cells in culture, Cell 1:9-21 (1974).

21. C. F. Arlett, D. Turnbull, S. A. Harcourt, A. R. Lehmann and C. M. Colella, A comparison of the 8-azaguanine and ouabain-resistance systems for the selection of induced mutant Chinese hamster cells. Mutat. Res. 33:261-278 (1975).

22. R. Cox and W. K. Masson, X-ray induced mutation to 6-thioguanine resistance in cultured human diploid fibroblasts, Mutat. Res. 37:125-136 (1976).

23. C. F. Arlett and S. A. Harcourt, Expression time and spontaneous mutability in the estimation of induced mutation frequency following treatment of Chinese hamster cells by ultraviolet light. Mutat. Res. 16:301-306 (9172).

DISCUSSION:

Q. Lewtas (EPA): I would like to make a few comments. I am very pleased to have heard Dr. Mermelstein and Dr. Arlett present their research on nitropyrenes. At EPA we are finding more evidence that these compounds are important environmental emissions resulting (at least partially) from the interaction of NO and NO_2 with aromatics in the presence of water. Nitropyrenes may be more important than the polycyclic aromatic hyrocarbons which have received so much attention over the last few years. I think that the different responses observed in the microbial and mammalian in vitro systems are very important and interesting. We have also examined some mononitropyrenes and trinitrofluoronone in mammalian cells in culture and have obtained either negative or extremely weak positive responses. The assays we have used include oncogenic transformation in 10T1/2 cells, transformation in the mouse lymphoma system and mutagenesis in the Chinese hamster ovary cell system.

GENOTOXIC PROPERTIES OF RADON AND ITS DAUGHTERS

Naomi H. Harley, Stuart M. Altman, Bernard S. Pasternack

New York University Medical Center
Institute of Environmental Medicine
550 First Avenue
New York, NY   10016   (U.S.A.)

INTRODUCTION

Gaseous ^{222}Rn and its short-lived solid daughter products are present in the atmosphere due to its production from ^{226}Ra in crustal earth. The short-lived ^{222}Rn daughters emit alpha particles, beta rays and gamma radiation. The carcinogenic effects of radon daughters are well documented in epidemiological studies of miners. Radon itself is an alpha-emitting noble gas and has no major sinks in the environment. Its removal from the atmosphere is through radioactive decay and its 3.8 day half-life allows it to be transported thousands of kilometers from its point of origin. It is the two short-lived alpha emitting daughter ^{218}Po(RaA) and ^{214}Po(RaC') which deliver the radiation dose associated with the carcinogenic properties of radon. Their effective half-lives are only 3.05 and 19.7 minutes respectively, so the concentration of the parent must be studied as well as the daughter activity in order to understand their presence in the human environment.

So far, there are two genotoxic effects of ^{222}Rn which have been observed, namely, lung cancer and skin cancer. A third effect, genetic mutation, has not been observed directly but may be inferred as a possibility. Some radon is dissolved in body tissue including the reproductive organs and thus it and its daughters deliver a radiation dose directly to the gonads. Also, the gamma rays from the airborne daughters deliver a small dose to the reproductive organs. Some of the very first experiments to induce genetic mutations were with ionizing radiation (1,2).

411

Lung cancer has been observed in eighteen underground mining groups exposed to radon daughters at moderate to high levels (3). Lung cancer is now thought to be the most important consequence of environmental ^{222}Rn because the bronchial epithelium in the first few branching airways of the tracheobronchial tree receive the highest dose delivered from any naturally-occurring radioactive substance. Ordinary environmental exposures in the United States are commonly about 1 to 5 percent of that allowed in occupational practices and in some locations can approach the levels permitted occupationally.

Skin cancer, although not as serious an outcome of radon daughter exposure as lung cancer, has also been observed. One reported study in Czechoslovakian uranium mines (4), shows the relationship between alpha emitting radon daughters deposited on the skin and the appearance of predominantly basal cell carcinoma (BCC). Skin cancers are generally not fatal and follow-up in exposed populations is difficult for this reason.

It is the purpose of this paper to indicate the environmental levels of ^{222}Rn that exist in typical atmospheres and their potential consequences in terms of these three genotoxic effects.

LEVELS OF ^{222}Rn EXISTING IN THE ENVIRONMENT

Harley (5) has estimated the global inventory of ^{222}Rn from measurements of the average emanation rate from soils. He found the average outdoor concentration based on this calculation is 200 pCi ^{222}Rn/m^3 over the continents and this value is in good agreement with measurements. Indoor ^{222}Rn has been shown to depend primarily upon three parameters: the ^{226}Ra concentration in underlying soil (for single family dwellings), the materials used in construction and the ventilation rate (6,7). In special instances, the use of well water containing high ^{222}Rn levels can be a major factor in determining indoor concentrations (8,9).

One often overlooked factor in estimating individual exposure, is that both indoor and outdoor ^{222}Rn concentrations undergo seasonal variations by a factor of perhaps two or three (10-12). Thus, accurate documentation of individual exposure is difficult or impossible without extensive environmental monitoring preferably over a full year, since the year-to-year concentration appears to be reproducible (12,13).

Table 1 is a summary of some of the short-term ^{222}Rn measurements which have been made in the United States and abroad and the few long-term measurements where data were collected for at

Table 1. Summary of worldwide measurements of ^{222}Rn concentrations

pCi ^{222}Rn/m^2

Country	INDOORS				OUTDOORS				REFERENCE
	Min.	Avg.	Max.	Study Duration	Min.	Avg.	Max.	Study Duration	
AFRICA (North)					–	13	–	*	23
ANTARCTICA (Little America)					–	0.5	–	*	24
					–	2.5	–	*	24
AUSTRIA (Insbruck)	<50	1190	7460	**1 year	10	390	2580	*	25
					–	360	2600	*	26
(Salzburg)	750	–	3100	*	<50	220	1300	*	10
	750	1300	2500	**1 year					10
									10
BOLIVIA					–	40	–	*	27
CANADA (13 cities) (Montreal)	140	410	880	*	20	60	130	*	28
									25
CZECHOSLOVAKIA (Plzen)					0.70	70	280	**3 year	29
FINLAND (Helsinki)	500	1000	3000	*¹					9

Table 1 (Continued)

Country	INDOORS				OUTDOORS				REFERENCE
	Min.	Avg.	Max.	Study Duration	Min.	Avg.	Max.	Study Duration	
FRANCE					–	250	–	*	30
GERMANY (Frankfurt)					30	150	530	*	25
HOLLAND					–	85^2	–	*	25
					–	20^3	–	*	25
HUNGARY	700	2640	5800	*					31
INDIA					–	100	–	*	32
JAPAN					–	60	–	*	27
NORWAY	1000	1400	> 6000	*	–	60	–	*	34
OCEANS									
(All)					–	1.2	–	*	25
(Indian)					–	2	–	*	35
(North Atlantic)					–	6	–	*	35
(Pacific)					–	3.3	–	*	25
(South Pacific)					–	2	–	*	35
(Sub-Arctic)					–	0.4	–	*	25

Table 1. (Continued)

Country	INDOOR				OUTDOORS				REFERENCE
	Min.	Avg.	Max.	Study Duration	Min.	Avg.	Max.	Study Duration	
PACIFIC ISLANDS									
(Caroline Island)					–	0.4	–	*	23
(Guam)					–	1.3	–	*	23
(Marshall Island)					–	0.5	–	*	23
(Samoan)					–	2.1	–	*	23
PERU					–	40	–	*	27
PHILIPPINES									
(Manila)					10	71	150	*	25
					–	7.1	–	*	23
POLAND	80	330	2140	*					31
	20	230	5340	*					36
SWEDEN	30	1540	16000	*					31
	–	430	–	*House, 1941–1950					37
	–	1300	–	*House, 1951–1960					37
	–	1700	–	*House, 1961–1970					37
	–	2500	–	*House, 1971–1975					37
	–	1100	–	*Apt., 1941–1950					37
	–	2100	–	*Apt., 1951–1960					37
	–	2500	–	*Apt., 1961–1970					37
	–	3200	–	*Apt., 1971–1975					37
SWITZERLAND					54	131	305	*	25
(Freiburg)									

Table 1. (Continued)

Country	INDOORS Min.	INDOORS Avg.	INDOORS Max.	INDOORS Study Duration	OUTDOORS Min.	OUTDOORS Avg.	OUTDOORS Max.	OUTDOORS Study Duration	REFERENCE
THAILAND	5	205	1200	*	110	–	460	**	38
UNITED KINGDOM	60	1340	6900	**(19 months)					31
					10	90	300	**(19 months)	39
					35	110	350	*	39
					10	–	430	*	25
	10	–	1500	*					40
									40
U.S.S.R. (TashKent)	200	7480	36400	*					41
	100	800	4500	**					41
					–	170	–	*	33
					60	70	90	*	42
U.S.A. (Alaska)					–	3	–	*	24
(California)					2.5	6	10	*	24
					–	90	–	*	24
(Colorado–Grand Junction)	<400	900	2300	*	1100	7800	28000	**+	43
"	200	35000	240000		130	830	4400	**++	43
"									44
"									44

*Home on Tailings

Table 1. (Continued)

Country	INDOORS				OUTDOORS				REFERENCE
	Min.	Avg.	Max.	Study Duration	Min.	Avg.	Max.	Study Duration	
(Colorado—cont.)									
Durango					3800	16000	34000	**+	44
"					90	510	1300	**++	44
(Florida)	30	1300	3600	*	20	–	300	*	24
									24
Polk Co.	1500	4500	7700	*	–	1700	–	*	46
(Hawaii)					–	5	–	*	45
					1.1	30	60	*	47
(Illinois–Chicago)					45	95	200	*	25
					70	–	300	*	24
					–	25	–	*	24
					50	–	1000	*	24
(Mass.–Boston)	< 5	150	940	*	<10	20	40	*	24
									24
(New Jersey) (Chester)	80	2000	7400	**	86	240	400	**	6
					150	320	490	*	6
					100	200	370	** 1 year cont. 1977	48
					100	220	340	** 1 year cont. 1978	11
					110	190	290	** 1 year cont. 1979	12
									13

Table 1.   (Continued)

Country	INDOORS				OUTDOORS				REFERENCE
	Min.	Avg.	Max.	Study Duration	Min.	Avg.	Max.	Study Duration	
(New Mexico)					-	240	-	*	24
(New York)	80	750	4600	**	110	150	250	**	6
(New York City)	3600	6000	7800	*	10	97	481	*	25
					20	100	230	**	49
					100	170	220	*	54
	60	95	170	**1 year	15	100	200	**1 year	54
	140	250	390	**1 year	20	130	500	*	24
									24
									24
									24
(Tuxedo, NY)					160	210	260	*	54
"					100	185	260	**Cont. for 3 mo.	50
(Howe Caverns, NY)	1300	17140	32000	**3 months					51
(Lloyd, NY)					20	80	200	**Cont. 8 weeks	55
(Schenectady)	-	1110	-	*					52
(Ohio)					170	480	1040	*	24
					70	270	850	*	24
					-	260	-	*	24

Table 1. (Continued)

Country	INDOORS				OUTDOORS				REFERENCE
	Min.	Avg.	Max.	Study Duration	Min.	Avg.	Max.	Study Duration	
(Pennsylvania Canonsburg)	460	1450	5500	*Off-site	610	1720	4700	*On-site former Ra Processing Plant	53
					250	390	780	*Off-site	53
(Puerto Rico)					–	0.1	–	*	24
(Tennessee)	130	1400	4800	*	–	17	–	*	24
(Texas)					500	1500	2500	*	8
(Utah, Salt Lake City)					1600	7200	22000	**+	44
"					60	380	1400	**++	44
(Utah, Monticello)					890	3500	12000	**+	44
"					30	340	1300	**++	44
(Washington State)					–	2	–	*	24
(Washington, D.C.)					–	47	–	*	24
					–	122	–	*	24

*Short term measurements usually individual samples less than 1 month.
**Long term measurements longer than 1 month.
1=Based on median values for dwellings without (average) and with maximum bored wells.
2= Land breeze.
3= Sea breeze.
+ On-pile-tailings 1 yr.; ++ Off-pile-tailings 1 yr.

least one year. Figure 1 shows the most comprehensive set of
published measurements which indicate the annual variations over
a three year period at Chester, New Jersey.

For the United States an average outdoor concentration might
be selected as about 200 pCi ^{222}Rn/m^3, while an indoor concentration
of about 800 pCi ^{222}Rn/m^3 seems representative.

ESTIMATES OF GENETIC MUTATIONS

Radon and its daughters are capable of delivering a dose to
the reproductive organs primarily in three ways, through external
irradiation from the gamma-ray emitting daughters, through the
^{222}Rn dissolved in body tissues or through translocation of the
inahaled daughters from the lung.

From the air dose gamma radiation calculations of Beck (14),
it is possible to estimate the annual gonadal tissue, gamma-ray
dose from ^{222}Rn daughters:

$$\text{annual gonadal gamma dose} = 9 \times 10^{-6} \text{ (C)} \quad \frac{\text{rad}}{\text{yr}} \qquad (1)$$

where C = ^{222}Rn concentration in pCi/m^3.

From data reported by Harley (15) it can be estimated that
the annual genetically significant alpha dose due to ^{222}Rn
solubility in tissue is:

$$\text{annual alpha dose} = 0.1 \times 10^{-6} \text{ (C)} \quad \frac{\text{rad}}{\text{yr}} \qquad (2)$$

Translocation of inhaled radon daughters from the lung to the
reproductive organs could about double the alpha dose due to
^{222}Rn solubility alone. Assuming a quality factor of 20, the
annual dose equivalent to the gonads

$$\text{annual gonadal dose equivalent} = 13 \times 10^{-6} \text{ (C)} \quad \frac{\text{rem}}{\text{yr}} \qquad (3)$$

ICRP (15) adopts a figure for the risk of hereditary ill
health within the first two generations, following irradiation of
either parent, of about 10^{-4} per rem. For environmental exposure
to ^{222}Rn and its daughters this is equivalent to a risk of 2 x 10^{-4}
per rem for domanant mutations since both parents are irradiated.

Assuming an average exposure in the U.S. of about 700 pCi/m^3,
the gonadal dose equivalent would be about 10^{-2} rem/year. The
risk would thus be very small.

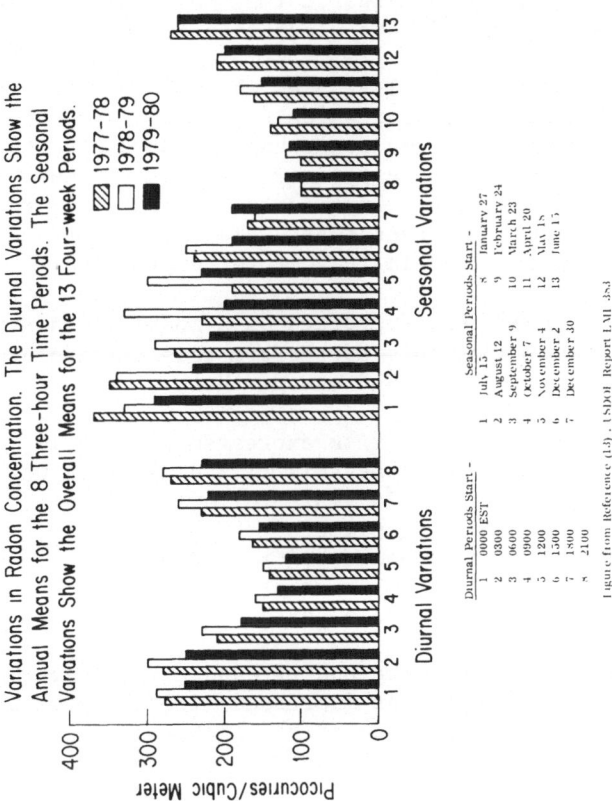

Figure 1. Variations in radon concentration. The diurnal variations show the annual means for the 8 three-hour time periods. The seasonal variations show the overall means for the 13 four-week periods. From reference 13: USDOE Report EML 383.

LUNG CANCER

    Lung cancer is a well-documented effect following exposure
to elevated levels of radon daughters in underground mines. A
statistically significant number of lung cancers above those expected have been reported at 100 working level months*. The current
U.S. occupational limit for exposure to radon daughters is 4 working
level months per year, equivalent to an exposure to approximately
3×10^4 pCi^{222} Rn/m^3 in equilibrium with its daughters for a
period of one working year or 2,000 hours.

    From Table 1 it can be seen that normal environmental levels
are routinely a few percent of this occupation limit.

    Although the epidemiological studies which show with reasonable
certainty that lung cancer is attributable to radon daughters involve exposures near 100 WLM, environmental exposures are typically
near 0.26 WLM per year**. Thus a typical lifetime exposure of about
$0.26 \times 50 = 13$ WLM is possible. In order to estimate environmental
effects, it is necessary to accept that a linear, nonthreshold,
stochastic process is involved in lung cancer expression.

    Harley and Pasternack have reported on a model which utilizes
the underground mining experience to perform this extrapolation
to environmental levels (17). They assume that:

    1. The attributable cancer risk expressed is 10×10^{-6}
        per year per WLM, but that this risk diminishes with an
        effective half life of 20 years following the year in
        which the exposure took place.
    2. That no lung cancer appears before age 40.
    3. That a minimum latent interval of at least 5 years exists
        after exposure before the appearance of lung cancer.

    This model appears to fit the underground mining experience
well and accounts for the apparent increased lifetime risk of
lung cancer for older age groups at first exposure.

----

*The working level (WL) is defined as the total potential alpha
 energy per liter of air of radon daughters considering decay through
 RaC' and is equivalent to 1.3×10^5 MeV/l. The working level is
 equivalent to 10^5 pCi/m^3 of ^{222}Rn in equilibrium with its
 daughters although equilibrium is not a necessary criterion. Exposure to one WL for one working month (170 hours) is the cumulative
 exposure in working level months.
**This cumulative exposure is attained from a typical environment
 containing 700 pCi ^{222}Rn/m^3 with ^{222}Rn to daughter ratios of
 1/0.9/0.7/0.7.

This risk model can also be related to ^{222}Rn concentration
rather than WLM through the alpha dose to the bronchial epithelium
per WLM. The lifetime risk of lung cancer per unit atmosphere concentration of ^{222}Rn (100 pCi^{222} Rn/m^3) has been calculated on this
basis and is shown in Table 2. It can be seen that an exposure to
500 pCi ^{222}Rn/m^3 would result in a lifetime lung cancer risk of
0.0018 or 0.2 percent.

Enstrom and Gadley (18) and Garfinkel (19) have reported the
annual age adjusted lung cancer rates in the United States for
nonsmokers, that is, spontaneous lung cancer. From their data a
lifetime risk (age 35-85) of lung cancer to men and women may be
calculated. These risks are 0.01 and 0.005 and 0.006 and 0.004
respectively for the two studies. Hirayama (20) in a study in
Japan showed that nonsmoking wives of smoking husbands have a higher
risk of lung cancer by a factor of two than nonsmoking wives of
nonsmoking husbands.

Although the spontaneous lifetime risk estimated here in non-
smokers is probably elevated by the effects of passive smoking it
indicates that typical environmental levels could generate about
20% of the spontaneous lung cancer.

SKIN CANCER (BASAL CELL CARCINOMA)

Because of the difficulty in follow-up studies of nonmelanotic
skin cancer (nonfatal cancer), the risk of basal cell (BCC) and
squamous cell carcinoma following radiation exposure is not well
documented. Only one study, the Czechoslovakian follow-up of
underground uranium miners, provides relevant risk estimates.
Although it is difficult to derive risk estimates directly from
their data, the following approach has been adopted in order to
estimate this effect. Sevcova et al. (13) report that in eight
years of follow-up there were 90 attributable skin cancers (predominantly BCC) per 10^4 miners. The average cumulative exposure
in this group was about 100 WLM. In the control population matched
for age there were 12 BCC per 10^4. We assume that the lifetime
risk of BCC in their control population is approximately that
reported in the U.S. by Scotto et al (21) for San Francisco, Iowa
and Minnesota (20% for males). Their control population has thus
expressed 12/2000 or about 1% of their expected lifetime skin
cancers. Since the oldest individual in both the mining and control
groups was only 55, such a low total yield is anticipated. Kopf
has documented that the majority of BCC appear after age 55 (22).
If the value of 1% of the total expected is assumed to apply to
the exposed miners as well, then

Lifetime estimates of BCC =

$$\frac{90 \text{ BCC}}{(100 \text{ WLM}) \, (10^4 \text{ persons}) \, (.01)} = 9 \times 10^{-3} \text{ per WLM}$$

A continuous exposure to 700 pCi ^{222}Rn/m^3 at environmental conditions results in an annual exposure of about 0.26 WLM/year. Thus,
the cumulative exposure of 0.26 WLM/year for an assumed effective
lifetime exposure of 50 years is 13 WLM. The estimate of lifetime
risk of BCC is thus $(9 \times 10^{-3}$ BCC WLM$^{-1})$ (13 WLM) $\simeq 0.10$. This
crude estimate of BCC risk,

Lifetime exposure to 700 pCi^{222} Rn/m^3 $\simeq$ 0.10 BCC

if correct, accounts for about half of the total observed spontaneous BCC at moderately high latitudes such as San Francisco. However, it must be pointed out that this prediction is based on very
limited data and the error associated with the estimate is extremely
high.

DISCUSSION AND SUMMARY

Three genotoxic effects of ^{222}Rn and its daughters have been
described. In order of importance these are lung cancer, skin
cancer and genetic mutation. Although the doses delivered from
^{222}Rn and its daughters are the highest received from natural
sources, the environmental data are not yet available to give a
good estimate of risk for the three effects. Unfortunately, the
risk is the quantity of interest to those individuals exposed to
elevated levels of ^{222}Rn.

An illustration, the lifetime risks of lung cancer for exposure
to three different ^{222}Rn levels from birth can be calculated using
the factors developed in Table 2.

	^{222}Rn concentration	Lifetime Lung Cancer Risk
Typical U.S.	500 pCi/m^3	0.002
Short-term measurements in one home in Polk Co., Florida	4500	0.02
Short-term measurements in one home in Grand Junction, Colorado built on Uranium mill tailings	35000	0.13

Table 2. Estimates of expected lifetime lung cancer risk from environmental radon daughter exposure based on epidemiological studies of underground miners

Lifetime Lung Cancer Risk for Environmental Conditions per 100 pCi ^{222}Rn/m^3*

Exposure Duration	Age at First Exposure			
	Birth	20	40	60
1 Year	2.5×10^{-6}	5.0×10^{-6}	8.3×10^{-6}	4.8×10^{-6}
10 Years	2.0×10^{-5}	5.8×10^{-5}	7.5×10^{-5}	3.6×10^{-5}
Lifetime	3.6×10^{-4}	3.0×10^{-4}	1.7×10^{-4}	4.8×10^{-4}

*Radon to radon daughter ratio 1/0.9/0.7/0.7 Unattached RaA/^{222}Rn = 0.07.

These high risk estimates indicate it is necessary to have accurate measurements of annual average exposures. The figures assumed for Polk County and Grand Junction are for individual houses were high concentrations were found.

An additional consideration is the number of persons in the exposed populations. Polk County, Florida, for example, has a population of about 32,000. Perhaps one tenth of these persons has an enhanced exposure and a few individuals have exposures to the levels indicated above. In any case, the number of attributable lung cancers would be impossible to detect against a background of those expected in the general population including smokers. Our risk estimates must be based on calculations and it would be desirable to have better information for a basis.

Although the calculations concerning the numbers of lung cancers produced are imprecise, those for skin cancer and genetic mutations have much larger associated errors. The risks per unit exposure (or dose) are not as well established and the problem of sparse exposure data persists.

ACKNOWLEDGEMENTS

The authors would like to thank Dr. John H. Harley for his many helpful suggestions during the preparation of this manuscript and Aimee Miranda for the arrangement and typing of the manuscript.

REFERENCES

1. H. J. Muller, Artificial transmutation of the gene, _Science_ 66:84-87 (1927).
2. E. A. Martell, V. H. Frahm, Jr., and S. E. Poet, Evidence that alpha radiation may be the primary agent of natural mutations: Preliminary experimental results for _Drosophila melanogaster_, _National Center for Atmospheric Research Report_, December, (1978).
3. V. E. Archer, E. P. Radford, and O. Axelson, Radon daughter cancer in man: Factors in exposure response relationship, _Conf. Workshop on Lung Cancer Epidemiology and Industrial Applications of Sputum Cytology_, Colorado School of Mines Press, Golden, CO, 324-367 (1979).
4. M. Sevcova, J. Sevc, and J. Thomas, Alpha irradiation of the skin and the possibility of late effects, _Health Physics_ 35:803-806 (1978).
5. J. Harley, Environmental Radon, _in_: "Nobel Gases", USEPA Symposium, Conf. 730915 September, (1973).

6. A. George and A. J. Breslin, The distribution of radon and radon daughters in residential buildings in the New York, New Jersey area, in: "The Natural Radiation Environment III", T. F. Gesell and W. L. Lowder, eds., Conf. 780422, DOE Symposium Series 51, April, (1978).

7. K. D. Cliff, Assessment of airborne radon daughters concentrations in dwellings in Great Britain, Phys. Med. Biol. 23: 696 (1978).

8. T. F. Gessell, and R. Prichard, The contribution of radon in tap water to indoor radon concentrations, in: "The Natural Radiation Environment III", T. F. Gesell and W. M. Lowder, eds., Conf. 780422, DOE Symposium Series 51, April, (1978).

9. O. Castren, The contribution of bored wells to respiratory radon daughter exposure in Finland, in: "The Natural Radiation Environment III", T. F. Gesell and W. M. Lowder, eds., Conf. 780422, DOE Symposium Series 51, April, (1978).

10. F. Steinhausler, W. Hofmann, E. Pohl, and J. Pohl-Ruling, Local and temporal distribution pattern of radon and daughters in an urban environment and determination of organ frequency distribution with demoscopical methods, in: "The Natural Radiation Environment III", T. F. Gesell and W. M. Lowder, eds., Conf. 780422, DOE Symposium Series 51, April, (1978).

11. J. Harley, Radon-222 measurements at Chester, USDOE Report EML-347, Environmental Measurements Laboratory, October, (1978).

12. J. Harley, Radon-222 measurements at Chester, USDOE Report EML-367, Environmental Measurements Laboratory, October, (1979).

13. I. M. Fisenne, Radon-222 measurements at Chester, USDOE Report EML-383, Environmental Measurements Laboratory, October, (1980).

14. H. L. Beck, Gamma radiation from radon daughters in the atmosphere, J. Geo. Res. 79:2215-2221 (1974).

15. J. H. Harley, E. S. Jetter, and N. Nelson, Elimination of radon from the body, USAEC Report HASL-32, Health and Safety Laboratory, March, (1958).

16. Recommendations of the International Commission on Radiological Protection, ICRP Publication 26, Annals of the ICRP, Vol. 1, No. 3, (1977).

17. N. H. Harley and B. S. Pasternack, A model for predicting lung cancer risks induced by environmental levels of radon daughters, Health Physics, 40:307-316 (1981).

18. J. E. Enstrom and F. H. Gadley, Cancer mortality among a representation sample of nonsmokers in the United States during 1966-68, J. Nat. Cancer Inst. 65:1175-1183 (1980).

19. L. Garfinkel, Cancer mortality in nonsmokers: Prospective study by the American Cancer Society, J. Nat. Cancer Inst. 65:1169-1173 (1980).

20. T. Hirayama, Nonsmoking wives of heavy smokers have a higher risk of lung cancer: A study from Japan, Brit. Med. J. 282: 183–185 (1981).

21. J. Scotto, A. W. Kopf, and F. Urbach, Non-melanoma skin cancer among caucasians in four areas of the United States, Cancer 34:1333–1338 (1974).

22. A. W. Kopf, Computer analysis of 3531 basal cell carcinomas of the skin, J. Dermatol. 6:267–282 (1979).

23. J. H. Blifford, Jr., J. Friedman, and L. B. Lockhart, Jr., Geographical and time distribution of radioactivity in the air, J. Atmos. Terr. Phy. 9:1–17 (1956).

24. N.C.R.P. #45, National Background Radiation in the United States, National Council on Radiation Protection and Measurements, November, (1975).

25. H. Israel, Radioactivity of the atmosphere, in: "Atmospheric Electricity", Akadem. Verl. Ges., Leipzig, 155–161 (1957).

26. F. Steinhausler, Long term measurement of ^{222}Rn, ^{220}Rn, ^{214}Pb, and ^{212}Pb concentrations in the air of private and public buildings and their dependence on meterological parameters, Health Physics 29:705–713 (1975).

27. L. B. Lockhart, Jr., Radioactivity of the radon–222 and radon–220 series in the air at ground level, in: "The Natural Radiation Environment", J. A. S. Adams and W. M. Lowder, eds., (1968).

28. R. G. McGregor, P. Vasudev, E. G. Letourneau, R. S. McCullough, F. A. Prantl, and H. Taniguchi, Background concentrations of radon and radon daughters in Canadian homes, Health Physics 39:285–289 (1980).

29. V. Hav lovic, Natural radioactive aerosols in the ground-level air of a Czechoslovak locality with respect to the radiological exposure of its population, Health Physics 11:553–566 (1965).

30. J. Fontan, A. Birot, and D. Blanc, Measurement of the diffusion of radon, thoron, and their radioactive daughter products in the lower layer of the Earth's atmosphere, Tellus 18: 623 (1966).

31. "Sources and Effects of Ionizing Radiation", United Nations Scientific Committee Report on The Effects of Atomic Radiation, pp. 77–78 (1977).

32. C. Rangarajan, S. S. Gopalakrithnan, and C. D. Eapen, The diurnal and seasonal changes in short-lived radon-thoron daughters' concentrations in the coastal and inland regions of India and their possible relations to regional climatology, Bhabha Atomic Center Report, Bombay (1974).

33. S. G. Malakhov, Diurnal variations of radon and thoron decay product concentrations in the surface layer of the atmosphere and their washout by precipitation, Tellus 18: 643 (1966).

34. E. Stranden, Radon in dwellings and lung cancer - A discussion, Health Physics 38:301–306 (1980).

35. J. Servant, Temporal and spatial variations of the concentration of short-lived decay products of radon in the lower atmosphere, _Tellus_ 18:663 (1966).

36. B. Jawarowski, K. Mamont-Ciesla, and M. Biernacka, The assessment of indoor exposure from γ-emitters and radon-222 in Poland, Central Laboratory for Radiological Protection, Warsaw, Poland, Specialist Meeting on the Assessment of Radon and Daughter Exposure and Related Biological Effects, CNEN, March, (1980).

37. G. A. Swedjemark, Radon in dwellings in Sweden, _in:_ "The Natural Radiation Environment III", T. F. Gesell and W. M. Lowder, eds., Conf. 780422, DOE Symposium Series 51, April, (1978).

38. P. Chittaporn, Work in progress, Thai Atomic Energy for Peace, Bangkok, Thailand, February, (1981).

39. A. K. M. M. Haque, A. J. L. Collinson, and C. O. S. Blyth-Brooke, Radon concentration in different environments and factors influencing it, _Phys. Med. Biol._ 10:505-514 (1965).

40. J. H. Harley, Radioactivity in building materials, _in:_ "Radioactivity in Consumer Products", A. A. Moghissi, P. Parks, M. W. Carter, and R. F. Barker, eds., NUREG/CP 0001 (1978).

41. A. E. Shemizade, Radioactivity of the air in the buildings of Tashkent, UDS 614.73:613.5(571.12), April, (1970).

42. S. G. Malakhov and P. G. Chernysheve, On the seasonal variations in the concentration of radon and thoron in the surface layer of the atmosphere, _in:_ "Radioactive Isotopes in the Atmosphere and Their Use in Meteorology", J. L. Karol, _et al._, eds., USAEC Report AEC-Tr-6711, 60-68 (1964).

43. H. Spitz, M. Wrenn, and N. Cohen, Diurnal variation of Rn measured indoors and outdoors in Grand Junction, Colorado and Teaneck, New Jersey and the influence that ventilation has on the buildup of Rn, _in:_ "The Natural Radiation Environment III", T. F. Gesell and W. M. Lowder, eds., Conf. 780422, DOE Symposium Series 51, April, (1978).

44. S. D. Shearer, Jr., and C. W. Sill, Evaluation of atmospheric radon in vicinity of uranium mill tailings, _Health Physics_ 17:77-88 (1969).

45. R. E. Larson, Radon profiles over Kilauea, the Hawaiian Islands and Yukon snow cover, _J. Geophys. Res._ 79:3432-3435 (1974).

46. W. M. Lowder and P. D, Raft, Radiation measurements in Polk County Florida, June, 1976, USERDA Report HASL 76-7 Health and Safety Laboratory, August, (1976).

47. H. E. Moore, S. E. Poct, and E. A. Martell, Origin of ^{222}Rn and its long-lived daughter in air over Hawaii, _J. Geophys. Res._ 29:5019-5024 (1974).

48. S. Watnick and V. Negro, Radon measuring instruments - Electrostatic field collection, USDOE Report EML 78-3, Environmental

Measurements Laboratory, November, (1978).

49. J. Thomas and I. M. Fisenne, Continuous monitoring of urban radon-222 concentrations, USDOE Report EML-298, January (1976).

50. P. Chittaporn, The development of a continuous monitor for the measurement of environmental radon, New York University Doctoral Thesis, June, (1980).

51. F. W. Seymore, R. M. Ryan, and J. C. Corelli, Radon and radon daughter levels in Howe Caverns, N.Y., Health Physics. 38: 858-859 (1980).

52. R. L. Fleischer, W. R. Giard, A. Mogro-Campero, L. G. Turner, H. W. Alter, and J. E. Gingrich, Dosimetry of environmental radon: Methods and theory for low dose, integrated measurements, Health Physics 39:957-962 (1980).

53. A. J. Breslin and A. C. George, Radon monitoring at Middlesex, Cannonsburg, and Lewiston, USDOE Internal Memorandum, September, (1979).

54. A. C. George, Indoor and outdoor measurements of natural radon and radon daughter decay products in New York City air, in: "The Natural Radiation Environment II, J. A. S. Adams and W. M. Lowder, eds., (1975).

55. J. H. Harley, Radon levels at the Lloyd, N.Y. regional station, USERDA Report HASL 306, July, (1976).

DISCUSSION

Q. Raabe (UC, Davis): First, a comment--I believe that in the United States, the states with the highest background radiation and with the highest level of radon in the air are also the states with the lowest incidence of lung cancer and, for that matter, other types of cancer. There isn't an obvious increase in cancers in the Rocky Mountain states versus the coastal industrial states.

My question relates to the influence of cigarette smoking on the calculation of the risk estimates you made. As I recall, cigarette smoking is a very important factor in determining the incidence of lung cancers among uranium miners. Could you please comment on how cigaerette smoking is considered separately in that risk-calculation?

A. Harley (NYU): A comment on your first point. This type of risk is very low--a .2% lifetime risk of lung cancer--55 lung cancers per million persons per year. By extrapolation, radon daughters account for one-fifth of that, or approximately 11 lung cancers per million persons per year. I don't think any epidemiological study will ever be able to see the difference in lung cancer incidences between two areas differing in radon concentrations.

With regard to smoking, this rate of 10 per million persons per year is supposedly the attributable rate to radon daughters independent of smoking. A recent report suggests that smoking may really confound the data in that the total lifetime risk to radon may be lower for smokers exposed to radon. It's just that the appearance of lung cancers may be sooner. Whether or not this is true remains to be seen when all the lung cancer data are evaluated. The most plausible explanation for this decreased risk in smokers is that smokers generate a lot of mucus in the bronchial tree, which is where the lung cancers really occur and particles don't really reach the bronchial epithelium through this mucus layer. Smoking may not be the problem in the underground mines as it was first thought.

Q. Costa (BNL): What is the concentration of radon daughters in cigarette smoke?

A. Harley (NYU): Radon daughters in cigarette smoke would be essentially nonexistant because they would be trapped by the filtering properties of the cigarette. The amount of radium in cigarette tobacco has been measured and its about equivalent to whats in normal soil - about one-half a picocurie per gram. So, you could possible be exposed to some radium but the levels would be really trivial.

The element which may be of concern is polonium-210. This nuclide is volatile at the temperature of burning tobacco and you do inhale a fair amount of polonium-210 per cigarette--something in the order of .1 picocurie. With each pack you would inhale 2 picocuries per day for a one pack-a-day smoker. This material is not short-lived; it has a 138-day half-life and may remain on the bronchial tree long enough to cause some prolems. The cancer risk from this nuclide is still unknown.

MULTIPLE TISSUE COMPARISONS OF SISTER CHROMATID EXCHANGES INDUCED BY INHALED STYRENE

M.K. Conner, Y. Alarie and R.L. Dombroske

Department of Biostatistics and IEHS,
Graduate School of Public Health
University of Pittsburgh, Pittsburgh, PA 15261 (U.S.A.)

INTRODUCTION

We have developed a multicellular in vivo assay for simultaneous assessment of relative mutagen-induced sister chromatid exchanges (SCE) in bone marrow, alveolar macrophages, and regenerating liver cells of hepatectomized mice and in bone marrow and macrophage cells of intact mice (1). Hepatectomy stimulates cell division thereby facilitating SCE analysis in an organ of primary metabolic importance and provides a means for assessing the effect of partial impairment of liver function on SCE responses in extrahepatic tissues (2). The specific cell types, while potentially sharing common hematopoietic origins (3), permit evaluation of relative susceptibilities unique to their residences in critical body tissues. In particular, analysis of induced SCE in alveolar macrophages promises to be a valuable method for assessment of indirect as well as direct acting hazardous airborne agents. We have recently reported our findings using this multicellular assay to study SCE induction and cytotoxicity of inhaled styrene (4,5).

Styrene is an important industrial monomer used in the manufacture of a variety of synthetic resins and plastics including packaging materials, containers, appliances, disposable food containers, and utensils. Although a slight elevation in the incidence of styrene-induced lung tumors in male mice (6,7) and in leukemia-lymphosarcoma type neoplasias in rats (8) have been reported, a recent NCI bioassay (7) provided no conclusive evidence for its carcinogenicity. However, the carcinogenicity of its presumed active metabolite, styrene oxide, has been confirmed in laboratory experiments (9).

Table 1. SCE induction of inhaled styrene oxide (5 hr. exposure)
          in bone marrow (BM), alveolar macrophages (AM) and
          regenerating liver (RL) of hepatectomized mice. Means
          are based on 3 mice (20 cells/mouse) unless otherwise
          indicated.

Dose (ppm)	BM	$\overline{AM}$	RL
0	3.1 ± 0.1	4.2 ± 0.6	3.6 ± 0.7
50	3.3 ± 0.4	6.3 ± 0.2	6.6 ± 0.6
70	3.4 (1 mouse)*	3.9 (1 mouse)*	4.9 (2 mice)*

*Very low yield of mitotic cells.

RESULTS

Styrene-Induced SCE

     We found that inhaled styrene in concentrations of 387 ppm
and above (6 hrs/d; 4 d) induced concentration dependent increases
in SCE frequencies in all cell types examined (4) (Fig 1).
Although dose-response relationships were apparent, styrene was
only weakly effective as an SCE inducer, producing maximum
frequencies of approximately 3-fold higher than baseline levels at
the highest tolerated dose of 922 ppm (6 hrs/d; 4 d). The acute
toxicity of styrene prevented investigation of its ability to
induce higher SCE frequencies. At 922 ppm, significant SCE
increases were also observed following 2 day exposures but not
after only one day exposure (Fig 2). In general, an increase in
the number of day exposures to styrene produced more cells with
higher levels of SCE as illustrated by SCE frequency distribution
plots for alveolar macrophages in Figure 3. It is of interest to
compare the SCE distribution plots of two days exposure to 922
ppm (Fig 3b) with that of four days·exposure to 387 ppm (Fig 3c).
Both protocols produced comparable mean SCE frequencies per cell.
However, in the latter case (actually a lower total dose) more
pronounced skewing of the distribution was apparent in the higher
mean SCE/cell range. This suggestive cummulative effect is consistent with pharmacokinetic clearance studies in rats and in humans
wherein styrene accumulated in fatty tissues only after repeated
exposures to concentrations greater than 200 ppm, levels at which
effective saturation of normal metabolic clearance pathways occurs
(10-12).

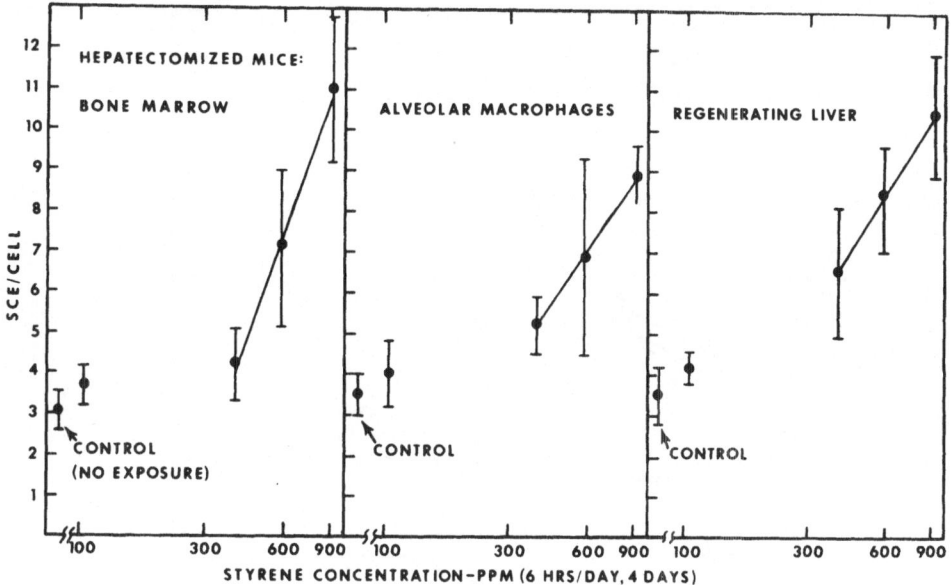

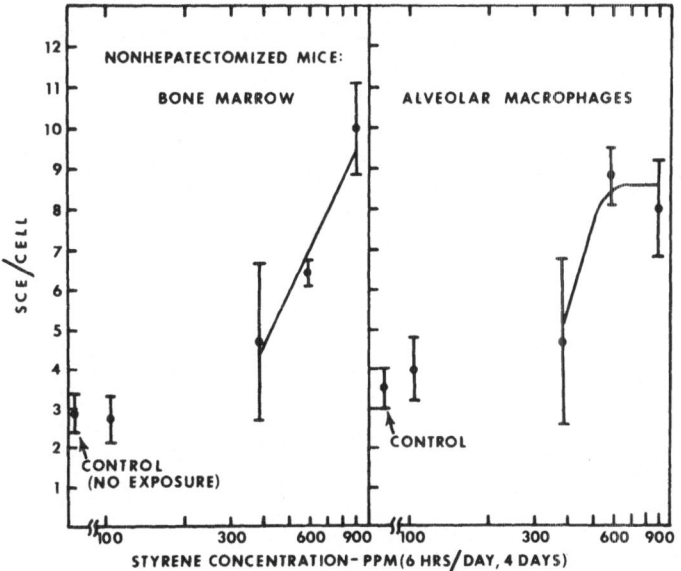

Figure 1: SCE frequencies in bone marrow, alveolar macrophage, and regenerating liver cells in hepatectomized mice and in bone marrow and alveolar macrophage cells of intact mice exposed to styrene (6 hr/d; 4 d) concentrations of 104 + 7.5, 387 ± 38, 591 ± 46, and 922 ± 52 ppm. (Published with permission of Academic Press, Inc. as it appeared in Conner et al., (4)).

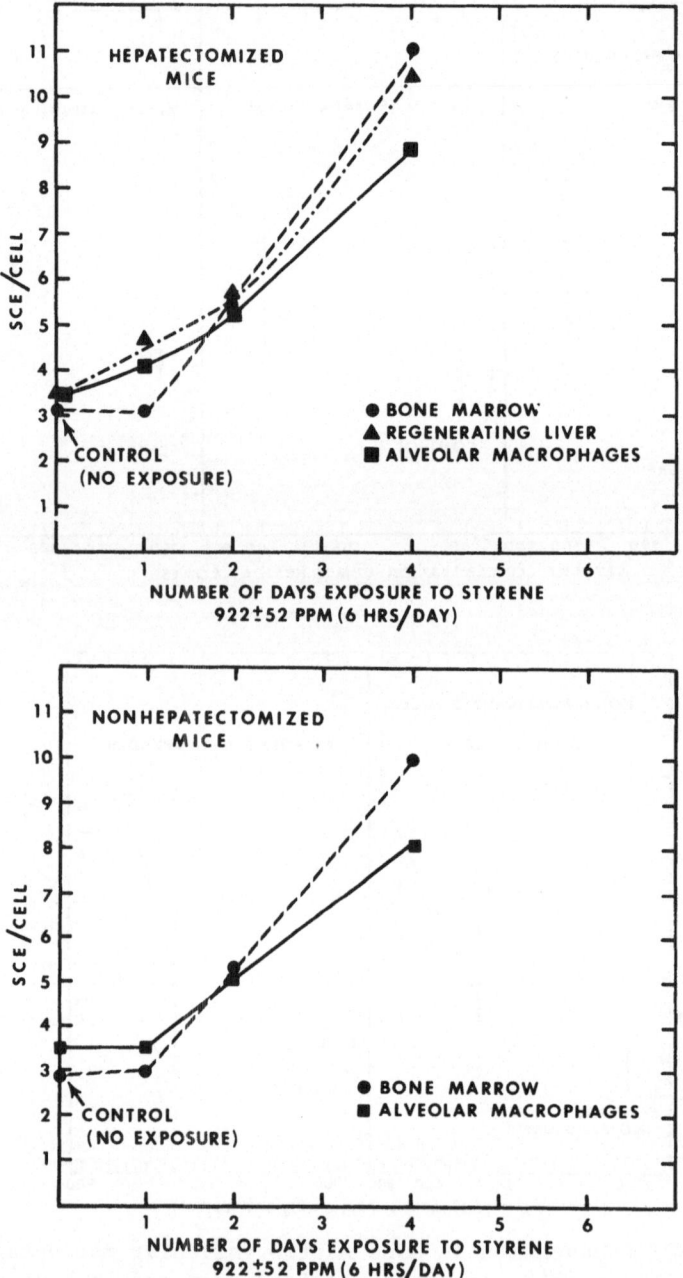

Figure 2: SCE frequencies in bone marrow, alveolar macrophages, and regenerating liver cells in hepatectomized mice and in bone marrow and alveolar macrophage cells of intact mice exposed to 922 ± 52 ppm styrene for the duration of 1,2 and 4 days (6 hr/d).

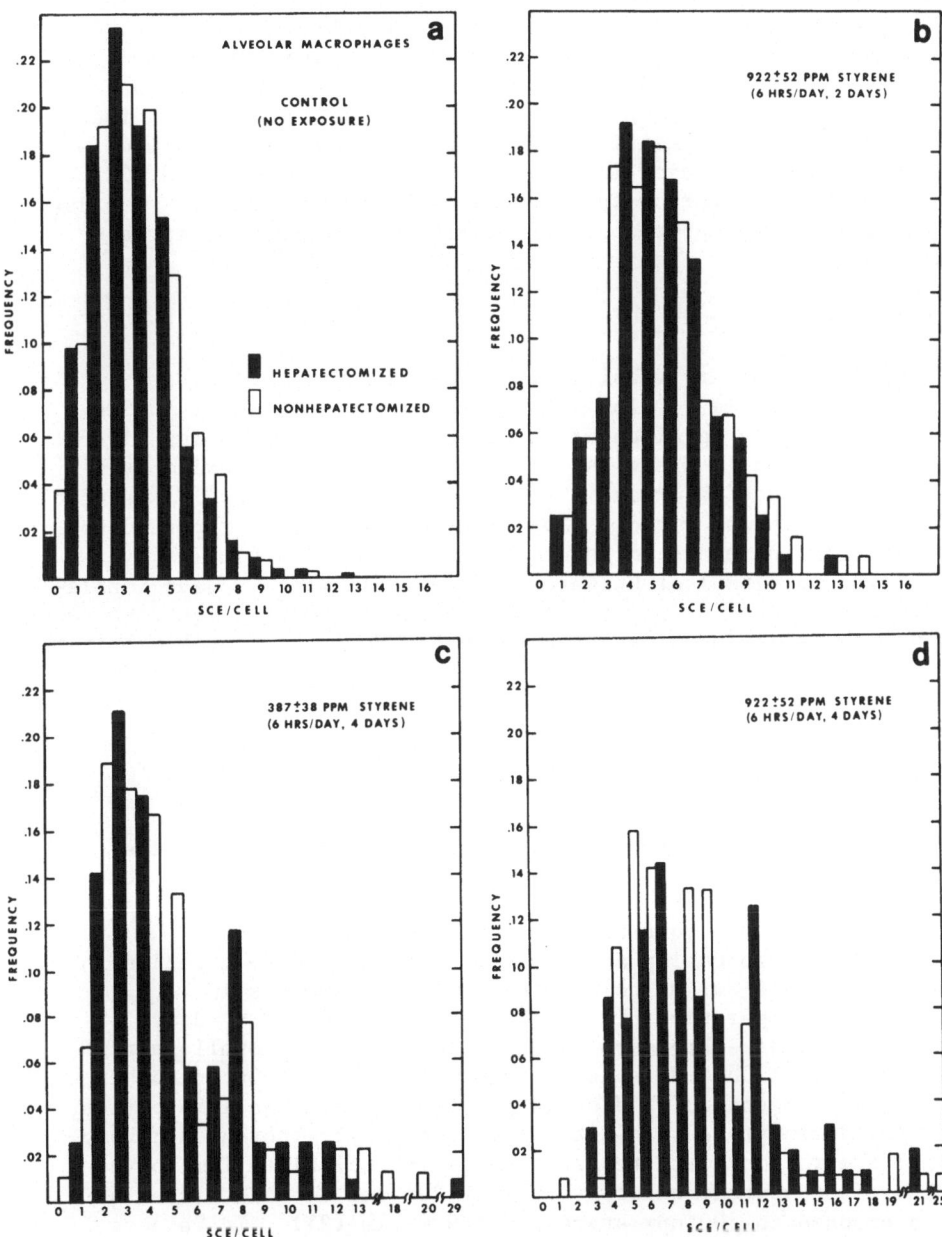

Figure 3: Frequency distribution plots of SCE in alveolar macrophage cells of mice exposed to styrene. a) No exposure controls
b) 922 + 52 ppm styrene (6 hr/d; 2 d) c) 387   38 ppm styrene
(6 hr/d; 4 d) d) 922 + 52 ppm styrene (6 hr/d; 4 d).

Hepatectomy affected styrene-induced cellular toxicity in all cell types in hepatectomized mice at the highest concentration (922 ppm; 4 d) employed and in regenerating liver cells at 591 ppm (4 d) as evidenced by an obvious decrease in proporation of second division relative to first division metaphases. However, hepatectomy did not alter cellular responses in extrahepatic tissues. The absence of tissue specific effects in SCE induction at all concentrations employed in our study suggest that induced SCE observed in bone marrow and alveolar macrophage cells are not due to hepatic biotransformation of styrene and release of styrene oxide.

Styrene Oxide-Induced SCE

Our preliminary results with inhaled styrene oxide are presented in Table 1. The data reveal slight increases in SCE frequencies only in alveolar macrophages and regenerating liver cells at 50 ppm (5 hrs). At 70 ppm (5 hrs) styrene oxide cytotoxicity produced a dramatic reduction in total as well as second division metaphase yields. Norppa et al. (13) also reported no significant increases in induced SCE in bone marrow cells (cultured in vitro) of Chinese hamsters following 20 days inhalation exposure to 25-100 ppm styrene oxide and only slight increases following injection of a lethal amount. However, in vitro culture of bone marrow cells may be complicated by selection of minimally damaged cell populations.

CONCLUSION

Other mutagenicity studies of styrene and styrene oxide have raised the concern of potential genetic risk. It is generally agreed that styrene is directly inactive in the Ames assay (14-16). However, contradictory evidence regarding the ability of microsomally activated styrene to induce mutation in bacteria (14-16) might be attributed to differences in microsomal preparations (17). Styrene produced no evidence of induction of sex-chromosome nondisjunction in Drosophila nor of micronuclei in Chinese hamster bone marrow (18). On the other hand, in cytogenetic studies in plants and in human lymphocytes, styrene demonstrated a c-mitotic effect (19). Increased chromosomal aberrations were reported in peripheral blood lymphocyte cultures of styrene exposed workers (20) and in bone marrow of rats exposed to 300 ppm styrene for 9 weeks (21). In Chinese hamster ovary (CHO) cells, styrene failed to elicit an increase in SCE even after microsomal activation whereas styrene oxide was a potent SCE inducer whose activity was diminished in the presence of microsomal fractions (22). In human lymphocytes, styrene oxide was found to be a potent SCE inducer while styrene (over the same concentration range employed in CHO cells) demonstrated approximately 1/10 the activity of styrene oxide (23). Furthermore, the SCE inducing activity of styrene in metabolically capable lymphocytes was correlated with its conversion to styrene oxide as monitored by gas chromatographic analysis (23).

The previously described in vitro SCE studies provide further insight into possible SCE-inducing activity of styrene and styrene oxide demonstrated in our studies. The limited ability of styrene oxide, a potent in vitro SCE inducer, to produce elevated SCE in vivo, as well as the demonstrated metabolic capability of lymphocytes to activate styrene to styrene oxide in vitro, suggest the possibility of in situ activiation of accumulated styrene to styrene oxide in vivo and relatively facile detoxification of systemically administered cytotoxic styrene oxide. Although the biological significance of the SCE phenomenon remains to be clarified, the striking parallels observed in our studies relative to pharmacokinetic and metabolic studies of styrene (10-12) are indicative of the potential utility of multicellular SCE analysis for evaluating hazardous airborne chemicals.

REFERENCES

1. M. K. Conner, Y. Alarie, and R. L. Dombroske, Sister chromatid exchange in murine alveolar macrophages, regeerating liver and bone marrow cells – A simultaneous multicellular in vivo assay. Chromosoma (Berl.) 74:51-55 (1979).
2. M. K. Conner, S. S. Boggs, and J. H. Turner, Comparison of in vivo BrdU labeling methods and spontaneous sister chromatid exchange frequencies in regenerating murine liver and bone marrow cells. Chromosoma (Berl.) 68:303-311 (1978).
3. R. Van Furth, Origin and kinetics of monocytes and macrophages. Sem. in Hematol. 7:125-141 (1970).
4. M. K. Conner, Y. Alarie, and R. L. Dombroske, Sister chromatid exchange in murine alveolar macrophages, bone marrow, and regenerating liver cells induced by styrene inhalation. Toxicol. Appl. Pharmacol. 55:37-42 (1980).
5. M. K. Conner, Y. Alarie, and R. L. Dombroske, Sister chromatid exchange in regenerating liver and bone marrow cells of mice exposed to styrene. Toxicol. Appl. Pharmacol. 50:365-367 (1979).
6. V. Ponomarkov, and L. Tomatis, Effects of long-term oral administration of styrene to mice and rats. Scan. J. Work Environ. Health, 4, Suppl. 2:127-135 (1978).
7. National Cancer Institute, Bioassay of styrene for possible carcinogenicity. Tech. Rep. Series. No. 185, D.H.E.W. Pub. No. (NIH) 79-1741, Washington, D.C. (1979).

8. N. C. Jersey, M. F. Blamer, J. F. Quast, C. N. Park, D. J.
    Schuetz, J. E. Beyer, K. J. Olson, S. B. McCollister,
    and L. W. Rampy, Two-year chronic inhalation toxicity
    and carcinogenicity study in monomeric styrene in rats,
    Dow Chemical U.S.A., MCA No: STY 1.1-TOX-INH (2 yr),
    Dec. 6 (1978).

9. C. Maltoni, G. Failla, and G. Kassapidis, First experimental
    demonstration of the carcinogenic effects of styrene
    oxide. Med. Lavoro 5:358-362 (1979).

10. J. C. Ramsey, J. D. Young, R. J. Karbowski, M. B. Chenoweth,
    L. P. McCarty, and W. H. Braun, Pharmacokinetics of
    inhaled styrene in human volunteers. Toxicol. Appl.
    Pharmacol. 53:54-63 (1980).

11. J. C. Ramsey, and J. D. Young, Pharmacokinetics of inhaled
    styrene in rats and humans. Scand. J. Work Health 4,
    suppl. 2:84-91 (1978).

12. J. D. Young, J. C. Ramsey, G. E. Blau, R. J. Karbowski, K. D.
    Nitche, R. W. Slauter, and W. H. Braun, Pharmacokinetics
    of inhaled or intraperitoneally administered styrene in
    rats. Proceedings of the Tenth Inter-American Conference
    on Occupational Medicine, pp. 297-309. Elseview/North-
    Holland, N.Y. (1979).

13. H. Norppa, E. Elovoora, K. Husgafuel-Pursiainen, M. Sorsa,
    and H. Vainio, Effects of styrene oxide on chromosome
    aberrations, sister chromatid exchange and hepatic drug
    biotransformation in chinese hamsters in vivo. Chem-
    Biol. Interact. 26:305-315 (1979).

14. H. Vainio, R. Paakkonen, K. Konnholm, V. Paminio, O. Pelkonen,
    A study on the mutagenic activity of styrene and styrene
    oxide. Scand. J. Work Environ. Health 3:147-151 (1976).

15. C. de Meester, F. Poncelet, M. Robertfroid, J. Rondelet,
    and M. Mercier, Mutagenicity of styrene and styrene
    oxide. Mutat. Res. 56:147-152 (1977).

16. L. Busk, Mutagenic effects of styrene and styrene oxide.
    Mutat. Res. 67:201-208 (1970).

17. K. Yoshikawa, Studies on metabolism and toxicity of styrene.
    III. The effect of metabolic inactivation by rat -
    liver S9 on the mutagenicity of phenyloxirane toward
    Salmonella typhimurium. Mutat. Res. 78:219-226 (1980).

18. M. Penttila, M. Sorsa, and H. Vainio, Inability of styrene
    to induce nondisjunction in Drosophila or a positive
    micromucleus test in the Chinese Hamster. Toxicol.
    Letters 6:119-123 (1980).

19. K. Linnainmaa, T. Meretoja, M. Sorsa, and H. Vianio, Cyto-
    genetic effects of styrene and styrene oxide. Mutat.
    Res. 58:277-286 (1978).

20. T. Meretoja, H. Vainio, M. Sorsa and H. Harkonen, Occupa-
    tional styrene exposure and chromosomal abberations.
    Mut. Res. 56:193-197 (1977).

21. T. Meretoja, H. Vainio, and H. Jarventous, Clastogenic effects
     of styrene exposure on bone marrow cells of rat. Toxicol.
     Lett. 1:315-318 (1978).
22. W. K. de Ratt, Induction of sister chromatid exchanges by
     styrene and its presumed metabolite styrene oxide in the
     presence of rat liver homogenate. Chem.- Biol. Interact.
     20:163-170.
23. H. Norppa, M. Sorsa, P. Pfaffle, H. Vainio, Styrene and
     styrene oxide induce SCE's and are metabolized in human
     lymphocyte cultures. Carcinogenesis 1:357-361 (1980).

AN OVERVIEW OF ETHYL CARBAMATE (URETHANE) AND ITS GENOTOXIC

ACTIVITY

James W. Allen, Yousuf Sharief*, and Robert J. Langenbach

Genetic Toxicology Division
Health Effects Research Laboratory,
U. S. Environmental Protection Agency
Research Triangle Park, NC 27711 (U.S.A.)

INTRODUCTION

During the last forty years ethyl carbamate, or urethane, has been the subject of many toxicology studies. Particular importance has been attached to this chemical because of widespread human exposure, experimental animal data indicating carcinogenic and teratogenic properties and an alluringly simple chemical structure. The simple chemical structure suggests an increased possibility of unveiling mechanistic pathways leading to its pathology. Consequently, a variety of studies have been concerned with ethyl carbamate's potential for interacting with DNA. In fact, ethyl carbamate became one of the earliest examples of a chemical mutagen when it was demonstrated to cause chromosome translocations in plants (1). However, the increasing appreciation of its carcinogenic capability has not been accompanied by very revealing patterns of mutagenic potential. While in vivo mammalian cytogenetic assays are clearly sensitive to this agent, urethane is negative in most short-term genetic tests for carcinogens. Thus, a fundamental problem is whether the relevant biological activity is due to ethyl carbamate itself or to a more potent host-generated metabolite. The enigmatic nature of ethyl carbamate has been alluded to in several excellent reviews (2-5). The present treatise provides an abbreviated perspective of ethyl carbamate, with updated genetic and, in particular, mammalian cytogenetic findings.

*Northrop Services Incorporated, Research Triangle Park, NC 27709 (U.S.A.)

HUMAN EXPOSURE

Ethyl carbamate, $CH_3-CH_2-O-CO-NH_2$, has been widely used in medicine as a hypnotic, anesthetic, and cancer chemotherapeutic agent, and in textile, pesticide, cosmetic, and various other industries. Another source of human exposure is through its occurrence, either naturally or as a reaction-product contaminant, in beverages (wine, beer, and orange juice), tobacco and tobacco smoke. Little data is available pertaining to commercial production of ethyl carbamate, although it is believed to have been produced in the U.S. for more than thirty years (4,6). Public health consequences have been of concern since 1943, when a very high incidence of lung tumors was discovered in mice that had been injected with ethyl carbamate as an anesthetic treatment (7). To our knowledge, there are no reported cases in which ethyl carbamate has been linked with the induction of human cancer; however, its action as a hepatocarcinogen in nonhuman primates has been reported (8).

CARCINOGENESIS AND TERATOGENESIS

Numerous rodent (mouse, rat, and hamster) studies employing various routes of exposure (oral, inhalation, intraperitoneal, and subcutaneous) have led to the recognition of ethyl carbamate as a multipotential procarcinogen of moderate potency (3,9,10). It is an initiator of mouse skin carcinogenesis, while with x-ray, it acts as a promoter or co-inducing agent to produce leukemia (3). Other agents observed to alter ethyl carbamate's expected yield of induced animal tumors include oxygen (11), thymidine (12), butylated hydroxytoluene (13), estradiol (14), diethylstilbestrol (15), zearalenone (16), growth hormone (17) and caffeine (18-20). Species, strain, age, sex, and tissue are also often important to the type and incidence of tumor produced (3). In vitro mammalian cell transformation has been demonstrated after in vivo (21,22) and in vitro (23) exposures to ethyl carbamate.

Ethyl carbamate is well known for its production of hepatomas, liver hemangiomas, thymic lymphomas and various other tumors. However, particular toxicological notoriety is derived from its strong, strain-specific induction of lung adenomas in mice. Presumably, these tumors arise from epithelial type II cells (13). Mice that are characterized by a high predisposition to spontaneous tumor development, e.g. strain A, respond to ethyl carbamate exposure with a much higher incidence of lung adenoma than do strains with a rate of low spontaneous pulmonary tumors, e.g. C57BL/6. Thus, it has been suggested that metabolic features may be relatively unimportant (3). That single-gene differences among strains may account for the relative levels of pulmonary tumor expression is supported by breeding trials (24). Strain-characteristic tumor induction susceptibilities are maintained in lung tissues grafted to hybrid mice prior to ethyl carbamate exposure, suggesting that the determinants are in the lung--the target tissue of the donor animals (25). Strain differences in lung enzyme levels have been noted, although without impressive correspondence to tumor susceptibility (26).

Radiolabeled ethyl carbamate is rapidly metabolized--more than 90% is expired as CO_2 within 24 hours and much of the remainder is excreted (3). As a low-molecular-weight, water- and lipid-soluble compound, it is rather evenly distributed throughout the body and is known to reach the developing fetus as well. It is both a teratogen and transplacental carcinogen, with these effects manifested in different species and among various tissues (3,27). Interestingly, excessive lung tumor incidence was detected in six-month-old mice exposed in utero to ethyl carbamate for only five minutes prior to cesarean delivery. This observation led to the suspicision that the parent molecule, rather than a metabolite, was the tumorigenic agent (28). Diverse information has come to bear on this point. However, the fundamental question as to whether ethyl carbamate or a metabolite is primarily responsible for car-cinogenic activity remains unresolved.

METABOLITES

Numerous compounds structurally related to ethyl carbamate and regarded as potential metabolites have been tested for carcino-genic potency (29-31). Only two of these, ethyl N-hydroxycarbamate and vinyl carbamate, have shown a marked propensity for lung and skin tumor induction and are seriously suspected to account for the carcinogenicity attributed to ethyl carbamate. Ethyl N-hydroxy-carbamate, although less carcinogenic than ethyl carbamate, is decidedly more potent in other biological activities including teratogenesis (32,33), inhibition of DNA synthesis (34,35), and the production of chromosome aberrations (36,37), DNA damage (38) and gene mutations (39). It is a detectable in vivo metabolite of ethyl carbamate (3). Of related interest, N-hydroxylation of aromatic amines is known to enhance carcinogenic activity (40).

In contrast to the long-standing interest in ethyl N-hydroxy-carbamate, vinyl carbamate has only recently come under investiga-tion (20,31). When injected into mice, vinyl carbamate is reportedly 10 to 50 times more potent than ethyl carbamate as an inducer of skin tumors and lung adenomas. Radiochemical evidence of ethyl group "desaturation" is consistent with the suggestion that ethyl carbamate is dehydrogenated in vivo to form the proxi--mate carcinogen vinyl carbamate. Vinyl carbamate in turn is pro-posed to undergo metabolic conversion to the ultimately reactive vinyl carbamate epoxide. Vinyl carbamate is not mutagenic for Salmonella typhimurium except with metabolic activation (31).

While both possible metabolites are of obvious biological
significance, their actual contribution to ethyl carbamate-assigned
carcinogenesis is not clear. Ethyl N-hydroxycarbamate is, to a
large extent, rapidly reduced in vivo to ethyl carbamate (3).   Yet
when this process is slowed by the enzyme inhibitor SKF 525A, ethyl
N-hydroxycarbamate carcinogenicity is reduced (3,41). The tumori-
genicity of ethyl N-hydroxycarbamate is reported to be more strongly
inhibited by another cytochrome P-450 inhibitor, DPEA (20). In
contrast, no similar inhibitions of tumor induction by ethyl carbamate are seen. Vinyl carbamate carcinogenicity is also not affected
by the aforementioned enzyme inhibitors; however, mutagenicity in
the S. typhimurium assay (with activation) is reduced (20,31). Also
vinyl carbamate, as a potentially carcinogenic intermediate, has
not been detected as a metabolite of ethyl carbamate with in vivo
or in vitro analytical techniques (20, 31). These observations
cast some doubt on the proposed role of vinyl carbamate and its
epoxide. Thus, the idea that ethyl carbamate itself is the proximate
carcinogen remains credible. In accord with this, only ethyl
carbamate induction of lung tumors has been shown to be repressible
by caffeine. This substance has little or no such effect on ethyl
N-hydroxycarbamate or vinyl carbamate (18-20). While the above
carcinogens are clearly distinguishable by some of their actions,
the degree of commonality in pathways leading to carcinogenesis
remains uncertain.

MUTAGENESIS

Although the significance of interim metabolic steps is not
clear, there is little question that in vivo ethyl carbamate
exposure will ultimately affect the DNA. Studies have shown that
in rodents injected with radiolabeled ethyl carbamate, nucleic acids
with bound radioactivity are recovered (42). The specific radioactivity is reportedly greater for lung DNA than for liver DNA,
except when liver DNA synthetic activity is heightened during liver
regeneration (43,44). It appears that the important molecule is a
covalently bound ethyl group residue (43,45). Evidence of hydrogen
loss from the ethyl group during metabolism has been presented (31).
Cytosine is known to be a target base for carboxyethylation after
in vivo exposure to ethyl carbamate (46) and after in vitro exposure
to ethyl N-hydroxycarbamate (but not ethyl carbamate under these
conditions) (47). Some interference with enzyme steps in pyrimidine
biosynthesis is also suggested by evidence for protective effects
of thymidine against chromosome damage (48) and carcinogenesis (12).
At present, it is not clear what genetic effects of ethyl carbamate
may relate to carcinogenesis. However, ethyl carbamate appears to
fit the general pattern of chemical carcinogenesis via an ultimate
covalent reaction with DNA (49).

Bateman (5) has organized much of the diverse literature on mutagenic actions of ethyl carbamate into a very useful review that is both chronological and comprehensive in its coverage up to the mid-1970's. From his summarizations, there is evidence of positive responsiveness in various plant systems, although without dose dependence. Gene mutations, aberrant meiotic configurations, and breakage in meiotic and mitotic chromosomes of various species are reported. However, in a study of Vicia faba root meristems, ethyl N-hydroxycarbamate induced chromosome aberrations, while ethyl carbamate did not. Experiments with Drosophila gave weakly positive results, showing sex-linked lethals and other gene and chromosomal anomalies, although again without dose responsiveness. In prokaryotes, Escherichia coli and Bacillus subtilis bacterial systems gave, respectively, weakly positive and negative results for mutagenesis. Neurospora and Schizosaccharomyces fungal mutation tests gave uniformly negative results. Clastogenesis in mammalian systems is discussed below.

Recent tests for DNA damage and mutations have shown variable, although predominantly negative, responses to ethyl carbamate. Some examples are given in Tables 1 and 2. For several of the nonmammalian cell tests, concern over the adequacy of the added activation system has been expressed. Consistent with this possible limitation is the apparent inability (in most tests) of ethyl carbamate to give a positive response. In Salmonella with metabolic activation, mostly negative results were obtained, even though ethyl N-hydroxycarbamate is mutagenic under these circumstances (39). Yet it is notable that Salmonella and Saccharomyces also gave negative results in the host-mediated bioassay using mice. Mammalian cell mutagenesis systems may be more sensitive to ethyl carbamate. Some support for this comes from unequivocally positive results obtained in unscheduled DNA synthesis and thioguanine/ouabain/fructose mutagenesis tests. More extensive coverage of various short-term test results with ethyl carbamate is given in a thorough review by Hollstein et al. (60) and in the forthcoming work "Short-Term Tests for Carcinogens: Report of the International Collaborative Program," edited by F.J. de Serres and J. Asbhy (61). In the latter, largely negative results (exceptions include yeast reversion, BHK transformation, and micronucleus assays) account for the designation of ethyl carbamate as a "problem carcinogen."

Mammalian test systems that involve effects at the chromosome level have also given variable results with ethyl carbamate (see Table 3). However, as a class, these tests have provided perhaps the most unequivocal evidence of genetic damage and the most predictable circumstances under which strongly positive results are obtained. Dominant lethal and micronucleus assays, which measure secondary manifestations of chromosome breakage, have given either completely negative or conflicting results, respectively. Chromosomal aberration and sister chromatid exchange (SCE) analyses are

Table 1. Representative Tests for Ethyl Carbamate Induction of DNA Damage

Test	Conditions (activation)	Organism Tissue/Cell Type	Results	References
Unscheduled DNA synthesis	in vitro (metabolic activation)	HeLa	+	50
Alkaline elution	in vivo (newborn rat)	rat brain rat liver	+ −	51 51
DNA repair (BrdUrd photolysis assay)	in vitro (no activation)	human fibroblasts	−	52
E. coli pol A₁⁻	(with or without metabolic activation)	DNA polymerase I deficient E. coli	−	38
Prophage induction	(no activation)	Salmonella phages	−	53

Table 2. Representative Tests for Ethyl Carbamate Induction of Mutations

Test	Condition (activation)	Organism Tissue/Cell Type	Results	References
Ames test (histidine reversion)	(with & without metabolic activation)	S. typhimurium (TA 100)	+,-	54,55
		(others)	-	55,56
Yeast (mitotic recombination)	(with & without metabolic activation)	Saccharomyces cerevisiae D$_3$	-	57
Host-mediated bioassay	in vivo (mouse)	S. typhimurium (TA 1535)	-	58
		Saccharomyces cerevisiae D$_3$	-	58
Mammalian cell mutagenesis				
(thioguanine/ ouabain/fructose)	in vitro (no activation)	human lung fibroblasts & Chinese hamster lung fibroblasts	+	23
(BrdUrd resistance)	in vitro (metabolic activation)	mouse lymphoma TK+/- cells	-	59

more direct assessments of chromosome alteration and, at least
with in vivo trials, appear to have consistently high sensitivity
for revealing effects of this particular chemical. While the
possible role of aberrations or SCEs in a carcinogenic process is
debatable, both phenomena are inducible by many potential mutagens
and carcinogens and thus serve in their detection (62,63). Recent
studies have suggested that aberrations and SCEs arise from
basically independent mechanisms (64).

When no activation system is included, in vitro chromosomal
aberration studies suggest a slight tissue-dependence in responsiveness to ethyl carbamate (Table 3). for the most part, weak positive responses have been obtained in lung-derived cells, e.g.
Chinese hamster lung-derived fibroblasts and mouse embryo lung
cells, while completely negative responses have characterized non-
lung-derived cells, e.g. rat embryo cells and Djungarian hamster
kidney cells. The data suggest the possibility that some level
of ethyl carbamate metabolism unique to lung tissue may be occurring.
However, this interpretation is compromised by the uncertainty of
detail concerning lung-cell derivation and by influences stemming
from species differences. Also, in vitro studies of SCE have not
revealed any special sensitivities of lung cells. Negative results
were obtained in Chinses hamster V-79 cells. However, this finding
may be due to the relatively low dose of ethyl carbamate (25 µg/mL)
used and/or to the distant lung derivation of V-79 cells. Although
lacking a clear dose-dependent response, Chinese hamster (DON)
fibroblasts exposed to ethyl carbamate have revealed up to a
doubling of the baseline SCE level. Both ethyl carbamate and ethyl
N-hydroxycarbamate have proven efficient inducers of SCE in human
lymphocytes exposed in culture. Interestingly, increases of
approximately three to four times baseline, obtainable in the
absence of activating systems, are drastically diminished when S-9
mix is added to the system.

In vivo chromosomal aberration and SCE studies have clearly
illustrated rodent capabilities to activate ethyl carbamate and
disperse a genetically active substance to diverse tissues (Table
3). Considering the former cytogenetic endpoint, a frequently
used dose of 1 gm/Kg body weight in mice typically results in half
the cells under analysis showing breakage phenomena. Species
(73) and strain (74,77) influences on the degree of damage are
also demonstrable. In keeping with the recognition of SCE as a
generally more sensitive cytogenetic indicator of chemically induced damage than aberrations (62-64, 79), relatively lower doses
of ethyl carbamate are needed to show a strong SCE effect. Depending on the cell type examined, intraperitoneal exposure to 50 mg/Kg
body weight leads to approximately a doubling of the control SCE
frequency while higher doses result in frequencies up to seven
times that of the control. When ethyl carbamate is administered
to mice either intraperitoneally (79) or by inhalation (83), re-

Table 3. Representative Ethyl Carbamate Results in Tests Involving Mammalian Chromosome Effects

Test	Condition (activation)	Organism Tissue/Cell Type	Results	References
Dominant lethal	in vivo	mouse	−	5
Micronucleus	in vivo	mouse bone marrow	+,−	61,65,66
Chromosomal aberrations	in vitro (no activation)	hamster, mouse (lung-derived cells)	+,−	67−70
		hamster, mouse, rat (non-lung-derived cells)	−	71,72
	in vivo	mouse, rat (bone marrow, thymus, spleen)	+	73−78
Sister chromatid exchange	in vitro (no activation)	mouse bone marrow	−	79
		Chinese hamster (V-79)	−	80
		Chinese hamster (DON)	+	81
		human lymphocytes	+	82
	in vivo	mouse liver, bone marrow, macrophages and spermatogonia	+	79,83

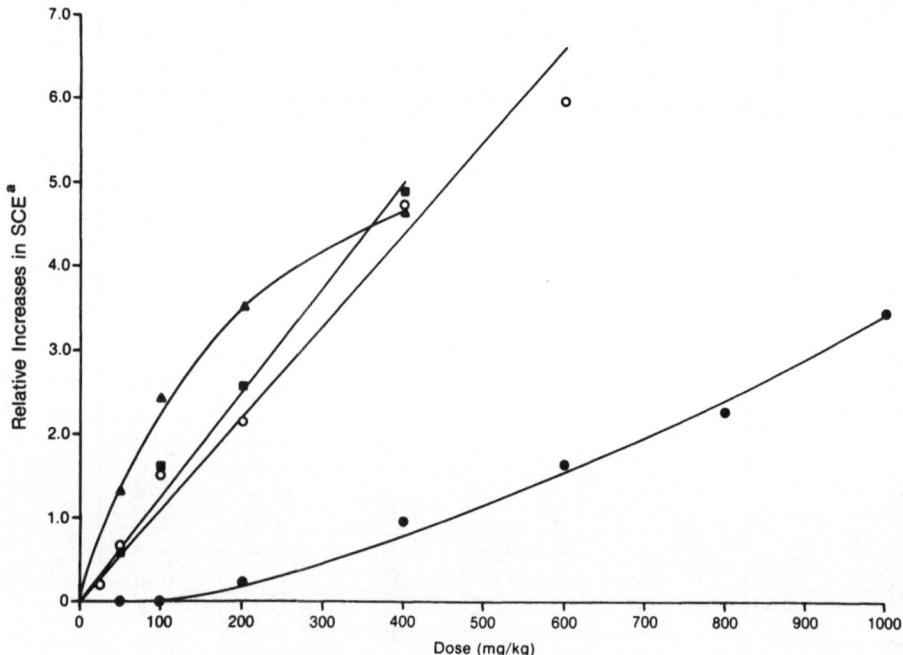

Figure 1.   Induction of SCE by ethyl carbamate in hepatocytes from
partially hepatectomized mice (▲), bone marrow from partially hepatectomized mice (■), bone marrow from intact mice (o), and spermatogonial cells from intact mice (●). [a]Relative increases in SCE
over respective control level.   (79, Published with permission from
Alan R. Liss, Inc)

generating liver cells reveal higher levels of SCE induction than
do bone marrow cells.   Alveolar macrophages also show higher SCE
levels (83), while spermatogonia reveal a lower increase in SCE than
other somatic tissues examined (79) (see Figure 1).   Additional
studies with ethyl carbamate are indicated in order to further
assess the degree of correspondence between sensitivities to SCE
formation and to tumorigenesis within a tissue.

From the aforementioned in vivo and certain in vitro mammalian
cell studies, it is clear that ethyl carbamate can potentially
exert strong genetic effects.   However, as with carcinogenesis, the
assignment of specific genetic activities to the parent compound
and/or to suspect metabolites has not been clear.   In the authors'
laboratories, SCE and gene mutation are being evaluated in Chinese
hamster V-79 cells exposed in culture to ethyl carbamate, ethyl
N-hydroxycarbamate, and vinyl carbamate.   Only preliminary data is
now available.   However, it appears that within this system, ethyl

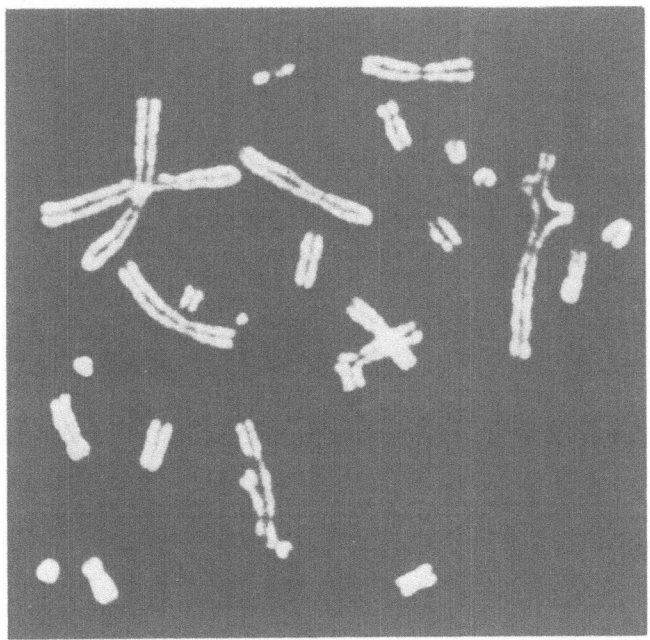

Figure 2. Chromosomal aberrations evident in a Chinese hamster V-79 cell exposed in vitro (without S-9) to 1 mg/mL ethyl N-hydroxycarbamate.

carbamate is largely unreactive by these criteria, both with and without S-9 activation. Doses of up to 10 mg/mL ethyl carbamate caused little or no mutagenesis for 6-thioguanine resistance or elevation in SCE frequency. The metabolite ethyl N-hydroxycarbamate has, thus far, proven inactive in V-79 cells at doses of up to 1 mg/mL in the presence of S-9 activation. Some elevation in SCE frequency was seen at these doses when S-9 activation was omitted, although this effect was overshadowed by evidence of powerful clastogenic action (see Figure 2). Similarly, in separate experiments with cultures of rat primary lung cells, exposure to 0.5-1.0 mg/mL ethyl N-hydroxycarbamate resulted in a dramatic incidence of chromosomal abberations while SCE increments were not very significant.

Vinyl carbamate, on the other hand, was a potent dose-dependent inducer of SCEs in V-79 cells. In the presence of S-9 activation, mean SCE frequencies were approximately 50 per cell, representing levels five times that of the control. Surprisingly, preliminary results suggest that comparable increases can be obtained (at higher doses) when the S-9 fraction is omitted (see Figure 3). In the presence of S-9 activation, vinyl carbamate doses of 0.5 and

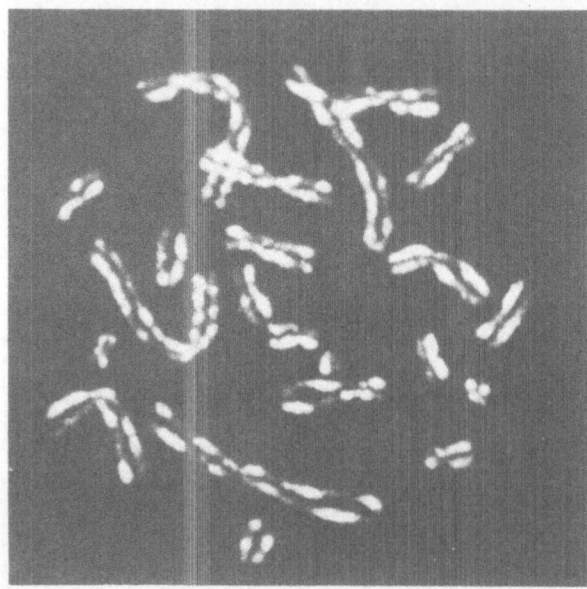

Figure 3.   Extensive sister chromatid exchange induction in a
Chinese hamster V-79 cell exposed in vitro (without S-9) to 1 mg/mL
vinyl carbamate.

1.0 mg/mL were toxic. Without S-9, these doses elevated the
control average SCE frequency of approximately 10/cell to more
than 30 and 60 SCEs/cell, respectively. Thus, in contrast with
the Ames test results (31), where the addition of a metabolic
activation system was required to detect genetic activity by vinyl
carbamate, SCEs are directly inducible. Perhaps metabolic capabilities of V-79 cells and/or differences in the nature of the
two genetic endpoints account for the disparate test system
results.

    A more complete interpretation of these data awaits final
tabulation of dose-range responses for gene mutagenesis and SCE.
However, it is noteworthy that strong activities for the induction
of chromosomal aberrations and SCEs following in vivo exposure to
ethyl carbamate appear, to some extent, to be assignable to different proposed carcinogenic metabolites. Aberrations and SCEs are
not mutually exclusive properties of ethyl N-hydroxycarbamate and
vinyl carbamate, respectively. Yet sensitivity differences measured
by these endpoints may facilitate the determination and identification of biologically relevant metabolites in carcinogenic
processes. Experiments to more fully characterize the genetic
actions of these chemicals are under way.

ACKNOWLEDGEMENTS

Vinyl carbamate was generously provided by Dr. James Miller, McArdle Laboratory for Cancer Research, University of Wisconsin, through the courtesy of Dr. Stephen Nesnow, Health Effects Research Laboratory, U.S. Environmental Protection Agency. We thank Karen Sasseville and Sharon Leavitt for their able technical assistance.

REFERENCES

1. F. Oehlkers, The production of chromosome mutations in meiosis by the action of chemicals, Z. Vererb. 81:313-341 (1943).

2. A. Haddow, "Professor Khanolkar Felicitation Volume", Indian Cancer Research Center, Bombay University Press, Bombay, India, pp. 158-181 (1963).

3. S. S. Mirvish, The carcinogenic action and metabolism of urethane and N-hydroxyurethane, Adv. Cancer Res. 11:1-42 (1968).

4. IARC, Some antithyroid and related substances, nitrofurans, and industrial chemicals, in: "IARC Monograph on the Evaluation of the Carcinogenic Risk of Chemicals to Man", Urethane 7:111-140 (1974).

5. A. J. Bateman, The mutagenic action of urethane, Mutat. Res. 39:75-96 (1976).

6. I. Schmeltz, K. G. Chiong, and D. Hoffmann, Formation and determination of ethyl carbamate in tobacco and tobacco smoke, J. Anal. Toxicol. 2:265-268 (1978).

7. A. Nettleship, P. S. Henshaw, and H. L. Meyer, Induction of pulmonary tumors in mice with ethyl carbamate (urethane), J. Natl. Cancer Inst. 4:309-319 (1943).

8. S. M. Sieber, P. Correa, D. W. Dalgard, and R. H. Adamson, Carcinogenic effects of urethane in nonhuman primates, Proc. Am. Assoc. Cancer Res. 21:78 Abstract (1980).

9. A. Tannenbaum, Contribution of urethane studies to the understanding of carcinogenesis, Natl. Cancer Inst. Monograph 14:341-356 (1964).

10. J. A. Miller and E. C. Miller, Metabolic activation of chemicals to reactive electrophiles: An overview, in: "Toxicology, Vol. 9: Advances in Pharmacology and Therapeutics", Y. Cohen, ed., Pergamon Press, New York, N.Y., 3-27 (1979).

11. J. A. DiPaolo, Effects of oxygen concentration on carcinogenesis induced by transplacental exposure to urethane, Cancer Res. 22:299-304 (1962).

12. G. B. Elion, S. Bieber, H. Nathan, and G. H. Hitchings, Uracil

antagonism and inhibition of mammary adenocarcinoma 755,
Cancer Res. 18:802-827 (1958).

13. H. Witschi and S. Lock, Butylated hydroxytoluene: a possible
     promoter of adenoma formation in mouse lung, in:
     "Carcinogenesis, Vol. 2: Mechanisms of Tumor Promotion
     and Cocarcinogenesis", T. J. Slaga, A. Sivak, and R. K.
     Boutwell, eds., Raven Press, New York, N.Y., 465-474 (1978).

14. S. Kawamoto, N. Ida, A. Kirschbaum, and G. Taylor, Urethane
     and leukemogenesis in mice, Cancer Res. 18:725-729 (1958).

15. F. Bojan, I. Redai, and Sz. Gomba, Induction of lymphomas by
     urethane in combination with diethylstilboestrol in CFLP
     mice, Experientia 35:378-379 (1979).

16. F. Bojan and P. Kertai, Induction of lymphomas by urethane in
     combination with zearalenone, presented at the Fifth Meeting
     of The European Association for Cancer Research (1979).

17. A. Robert and P. Klarner, Wirkung von Adaptationshormonen auf
     Urethan-Lungentumoren in der intakten Maus, Arch.
     Geschwulstforsch. 12:28-30 (1958).

18. J. C. Theiss and M. B. Shimkin, Inhibiting effect of caffeine
     on spontaneous and urethane-induced lung tumors in strain A
     mice, Cancer Res. 38:1757-1761 (1978).

19. T. Nomura, Timing of chemically induced neoplasia in mice
     revealed by the antineoplastic action of caffeine, Cancer
     Res. 40:1332-1340 (1980).

20. G. A. Dahl, E. C. Miller and J. A. Miller, Comparative carcino-
     genicities and mutagenicities of vinyl carbamate, ethyl
     carbamate, and ethyl N-hydroxycarbamate, Cancer Res. 40:
     1194-1203 (1980).

21. J. A. DiPaolo, R. L. Nelson, P. J. Donovan, and C. H. Evans,
     Host-mediated in vivo-in vitro assay for chemical carcino-
     genesis, Arch. Pathol. 95:380-385 (1973).

22. J. M. Quarles, M. W. Sega, C. K. Schenley, and M. Lijinsky,
     Transformation of hamster fetal cells by nitrosated
     pesticides in a transplacental assay, Cancer Res. 39:4525-
     4533 (1979).

23. D. A. Spandidos and L. Siminovitch, The relationship between
     transformation and somatic mutation in human and Chinese
     hamster cells, Cell 13:651-552 (1978).

24. D. S. Falconer and J. L. Bloom, A genetic study of induced
     lung-tumors in mice, Br. J. Cancer 16:665-685 (1962).

25. J. R. Shapiro and A. Kirschbaum, Intrinsic tissue response to
     induction of pulmonary tumors, Cancer Res. 11:644-647 (1951).

26. D. W. Nebert and H. V. Gelboin, The in vivo and in vitro
     induction of aryl hydrocarbon hydroxylase in mammalian cells
     of different species tissues, strains, and developmental
     and hormonal states, Arch. Biochem. Biophys. 134:76-89
     (1969).

27. T. Nomura, An analysis of the changing urethane response of the
     developing mouse embryo in relation to mortality, malforma-
     tion, and neoplasm, Cancer Res. 34:2217-2231 (1974).

28. M. Klein, Induction of lung adenomas following exposure of pregnant, newborn, and immature male mice to urethane, Cancer Res. 14:438–440 (1954).

29. C. D. Larsen, Evaluation of the carcinogenicity of a series of esters of carbamic acid, J. Natl. Cancer Inst. 8:99–101 (1947).

30. I. Berenblum, D. Ben-Ishai, N. Haran-Ghera, A. Lapidot, E. Simon, and N. Trainin, Skin initiating action and lung carcinogenesis by derivates of urethane (ethyl carbamate) and related compounds, Biochem. Pharmacol. 2:168–176 (1959).

31. G. A. Dahl, J. A. Miller, and E. C. Miller, Vinyl carbamate as a promutagen and a more carcinogenic analogue of ethyl carbamate, Cancer Res. 38:3793–3804 (1978).

32. S. Chaube and M. L. Murphy, The effects of hydroxyurea and related compounds on the rat fetus, Cancer Res. 26:1448–1457 (1966).

33. J. A. DiPaolo and J. Elis, The comparison of teratogenic and carcinogenic effects of some carbamate compounds, Cancer Res. 27:1696–1701 (1967).

34. J. Perpich, J. W. Yarbro, and B. J. Kennedy, Effect of urethane, hydroxyurethane, and hydroxyurea on synthesis of nucleic acid, Cancer 21:456–460 (1968).

35. P. Perocco, A. M. Ferreri, C. Franceschi, S. Grilli, P. Rocchi, G. Arfellini, and G. Prodi, Compared effects of N-hydroxy-urethane, urethane, and hydroxyurea on DNA synthesis, In vivo and in vitro studies, Z. Krebsforsch 89:99–106 (1977).

36. E. Borenfreund, M. Krim, and A. Bendich, Chromosomal aberrations induced by hyponitrite and hydroxylamine derivatives, J. Natl. Cancer Inst. 32:667–672 (1964).

37. E. Boyland, R. Nery, and K. S. Peggie, The induction of chromosome aberrations in Vicia faba root meristems by N-hydroxyurethane and related compounds, Br. J. Cancer 19:878–882 (1965).

38. H. S. Rosenkranz and Z. Leifer, Determining the DNA-modifying activity of chemicals using DNA-polymerase-deficient Escherichia coli, in: "Chemical Mutagens, Vol. 6", F. J. de Serres and A. Hollaender, eds., Plenum Press, New York, N.Y., 109–147 (1980).

39. T. Sugimura, S. Sato, M. Nagao, T. Yahagi, T. Matsushima, Y. Seino, M. Takeuchi, and T. Kawachi, Overlapping of carcinogens and mutagens, in: "Fundamentals in Cancer Prevention", P. N. Magee, S. Takayama, T. Sugimura, and T. Matsushima, eds., University Park Press, Baltimore, MD, 191–215 (1976).

40. J. A. Miller, Carcinogenesis by chemicals: An overview, G. H. A. Clowes Memorial Lecture, Cancer Res. 36:559–576 (1970).

41. A. M. Kaye and N. Trainin, Urethane carcinogenesis and nucleic acid metabolism: factors influencing lung adenoma induction Cancer Res. 26:2206–2212 (1966).

42. G. Prodi, P. Rocchi, and S. Grilli, In vivo interaction of urethane with nucleic acids and proteins, Cancer Res. 30: 2887-2892 (1970).

43. T. A. Lawson and A. W. Pound, The interaction of [3H] ethyl carbamate with nucleic acids of regenerating mouse liver, Chem.-Biol. Interact. 4:329-341 (1971).

44. S. V. Bhide, Urethane interaction with nucleic acids and proteins from non-malignant fast growing tissues, Chem.-Biol. Interact. 8:19-23 (1974).

45. A. W. Pound, F. Franke, and T. A. Lawson, The binding of ethyl carbamate to DNA of mouse liver in vivo: the nature of the bound molecule and the site of binding, Chem.-Biol. Interact. 14:149-163 (1976).

46. E. Boyland and K. Williams, Reaction of urethane with nucleic acids in vivo, Biochem. J. 111:121-127 (1969).

47. R. Nery, Acylation of cytosine by ethyl N-hydroxycarbamate and its acyl derivatives and the binding of these agents to nucleic acids and proteins, J. Chem. Soc. C:1860-1865 (1969).

48. E. Boyland and P. C. Koller, Effects of urethane on mitosis in the Walker rat carcinoma, Br. J. Cancer 8:667-684 (1954).

49. P. Brookes, Covalent interaction of carcinogens with DNA, Life Sci. 16:331-344 (1975).

50. C. N. Martin, A. C. McDermid, and R. C. Garner, Testing of known carcinogens and noncarcinogens for their ability to induce unscheduled DNA synthesis in HeLa cells, Cancer Res. 38:2621-2627 (1978).

51. G. L. Petzold and J. A. Swenberg, Detection of DNA damage induced in vivo following exposure of rats to carcinogens, Cancer Res. 38:1589-1594 (1978).

52. J. D. Regan, Interactions of carcinogens with human cell DNA: damage and repair, in: "Biological Reactive Intermediates", D. J. Jollow, J. J. Kocsis, R. Snyder, and H. Vaino, eds, Plenum Press, New York, N.Y., 483-491 (1977).

53. N. Yamamoto, M. D. Anderson, and J. A. DiPaolo, Phage and bacterial inactivation and prophage induction by chemical carcinogens, Molec. Pharmacol. 10:640-647 (1974).

54. D. Anderson and J. A. Styles, An evaluation of 6 short-term tests for detecting organic chemical carcinogens, Appendix 2, The bacterial mutation test, Br. J. Cancer 37:924-930 (1978).

55. V. F. Simmon, In vitro mutagenicity assays of chemical carcinogens and related compounds with Salmonella typhimurium, J. Natl. Cancer Inst. 62:893-899 (1979B).

56. J. McCann, E. Choi, E. Yamasaki, and B. N. Ames, Detection of carcinogens as mutagens in the Salmonella/microsome test: Assay of 300 chemicals, Proc. Natl. Acad. Sci. USA 72:5135-5139 (1975).

57. V. F. Simmon, In vitro assays for recombinogenic activity of chemical carcinogens and related compounds with Saccharomyces cerevisiae D3, J. Natl. Cancer Inst. 62:901-909 (1979A).

58. V. F. Simmon, H. S. Rosenkranz, E. Zeiger, and L. A. Poirier,
     Mutagenic activity of chemical carcinogens and related
     compounds in the intraperitoneal host-mediated assay, J.
     Natl. Cancer Inst. 62:911-918 (1979).
59. D. Clive, A linear relationship between tumorigenic potency
     in vivo and mutagenic potency at the heterozygous thymidine
     kinase (TK+/-) locus of L5178 mouse lymphoma cells coupled
     with mammalian metabolism, Mutat. Res. 53:95 Abstract (1978).
60. M. Hollstein, J. McCann, F. Angelosanto, and W. Nichols, Short-
     term tests for carcinogens and mutagens, Mutat. Res. 65:133-
     226 (1979).
61. F. J. de Serres and J. Ashby, eds., "Short-term Tests for
     Carcinogens: Report of the International Collaborative
     Program", Elsevier/North Holland, Amsterdam (in press).
62. S. A. Latt, Sister chromatid exchanges, indices of human
     chromosome damage and repair: detection by fluorescence and
     induction by mitomycin C, Proc. Natl. Acad. Sci. USA
     71:3162-3166 (1974).
63. P. Perry and H. J. Evans, Cytological detection of mutagen-
     carcinogen exposure by sister chromatid exchange, Nature
     258:121-125 (1975).
64. S. M. Galloway and S. Wolff, The relationship between chemically
     induced sister-chromatid exchanges and chromatid breakage,
     Mutat. Res. 61:297-307 (1979).
65. D. Wild, Cytogenetic effects in the mouse of 17 chemical
     mutagens and carcinogens evaluated by the micronucleus test,
     Mutat. Res. 56:319-327 (1978).
66. R. J. Trzos, G. L. Petzold, M. N. Brunden, and J. A. Swenberg,
     The evaluation of sixteen carcinogens in the rat using the
     micronucleus test, Mutat. Res. 58:79-86 (1978).
67. M. Ishidate and S. Odashima, Chromosome tests with 134 compounds
     on Chinese hamster cells in vitro--a screening for chemical
     carcinogens, Mutat. Res. 48:337-354 (1977).
68. G. M. Platonova, Comparative study of the mutagenic action of
     urethane and N-hydroxyurethane on the chromosomes of
     embryonic lung cells of A/Sn and C57 Black mice, Genetika
     5:168-169 (1969).
69. E. E. Pogosyants, G. M. Platonova, E. N. Tolkacheva, and
     L. F. Ganzenko, Effect of urethane on mammalian chromosomes
     in vitro, Genetika 4:60-72 (1968).
70. E. N. Tolkacheva, G. M. Platonova, and N. S. Vishenkova,
     Mechanism of the mutagenic action of urethane in mammalian
     cells in vitro, Genetika 6:108-115 (1970).
71. C. C. Huang, Induction of a high incidence of damage to the
     X chromosomes of Rattus (Mastomys) natalensis by base
     analogues, viruses, and carcinogens, Chromosoma,
     23:162-179 (1967).
72. O. I. Sokova, O. E. Kulagina, and E. E. Pogosyants, A study of
     the action of gamma rays, hydroxylamine, and urethane on
     the chromosomes of the Djungarian hamster in vitro,

Genetika 6:58–62 (1970).

73. B. J. Dean, Chemical-induced chromosome damage, Lab. Anim. 3: 157–174 (1969).

74. R. Wakonig-Vaartaja, The effect of urethane on mitotic cells of mice of different ages and strains, Aust. J. Exp. Biol. Med. Sci. 42:165–172 (1964).

75. T. P. Stromskaya, and E. E. Pogosyants, Effect of Rubomycin C and urethane on chromosomes of normal and leukemic mouse cells, Genetika 7:57–63 (1971).

76. M. I. Colnaghi, G. Della Porta, G. Parmiani, and G. Caprio, Chromosomal changes associated with urethane leukemogenesis in mice, Int. J. Cancer 4:327–333 (1969).

77. Y. Kurita, H. Shisa, M. Matsuyama, Y. Nishizuka, R. Tsuruta, and T. H. Yoshida, Carcinogen-induced chromosome aberrations in hematopoietic cells of mice, Gann 60:91–95 (1969).

78. N. Ueda, Action of chemical carcinogens at the chromosomal level, Kobe Ika Daigaku Kiyo 37:127–135 (1977).

79. G. T. Roberts and J. W. Allen, Tissue-specific induction of sister chromatid exchanges by ethyl carbamate in mice, Environ. Mutagen 2:17–26 (1980).

80. N. C. Popescu, D. Turnbull, and J. A. DiPaolo, Sister chromatid exchange and chromosome aberration analysis with the use of several carcinogens and noncarcinogens, J. Natl. Cancer Inst. 59:289–293 (1977).

81. S. Abe and M. Sasaki, Chromosome aberrations and sister chromatid exchanges in Chinese hamster cells exposed to various chemicals, J. Natl. Cancer Inst. 58:1635–1641 (1977).

82. I. Csukas, E. Gungl, I. Fedorcsak, G. Vida, F. Antoni, I. Turtoczky, and F. Solymosy, Urethane and hydroxyurethane induce sister-chromatid exchanges in cultured human lymphocytes, Mutat. Res. 67:315–319 (1979).

83. M. Cheng, M. K. Conner, and Y. Alarie, Multicellular in vivo sister-chromatid exchanges induced by urethane, Mutat. Res. 88:223–231 (1981).

SESSION VIII: MONITORING OF HUMAN POPULATIONS AT RISK

R. R. Tice, Moderator

Medical Department
Brookhaven National Laboratory
Upton, NY 11973 (U.S.A.)

The identification of carcinogenic and/or mutagenic activity in chemicals and the extrapolation of the experimental data to the human situation is the primary approach to controlling possible adverse health effects in human populations. Another useful approach, and one negating the difficulties inherent in extrapolating cell and animal data to humans, is to actually monitor human populations for genotoxic damage, e.g cancer, heritable diseases, during and after their exposure to hazardous agents. Epidemiological studies, the classical approach to population surveillance, are limited in that the ill-health effect must be readily apparent before an adverse health effect situation can be inferred. For epidemiological studies directed at assessing cancer incidence, a 15 to 20 year delay between the exposure to the causitive agent and the appearance of the cancers is not uncommon. For heritable diseases the delay in appearance is at least one generation. Except for an unrelated decrease in agent usage dictated by economic or for more immediate health effects, the burden of ill-health will increase throughout this delayed expression time. To circumvent this lag time associated with cancer or heritable diseases, several approaches have been developed or are being developed to assess several "indicators" of potential genotoxic ill-health. These indicators include: cytogenetic and mutagenic endpoints in human peripheral blood lymphocytes, mutagenic or clastogenic substances in body fluids (urine, seminal fluid), nondisjunctional events or abnormal morphology in sperm, and the alkylation of hemoglobin.

Although monitoring is generally agreed as useful, several difficulties are inherent in the process. First, the relevance of specific endpoints, e.g. alkylation of hemoglobin, abnormal sperm, to the exposure situation is nearly always confounded by extraneous

461

exposures to other agents. The resulting data is further confounded by the heterogeneity of the human population with regard to innate sensitivity to the genotoxic agent. Finally, positive findings are not interpretable on an individual basis, but can only be used to infer possible future ill-health for the population in general.

In this session three methodologies for monitoring human populations or with potential use for monitoring will be discussed. Two of the approaches will assess somatic manifestations of genotoxicity while the third presentation will address germinal effects. The first presentation, by S. M. Galloway and R. R. Tice, will discuss the cytogentic techniques--chromosomal aberrations and sister chromatid exchanges--for monitoring humans. These techniques have been used the most extensively, especially in occupational settings. The next presentation, by R.J. Albertini and D. J. Sylvester, will examine the utility of a mutagenic assay using human peripheral blood lymphocytes. This assay is very much in the developmental stages and has not been used in any occupational setting. The last presentation, by H. V. Malling and his coworkers, will discuss a mutational assay in sperm. This assay has been developed for rodent populations exposed to specific genotoxic chemicals. However, it does have future applicability to humans. These three approaches are not all inclusive, but they do offer a spectrum of methodologies for monitoring human populations at risk for both somatic and germ cell effects.

CYTOGENETIC MONITORING OF HUMAN POPULATIONS

Sheila M. Galloway and Raymond R. Tice*

Litton Bionetics, Inc.
Kensington, MD   20795 (U.S.A.)

INTRODUCTION

It is clear from examining the impact of modern technology
on the environment that many individuals are being exposed to genotoxic agents, i.e. agents which may increase the risk of cancer, of
mutations that might adversely affect future generations, and of
damage to unborn children. It takes lengthy epidemiological
research to establish links between exposure to specific agents
and increases in incidence of cancer or birth defects. Biochemical
methods for monitoring children for inherited genetic changes are
not yet widely available. However, one practical way to detect
damage to genetic material is by studies of lymphocyte chromosomes.

Cytogenetic monitoring has been increasingly used during the
last two decades, and in the last 5 or 6 years its scope has been
expanded considerably by the addition of sister chromatid exchange
studies. We now have the technical information needed to plan
reliable, well-standardized cytogenetic studies. Interpretation
of the findings is more difficult and will require additional
information, but the presence of chromosomal damage is an acknowledgement indicator of exposure to "genotoxic" agents.

THE CYTOGENETIC ENDPOINTS

In studies of chromosomes of peripheral blood lymphocytes,
we can evaluate chromosomal aberrations, sister chromatid exchanges

*Brookhaven National Laboratory, Upton, NY 11973 (U.S.A.)

(SCEs), the frequency of dividing cells and the rate of cell
proliferation.

Chromosomal Changes: Morphology and Mechanism of Formation

Chromosomal aberrations: Aberrations are changes in overall
shape and morphology of chromosomes which result from damage to
the genetic material. Many can be detected by light microscopy
and conventional staining, e.g. with Giemsa or orcein.

It is thought that when breaks are induced in chromatin they
may be repaired normally (restitution), remain open or rejoin in
abnormal configurations to give aberrations. The classical theory
of chromosomal aberrations was based on observations in cells exposed to ionizing radiation. If the cells were damaged during or
after the DNA synthesis period (in S phase or G_2), most ionizing
damage affected only one chromatid (chromatid-type damage), e.g.
chromatid breaks. However, if the cells were treated before DNA
synthesis (in G_1), when each chromosome consisted of a single
chromatid, unrepaired breaks would be duplicated during replication
and would result in damage affecting the whole chromosome (chromosome-type damage). Chromosomal aberrations are of two major types:
breaks, such as simple deletions, which may result from a break
through a single chromosome, and exchanges which involve more than
one break. An exchange between two chromosomes is called an
interchange, and within a chromosome, an intrachange. Detailed
descriptions of chromosomal aberrations and the theories about
their formation can be found in references 1-4.

Because the mechanisms of aberration induction by chemicals
and ionizing radiation differ, the patterns of aberration formation
also differ (5). Damage induced by most chemical agents requires
a period of DNA synthesis to elicit formation of aberrations, and
most of the damage seen at the first mitosis after exposure is of
the chromatid-type. After the cells undergo a further DNA synthesis
period, the chromatid-type damage can be duplicated to form chromosome type aberrations (1-5). For example, if stem cells in bone
marrow were damaged by a chemical, and then underwent one division
cycle to produce circulating lymphocytes, the damage seen at the
first metaphase would resemble chromosome-type aberrations but would
actually be of the derived chromatid-type.

Chromosomal aberrations can be detected by simple staining of
metaphase chromosomes, as many of them involve conspicuous changes
in chromosome morphology. Examples of these aberrations are dicentrics, rings and quadriradials (Fig 1). However, more subtle changes
such as reciprocal translocations, small deletions and inversions
are less easy to recognize by means of simple staining methods.
These stable aberrations may have important genetic consequences
because of their ability to persist through successive cell

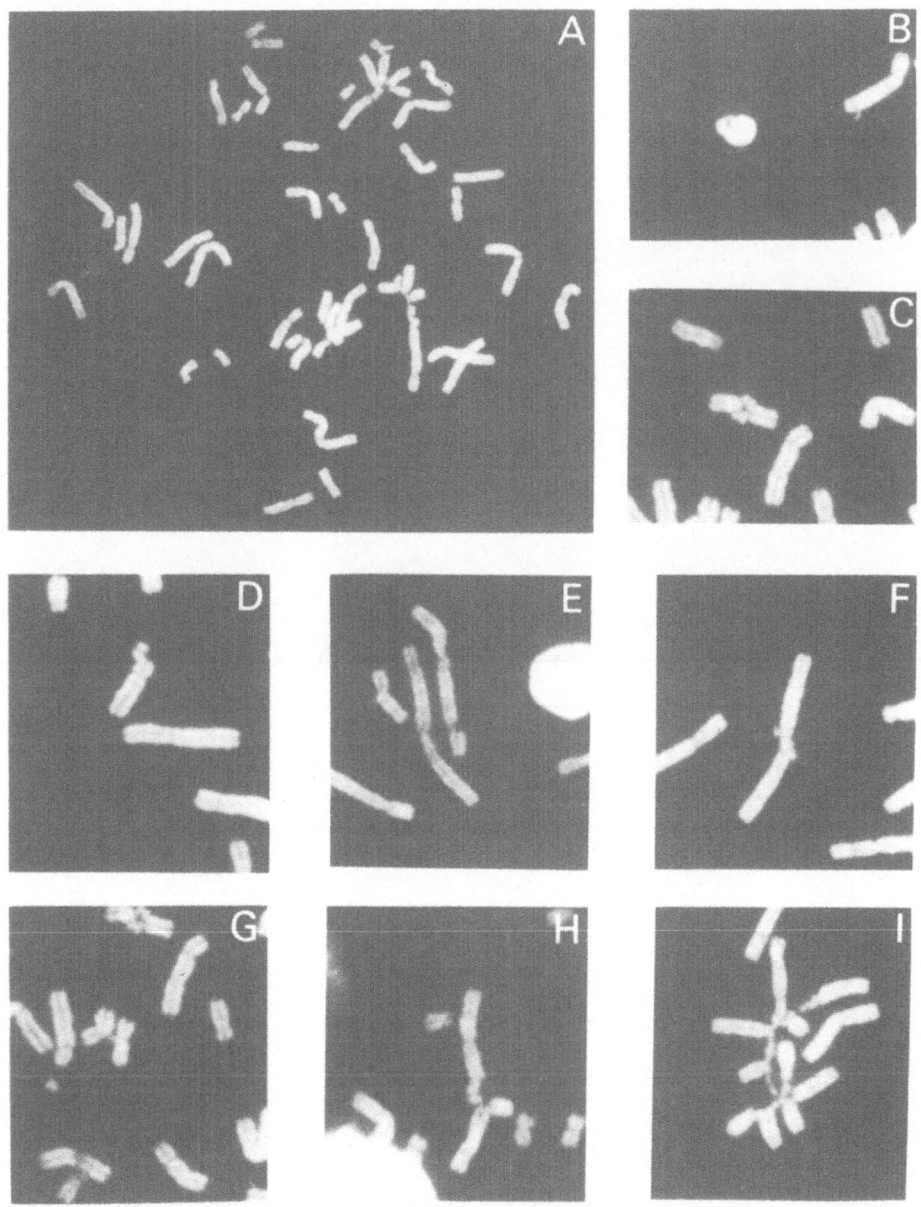

Figure 1: Fluorescent photomicrographs of chromatid-type aberrations in human peripheral blood lymphocytes exposed to 100 ng/mL Mitomycin C. A. Metaphase cell exhibiting extensive rearrangements; B. Ring; C. Quadraradial; D. Isochromatid Break; E. Break; F. Interstitial Break; G. Interstrand Rearrangements; H. Incomplete Quadraradial; I. Complex Chromatid Rearrangements.

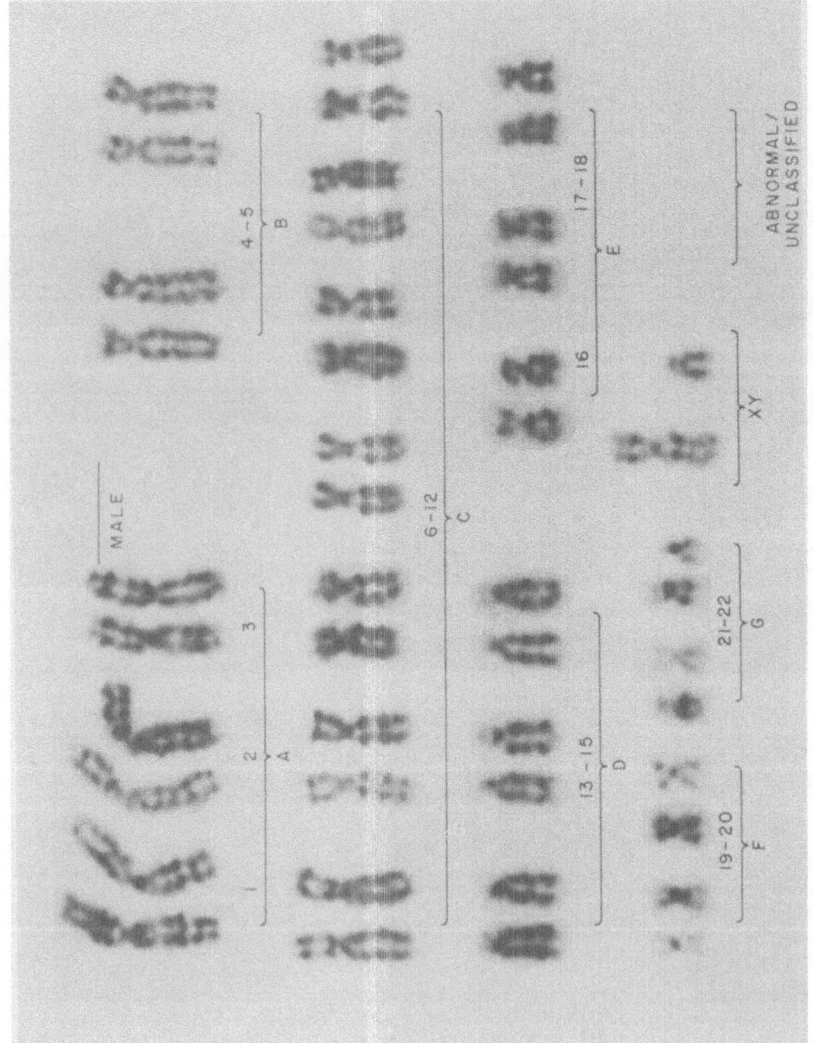

Figure 2: G-banded karyotype of a normal human male (photograph and karyotype by Mr. R. N. Ruffing)

generations. In contrast, unstable damage such as dicentrics and
fragments interfere with the division process, resulting in a loss
of genetic material and cell death. In the last decade, several
staining techniques have been developed, such as Q (Quinacrine fluorescence) and G (Giemsa) bands (see 6), which reveal more detailed
features of chromosome morphology. Each chromosome pair has a characteristic banding pattern that allows the detection of minor changes and stable-type aberrations (Fig 2). Some laboratories prefer
to use banding techniques in studying chromosomal damage, and, for
highly trained individuals, analysis can be as fast as with conventional staining techniques. Although banding provides better reso-
lution of chromosomal rearrangements, analysis using traditional
staining techniques is quite efficient and the frequency of abnormal
cells detected may not be greatly different (7,8).

     Sister chromatid exchanges: It is now more than twenty years
since Taylor and his associates used autoradiography to demonstrate
differential labelling of the sister chromatids and detected exchanges between them (9). The stimulus to study SCEs intensively was
the development of staining techniques that gave clear differerent-
iation between chromatids, and extremely accurate localization of
SCEs (10-12). In place of tritiated thymidine labelling, chromatid
differentiation is achieved by incorporating the thymidine base analogue 5-bromodeoxyrudine (BrdUrd) during DNA synthesis (10) (Fig 3).
SCEs were quickly shown to be a very sensitive indicator of exposure
to chemicals (13,14), but not to ionizing radiation (14). Dramatic
increases in SCEs were demonstrated in cultured cells exposed to
chemical mutagens (Fig 4) and in blood lymphocytes from individuals
treated with chemotherapeutic agents (14 and see below).

     Like chemically-induced chromosomal aberrations, SCEs require
a period of DNA synthesis for their expression. Damage induced in
cells that have duplicated their DNA does not elicit an increase
in SCE at the subsequent metaphase. Only if DNA synthesis occurs
in the presence of DNA lesions does the SCE frequency increase
(15). Although the precise molecular mechanism(s) of SCE formation
are not yet known, SCE formation probably involves events at or near
the replication fork leading to a reciprocal double-stranded exchange of DNA. In many ways SCEs behave differently from chromosomal
aberrations and the derivations of the two types of effects appear
to be distinct (16-18). Reviews on SCE include references 19-21.

     Cell cycle kinetics: Circulating lymphocytes are normally
quiescent in vivo, but can be stimulated to divide in culture.
The several subpopulations of lymphocytes (22-24) begin to grow
and divide at different times (25-27). With the expectation of
obtaining primarily first generation metaphase cells, cytogenetic
investigators have usually fixed cultures after 45-48 hours of
incubation (28). The yield of chromosomal aberrations is drastically influenced by the time in culture (29-31). Up to half of

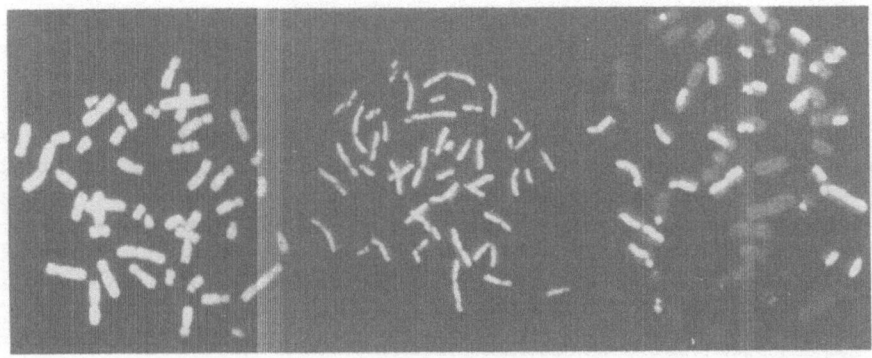

Figure 3: A schema outlining the incorporation of bromodeoxyuridine
into DNA during three consecutive rounds of DNA synthesis in the
presence of this thymidine analogue. Fluorescence photomicrographs
of the corresponding metaphase cells are presented below the schema.
The solid lines indicate DNA strands unsubstituted with BrdUrd,
the dashed lines indicate DNA strands substituted with BrdUrd.
White chromatids indicate bright fluorescent staining, dark chromatids indicate dull fluorescent staining.

the unstable aberrations (asymmetrial exchanges) are lost at each
cell division, and chromatid aberrations are converted to derived
types. The BrdUrd labelling technique allows the identification of
those cells which have undergone one, two, and three or more cell
cycles in culture (M1, M2 and M3 cells) at the time of fixation
(Fig 3) (11). This technique has been used to show that in many
individuals up to 50% of the cells have reached their second metaphase in vitro by 48 hours (32). This observation is of great practical importance in cytogenetic monitoring. Since first generation
metaphase cells must be used to obtain a true estimate of in vivo
aberration frequencies it is necessary to use earlier fixation times
to ensure the absence of second generation metaphase cells. Alternatively, cultures containing BrdUrd can be used to permit unequivocal identification of first generation metaphase cells for aberration
assessments.

There is a great deal of individual variation in the rate of growth in lymphocytes of PHA-stimulated cultures (32). The rate of growth is affected by age (Fig 4; 33), and while we do not yet know if alterations in growth rate are an indication of immunological status and/or related in any way to exposure to harmful agents, cell cycle kinetics offers additional information about cytotoxic effects of chemical agents (e.g. 34). Assessment of cell cycle kinetics may also prove useful in human monitoring studies. The cell proliferation rate has already been studied in bone marrow of human cancer patients, and for acute myeloid leukemia, proliferation data may be helpful in monitoring the stage of the disease (35).

APPLICATIONS OF CYTOGENETIC STUDIES TO HUMAN POPULATIONS

Chromosomal aberrations

Significance: Structural aberrations were among the very first types of mutations studied in the early mutagenesis studies using Drosophila and mouse. In the mouse more than 50% of the specific locus mutations induced in oocytes or in the post meiotic male germ cells are the result of small chromosomal deficiencies (36). In contrast to these small defects, a great deal of the gross chromosomal damage observed at the light microscope level is lethal to the cells. However, measurement of such damage reflects an underlying rate of less obvious non-lethal damage that must be deleterious. We do not yet have for humans accurate estimates of the risks of acquired chromosome damage resulting from, for example, occupation exposure. The effects of inherited chromosomal abnormalities are seen in humans as congenital anomalies and in reduced fertility. For example, while the carrier of a balanced translocation is phenotypically normal, the offspring are affected drastically with a high chance of being stillborn or spontaneously aborted. Also, a proportion of the survivors has the translocation in the unbalanced form with serious consequences such as congenital abnormalities and mental retardation (cited in 37).

Population studies: Chromosomal aberrations are well established as a very accurate biological dosimeter of exposure to radiation, and have been used in numerous surveys--from the drastic situation after the atomic bomb explosions at Hiroshima and Nagasaki (38,39) to the nuclear dockyard workers in Scotland (40). In the latter study it was shown that this technique is sensitive enough to detect chromosome damage in workers exposed for long periods to very low doses of radiation, i.e. doses below the supposedly safe standard level (40).

Aberrations have also been used in monitoring populations exposed to various chemicals, e.g. the well-documented case of polyvinyl chloride (PVC) workers. Many studies have demonstrated

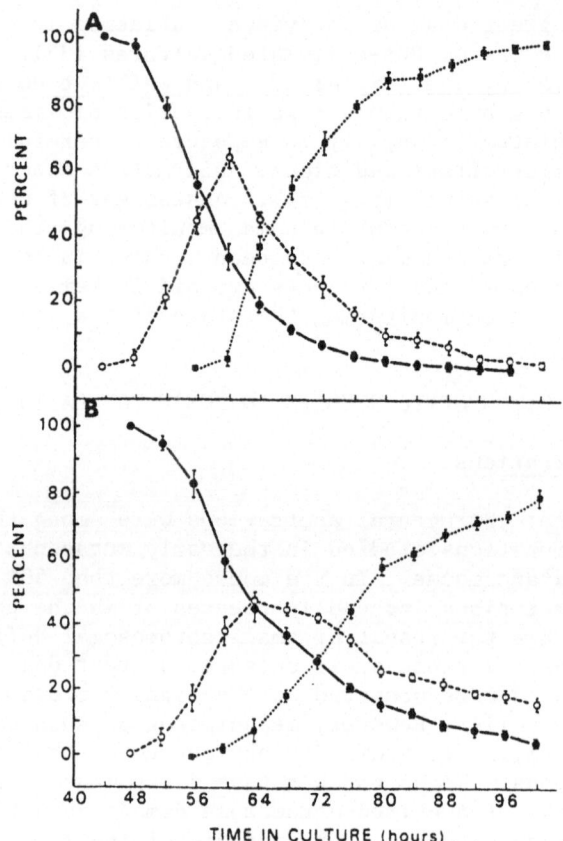

Figure 4: The distribution of first generation (●), second generation (0) and third or subsequent generation (■) metaphase cells in young (A0 and aged (B) lymphocyte cultures as a function of time (40-100 hr) after PHA addition, Each data point indicates the mean and standard error of the mean for four individuals. The young individuals averaged 25 years of age, the aged individuals 75 years of age. (reproduced with permission from ref. 33)

aberrations in PVC workers exposed to vinyl chloride monomer (41-49), a known carcinogen and a possible teratogen (50).

The study of the effects of lead is a good example of a topic that stimulated a great deal of work, many publications, and no agreement among investigators. Of 15 papers listed in the IARC Monograph on carcinogenic risk from metals and metallic compounds (51), nine showed an increase in chromosomal aberrations with lead exposure, but six did not. These results illustrate the importance of having an estimate of the dose when studying exposure.

The individuals with chromosomal damage included workers in lead
oxide, chemical, battery and ceramic factories, smelters and blast
furnaces. The workers with no apparent damage include shipyard
workers burning lead-painted metal on ships, policemen exposed
to automobile exhaust fumes and children living near lead smelters.
Why were some studies positive and some negative? A study of
Swedish bus drivers (52) showed a correlation between blood lead
levels and the frequency of chromosomal damage. If reported
blood lead levels from all lead studies are compared, it is clear
that aberrations were found in workers when lead blood levels
exceed 600 µg/L (51). However, it is interesting that in a study
which included three individuals with acute lead poisoning, chromosomal damage was not detected despite a maximum reported blood level of 590 mg/L (53).

    Laboratory variability: A recent paper on radiation damage
(54) summarized the background levels from many published studies
on chromosomal aberrations in human populations and pointed out
the large variation among laboratories. The major differences
are probably due to the quality of cells selected for analysis and
to differences among scorers. Some of the variation may also be
due to differences in numbers of cells in their first and second
division in culture. These frequencies could vary among laboratories as a result of fixation times and of the types of reagents used,
e.g. tissue culture medium, phytohaemagglutinin and serum. The
choice of controls and the information available about them is also
important, because a small number of people inadvertantly included
as controls may be actually exposed to the agent in question or to
some other damaging agent, affecting "baseline" levels. In cytogenetic surveys, the small but significant effect of age must also
be taken into account (40,55).

    Aberration persistence: Aberrations persist long after exposure.
For example, many years after radiation therapy, cells can still be
found which contain dicentrics with associated fragments, an indicator that the cell is dividing for the first time since exposure
(56). This observation has even been used to estimate the average
lifespan of human lymphocytes at over 4 years (56). The frequency ·
of aberrations does decrease with time post exposure. In studies
of agricultural workers exposed to pesticides (57) and of vinyl
chloride workers (45,48), a reduction in the number of unstable
aberrations in lymphocytes was found subsequent to a reduction in
exposure. All damage has not necessarily been "cleared from the
system" since there is likely to be considerable residual damage
in non-dividing cells in other tissues. Also, stable aberrations
were not investigated in detail in these studies. In cells from
patients given cytostatic drugs, the levels of unstable aberrations
decreased with time after the end of treatment, but stable aberrations detected by the more detailed technique of G banding persis-

ted for years (58). Similar observations were made on atomic bomb
survivors: One study showed the persistence of stable aberrations
for three decades in atomic bomb survivors while unstable aberrations
became rare (39).

Sister Chromatid Exchanges

Significance: The significance of SCE is not known. Their
frequency increases in parallel with mutations induced by some
chemicals (59) but for other agents, e.g. light, SCEs are not
related to mutational events (60). It is possible that there
exists more than one mechanism of SCE, with or without associated
mutational events. On the other hand, SCEs show an excellent
correlation with exposure to most mutagens and carcinogens and can
be more sensitive than chromosomal aberrations as an indicator of
genotoxic activity. Because SCEs are more frequent than aberrations,
it is possible to establish statistically significant increases by
scoring fewer cells, and to detect, under controlled conditions,
very small differences in SCE frequencies.

Population studies: Measurement of SCEs is not a good indicator of radiation exposure; very little increase is seen in vitro
(14,61,62) and in vivo (63). In contrast to chromosomal aberration
effects, atomic bomb survivors (64) do not have increased SCE levels.
It is clear that agents which produce predominantly DNA strand breakage rather than DNA adducts, e.g. ionizing radiation and the cytotoxic drug bleomycin (14), are equally poor inducers of SCEs.

SCEs have been examined extensively in studies of chemotherapy
patients (14,65-71), and to a lesser extent in industrial workers.
Increased SCE frequencies have been found in laboratory workers exposed to a range of solvents and other chemicals (72), in steriliz-
ing process workers inhaling ethylene oxide gas (73) and in vinyl
chloride workers (47).

As with chromosomal aberration studies, there are discrepancies
among studies of SCEs. A good example is the issue of the effect
of cigarette smoking on SCE frequency. Several studies show slight
increases in SCEs in smokers compared with non-smokers, (see 74),
with evidence for a dose relation to numbers of cigarettes smoked
(75). Other investigators have reported no effect (70,76). Numerous
reports give data on smoking as a secondary issue to the chemical
under investigation, and controversy about the effect on SCE arises
to a large extent from the difficulty of interpretation of results
where the increases are small and where there is considerable variation among individuals. Statistical analysis of SCE in one study
of smokers versus non-smokers has shown that while there was little
difference between the overall mean SCE frequencies for the two
groups, the smokers had more "outlier cells" with high SCEs (77).

Thus, it is important to examine the distribution of SCE among cells. It is possible that exposure to genotoxic agents may be best detected through subpopulations of cells with elevated SCE frequencies, partially obscured by the majority of normal cells.

Variability: There are several clearcut reasons for variations in SCE frequencies among laboratories. The amount of BrdUrd in cultures (10-13,78,79), exposure to light (80) and the number of responding lymphocytes (79) are all important variables, but variables which can be standardized. The time of fixation may also be important bcause there are conflicting data on whether baseline SCE vary among faster- or slower-growing populations of lymphocytes (79,81-83). Even the type of serum used may affect SCE frequenices (84). Scoring SCE is much less subjective than is scoring chromosomal aberrations and with good cytological preparations scoring can be done without large inter-laboratory variation. As in any cytogenetic study, a good medical history is necessary to aid in interpretation of results. Besides smoking and alcohol consumption, age may be important since there is some evidence from in vivo studies for reduced inducibility of SCE in older animals (85). Also there are reports of slight increases in SCE associated with use of oral contraceptives (86) or of antibiotics such as nalidixic acid (87). Virus infection (88) or vaccination (89,90) may also affect SCE levels.

Persistence: SCEs result when lesions in the DNA somehow affect DNA replication. Since most eukaryote cells can repair many lesions in DNA, a long time-lag between induction of the lesions and examination of the cells may result in a reduced yield of SCEs (91). This has been clearly demonstrated after in vivo exposure. The first studies in experimentally exposed animals and in humans on chemotherapy showed that there was a sharp rise in SCE in lymphocytes within a day or two of exposure, followed by a rapid decline (14, 65,66). The decline in SCE frequency may also be due to dilution of damaged cells by new cells. The pattern of SCE frequencies at different times after exposure depends on the nature of the induced lesion, the magnitude of the dose of chemical and on the manner in which it is administered, i.e. whether by an acute treatment of by repeated exposures. It was found in rabbits given "split doses" of Mitomycin C (MMC), that within a few days the induced frequencies of SCEs sharply declined. However, after repeated exposures, the SCE rate climbed to a plateau and remained high for weeks after the end of treatment (92). Similar effects have now been observed in human chemotherapy patients (68,71).

SCE frequencies depend on the nature of the chemical agent since chemicals differ in the spectrum of DNA lesions induced. The results also reflect the type of cell under investigation since cells differ in their capacity for DNA repair. Some lesions are more efficiently removed than others. In particular, the range in capacities for removal of the O^6 alkyl-guanine (93) or 3-adenine adducts (94) seems to be a critical factor affecting the persistence of lesions leading to SCE formation.

Peripheral blood lymphocytes, a largely quiescent cell population in vivo, are comparatively inefficient in the ability to excise DNA damage compared to proliferating lymphocytes or undifferentiated cell populations (95,96). This observation suggests increased opportunity for detecting in vivo DNA damage occurring over time in this cell population.

POSSIBLE CYTOGENETIC DETECTION OF INDIVIDUALS SUSCEPTIBLE TO CANCER

Chromosomal Damage and Cancer

Certain well-known hereditary syndromes known as the Chromosome Instability Syndromes are characterized by high rates of "spontaneous" chromosomal breakage. Patients with thes diseases, e.g. ataxia telangiectasia (AT), Fanconi's anemia (FA) and Bloom's syndrome (BS), also have high risks of developing various sorts of cancer. Chromosomal aberrations are more frequent than normal in lymphocytes from most of the affected individuals. While SCE levels in AT and FA cells are normal (97,98), they are exceptionally high in BS cells (99). SCE frequency is also reported to be slightly increased in another cancer-prone syndrome, dyskeratosis congenita (100). Another form of "chromosome imbalance" is Down's syndrome (DS) where some or all of the cells of the individual contain an extra chromosome number 21. Individuals with DS are also more prone than the general population to developing cancer. However, these individuals do not exhibit chromosome instability or elevated SCE frequencies (101, Galloway and Tice, unpublished data).

Cells from cancer patients often contain chromosome abnormalities. One example is the "marker" known as the Philadelphia chromosome, a translocation between chromosomes 9 and 22 which is associated exclusively with chronic myeloid leukemia (CML). There is now a highly convincing body of evidence for the association of particular chromosome rearrangements with specific cancers (for a recent review, see 102).

SCE has also been studied in cancer patients, with some conflicting results. Some of the discrepancies among studies may result from the inclusion of patients who have had chemotherapy, a process known to induce SCE. Table 1 summarizes the published data on leukemia and lymphoma, and illustrates the slight increases in SCE reported in untreated patients with malignant lymphoma (108) or acute lymphoid leukemia (ALL) (109). In ALL, the high SCE frequencies in patients before treatment were seen to decrease to the normal range during remission of the disease (109).

Table 1. SCE in human leukemia and lymphoma

Cancer Type	Cell Type	SCE Frequency*		Comments	Reference
		Patients	Controls		
CML	BM	Normal		Ph' Positive Cells Normal	103
CML	BM	Normal (9) Blastic Phase (2) 8.2	3.3	Patients Had Previous Chemotherapy	104
CML	BM	Low (12)		Untreated Patients	105
CML	PBL	Normal			106
AML	BM	Low (17)		Cells Also Grow More Slowly	35
ANLL	BM	Low (12)		Aneuploid Cells In Marrow Had Low SCE cf Diploid Cells Or Cells In Remission	107
ML	PBL	12.7 ± 0.9 (47)	6.1 ± 0.3 (40)	Untreated Patients; Active Phase Slightly Higher Than Remission	108
ALL	PBL	12.2 ± 0.2 (16) (before treatment)	7.6 ± 0.2 (14)	SCE Increased further after Chemotherapy But Returned To Control Levels In Remission	109
CLL	PBL	Normal			110

* Number of SCEs per cell ± the standard deviation. Number of individuals in ( ).
CML = chronic myelogenous leukemia; AML = acute myelogenous leukemia; CLL = chronic lymphocytic leukemia; ALL = acute lymphocytic leukemia; ANLL = acute nonlymphocytic leukemia
BM = bone marrow; PBL = peripheral blood lymphocye

In chronic myelogenous leukemia (CML), SCE levels were normal in bone marrow cells including the leukemic cells themselves, i.e. those cells which contained a Philadelphia chromosome (103). It is not clear whether the increased SCE in the two CML patients in blastic phase (104) result from earlier chemotherapy. In another study of CML, where no treatment had been given (105) SCE levels were lower than normal. The picture for SCE in cancer patients is thus confusing, giving low, high or normal SCE frequencies (111). Clearly SCE measurements cannot currently help to predict the occurrence of cancer.

Detection of Cancer Prone Individuals

We know that individuals with inherited chromosome instability are susceptible to cancer, and that cancer patients may show chromosome rearrangements, and possibly altered SCE frequencies. Is it possible then to predict from cytogenetic properties whether an individual is likely to develop cancer?

The genes for the chromosomal instability syndromes are autosomal recessives. Although the affected individuals are rare, the gene frequencies in the population are surprisingly high. For example, it has been estimated that up to 1% of the general population may be heterozygous for the gene for ataxia telangiectasia. AT patients have a very high risk of developing cancer, and it is now clear from family studies of blood relatives of AT homozygotes that the heterozygotes also have an increased cancer rate. For AT heterozygotes less than 45 years old, the risk of dying from cancer may be as much as five times that for the general population (112). Thus, the impact of such genes on the "cancer burden" may be very large. Since there may well be other genes as yet uncharacterized which predispose to cancer, it would be highly desirable to have a method of detecting such susceptible individuals. Unfortunately, AT heterozygotes do not show chromosome instability. Similarly, in a survey of patients and their relatives in "cancer families" no difference in SCE frequencies from normal individuals has been detected (106,115).

In another chromosome instability syndrome, Fanconi's anemia, there is no clear increased cancer risk in heterozygotes (113). However, there is increased cancer susceptibility in patients with Xeroderma Pigmentosum (XP) in which DNA repair is abnormal (114). Heterozygotes also have a clearly elevated risk of skin cancer (114), but there is as yet no laboratory test that reliably distinguishes the heterozygote. While abnormally high rates of SCEs are induced by chemicals in XP lymphocytes (116) and in some lines of transformed fibroblasts (178) from XP patients, no such distinguishing feature has been found in heterozygotes. Even within XP the hypersensitivity to chemical mutagens is not found in all cell lines (118).

The degree of chromosome damage induced by chemicals in vitro may correlate with DNA repair capacity. In XP cells, it is thought that SCE levels induced by chemical mutagens are increased because the cells are unable to repair damage before the DNA lesions induce SCE during DNA replication (117). This conclusion is supported by studies on certain strains of human tumor cells (93) in which there is deficient repair of damage induced by methylnitronitro-guanidine (MNNG); the cells are highly susceptible to SCE induction by MNNG compared with repair-proficient cells.

As standardized information accumulates on normal individual variability, it may be possible to detect slight differences in SCE inducibility that are linked to disease susceptibility. Several in vitro studies have been published which state that individuals vary in the amounts of DNA damage and chromosome aberrations accumulated in lymphocytes and that the effects of DNA induced in vitro correlate with the amount of damage seen in vivo (119).

Another difference among individuals which may affect susceptibility to cancer is variations in metabolism. SCE may prove useful in detecting the ability to metabolize chemicals to their active forms. There was great excitement some years ago when it was reported that there might be a link between the presence of absence of a certain enzyme system known as AHH (see below) and susceptibility to lung cancer among smokers (120). Many chemicals undergo metabolic conversion in the body; a process which often leads to the formation of highly reactive metabolites that are often carcinogenic.

One well-characterized enzymatic system is the cytochrome P_1 · 450-mediated monooxygenase series, which includes the enzyme aryl hydrocarbon hydroxylase (AHH). This enzyme activates polycyclic hydrocarbons such as methylcholanthrene and benzo(a)pyrene (BaP) (one of the components of cigarette smoke) to carcinogenic metabolites. In mice the ability to metabolize BaP is genetically determined and is under control of the Ah locus (reviewed in 121). It was thought that humans could also be classified as Ah positive, negative or intermediate, as one would expect if this trait were determined in a simple Mendelian fashion (122). It was postulated that if an individual responded to a polycyclic aromatic hydrocarbon, e.g. BaP, by producing enzymes to convert it into its carcinogenic form, there would be far greater risk of developing cigarette-induced cancer than in someone who did not activate the carcinogen. Unfortunately, although the Ah locus does exist in humans, the association with cancer risk has not been confirmed (123). However, in mice the link with cancer susceptibility is definitive enough to encourage further investigation of the relation between metabolic capacity and cancer.

In inbred strains of mice one can identify by means of SCE
induction by BaP those strains which have, or lack, inducible AHH
(124).   In embryos of Ah positive strains, BaP induces high frequencies of SCE not found in "non-inducible" strains.   This relation of
BaP-induced SCE to AHH induction is not always straight-forward.
For example, other studies showed no correlation between AHH inducibility and SCE in partially hepatectomized mice (125).   In such
mice, the balance of activation and detoxification is altered.
Similar complexities of metabolic pathways may also explain the
reported lack of correlation between AHH "status" and SCE induction
in two human subjects (126).   The interpretation of results of
these experiments clearly requires a detailed knowledge of the
relevant biochemical pathways.

It remains possible that SCE induction might detect differences
among individuals in their response to chemicals.   For example
induction of SCE in cultured lymphocytes by cigarette smoke condensate (CSC) is higher in smokers than in non-smokers and higher still
in lung cancer patients (75).   This is not related to BaP metabolism
because the main SCE-inducing component of CSC is not BaP (127) and
BaP-induced SCE reportedly do not differ between lung cancer patients
and controls (128).

The cytogenetic effects of inhaled benzene in mice depend on
not only the mouse strain, but also on the age and sex (34, see Tice,
et al., these proceedings).   There are also differences among mouse
strains in induction of SCE by direct-acting chemicals that do not
require complex metabolizing systems, e.g. MMC (129) and bischloro-
ethylnitrosourea (130).

Clearly, differences exist among individuals in cancer
susceptibility and in chromosomal responses to genotoxic agents.
The relation between chromosomes and cancer is a fascinating and
complex area of investigation.

CONCLUSIONS

Cytogenetic monitoring is our only established means of detecting
damage to the human genetic material.   Although we need long-term
epidemiological studies before the information can have predictive
value for assessing risk to individuals or populations, chromosomal
damage appears to be associated with deleterious health effects.
The most useful studies would include detailed dose/exposure information.

It is possible that the use of cytogenetics tests for in
vitro susceptibility to chemical damage may be developed to identify
individuals at particular risk from chemicals.   This will require
more information on individual variability and drug metabolism.

ACKNOWLEDGEMENT

RRT is supported by the Department of Energy and the Environmental Protection Agency.

REFERENCES

1. H. J. Evans, Effects of ionizing radiation on mammalian chromosomes, in; "Chromosomes and Cancer", J. German, ed., John Wiley and Sons, Inc., New York, N.Y., 191-237 (1974).
2. M. A. Bender, H. G. Griggs, and J. Bedford, Mechanisms of chromosomal aberration production. III. Chemicals and ionizing radiation, Mutat. Res. 23:197-212 (1974).
3. S. H. Revell, The breakage-and-reunion theory and the exchange theory for chromosomal aberrations induced by ionizing radiations: a short history, in: "Adv. Radiat. Biol. IV", J. Y. Lett, H. Adler, and M. Zelle, eds., Academic Press, New York, N.Y., 367-416 (1974).
4. J. R. K. Savage, Classification and relationships of induced chromosomal structural changes, J. Med. Genet. 12:103-122 (1975).
5. H. J. Evans and D. Scott, The induction of chromosome aberrations by nitrogen mustard and its dependence on DNA synthesis, Proc. Roy. Soc. B 173:491-512 (1969).
6. Paris Conference, Standarization in human cytogenetics. Birth Defects Original Article Series VIII, No. 7, The National Foundation (1972).
7. K. E. Buckton, Classification with G and R banding of the position of breakage points induced in human chromosomes by in vitro irradiation, Int. J. Radiat. Biol. 29:475-488 (1976).
8. K. E. Buckton, G. E. Hamilton, L. Paton, and O. Langlands, Chromosome aberrations in irradiated ankylosing spondylitis patients, in: "Mutagen-induced chromosome damage in Man", H. J. Evans and D. C. Lloyd, eds., Edinburgh University Press, Edinburgh, Scotland, 142-150 (1978).
9. J. H. Taylor, Sister chromatid exchanges in tritium labelled chromosomes, Genetics 43:515-529 (1958).
10. H. Kato, Spontaneous sister chromatid exchanges detected by a BUdR labelling method, Nature 251:70-72 (1974).
11. S. A. Latt, Microfluorimetric detection of deoxyribonucleic acid replication in human metaphase chromosomes, Proc. Natl. Acad. Sci. USA 70:3395-3399 (1973).
12. P. Perry and S. Wolff, New Giemsa method for the differential staining of sister chromatids, Nature 251:156-158 (1974).
13. S. A. Latt, Sister chromatid exchanges, indices of human chromosome damage and repair: detection by fluorescence and induction by Mitomycin C, Proc. Natl. Acad. Sci. USA 71:3162-3166 (1974).

14. P. Perry and H. J. Evans, Cytological detection of mutagen-carcinogen exposure by sister chromatid exchange, Nature 258:121-125 (1975).

15. S. Wolff, J. Bodycote, and R. B. Painter, Sister chromatid exchanges induced in Chinese hamster cells by UV irradiation at different stages of the cell cycle: The necessity for cells to pass through S, Mutat. Res. 25:73-81 (1974).

16. S. M. Galloway, What are sister chromatid exchanges?, in: "DNA Repair Processes", W. W. Nichols and D. G. Murphy, eds., Symposia Specialists, Florida, 191-201 (1977).

17. P. E. Perry, Chemical mutagens and sister-chromatid exchanges, in: "Chemical Mutagens - Principles and Methods for their Detection, Vol. 6", F. J. de Serres and A. Hollaender, eds., Plenum Press, New York, N.Y., 1-39 (1979).

18. S. Wolff, Sister chromatid exchange, Ann. Rev. Genet. 11:183-201 (1977).

19. H. J. Evans, What are sister chromatid exchanges?, in; "Chromosomes Today, Vol. 6" A. de la Chapelle and M. Sorsa, eds., Elsevier/North Holland Biomedical Press, Amsterdam, 315-326 (1977).

20. S. A. Latt, J. W. Allen, L. Shuler, K. S. Loveday and S. H. Munroe, The detection and induction of sister chromatid exchanges, in: "Molecular Human Cytogenetics", R. S. Sparkes, D. E. Comings, and C. F. Fox, eds., Academic Press, New York, N.Y., 315-334 (1977).

21. E. L. Schneider, R. R. Tice, and D. Kram, Bromodeoxyuridine differential chromatid staining techniques: a new approach to examining sister chromatid exchange and cell replication kinetics, in: "Methods in Cell Biology" D. M. Prescott, ed., 20:380-407 (1978).

22. A. Michalowski, Time-course of DNA synthesis in human leukocyte cultures, Exptl. Cell Res. 32:609-612 (1963).

23. G. Dudin, B. Beek, and B. Obe, The human leukocyte test system. I. DNA synthesis and mitoses in PHA-stimulated 2-day cultures, Mutat. Res. 23:279-281 (1974).

24. J. A. Heddle, H. J. Evans, and D. Scott, Sampling time and the complexity of the human leukocyte culture system, in: "Human Radiation Cytogenetics", H. J. Evans, W. M. Court-Brown, and A. S. McLean, eds., North-Holland, Amsterdam, 6-19 (1967).

25. P. E. Crossen and W. F. Morgan, Occurrence of 1st division metaphases in human lymphocyte cultures, Human Genet. 41: 97-100 (1978).

26. S. Wolff and R. Arutunyan, The apparent decrease in thiotepa-induced chromosome aberrations in human lymphocytes caused by an effect of WR 2721 on the cell cycle as found by the definitively determined division method, Environ. Mutagen 1:5-13 (1979).

27. J. A. Steffen, K. Swierkowska, A. Michalowski, E. Kling, and A. Nowakowska, In vitro kinetics of human lymphocytes

activated by mitogens, in; "Mutagen Induced Chromosome Damage in Man", H. J. Evans and D. C. Lloyd, eds., Edinburgh University Press, Edinburgh, Scotland, 89-107 (1978).

28. M. A. Bender and J. G. Brewen, Factors influencing yields in the human peripheral leukocyte system, Mutat. Res. 8:383-389 (1969).

29. M. S. Sasaki and A. Norman, Selection against chromosome aberrations in human lymphocytes, Nature 214:502-503 (1967).

30. A. V. Carrano, Chromosome aberrations and radiation-induced cell death, I. Transmission and survival parameters of aberrations, Mutat. Res. 17:341-353 (1973).

31. K. E. Buckton and M. C. Pike, Chromosome investigations on lymphocytes from irradiated patients: effect of time in culture, Nature 202:714-715 (1964).

32. P. E. Crossen and W. F. Morgan, Analysis of human lymphocyte cell cycle time in culture measured by sister chromatid differential staining, Exptl. Cell Res. 104:453-457 (1977).

33. R. R. Tice, E. L. Schneider, D. Kram, and P. Thorne, Cytokinetic analysis of the impaired proliferative response of peripheral lymphocytes from aged humans to phytohemagglutinin, J. Exp. Med. 149:1029-1041 (1979).

34. R. R. Tice, D. L. Costa, and R. T. Drew, Cytogenetic effects of inhaled benzene in murine bone marrow: induction of sister chromatid exchanges, chromosomal aberrations and cellular proliferation inhibition in DBA/2 mice, Proc. Natl. Acad. Sci. USA 77:2148-2152 (1980).

35. S. Abe, S. Kakati, and A. A. Sandberg, Growth rate and sister chromatid exchange (SCE) incidence of bone marrow cells in acute myeloblastic leukemia (AML), Cancer Genet. Cytogenet. 1:115-130 (1979).

36. L. B. Russell, Definition of functional units in a small chromosomal segment of the mouse and its use in interpreting the nature of radiation - induced mutations, Mutat. Res. 11: 107- (1971).

37. C. P. Swanson, T. Merz, and W. J. Young, in: "Cytogenetics: The Chromosome in Division, Inheritance, and Evolution", 2nd edition, Prentice Hall (1981).

38. A. A. Awa, S. Neriishi, J. Honda, M. C. Yoshida, T. Sofuni, and T. Matsui, Chromosome aberration frequency in cultured blood-cells in relation to radiation dose of A-bomb survivors, Lancet 2:903-905 (1971).

39. T. Sofuni, H. Shimba, K. Ohtaki, and A. A. Awa, A cytogenetic study of Hiroshima atomic bomb survivors, in: "Mutagen-Induced Chromosome Damage in Man", H. J. Evans and D. C. Lloyd, eds., Edinburgh University Press, Edinburgh, Scotland 108-114 (1978).

40. H. J. Evans, K. E. Buckton, G. E. Hamilton, and A. Carothers, Radiation-induced chromosome aberrations in nuclear-dockyard workers, Nature 277:531-534 (1979).

41. A. Ducatman, K. Hirschron, and I. J. Selikoff, Vinyl chloride
     exposure and human chromosome aberrations, Mutat. Res. 31:
     163-168 (1975).
42. F. Funes-Cravioto, B. Lambert, J. Linsten, L. Ehrenburg,
     A. T. Natarajan, and S. Osterman-Golkar, Chromosome
     aberrations in workers exposed to vinyl chloride, Lancet
     1:459 (1975).
43. I. F. H. Purchase, C. R. Richardson, and D. Anderson,
     Chromosome and dominant lethal effects of vinyl chloride,
     Lancet 2:410-411 (1975).
44. I. Szentesi, E. Hornyak, G. Ungvary, A. Czeizel, Z. Bognar,
     and M. Timar, High rate of chromosomal aberrations in PVC
     workers, Mutat. Res. 37:313-316 (1976).
45. I.-L. Hansteen, L. Hillestad, E. Thiis-Evensen, and
     S. S. Heldaas, Effects of vinyl chloride in man. A cyto-
     genetic follow-up study, Mutat. Res. 51:271-278 (1978).
46. I. F. H. Purchase, C. R. Richardson, D. Anderson, G. M. Paddle,
     and W. G. F. Adams, Chromosomal analysis in vinyl chloride
     exposed workers, Mutat. Res. 57:325-334 (1978).
47. M. Kucerova, Z. Pilivkova, and J. Balova, Comparative evaluation
     of the frequency of chromosomal aberrations and the SCE
     numbers in peripheral lymphocytes of workers occupationally
     exposed to vinyl chloride monomer, Mutat. Res. 67:97-100
     (1979).
48. D. Anderson, C. R. Richardson, T. M. Weight, I. F. H. Purchase,
     and W. G. F. Adams, Chromosomal analysis in vinyl chloride
     exposed workers. Results from analysis 18 and 42 months
     after an initial sampling, Mutat. Res. 79:151-162 (1980).
49. P. Rossner, R. J. Sram, J. Novakova, and V. Lambl, Cytogenetic
     analysis in workers occupationally exposed to vinyl chloride,
     Mutat. Res. 73:425-427 (1980).
50. P. F. Infante, J. K. Wagoner, and R. J. Waxweiler, Carcinogenic
     mutagenic and teratogenic risks associated with vinyl
     chloride, Mutat. Res. 41:131-142 (1976).
51. IARC Mongraphs, IARC working groups on the evaluation of
     carcinogenic risks of chemicals to humans; some metals and
     metallic compounds, V. 23, 325-415 (1979).
52. B. Hogstedt, A.-M. Kolnig, F. Mitelman, and A. Schutz,
     Correlation between blood-lead and chromosomal aberrations,
     Lancet 2:2629 (1979).
53. E. Schmid, M. Bauchinger, S. Pietruk, and G. Hall, Cytogenetic
     action of lead in human peripheral lymphocytes in vitro
     and in vivo, Mutat. Res. 16:401-406 (1972).
54. D. C. Lloyd, R. J. Purrott, and E. J. Reeder, The incidence of
     unstable chromosome aberrations in peripheral blood
     lymphocytes from unirradiated and occupationally exposed
     people, Mutat. Res. 72:523-532 (1980).
55. N. P. Bochkov, V. M. Kozlov, R. A. Pilsov, and A. V. Sevenkaev,
     The level of spontaneous chromosome aberrations in the
     cultures of human leucocytes, Genetika 4:93-98 (1968).

56. K. E. Buckton, P. G. Smith, and W. M. Court-Brown, The estima-
     tion of lymphocyte lifespan from studies on males treated
     with X-rays for ankylosing spondylitis, in: "Human Radiation
     Cytogenetics", H. J. Evans, W. M. Court-Brown, and
     A.S. McLean, eds., North Holland, Amsterdam, 106-114 (1967).

57. J. Yoder, M. Watson, and W. W. Benson, Lymphocyte chromosome
     analysis of agricultural workers during extensive occupa-
     tional exposure to pesticides, Mutat. Res. 21:335-340 (1973).

58. R. C. Miller, R. B. Hill, W. W. Nichols, and A. T. Meadows,
     Acute and long-term cytogenetic effects of childhood cancer
     chemotherapy and radiotherapy, Cancer Res. 38:3241-3246
     (1978).

59. A. V. Carrano, L. H. Thompson, P. A. Lindl, and J. L. Minkler,
     Sister chromatid exchange as an indicator of mutagenesis,
     Nature 271:551-553 (1978).

60. M. O. Bradley, I. C. Hsu, and C. C. Harris, Relationships
     between sister chromatid exchange and mutagenicity toxicity
     and DNA damage, Nature 282:318-320 (1979).

61. S. M. Galloway, Ataxia telangiectasia: The effects of chemical
     mutagens and x-rays on sister chromatid exchanges in blood
     lymphocytes, Mutat. Res. 45:343-349 (1977).

62. G. K. Livingston and L. A. Dethlefsen, Effects of hyperthermia
     and x-irradiation on sister chromatid exchange frequency in
     Chinese hamster ovary (CHO) cells, Radiat. Res. 70: A 611
     (1977).

63. Y. Nakanishi and E. L. Schneider, In vivo sister chromatid
     exchange: A sensitive measure of DNA damage, Mutat. Res.
     60:329-337 (1979).

64. G. S. Pant, N. Kamada, and R. Tanaka, Sister chromatid ex-
     changes in peripheral lymphocytes of atomic bomb survivors
     and of normal individuals exposed to radiation and chemical
     agents, Hiroshima J. Med. 25:99-105 (1976).

65. N. P. Nevstad, Sister chromatid exchanges and chromosomal
     aberrations induced in human lymphocytes by the cytostatic
     adriamycin in vivo and in vitro, Mutat. Res. 57:253-258
     (1978).

66. T. Raposa, Sister chromatid exchange studies for monitoring
     DNA damage and repair capacity after cytostatics in vitro
     and in lymphocytes of leukaemic patients under cytostatic
     therapy, Mutat. Res. 57:241-251 (1978).

67. B. Lambert, U. Ringborg, E. Harper, and A. Lindblad, Sister
     chromatid exchanges in lymphocyte cultures of patients
     receiving chemotherapy for malignant disorders, Cancer
     Treat. Rep. 62:1413-1419 (1978).

68. B. Lambert, U. Ringborg, and A. Lindblad, Prolonged increase
     of sister chromatid exchanges in lymphocytes of melanoma
     patients after CCNU treatment, Mutat. Res. 59:295-300
     (1979).

69. J. Musilova, K. Michalova, and J. Urban, Sister-chromatid

exchanges and chromosomal breakage in patients treated with cytostatics, Mutat. Res. 69:289-294 (1979).

70. D. H. Hollander, M. S. Tockman, Y. W. Liang, D. S. Borgaonkar, and J. K. Frost, Sister chromatid exchanges in the peripheral blood of cigarette smokers and in lung cancer patients; and the effect of chemotherapy, Human Genet. 44:165-171 (1978).

71. M. Ohtsuru, Y. Ishii, S. Takai, H. Higashi, and G. Kosaki, Sister chromatid exchanges in lymphocytes of cancer patients receiving mitomycin C treatment, Cancer Res. 40:477-480 (1980).

72. F. Funes-Cravioto, C. Zapata-Gayon, B. Kolmodin-Hedman, B. Lambert, J. Lindsten, E. Norberg, M. Nordenskjold, R. Olin, and A. Swensson, Chromosome aberrations and sister chromatid exchange in laboratory and factory workers and their children, in: "Mutagen-induced Chromosome Damage in Man", H. J. Evans and D. C. Lloyd, eds., Edinburgh University Press, Edinburgh, Scotland, 275-289 (1978).

73. V. F. Garry, J. Hozier, D. Jacobs, R. L. Wade, and D. G. Gray, Ethylene oxide: Evidence of human chromosomal effects, Environ. Mutagen. 1:375-382 (1979).

74. J. M. Hopkin and H. J. Evans, Cigarette-smoke-induced damage and lung cancer risks, Nature 283:388-390 (1980).

75. B. Lambert, A. Lindblad, M. Nordenskjold, and B. Werelius, Increased frequency of sister chromatid exchanges in cigarette smokers, Hereditas 88:147-149 (1978).

76. G. Ardito, L. Lamberti, E. Ansaldi, and P. Ponzetto, Sister chromatid exchanges in cigarette-smoking human females and their newborn, Mutat. Res. 78:209-212 (1980).

77. A. V. Carrano and D. H. Moore, II, The rationale and methodology for quantifying sister chromatid exchange in humans, in: "Mutagenicity from Bacteria to Man", J. A. Heddle, ed., Academic Press, N.Y., in press (1981).

78. H. Kato, Spontaneous and induced sister chromatid exchanges as revealed by the BUdR labelling method, Int. Rev. Cytol. 49: 55-97 (1977).

79. A. V. Carrano, J. L. Minkler, D. G. Stetka, and D.H. Moore, Variation in the baseline sister chromatid exchange frequency in human lymphocytes, Environ. Mutagen. 2:325-338 (1980).

80. T. Ikushima and S. Wolff, Sister chromatid exchanges induced by light flashes to 5-bromodeoxyuridine and 5-iododeoxy-uridine substituted Chinese hamster chromosomes, Exp. Cell Res. 87:15-19 (1974).

81. W. F. Morgan and P. E. Crossen, The incidence of sister chromatid exchanges in cultured human lymphocytes, Mutat. Res. 42:305-312 (1977).

82. B. Beek and G. Obe, Sister chromatid exchanges in human leukocyte chromosomes: Spontaneous and induced frequencies in early and late-proliferating cells in vitro, Human Genet. 49:51-61 (1979).

83. A. J. Snope and J. M. Rary, Cell-cycle duration and sister

chromatid exchange frequency in cultured human lymphocytes, Mutat. Res. 63:345–349 (1979).

84. H. Kato and A.A. Sandberg, The effect of sera on sister chromatid exchanges in vitro, Exptl. Cell Res. 109:445–448 (1977).

85. E. L. Schneider, D. Kram, Y. Nakanishi, R. E. Monticone, R. R. Tice, B. A. Gilman, and M. L. Nieder, The effect of aging on sister chromatid exchange, Mech. Aging Develop. 9:303–311 (1979).

86. P. Bala Krishna Murthy and K. Prema, Sister chromatid exchanges in oral contraceptive users, Mutat. Res. 68:149–152 (1979).

87. J. Kowalczyk, Sister chromatid exchanges in children treated with nalidixic acid, Mutat. Res. 77:371–375 (1980).

88. K. Kurvink C. D. Bloomfield, and A. J. Cervenka, Sister chromatid exchange in patients with viral disease, Exptl. Cell Res. 113:400–453 (1978).

89. S. Knuutila, J. Maki-Paakemen, M. Kahkonen, and E. Hokkanen, An increased frequency of chromosomal changes in SCEs in cultured blood lymphocytes of 12 subjects vaccinated against smallpox, Human Genet. 41:89–96 (1978).

90. B. Lambert, A. Ehrnst, K. Hansson, A. Lindblad, M. Morad, and B. Werelius, Sister chromatid exchange in peripheral lymphocytes of subjects vaccinated against measles, Human Genet. 50:291–296 (1979).

91. J. B. Schvartzman and C. Gutierrez, The relationship between the cell time available for repair and the effectiveness of a damaging treatment in provoking the formation of sister chromatid exchange, Mutat. Res. 72:483–489 (1980).

92. D. G. Stetka, J. Minkler, and A. V. Carrano, Induction of long-lived chromosome damage as manifested by sister chromatid exchange in lymphocytes of animals exposed to mitomycin C, Mutat. Res. 51:383–396 (1978).

93. R. S. Day III, C. H. J. Ziolkowski, D. A. Scudiero, S. A. Meyer, A. S. Lubiniecki, A. J. Girardi, S. M. Galloway, and G. D. Bynum, Defective repair of alkylated DNA by human tumor and SV40-transformed human cell strains, Nature 288:724–727 (1980).

94. R. H. Heflich, S. M. Morris, D. T. Beranek, R. L. Kodell, and D. A. Casciano, Induction of mutations and sister chromatid exchanges in CHO cells by ethylating agents: Relationship to specific DNA adducts, 12th Annual Environmental Mutagen Society, San Diego, California, abstract (1981).

95. D. Scudiero, A. Noun, P. Karran, and B. Strauss, DNA excision-repair deficiency of human peripheral blood lymphocytes treated with chemical carcinogens, Cancer Res. 36:1397–1403 (1976).

96. Z. Darzynkiewicz, Radiation induced DNA synthesis in normal and stimulated human lymphocytes, Exp. Cell Res. 69:356–358 (1971).

97. S. M. Galloway and H. J. Evans, Sister chromatid exchanges in

human chromosomes from normal individuals and patients with ataxia telangiectasia, Cytogent. Cell Genet. 15:17–29 (1975).

98. K. Sperling, R.-D. Wegner, H. Riehm, and G. Obe, Frequency and distribution of sister chromatid exchanges in a case of Fanconi's anemia, Humangenetik 27:227–230 (1975).

99. R. S. K. Chaganti, S. Schonberg, and J. German, A manyfold increase in sister chromatid exchanges in Bloom's syndrome lymphocytes, Proc. Natl. Acad. Sci. USA 71:4508–4512 (1974).

100. W. Burgdorf, K. Kurvink, and J. Cervenka, Sister chromatid exchange in dyskeratosis congenita lymphocytes, J. Med. Genet. 14:256–257 (1977).

101. E. A. Lezana, N. O. Bianchi, M. S. Bianchi, and J. E. Zabala-Suarez, Sister chromatid exchanges in Down's syndrome and normal human beings, Mutat. Res. 45:85–90 (1977).

102. J. D. Rowley, Chromosome abnormalities in cancer, Cancer Genet. Cytogenet. 2:175–198 (1980).

103. S. Knuutila, E. Helminen, P. Vuopio, and A. de la Chapelle, Sister chromatid exchanges in human bone marrow cells. I. Control subjects and patients with leukemia, Hereditas 88: 189–196 (1978).

104. S. Kakati, S. Abe, and A. A. Sandberg, Sister chromatid exchange in Philadelphia chromosome (Ph')-positive leukemia, Cancer Res. 38:2918–2921 (1978).

105. R. Becher, C. G. Schmidt, G. Theis, and D. K. Hossfeld, Sister chromatid exchange in Ph'-positive chronic myelocytic leuk, Int. J. Cancer 24:713–716 (1979).

106. W. S. Cheng, J. J. Mulvihill, M. H. Greene, L. W. Pickle, S. Tsai, and J. Whang-Peng, Sister chromatid exchanges and chromosomes in chronic myelogenous leukemia and cancer families, Int. J. Cancer 23:8–13 (1979).

107. S. Obe and A. A. Sandberg, Sister chromatid exchange and growth kinetics of marrow cells in aneuploid non-lympho-cytic leukemias, Cancer Res. 40:1292–1299 (1980).

108. K. Kurvink, C. D. Bloomfield, K. M. Kennan, S. Levitt, and J. Cervenka, Sister chromatid exchange in lymphocytes from patients with malignant lymphoma, Human Genet. 44:137–144 (1978).

109. M. Otter, C. G. Palmer, and R. L. Baehner, Sister chromatid exchange in lymphocytes from patients with acute lympho-blastic leukemia, Hum. Genet. 52:185–192 (1979).

110. M. A. McDonald and P. H. Fitzgerald, Sister chromatid exchange and cell cycle progression in cultured lymphocytes from patients with chronic lymphatic leukemia, J. Natl. Cancer Inst. 62:1169–1171 (1979).

111. Y. Shiraishi and A. A. Sandberg, Sister chromatid exchange in human chromosomes, including observations in neoplasia. Cancer Genet. Cytogenet. 1:363–380 (1980).

112. M. Swift, L. Sholman, M. Perry, and C. Chase, Malignant

neoplasms in the families of patients with ataxia telangiectasia. Cancer Res. 36:209-215 (1976).

113. M. Swift, R. J. Caldwell, and C. Chase, Reassessment of cancer
predisposition of Fanconi anemia heterozygotes. J. Natl.
Cancer Inst. 65:863-867 (1980).

114. M. Swift and C. Chase, Cancer in families with xeroderma
pigmentosum. J. Natl. Cancer Inst. 62:1415-1421 (1979).

115. R. R. Tice and E. L. Schneider, Spontaneous and induced sister
chromatid exchanges in peripheral blood lymphocytes from
individuals affected with Gardner's syndrome, in preparation
(1981).

116. P. E. Perry, M. Jager, and H. J. Evans, Mutagen-induced sister
chromatid exchanges in xerodema pigmentosum and normal
lymphocytes, in: "Mutagen-induced Chromosome Damage in Man",
H. J. Evans and D. C. Lloyd, eds., Edinburgh Univ. Press,
Edinburgh, Scotland, 201-207 (1978).

117. S. Wolff, B. Rodin, and J. E. Cleaver, Sister chromatid
exchanges induced by mutagenic carcinogens in normal and
xeroderma pigmentosum cells, Nature 265:347-349 (1977).

118. J. A. Heddle and C. V. Arlett, Untransformed xeroderma
pigmentosum cells are not hypersensitive to sister chromatid
exchange production by ethylmethane sulfonate: implications
for the use of transformed cell lines and for the mechanisms
by which SCE arise, Mutat. Res. 72:119-125 (1980).

119. R. W. Pero and F. Mitelman, Another approach to in vivo
estimation of genetic damage in humans, Proc. Natl. Acad.
Sci. USA 76:462-463 (1979).

120. G. Kellerman, C. R. Shaw, and M. Luyten-Kellerman, Aryl hydrocarbon hydroxylase inducibility and bronchogenic carcinoma,
N. Engl. J. Med. 289:934-937 (1973).

121. D. W. Nebert, S. A. Atlas, T. M. Guenthner, and R. E. Kouri,
The Ah locus: Genetic regulation of the enzymes which
metabolize polycyclic hydrocarbons and the risk for cancer,
in: "Polycyclic Hydrocarbons and Cancer 2", H. V. Gelboin
and P. O. P. Ts'o, eds., Academic Press, New York, NY, 346-
389 (1978).

122. G. Kellerman, M. Luyten-Kellerman, and G. R. Shaw, Genetic
variation of aryl hydrocarbon hydroxylase in human
lymphocytes, Am. J. Hum. Genet. 25:327-331 (1973).

123. B. Paigen, H. L. Gurtoo, J. Minowada, E. Ward, L. Houten,
K. Paigen, A. Reilly, and R. Vincent, Genetics of aryl
hydrocarbon hydroxylase in the human population and its
relationship to lung cancer, in: "Polycyclic Hydrocarbons
and Cancer 2", H. V. Gelboin and P. O. P. Ts'o, eds.,
Academic Press, New York, NY. 391-406 (1978).

124. S. M. Galloway, P. E. Perry, J. Meneses, D. W. Nebert, and
R. A. Pedersen, Cultured mouse embryos metabolize benzo(a)
pyrene during early gestation: Genetic differences detectable by sister chromatid exchange, Proc. Natl. Acad. Sci.
USA 77:3524-3528 (1980).

125. R. R. Schreck and S. A. Latt, Comparison of benzo(a)pyrene

metabolism and sister chromatid exchange induction in mice, Nature, in press, (1981).

126.   H. W. Rudiger, F. Kohl, W. Mangels, P. von Wichert, C. R. Bartram, W. Wohler, and E. Passarge, Benzpyrene induces sister chromatid exchanges in cultured human lymphocytes, Nature 262:290-292 (1976).

127.   J. M. Hopkin and P. E. Perry, Benzo(a)pyrene does not contribute to the SCEs induced by cigarette smoke, Mutat. Res. 77:377-381 (1980).

128.   A. D. Schonwald, C. R. Bartram, and H. W. Rudiger, Benzpyrene induced sister chromatid exchanges in lymphocytes of patients with lung cancer, Hum. Genet. 36:261-264 (1977).

129.   D. Kram and E. L. Schneider, Reduced frequencies of Mitomycin c-induced sister chromatid exchanges in AKR mice, Hum. Genet. 41:45-51 (1978).

130.   J. A. Biegel, S. S. Boggs, and M. K. Conner, Comparison of BCNU-induced SCE in bone marrow cells of AKR/J and BDF1 mice, Mutat. Res. 79:87-90 (1980).

DISCUSSION

Q. Albertini (Vermont): The lymphocytes you and I use in our respective assays are part of the immune system and many of the diseases we look at are characterized by altered immunological responses. Does an immune response alter the frequency of SCE and must these physiological responses be taken into account when examining environmental effects on SCE levels?

A. Galloway (Litton Bionetics): I don't know of any clear-cut information about the possible interaction between an altered immune response and SCEs. One possible effect would be to alter the relative frequencies of first, second and third cycle cells in your test population and there is some controversial evidence that suggests different base-line SCE levels in early versus late appearing cell populations. This possibility worries me, especially in studies showing a significant difference of two SCEs between a control and an exposed population. If the reports about different base-line levels of SCEs at different culture termination times are true then the small but stistically significant increase of two SCEs may merely be the result of different lymphocyt growth patterns.

THIOGUANINE RESISTANT LYMPHOCYTES AS INDICATORS OF

SOMATIC CELL MUTATION IN MAN

Richard J. Albertini and David L. Sylwester

University of Vermont, Burlington, VT   05405   (U.S.A.)

INTRODUCTION

This paper briefly reviews studies, which, over the past several years in our laboratory, have resulted in the development of a presumptive test for somatic cell mutation occurring in vivo in man. The test enumerates in vitro variant 6-thioguanine resistant (TG^r) peripheral blood lymphocytes (PBL's). Conditions of testing are such that variant-producing events (presumably mutations) antedate the assay and must occur in vivo. Tests of this kind are called direct mutagenicity tests (1) and have unique advantages for human mutagenicity monitoring.

The recognition of somatic cell mutations in cultured mammalian cells, and the concept of using cultured cells for tests of mutagenicity, are quite old (2). Although the genetic basis of phenotypic variation in cultured somatic cells was, for a time, in question, subsequent demonstrations have clearly shown that at least some variants are truly genetic mutatns (3,4). Several criteria have been proposed for the definition of mutant somatic cells (5).

Cultured diploid human cells, on the other hand, have been used for mutagenicity testing in vitro for only about ten years (6). At present, two human cell types--fibroblasts and lymphobasts-- are in routine use for this purpose. Several markers of mutation are available in human cells, one of which is hypoxanthine guanine phosphoribosyltransferase (HPRT) deficiency and purine analogue resistance resulting from the Lesch-Nyhan (LN) mutation (6,7). Mutation rates at this locus in fibroblasts and lymphoblasts in vitro have been determined by fluctuation tests to be of the order

489

of 10^{-6} per cell generation (8–12). However, mutation at other loci in cultured human fibroblasts may occur at a higher rate (1).

Direct mutagenicity testing for specific locus mutations occurring in vivo in man is conceptually (and technically) an out-growth of in vitro mutagenicity testing with cultured human cells. The rationale for such testing rests on certain assumptions: (i) specific locus mutations in human cells occur in vivo, (ii) mutant cells arising from mutations occurring in vivo are quantitatively detectable in vitro, and (iii) frequencies of mutant cells correlate with and may serve as markers for genotoxic exposures of the individuals tested.

The Lesch–Nyhan Mutation as as Prototype for Human Mutagenicity Testing

The LN mutation is a naturally occurring mutation of the X–linked gene for HPRT that results in a dramatic clinical syndrowm in man (13,14). Of importance for mutagenicity testing is the fact that this mutation produces a clear and unambiguous phenotype at the single cell level (6,7,15). LN cells, in contrast to normals, are resistant to the cytotoxicity of purine analogues such as 8–azaguanine (AG^r) and 6–thioguanine (TG^r), are unable to utilize hypoxanthine (the normal substrate of HPRT) for growth, and, of course, are deficient in HPRT activity. These phenotypic characteristics, as well as the location of the gene on the X–chromosome, render the LN mutation ideal as a prototype for somatic cell mutation in man.

The LN phenotype is expressed in peripheral blood lymphocytes (PBL's) as well as in cultured fibroblasts and lymphoblasts (16). Because of single X–chromosome inactivation, women who are hetero-zygous at X–linked loci are functionally mosaics with two popula-tions of somatic cells. One cell population will express the genes of one X–chromosome (including the normal allele for HPRT) while the other population will express the genes of the other X–chromo-some (including the LN mutation). Thus, LN heterozygotes are naturally occurring models for direct mutagenicity testing in that they possess a population of variant cells known to be mutant. Furthermore, the LN fraction of PBL's in heterozygotes is small, comprising approximately 1% or less of the total rather than the expected 50% (16,17). This difference probably results from negative selection against the LN phenotype. Not only do heterozygotes contain mutant PBL's for detection, but these cells are in the minority. A method capable of detecting these infrequent mutant PBL's in LN heterozygotes should be able also to detect LN–like variant PBL's arising in non–LN individuals, providing, of course, the method is sensitive enough. Negative selection in vivo against LN–like PBL's should serve to keep variant cells from accumulating in vivo in non–LN individuals.

A Method For Enumerating TG^r PBL's

Peripheral blood mononuclear cells (MNC's) which are easily obtainable and are polyclonal in origin, are sampled for TG^r variants. The peripheral blood lymphocyte (PBL) component of MNC's is largely in an arrested G_0 stage in the peripheral blood, but can be stimulated to DNA synthesis in vitro by phytohemagglutinin (PHA). DNA synthesis is conveniently measured by tritated (^{3}H) thymidine incorporation and is inhibited by several purine analogues, including TG, in normal but not LN cells.

The details of the method have been presented elsewhere (18-20) and are only summarized here. Blood is obtained in heparinized syringes and the MNC fraction separated by the Ficoll-Hypaque method (21). The MNC fraction is cryopreserved in medium RPMI 1640, human AB serum, dimethylsulfoxide (DMSO) and stored in the vapor phase of liquid nitrogen until used. Fresh MNC's may be tested directly but this often results in an over-estimation of LN-like PBL's due to failure of TG to inhibit DNA synthesis in all TG sensitive (TG^s) cells (see below). For use, frozen MNC's are suspended after rapid thawing in medium RPMI 1640 containing 25 mM Hepes, 2 mM glutamine, penicillin and streptomycin (100 units and 100 µg, respectively), and 30% human AB serum. Cells are suspended directly if tested fresh.

Cells from each individual tested are inoculated into a series of culture tubes so as to achieve final culture density of 10^6 MNC's/mL. All cultures are made in one or two mL volumes containing 5 µg/mL PHA-P (Burroughs Wellcome). Most cultures contain TG at a final concentration of 2×10^{-4} M (test cultures). However, at least two cultures lack TG but contain instead, an equal volume of medium RPMI 1640 with the pH adjusted to that of the TG solution (control cultures).

All cultures are incubated for 30 hours in a 37°C humidified atmosphere and labeled with 5 µCi ^{3}H-thymidine/mL. Cultures are incubated for an additional 12 hours and terminated. Termination consists of adding 4 mL 0.1 M citrate to each culture tube and centrifuging at room temperature at 600 x G for 10 minutes. This results in cytoplasmic stripping of PBL's and free nuclei. The supernatant is removed and the pellet containing the nuclei resuspended in 4 mL of fixative consisting of 7 parts methanol and 1.5 parts glacial acetic acid. Tubes are recentrifuged, supernatant is removed and the pellets are resuspended in 0.2 mL fixative. Tubes are tightly capped and nuclei left to fix in the cold (4°C) for at least one hour.

Suspensions of nuclei in each tube then are counted with a Coulter counter and added in toto in measured volumes to 18 x 18 mm coverslips fixed with a permount to glass microscope slides.

Slides are dried, stained, and autoradiographed according to published methods. After preparation, slides are scored by light
microscopy.

The number of cells on coverslips is determined from the volume
of suspension X the number of nuclei in that volume. Coverslips
containing nuclei from control cultures are scanned at high power
microscopy (970X) and 2500 nuclei on each of the two control coverslips are evaluated. The number of autoradiographically labeled
nuclei among these 5,000 is counted and the labeling index of the
control cultures determined:

$$\text{LI (control)} = \frac{\text{\# labeled nuclei/5000 nuclei}}{5000}$$

Coverslips containing nuclei for test cultures are scanned
in their entirety at 160X and all labeled nuclei on all coverslips
containing nuclei from test cultures of a given individual are
counted. Since the total number of nuclei on these coverslips
is known, the LI of the test cultures is determined:

$$\text{LI (test)} = \frac{\text{\# labeled nuclei on all coverslips}}{\text{total \# of nuclei on all test coverslips}}$$

The TG^r PBL variant frequency (V_f) of the individual is the
ratio of the two LI's:

$$V_f = \frac{\text{LI (test)}}{\text{LI (control)}}$$

SUMMARY OF RESULTS

Figures 1 and 2 show coverslips prepared from nuclei from
representative control (Fig 1) and test (Fig 2) cultures. The
control culture in this instance had a LI of 42%. A segment of
the control culture coverslip photographed under high power microscopy (970X, Fig 1b) shows that labeled nuclei may be unambiguously
differentiated from unlabeled nuclei. Coverslips made with nuclei
from test cultures (containing TG) show rare labeled nuclei. Here,
too, labeled nuclei are readily differentiated from unlabeled nuclei
when viewed at low (Fig 2a) and high (Fig 2b) power microscopy.

Using this method, we showed that 8 LN heterozygous females
had TG^r PBL's in their peripheral blood in frequencies that ranged
from 1.4×10^{-3} to 5.7×10^{-2} (17). Several LN males showed TG^r
PBL frequencies that ranged from 23 to 100% (19). Clearly, LN cells
in such individuals must comprise 100% of the total PBL's. These
results, therefore, reflect the range of error of the method.
This range for LN males is either due to the variability within

the procedure, e.g. the LI determined for test and control cultures
of a given individual, or to partial sensitivity of LN PBL's to 2 x
10^{-4} M TG. LN PBL's are clearly sensitive to the cytotoxicity of
2 x 10^{-3} M TG (18).

In early studies reported elsewhere, we used this method to
determine TG^r PBL V_f's for 98 fresh MNC samples from 63 healthy,
non-mutagen exposed, non-LN individuals (18). The median TG^r
PBL V_f for this group, determined at 2 x 10^{-4} M TG, was 1.1 x 10^{-4}
with a range for all test values of 2.5 x 10^{-5} to 3.18 x 10^{-4}. The
ages of the 63 individuals ranged from 11 to 75 years and TG^r
PBL V_f values showed no correlation with age. This is in accord
with expectations for selection in vivo against LN-like PBL's.

Fresh MNC samples were then obtained from several groups of
individuals who had been exposed therapeutically to known or suspected mutagens. These early studies are also reported in detail
elsewhere (18,19,22). The median TG^r PBL V_f determined at 2 x 10^{-4}
M TG for 12 cancer patients receiving chemotherapy was 8.5 x 10^{-4}
with a range of 4.0 x 10^{-5} to 9.9 x 10^{-3} . Ten of these 12 values
were higher than the highest V_f value determined for a group of
eight healthy non-mutagen exposed controls tested concurrently.
Similarly, fresh MNC samples obtained from 18 patients with
psoriasis who were being treated with 8-methoxypsoralen and UVA
light (PUVA) were tested. The median V_f of this group was 7.9 x 10^{-4}
with a range of 5.0 x 10^{-5} to 6.3 x 10^{-3}. Also, testing of fresh
MNC's from 10 patients with vitiligo receiving PUVA gave a median
TG^r PBL V_f value of 6.1 x 10^{-4} with a range of 1.0 x 10^{-4} to 4.3 x
10^{-3}. Again, V_f values determined concurrently for normal controls
were in the normal range. Cancer patients receiving X-irradiation
therapy showed elevated TG^r PBL V_f's when compared to normals, and
these values appeared to rise with therapy in a dose dependent
fashion (19).

Although elevated TG^r PBL V_f values in humans receiving
potentially mutagenic therapies might easily be ascribed to these
therapies, other results of these early studies were not so easily
explained. A group of 14 patients with breast cancer not yet
receiving chemotherapy also had elevated TG^r PBL V_f's as compared
to normal (median value of 5 x 10^{-4} , range of 5.8 x 10^{-5} to 5.0 x
10^{-3}) (19). Similarly, 16 psoriatic patients receiving conventional
therapy had elevated TG^r PBL V_f's (median value 1.4 x 10^{-3}, range
of 1.2 x 10^{-4} to 7.9 x 10^{-3}) (22). However, a group of 7 untreated
vitiligo patients had TG^r PBL V_f values indistinguishable from a
cohort of normal controls (median value of 1.4 x 10^{-4} for vitiligo
patients vs 1.1 x 10^{-4} for controls) suggesting that the elevated
V_f values in PUVA treated vitiligo patients were due to the treatments.

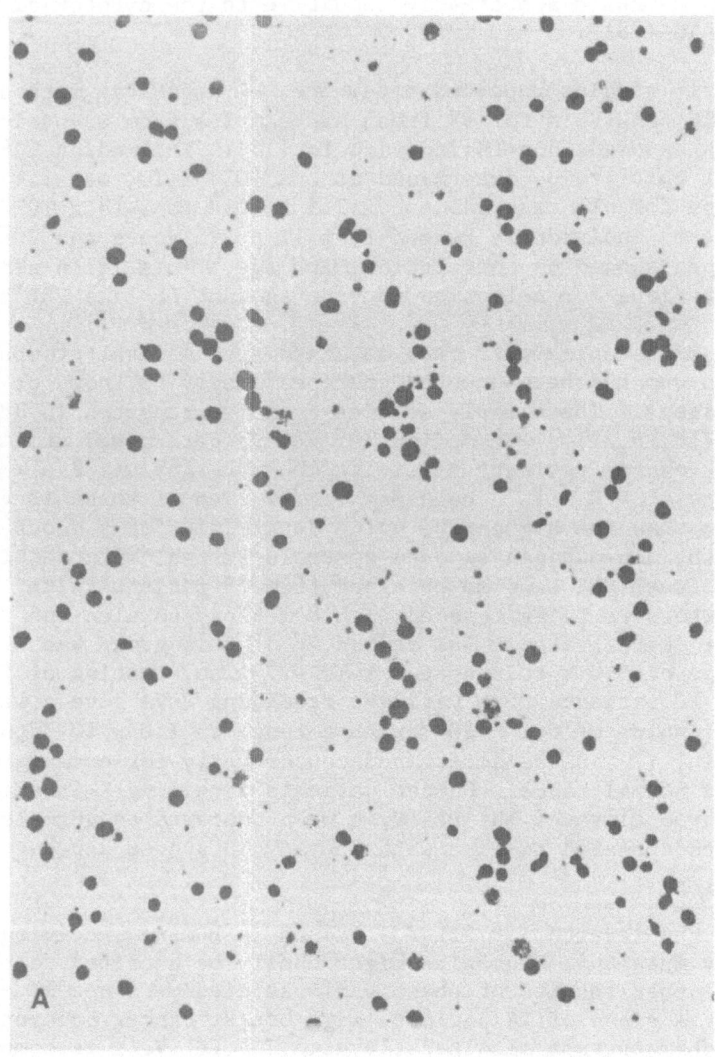

Figure 1. Autoradiograph of coverslip containing nuclei from
control culture shown at low power (160X) microscopy (Figure 1a)
and high power microscopy (Figure 1b). The LI of the control
culture was 42%. Eight of the fifteen nuclei shown in Figure 1b
are positive.

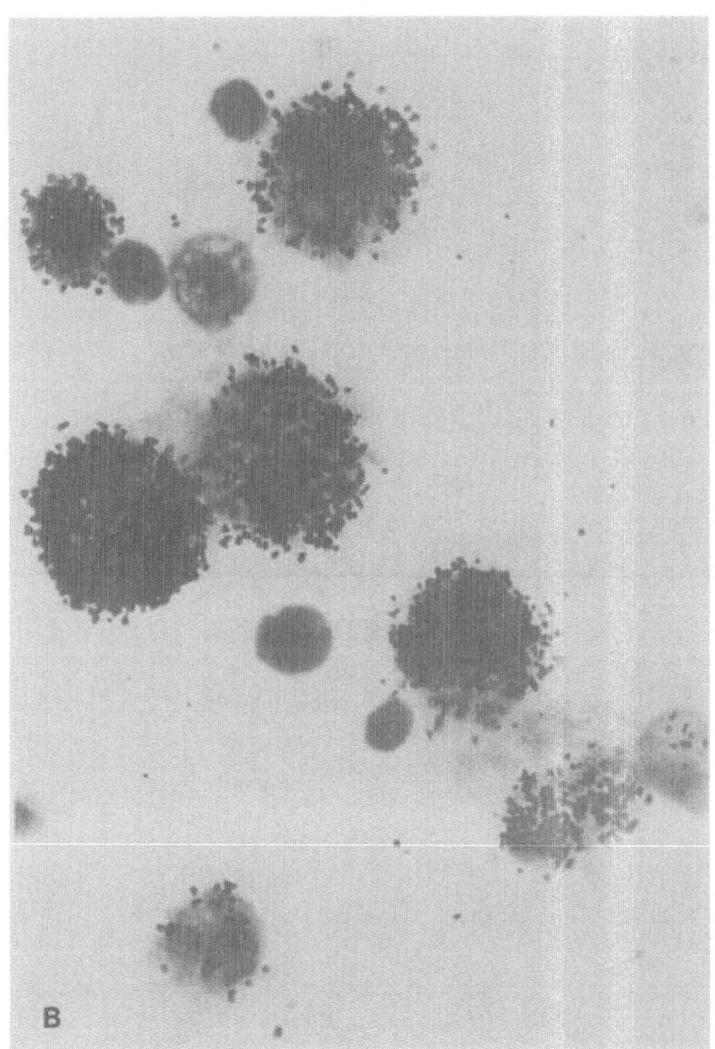

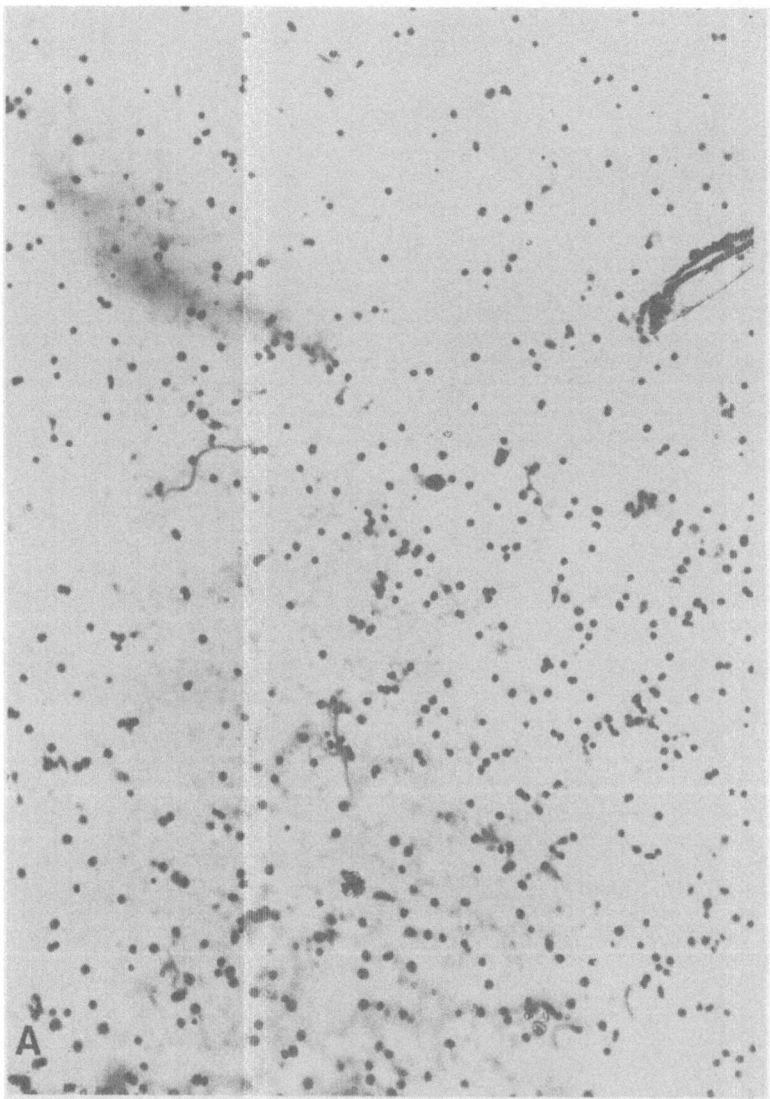

Figure 2. Autoradiograph of coverslip containing nuclei from test
culture shown at low power (160X) microscopy (Figure 2a) and high
power microscopy (Figure 2b). Only a single labeled nucleus is
seen in each figure.

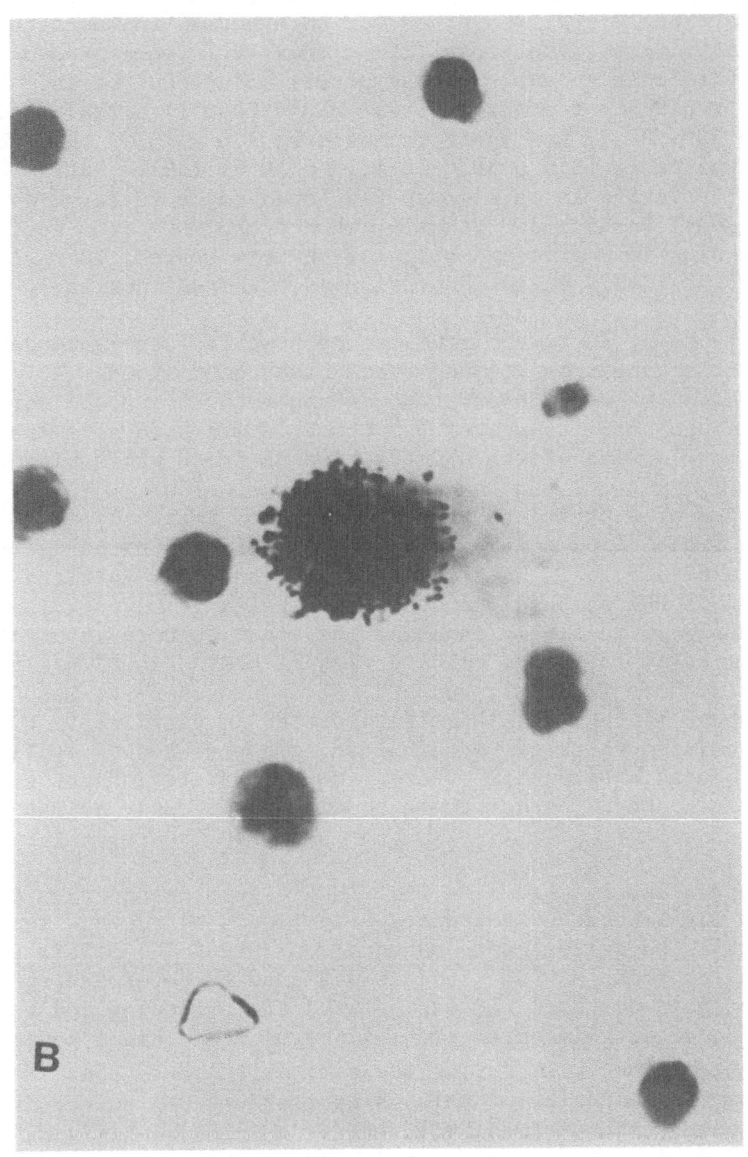

The results obtained with non-treated cancer and psoriatic
patients suggested that labeled nuclei, reflecting PBL's apparently
resistant to TG under conditions of testing, might arise by
mechanisms other than mutation. Also, we recognized an earlier
error in electronic counting of nuclei that probably resulted in an
underestimate of TG^r PBL V_f values for both normal and mutagen
exposed individuals given above (19). Thus, V 's were of a magnitude
that were difficult to explain on a purely mutational basis. In
subsequent replicate testing of fresh MNC's from 11 normal individuals,
the median TG^r PBL V_f was again found to be 1.6×10^{-4}, but the range
of values was large $(3.0 \times 10^{-5}$ to $1.15 \times 10^{-3})$ (19). Intraindividual variation in this group was found to be as large as
interindividual variation. Finally, when fresh MNC' from individuals undergoing immunological stimulation were tested, TG^r PBL V_f
values were decidedly elevated over those of normal adults.

Clearly then, in some circumstances some influences tended to
make TG^s PBL's appear as TG^r under our assay conditions. It
appears that a frequent observation in non-mutagen exposed normal
individuals who showed elevated TG^r PBL V_f's was some stimulus to
proliferation in vivo of the MNC fraction of fresh whole blood.

A series of experiments then demonstrated that cycling MNC's
present in fresh blood samples are not reliably inhibited during
the short exposure to TG in this assay and thus, may initiate their
first round of DNA synthesis and become labeled in vivo (23).
The rationale for initially choosing the short culture interval
used in our assay (30 hours culture plus 12 hours label) has been
detailed elsewhere (18). In summary, it is necessary to keep this
interval short so that all TG cells detected in vitro reflect
variation arising in vivo. Nonetheless, the presence of cycling
cells introduces substantial errors. We termed the TG^s PBL's
which appear to be TG^r under these conditions of assay as phenocopies.

At approximately this same time, MNC's were being cryopreserved
in DMSO in our laboratory for storage in liquid nitrogen. Testing
of these cells revealed lower and more consistant TG^r PBL V_f's than
did testing of fresh MNC's. Studies designed to determine spontaneous
DNA synthesis in cultured MNC's suggested that freezing and thawing
greatly reduced or eliminated the ability of these cells to incorporate ^3H thymidine in TG during our period of label (23). We
do not know the mechanism of this as spontaneous DNA synthesis
occurs in some frozen-thawed MNC's early in cultures but does not
occur in TG during the interval of label(30 hours culture followed
by 12 hours label). Thus, by cryopreservation procedures, apparent
phenocopies are eliminated or greatly reduced in number. We have
also shown that cooling of MNC's to 4°C for several hours, and
certain other procedures, tend to eliminate some or all of the
phenocopies.

Our more recent studies have been with cryopreserved MNC's. Results of these studies show that the phenocopy producing influences mentioned above fail to produce elevated TG^r PBL V_f's in normal individuals. We feel, therefore, that the TG^r PBL's now scored reflect, in large part, somatic cell mutation occurring in vivo. For example, recent preliminary studies with cryopreserved MNC's from seven normal non-mutagen exposed control adults gave TG^r PBL V_f values that ranged from 2×10^{-6} to 8.3×10^{-6}. We now consider this to be the range of normal values. Two newborn umbilical cord blood samples had V_f values within this range as did a sample obtained from a cancer patient prior to initiation of chemotherapy. By contrast, preliminary studies in ten cancer patients receiving either chemotherapy of X-irradiation therapy showed TG^r PBL V_f values that ranged from 13×10^{-6} to 306×10^{-6}. All but one of the assays were performed with cryopreserved MNC's, and the exceptional sample was cultured at 37°C for 4 days in order to remove cycling cells. These and other results obtained with cryopreserved MNC's are to be published in detail elsewhere. Results of TG^r PBL V_f determinations for the various groups of individuals made with fresh or cryopreserved MNC's are presented in Table 1.

Statistical Concerns with TG^r PBL V_f Assays:

As noted, with removal of potential phenocopies from MNC samples by freezing, the normal TG^r PBL V_f's are much lower than we originally reported. Therefore, in order to achieve useful precision in V_f estimates, experiments must be larger (i.e., more cells tested when frozen MNC's are used). Otherwise, estimates are based on counts of very few labeled nuclei recovered from test cultures.

In order to determine the precision of the assay and the optimum size of experiments, methods were devised to establish 95% confidence intervals (CI's) for the true variant frequency measured by V_f (20). Recall that

$$V_f = \frac{LI \text{ (test)}}{LI \text{ (control)}} = \frac{M/T}{LI} = \frac{M}{LI \cdot T} = \frac{M}{N}$$

where M equals number of labeled nuclei on all test coverslips, T equals the total number of nuclei on test coverslips, LI equals the labeling index of control cultures, and N equals the evaluatable nuclei in test cultures.

We assume that M is a Poisson variate and LI is a binomial fraction. Exact CI's for the true Poisson mean of labeled nuclei

Table 1. TG^r PBL V$_f$ Values determined for various groups of
individuals with fresh or cryopreserved mononuclear
cells (MNC's). (TG = 2 x 10^{-4} M)

FRESH MNC's		TG^r PBL V$_f$ (x10^{-6})	
	# TESTS	MEDIAN	RANGE
Normal 1[f]	98[a]	110[b]	25, 318[c]
Normal 2	46[d]	160[c]	30, 1150[c]
Vitiligo(untreated)[f]	7	140	74, 2000
Psorasis (conventional treatment)[f]	16	1400	120, 7900
Cancer (Pretreatment)[f]	14	500	58, 5000
Vitiligo - PUVA[f]	10	610	100, 4200
Psoriasis - PUVA[f]	18	790	50, 6300
Cancer - Chemotherapy[f]	12	850	40, 9900
CRYOPRESERVED MNC's			
Normal	9[e]	2.4	2.0, 8.3
Cancer chemotherapy or X-irradiation	10	44	13, 306

[a]98 tests in 63 individuals

[b]MEDIAN value for the 63 individuals

[c]MEDIAN value or range of values for all of the tests

[d]46 tests in 11 individuals

[e]includes two samples from newborn umbilical cord blood

[f]Value underestimated due to error in electronic counting

measured by M can be obtained from tables. Since N is a binomial
random variable with parameters T and p (the probability that a
nucleus from a control culture will be labeled), it can be shown
that the approximate standard deviation of N (s_N) is:

$$s_N = T \sqrt{pq/C}$$

where q = 1-p and C = # of nuclei evaluated in control cultures =
5000.

Two methods can be used to establish CI's for true variant
frequencies. A method which provides the narrower CI treats N as
a constant and divides the endpoints of the CI for the true Poisson
mean value measured by M (a_M and b_M obtained from tables) by N. Thus,
the CI is (a_M/N, b_M/N). A method which provides a wider CI is obtained
by dividing the lower C.I. bound of M, i.e. a_M, by the upper CI
bound for N, i.e. b_N, and the upper CI bound for M, i.e. b_M,
by the lower CI bond for N, i.e. a_N. Thus, the CI is

$$\left(\frac{a_M}{b_N}, \frac{b_M}{a_N} \right)$$

In comparing V_f's (V_{f1}, V_{f2}) we wish to determine if the ratio

$$\frac{V_{f1}}{V_{f2}}$$

is significantly different (p<0.05) from 1. This is true if the
approximate CI for the ratio, i.e. the lower CI bound for V_{f1}, divided
by the upper CI bound for V_{f1}, and vice versa, excludes the special
value 1.

We describe elsewhere a method for using the transformation
Y = ln X (where ln denotes the natural logarithm) to approximate
CI's. X may represent a Poisson variate such as M or a binomial
fraction such as LI. The method allows close approximation of
the CI for the true value measured by V_f for moderate values of
M (i.e., 10 or more). Using the transformation, an estimate of
the approximate variance of V_f (s^2) is

$$s^2 = \frac{1}{M} + \frac{1-LI}{C \cdot LI}$$

and the approximate 95% CI for the true variant frequency is

$$\left(V_f \cdot e^{-1.96s}, V_f \cdot e^{-1.96s}\right).$$

The precision of the estimate, V_f, is governed mostly by the size of M. The greater the M the greater the precision. Ideally, TG^r PBL V_f determinations should involve experiments of sufficient size so that two fold differences in V_f's are detectable. Elsewhere we show that V_f determinations must be based on a minimum of 14 labeled nuclei in test cultures in order to detect a true doubling of the V_f as statistically different (20). Therefore, if the median normal TG^r PBL V_f is approximately 5×10^{-6}, experiments must be of sufficient size so that between 2 and 3×10^6 evaluatable nuclei are recovered from test cultures.

DISCUSSION

The TG^r PBL V_f assay has undergone considerable evolution since its original description. Our initial estimate of V_f's for non-mutagen exposed normal individuals were clearly in error and far too high. We now feel that we know the cause of much of this error and realize that MNC's in cell cycle at the time of exposure to TG in vitro may not be inhibited from synthesizing DNA and incorporating ^3H thymidine by 30 hours (time of label). These cells are almost certainly TG^r non-mutant cells and thus are phenocopies.

Two modification of the assay method have been made: (i) measures (principally cryopreservation) and thawing) are now taken to eliminate or reduce greatly the phenocopies, and (ii) efforts are made to perform experiments of sufficient size so that more precise TG^r PBL V_f determinations can be made.

We do not know, of course, that all TG^r PBL's detected by this method are genetic mutants, since this will require characterization of variants, much as has been done for variant purine analogue resistant human fibroblasts derived in vitro (7,8). Our evidence that TG^r PBL's arise from mutation has been presented elsewhere (18,19,24). It is indirect but sufficient to indicate that at least some variants are true mutants. Currently available methods that will allow propagation of PBL's in vitro will be exploited to demonstrate directly the genetic basis of TG^r in PBL's.

Despite the errors made in our initial studies, the most critical conclusions drawn from them remain valid. Variant TG^r PBL's are present in low frequencies in the peripheral blood of normal non-mutagen exposed individuals. Age is not a factor in these frequencies. Most importantly, potential mutagenic therapies, such as cytotoxic chemotherapy and X-irradiation therapy, are associated with elevations of TG^r PBL V_f's. Thus, the method detects mutagenic exposures of at least this magnitude. With the removal of phenocopies, the method is more sensitive than originally described. Longitudinal studies of patients receiving X-irradiation therapy are in progress to determine if intraindividual TG^r PBL V_f elevations occur in a dose dependent fashion in man.

We have listed elsewhere the unique advantages provided by direct mutagenicity testing in man (19,24). Perhaps of greatest importance is that such testing may allow assessments of human health risks to be made. Individuals in whom direct mutagenicity tests have been performed may be followed to determine the late health effects of exposures to genotoxic agents. Mutagenicity test results may then be correlated with the occurrence of these late effects, e.g., malignancies or birth defects in the offspring. Thus, direct mutagenicity tests may be validated in terms of being predictors of human health risks. Once validated, test results may be substituted for the occurrence of illness as an endpoint for assessing the hazard of environmental agents. This is the goal of mutagenicity monitoring.

The TG^r PBL V_f assay continues to evolve. We feel, however, that the major non-mutant sources of error are mostly eliminated and that the method may be used for studies in man. It remains to test large numbers of individuals undergoing known or suspected mutagenic exposures, as well as individuals in industrial or other environments undergoing unknown exposures, so that the place of this system in human mutagenicity monitoring can be determined.

REFERENCES

1. R. DeMars, Suggestions for increasing the scope of direct testing for mutagens and carcinogens in intact humans and animals, in: "Banbury Report 2, Mammalian Cell Mutagenesis: The Maturation of Test Systems", A. W. Hsie, J. P. O'Neill, and V. K. McElheny, eds., Cold Spring Harbor Laboratory, Cold Spring Harbor, N.Y., 329-340 (1979).
2. T. T. Puck and H. W. Fisher, Genetics of somatic mammalian cells, J. Exp. Med. 104:427-433 (1956).
3. R. DeMars, Resistance of cultured human fibroblasts and other cells to purine and pyrimidine analogues in relation to mutagenesis detection, Mutat. Res. 24:335-364 (1974).
4. L. Siminovitch, On the nature of heritable variation in cultured somatic cells, Cell 7:1-11 (1976).
5. E. H. Y. Chu and S. S. Powell, Selective systems in somatic cell genetics, in: "Advances in Human Genetics", Vol. 7, Chapt. 5, Harris and Hirschhorn, eds., Plenum Press, N.Y., 189-258 (1976).
6. R. J. Albertini and R. DeMars, Diploid azaguanine-resistant mutants of cultured fibroblasts, Science 169:481-485 (1970).

7.   R. J. Albertini and R. DeMars, Detection and quantification of
     X-ray induced mutation in cultured, diploid human fibro-
     blasts, Mutat. Res. 18:199-244 (1973).
8.   R. DeMars and K. Held, The spontaneous azaguanine-resistant
     mutants of diploid human fibroblasts, Humangenetik 16:87-110
     (1972).
9.   M. Buchwald, Mutagenesis at the Ouabain resistance locus in
     human diploid fibroblasts, Mutat. Res. 44:401-412 (1977).
10. R. Gupta and L. Siminovitch, Isolation and characterization of
     mutants of human diploid fibroblasts resistant to diphtheria
     toxin, Proc. Nat. Acad. Sci. U.S.A 73:3337-3340 (1978).
11. D. Pious and C. Soderland, HLA variants of cultured human
     lymphoid cells: Evidence for mutational original and
     estimation of mutation rate, Science 197:769-771 (1977).
12. W. G. Thilly, J. G. DeLuca, E. E. Furth, H. Hoppeiv, D. A.
     Kaden, J. J. Krolewski, H. L. Liber, T. R. Skopek, S. A.
     Slapikoff, R. J. Tizard and B. W. Penman, "Gene-locus
     mutation assays in diploid human lymphoblast lines, in:
     Chemical Mutagens: Principles and Methods for their
     Detection", Vol. 6, A. Hollaender and F. de Serres,
     eds., Plenum Press, New York, NY, 331-364 (1980).
13. M. Lesch and W. L. Nyhan, A familial disorder of uric acid
     metabolism and central nervous system functions, Am. J.
     Med. 36:561-570 (1964).
14. J. E. Seegmiller, F. M. Rosenbloom, and W. N. Kelley, Enzyme
     defect associated with a sex-linked human neurological
     disorder and excessive purine synthesis, Science 155:1682-
     1684 (1967).
15. R. DeMars, Genetic studies of HG-PRT deficiency and the Lesch-
     Nyhan syndrome with cultured human cells, Fed. Proc. 30:
     944-955 (1971).
16. R. J. Albertini and R. DeMars, Mosaicism of peripheral blood
     lymphocyte populations in females heterozygous for the Lesch-
     Nyhan mutation, Biochem. Genet. 11:397-411 (1974).
17. G. H. Strauss, E. F. Allen, and R. J. Albertini, An enumerative
     assay of purine analogue resistant lymphocytes in women
     heterozygous for the Lesch-Nyhan mutation, Biochem. Genet.
     18:529-547 (1980).
18. G. H. Strauss and R. J. Albertini, Enumeration of 6-thioguanine
     resistant peripheral blood lymphocytes in man as a potential
     test for somatic cell mutation arising in vivo, Mutat. Res.
     61:353-379 (1979).
19. R. J. Albertini, Drug resistant lymphocytes in man as indicators
     of somatic cell mutation, Teratogenesis, Carcinogenesis, and
     Mutagenesis 1:25-48 (1980).
20. R. J. Albertini, D. L. Sylwester, and E. F. Allen, The 6-
     thioguanine resistant peripheral blood lymphocyte assay for
     direct mutagenicity testing in man, in: "Mutagenicity:
     From Bacteria to Man", J. A. Heddle, ed., Academic Press,
     N.Y., in press (1981).

21. A. Boyum, Separation of leukocytes from blood and bone marrow,
     Scand. J. Clin. Lab. Invest. 21(Supple. 97):51-76 (1968).
22. G. H. Strauss, R. J. Albertini, P. A. Krusinski, and
     R. D. Baughman, 6-Thioguanine resistant peripheral blood
     lymphocytes in humans following psoralen long-wave UV light
     (PUVA) therapy, J. Invest. Dermatol. 73:211-216 (1979).
23. R.J. Albertini, E. F. Allen, A. S. Quinn, and M. R. Albertini,
     Human somatic cell mutation: In vivo variant lymphocyte
     frequencies as determined by 6-thioguanine resistance, in:
     "Birth Defects Institute Symposium XI", Academic Press, New
     York, N.Y., in press (1981).
24. R. J. Albertini and E. F. Allen, Direct mutagenicity testing
     in man, in: "Health Risk Analysis", P. J. Walsh,
     C. R. Richmond, and E. D. Copenhaver, eds., Franklin Institute
     Press, in press (1981).

DISCUSSION

Q. Borg (BNL): I realize that this system is fairly new but have
you had the opportunity to survey any groups of people in the general
population who might be exposed to genotoxic agents?

A. Albertini (U. of Vermont): Other investigators are doing this.
I think until we validate this assay and really understand the
significance of the data I'm not really sure we should be using
this assay system in a general way.

Q. Galloway (Litton Bionetics): Do you have any information of
whether your variant level will change with the length of time
your cell samples are stored in liquid nitrogen?

A. Albertini (U. of Vermont): We have looked at cells from 5
individuals stored for up to 60 days and have found no variation.
We are thinking of using this consistancy as a technical control
over time.

DETECTION OF POINT MUTATIONS IN MAMMALIAN SPERM

H. V. Malling, J. G. Burkhart, M. A. Baig and
A. A. Ansari
Laboratory of Biochemical Genetics
National Instirute of Environmental Health Sciences
Research Triangle Park, NC   27709   (U.S.A.)

INTRODUCTION

Our ability to assess the extent and significance of genetic damage resulting from exposure to environmental agents is limited. For the human population the two most important types of transmissible genetic damage are chromosomal aberrations and gene mutations. Established techniques are use routinely to detect chromosomal aberrations, but there are only a few established methods for the detection of point mutations. Any assessment of mutation frequency in potentially exposed humans is also complicated by family size, privacy considerations, and other sampling difficulties. Transmissible mutation frequency after exposure to a chemical agent can be measured by analyzing the offspring of the exposed individuals, or the possibility that genetic damage has occurred can be estimated by analyzing the cells of the exposed person. Methods for the detection of mutations in single cells from exposed individuals are under development. One such method is based on detection of thioguanine-resistant lymphocytes (1; see Albertini, these proceedings), and another is based on detection of variant hemoglobin types in single red blood cells by use of fluorescent antibodies (2). However, since germinal cells have singular relevance for the next generation, it would be beneficial to have a system for detection of point mutations directly in sperm. We have begun to develop such a system in laboratory animals and out goal is to examine similar techniques that might be applicable to the human population.

There are several factors that must be considered in the development and of such a system. Among these are (a) choice of

507

a suitable biochemical marker for sperm, (b) discrete localization
of the marker in single sperm by whatever technique is to be used
for detection, (c) a genetic basis and definition of the alteration
that denotes a presumptive mutant, (d) detection of events with a
frequency of 10^{-6} or 10^{-7}, and (e) using the same chemical agents
correlation of results obtained in single cell mutation induction
experiments with results from transmissible germinal mutation
studies. Our initial efforts have been based upon the immunofluorescent detection of alterations in the sperm specific enzyme lactate
dehydrogenase-X (LDH-X).

LDH-X

Several lactate dehydrogenase (LDH) isozymes occur in mammalian
somatic tissues. LDH-X, however, is restricted in its distribution
to developing spermatids and mature sperm; LDH-X synthesis begins
during the midpachytene stage of spermatogenesis. In mammals the
somatic LDH's are coded by two genes, Ldh-a and Ldh-b; LDH-X is
coded by a third gene, Ldh-c. LDH-X is a tetramer and exists only
in one form in mice (C_4) (3).

We have found no detectable genetic variation in LDH-X among
inbred strains of mice. However, LDH-X from mice and rats are
different. The amino acid sequences of rat and mouse LDH-X have
been partially analyzed and there is approximately 10% difference
in the amino acid sequence (4). The rat and mouse LDH-X also
differ with respect to their antigenic properties. Although the
major portion of antibodies to rat and mouse LDH-X are cross
reactive, there is a fraction of the antibodies to rat LDH-X
which do not ract with the mouse LDH-X (5). The specific fraction
of anti-rat LDH-X can be isolated and used to detect, by immuno-
fluorescence, an altered antigenic site in the mouse LDH-X that
resembles the antigenic site of rat LDH-X. Theoretically, a mouse
sperm that binds the rat specific antibody contains mutant LDH-X;
therefore, such sperm are here defined as a presumptive mutant.

Localization of LDH-X

The various components of the seminal fluid can adhere to sperm
in the course of spermiogenesis. The possiblity exists, therefore,
that enzymes present in the fluid might bind to the surface of
the sperm during development. If this were the case for LDH-X,
then the enzyme localized to any given sperm would represent a
pool of molecules produced by many sperm cells and not the particular germ cell in which it was located. This problem would, of
course, prevent the use of LDH-X as a mutation system. A series
of preliminary experiments were performed to determine if LDH-X
could be discretely localized within single sperm.

(i) Sperm were stained with fluorescent antibodies with a double label technique. The sperm were first treated with rabbit anti LDH-X and then with fluorescein isothiocyanate (FITC)-conjugated goat anti-rabbit immunoglobin. The first attempt to localize LDH-X indicated that it covered the total sperm surface (Fig 1). Biochemical studies previously indicated that LDH-X was compartmentalized in the mitochondria of mature sperm and immunological staining of developing spermatids showed the enzyme also was located in the cytoplasm (6,7). If LDH-X were an important enzyme in energy metabolism in sperm, it is puzzling that the enzyme should be found in the cytoplasm of spermatids but limited to the mitochondria of mature sperm and seen adhering to the surface of sperm. In later experiments, when the antibodies were absorbed on Sepharose before staining, only the tail of the sperm was stained (Fig 2). This indicated that the serum we produced against LDH-X was contaminated with antibodies against polysaccharides and glycoproteins that were present on the surface of the sperm.

(ii) The localization of LDH-X in the sperm tail, the major cytoplasmic compartment, agreed with previous demonstrations of LDH-X in the cytoplasm of spermatids, but the question remained as to mitochondrial LDH-X. In subsequent experiments, the sperm were treated with Triton X-100 to remove the outer cell membrane. After this treatment the mitochondria stained heavily with antibody to LDH-X. The mitochondrial sheath can be removed by treatment with dithiothreitol (DTT). Immunofluorescent staining of sperm after DTT treatment revealed that LDH-X was located in the tail sheath but not the axonemal fibers; in addition, there was

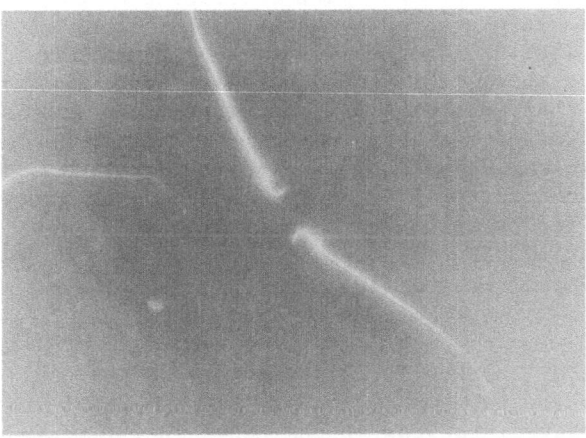

Figure 1. Rat sperm labeled with rabbit anti-LDHX and FITC conjugated goat antirabbit immunoglobin before absorption of the antibodies on Sepharose.

Figure 2.   Rat sperm labeled with double antibody technique after absorption of the antibodies on Sepharose.

fluorescence in the mitochondria. These experiments agreed with biochemical evidence for intramitochondrial LDH-X and confirmed that LDH-X was an integral part of sperm and not simply bound to the surface during development.

(iii) The questions of cross contamination between sperm and detection of rare fluorescent sperm were also important. When mouse sperm were mixed with rat seminal fluid, incubated for several hours, and then stained with monospecific antibodies to rat LDH-X, the mouse sperm did not stain, indicating that the rat LDH-X in the seminal fluid did not adhere to the mouse sperm. Therefore, single sperm cells could be stained for the LDH-X which was produced by that cell or its predecessors. Additionally, a few rat sperm mixed with a large number of mouse sperm could be specifically labeled for rat LDH-X (Fig 3). Model detection experiments with mixtures of rat and mouse sperm stained with antibodies specific for the rat antigenic sites gave a linear relationship of expected and observed frequencies down to the lowest ratio of rat to mouse sperm tested (10^{-5}).

Application of LDH-X system to Detect Mutations

The presumptive mutations in the LDH-X in the mouse sperm were detected by reaction with the monospecific antibodies to rat LDH-X. Theoretically, we can consider that in such a sperm a mutation has occurred which resulted in an exchange of an amino acid normally present in mouse LDH-X with one normally present in rat LDH-X with the simultaneous creation of an antigenic site

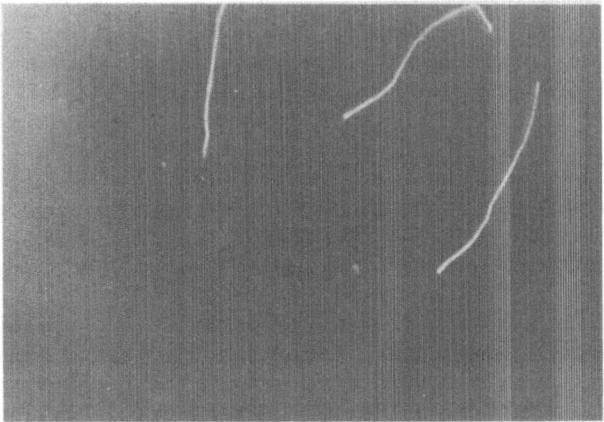

Figure 3.   Specific fluorescence of rat sperm in a mixture with large numbers of mouse sperm.

specific for rat LDH-X. Antigenic difference due to a single amino acid difference has been described for hemoglobin (8). The amino acid differences between variants of the same protein is often caused by a single base-pair substitution, which would then be the type of mutation expected to be detected in the LDH system.

The production of LDH-X is initiated during pachytene spermatogenesis and the gene probably does not have significant haploid expression. In general, expression of a newly induced mutation requires DNA replication. The induced mutant sperm should be detected among sperm which were treated with mutagens during the spermatogonid stage. After treatment with a chemical mutagen the animal may go through a sterile period due to the high sensitivity of early spermatids and differentiating spermatogonia to the drug. In our experiments with the LDH-X system, the analyzed sperm are treated as stem cell spermatogonia (10-15 weeks after treatment).

Detection of gene mutations in male germinal tissue of whole animals has mainly been accomplished using the visible specific locus mutation system (9) and the biochemical specific locus mutation system (10,11). Only a few compounds have been tested and four have been shown to induce mutations in spermatogonial stem cells (procarbazine, mitomycin C, triethylenemelamine, and n-ethyl-n-nitrosourea). The first agent we selected for testing in the LDH-X system was procarbazine, a monofunctional alkylating agent that is commonly used as an antineoplastic drug (2). Ehling (13), utilizing a mouse model, measured the frequency of induction of visible specific locus mutations by procarbazine and found that the

frequency increased linearly with doses up to 600 mg/Kg and decreased significantly at a dose of 800 mg/Kg. In the LDH-X mouse
sperm system, the frequency of presumptive mutant sperm increased
linearly with increasing dose up to 800 mg/Kg (Table 1). For a
mutation to be detected in the visible specific locus mutation
system, the sperm carrying the mutation must be capable of fertilization, and the offspring must survive. In the LDH-X system
sperm do not have to be functionally active to be recognized as
mutants. This may account for the difference in the dose response
of the two systems.

It is interesting to consider the mutation frequencies for the
systems. The spontaneous mutation frequency in the visible specific
locus system is 7.5×10^{-6} per locus and in the LDH-X system it is
0.43×10^{-6}, approximately a 17-fold difference. The frequency
of visible mutations after treatment with 600 mg/Kg procarbazine
is 50×10^{-6} per locus, and in the LDH-X system the frequency of
presumptive mutant sperm is 5.5×10^{-6}, approximately a 9-fold
difference. The type of mutations detected in the visible specific
locus mutation system should include deletions of the gene as well
as basepair substitutions, frameshifts, etc. In the LDH-X system
the mutations would most likely be basepair substitutions. If a
broader spectrum of mutations are detected in the specific locus
mutation system than in the LDH-X system, we might expect the LDH-X
system to have a considerably lower mutation rate; therefore, a
9 to 17-fold difference in mutation rates seems reasonable.

In the LDH-X sperm mutation system, mutations result in the
exchange of one amino acid for another specific amino acid because
each mutation is probably restricted to a change in one nucleotide

Table 1. Mutation data from procarbazine-treated
         DBA/2J mice

Procarbazine Dose (mg/Kg)	Total Sperm Screened	Fluorescent Sperm Found	Frequency of Fluorescent Sperm
0	167.0×10^6	72	0.43×10^{-6}
200	57.6×10^6	112	1.94×10^{-6}
400	32.5×10^6	119	3.66×10^{-6}
600	4.2×10^6	23	5.47×10^{-6}
800	2.5×10^6	19	7.60×10^{-6}

of the codon. One subunit of LDH-X contains approximately 330 amino acid residues and there are about 30 amino acid sequence differences between mouse and rat LDH-X (4). If there are 10 antigenic sites (we do not know the precise number) and the spontaneous mutation rate in the LDH-X system is 4.3×10^{-7}, this would correspond to a spontaneous mutation rate of 4.3×10^{-8} per nucleotide. Stamatoyannopoulos and Nute (14) have estimated the mutation rate per nucleotide per generation for β-hemoglobin in humans to be 7.4×10^{-9} and the rate derived using de novo α-hemoglobin mutants is 10×10^{-9}. Kimura and Ohta (15) have calculated the mutation rate per codon per generation for abnormal hemoglobin to be 9×10^{-8}. This may correspond to a nucleotide rate of 3.5×10^{-8}. Even though these are only estimates, there seems to be reasonably good agreement between the induced and spontaeous frequency in the LDH-X system and other mutation systems.

The second mutagen tested in the LDH-X mutation system was n-ethyl-n-nitrosourea (ENU) which was a potent mutagen in the visible specific locus mutation system (16). In our experiments, data obtained in the LDH-X system and in the biochemical specific locus mutation system (11) were directly related: the same animals were used in both tests. The LDH-X mouse sperm test was done almost a year after the initial treatment with ENU. The data for the mouse LDH-X system in Table 2 indicated an increase in the spontaneous mutation rate from 0.43×10^{-6} to 2.8×10^{-6}. We do not yet know if this represented an increase in the spontaneous rate or an

Table 2. Mutation data from ENU-treated mice

Mutation System	Mutation Frequencies Per 10^6	
	(Untreated)	(ENU)
LDH-X 15 weeks[a]	.42	---
50+ weeks[a]	2.9	54.3
ELECTROPHORESIS	0/300,000	420[b]
VISIBLE	7.5	530[c]

[a]Age of mice at time of sampling.

[b]Johnson and Lewis (11).

[c]Russell et al. (16).

unspecific reaction of the antibody in the older animals. Therefore, if we consider the spontaneous rate in 15 week-old animals and compare it with the induced rate after ENU treatment, there is a 100-fold increase. The increase above the spontaneous level is of the same order of magnitude of the exact same ENU treatment in the visible as well as in the biochemical specific locus test.

FUTURE COURSE OF POINT MUTATION SYSTEM IN SPERM

The LDH-X system detects sperm which bind heterologous antibody and the frequency of such sperm increases after mutagenic treatment. These sperm we presume to be mutants, but the validity of the system requires that the origin of the variant cell shown be an alteration in the LDH-X protein that has resulted from a change in the LDH-X gene. We are taking several approaches in seeking answers to this problem: (a) by use of a cell sorter, collecting enough variant sperm or spermatids for biochemical analysis of either the protein or the DNA. In the latter case we will use a normal probe and isolate the "mutant DNA" for further analysis. (b) Develop monoclonal antibodies which do not react with mutant sperm to enhance the frequency of mutant sperm in a sample. This enriched sperm sample would be used for in vivo or in vitro fertilization. (3) Induce LDH-X production in mouse somatic cell cultures, and by use of monospecific antibodies to rat LDH-X, isolate mutant clones which react with this antibody. Sufficient amounts of LDH-X can then be produced from these cell cultures for biochemical analyses.

The experience in measuring mutations that may be induced by chronic low level exposure is very limited. When these studies have been done, the systems have not given the expected accumulative effect of the chronic exposure. One exception is the visible specific locus mutation system which has shown an accumulative effect of prolonged exposure to ionizing radiation (17). Because many exposures of the human population occur over a prolonged period of time, we should consider methods that are applicable to measure genetic damage after this type of exposure. We have not shown that the LDH-X system is capable of detecting accumulative damage from low-dose chronic exposure, but the data in Table 2 suggests this may be a possiblity. The mice were analyzed one year after treatment and the mutation frequency was sufficiently high to believe that many LDH-X variants were capable of surviving selection. In our low level exposure studies the sensitivity of the system may become much more critical. Therefore, we are also in the process of developing additional markers for use in the sperm point mutation system. If a sperm system is to be applicable to detecting exposure to many types of chemical agents, then techniques are also needed that will pick up events such as deletions and frameshift mutations.

The LDH-X system might be adapted to humans following the same rationale as used for its development in the mouse. Human sperm react with the purified antimouse LDH-X such that only the tail is stained (Fig 4). It is perhaps true that there are some anti-mouse LDH-X antibodies that do not react with human sperm. If so, these antibodies, or non-reacting antibodies from other natural variations in LDH-X, might be used to measure the frequency of alteration of human LDH-X in single sperm.

Because the frequency of variant cells in the LDH-X system or other similar systems is low (10^{-4} to 10^{-7}), manual scoring of these events on microscope slides is a slow process. If this technology is to be routinely applied, automatic counting systems would be advantageous. This work is being explored in collaboration with Drs. Bigbee and Branscombe at the Lawrence Livermore Laboratory. The first experiments have been done with artificial mixtures of rat and mouse sperm and give sorting enrichment of 100-fold.

If a human sperm point mutation system can be developed there are several approaches to correlation with transmissible germinal mutation data that might be useful. One possibility we are pursuing is to identify several potential mutagens for which there are models of acute exposure in the human population such as chemotherapy patients, e.g., procarbazine. Laboratory animals can then be treated with similar doses and the same animals used in both the various germinal mutation systems and the sperm system. At the same time, the frequency of presumptive mutant sperm in the exposed humans can be compared with a control group. Then, with due consideration of the pharmacological differences between mice and humans the mutation rates obtained in the various sytems can be analyzed.

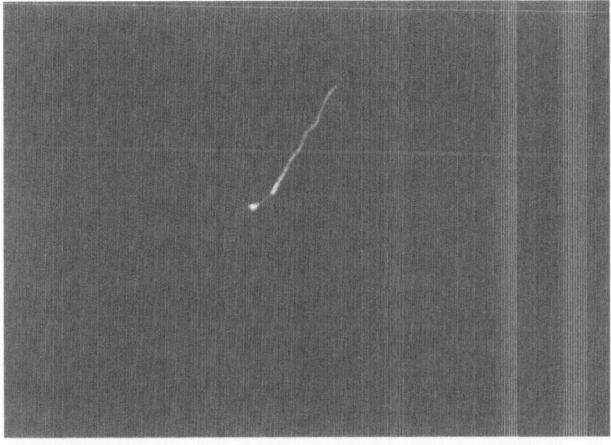

Figure 4. Human sperm labeled with purified mouse anti-LDHX.

The biochemical specific locus mutation system in laboratory
animals relates to the use of biochemical markers in the human
population. The markers used are the same in both system, so
mutation frequencies in humans and the laboratory animals can be
compared directly. Typically, human mutation monitoring has
required a population size that is too large to make it practical
to survey small subpopulations exposed to chemical mutagens. Neel
and his collarborators are working to improve their mutation
system by increasing the number of markers and the sensitivity of
biochemical tests, both of which will significantly reduce the
previously required number of offspring (19,19). The indirect
tests such as the lymphocyte chromosome aberration system, LDH-X
system, etc., might be validated by analyzing the offspring of
the groups which are tested with the indirect tests for induced
transmissible mutations. This direct comparison should continue
until two very important concerns have been clarified: (i) are
there groups of people who have higher transmissible mutation
rates when compared with matched controls?, and (ii) which of the
indirect tests or combination of tests is the best predictor for
damage?

Germinal mutation tests in laboratory animals are too costly
and time consuming to be used for all present and future potentially-
mutagenic agents. Only a few compounds have been tested in these
system. With a system such as the LDH-X system, the number of
chemicals able to be tested could be dramatically increased with
minimal time and cost investment. Then, when active compounds are
identified, these could be tested in the germinal mutation systems
to verify genetic alteration and to better understand the mechanisms
and consequences of those mutations. Therefore, the development of
a point mutation system in single germ cells is an important effort
that would allow us to screen more quickly the spectrum of chemicals
capable of inducing mutations in male germinal tissue. With this
information we can responsibly act to limit exposure of human
populations to those agents.

REFERENCES

1. G. H. Strauss and R. J. Albertini, Enumeration of 6-thioguanine-
    resistant peripheral blood lymphocytes in man as a potential
    test for somatic cell mutations arising in vivo, Mutat. Res.
    61:353-379 (1979).
2. G. Stamatoyannopoulos, Possibilities for demonstrating point
    mutations in somatic cells, as illustrated by studies of
    mutant hemoglobins, in: "Expert Conference on Genetic Damage
    Caused by Environmental Agents", K. Berg, ed., Academic
    Press, N.Y., 49-62 (1979).
3. E. Goldberg, Isozymes in testis and spermatozoa, in:

M. C. Rattazzi, et al., eds., Alan R. Liss, New York, Vol. 1, pp. 79-124 (1977).

4. Y.-C.E. Pan, M. Okabe, J. F. Sharie, and S.-L. Li, The amino acid sequence and immunological properties of testicular lactate dehydrogenase isozymes from rat and mouse, Fed. Proc. 40:1886, Abstract, (1981).

5. A. A. Ansari, M. A. Baig, and H. V. Malling, In vivo germinal mutation detection with "monospecific" antibody against lactate dehydrogenase-X, Proc. Natl. Acad. Sci. USA 7:7352-7356 (1980).

6. B. T. Storey and F. J. Kayne, Energy metabolism of spermatozoa, VI, Direct intramitochrondrial lactate oxidation by rabbit sperm mitochondria, Biol. Reprod. 16:549-556 (1977).

7. M. Hinz and E. Goldberg, Immunohistochemical localization of LDH-X during spermatogenesis in mouse testes, Dev. Biol. 57: 375-379 (1977).

8. Th. Papayannopoulou, T. C. McGuire, G. Lim, E. Garzel, P. E. Nute, and G. Stamatoyannopoulos, Identification of HbS in red cells and normoblasts using fluorescent anti-HbS antibodies, Brit. J. Haematol. 34:25-31 (1976).

9. W. L. Russell, X-ray induced mutations in mice, Cold Spring Harbor Symposium 16:327-366 (1951).

10. L. R. Valcovic and H. V. Malling, An approach to measuring germinal mutations in the mouse, Environ. Health Perspectives 6:201-205 (1973).

11. F. M. Johnson and S. E. Lewis, Electrophoretically detected germinal mutations induced by ethylnitrosourea in the mouse, Proc. Natl. Acad. Sci. USA,(in press) (1981).

12. U. W. Dold and H. Sack, Praktische Tumortherapie, Thieme, Stuttgart (1976).

13. U. H. Ehling, Induction of gene mutations in germ cells of the mouse, Arch. Toxicol. 46:123-138 (1980).

14. G. Stamatoyannopoulos and P. E. Nute, De novo mutations producing unstable hemoglobin or hemoglobin M, II. Direct estimation of minimum nucleotide mutation rates in man, Human Genet. (in press) (1981).

15. M. Kimura and T. Ohta, The age of a neutral mutant persisting in a finite population, Genetics 75:199-212 (1973).

16. W. L. Russell, E. M. Kelley, P. R. Hunsicker, J. W. Bangham, S. C. Maddux, and E. L. Phipps, Specific locus test shows ethylnitrosourea to be the most potent mutagen in the mouse, Proc. Natl. Acad. Sci. USA 76:5818-5819 (1979).

17. W. L. Russell, The effect of radiation dose and fractionation on mutation in mice, in: "Repair of Genetic Radiation Damage", F. H. Sobels, ed., Pergamon Press, Oxford, 205-217 (1963).

18. J. V. Neel, H. W. Mohrenweiser, and M. H. Meisler, Rate of spontaneous mutations at human loci encoding protein structure, Proc. Natl. Acad. Sci. USA 77:60376041 (1980).

19. J. V. Neel, In quest of better ways to study human mutation
     rates, in: "Human Mutations: Population and Biological
     Aspects", E. B. Hook, ed., Academic Press, N.Y. (in press)
     (1981).

DISCUSSION

Q. McCoy (New York Medical College): Is this enzyme coded for by the
mitochondrial genome?

A. Malling (NIEHS): The enzymes are clearly nuclear as shown by cell
hybridization studies.

SESSION IX: EXTRAPOLATION OF GENOTOXIC DATA TO HUMAN HEALTH EFFECTS

V. P. Bond, Moderator

Associate Director for Life Sciences and Safety
Brookhaven National Laboratory
Upton, NY 11973 (U.S.A.)

The detection of genotoxic activity in any agent, whether by
"short-term" bioassays or by chronic tumor studies in rodents, is
only the first step in the regulation of hazardous agents. The next
step, and one far more difficult, is the extrapolation of these experimentally derived data to a quantitative estimate of human health
risk. Experimental data are generally obtained at "unnaturally" high
exposure levels under "unreal" conditions. Consequently, these data
must be extrapolated to low exposure levels where a threshold for
effects may exist, between species certainly differing in innate cellular sensitivities and in metabolic profiles, and to multiple exposure situations where potential synergistic/antagonistic interactions
may be commonplace.

In this session, three approaches for extrapolating data from
high exposure to low exposure situations and from one species to
another will be presented. David Hoel will discuss mathematical
models for extrapolating laboratory data to human health effects.
William Lee will present a molecular dosimetric approach based on
the alkylation of DNA in germinal tissues for extrapolating mutagenic
effects between species. Finally, V. P. Bond will discuss the applicability of experimental data and models derived from radiation studies to interpret and estimate genotoxic effects resulting from low
chemical exposure levels.

EXTRAPOLATION OF LABORATORY DATA TO HUMAN HEALTH EFFECTS

David Hoel

NIEHS
Research Triangle Park, NC   27709   (U.S.A.)

Experimental dose-response data has traditionally been fit by various mathematical models in order to estimate parameters such as an LD_{50}. Although statisticians have debated the appropriateness of the mathematical functions, the resultant estimates have not depended to any great extent upon the choice of function, as long as responses are estimated within the experimental region. However, human risk estimation invariably involves estimating low-dose effects well below the normal experimental region. This extrapolation process then becomes a scientifically fragile activity. Needless to say, the choice of mathematical model is crucial. Unfortunately, the choice is often made without sufficient biological understanding of the mechanisms involved in the response being estimated.

To illustrate the issue of low dose extrapolation consider the dose-response data of Bliss as described by Prentice (1). The response consisted of the killing of flour beetles with carbon disulfide. For the six doses employed, with 60 beetles used per dose, the response ranged from 5% to 95%. Bliss fit the data with both a probit and a logit function. The fits were adequate and essentially equal so that neither function was considered to be statistically preferable. Since the work of Bliss, Prentice has constructed a generalized function which included as special cases the two functions used by Bliss and in doing so obtained a superior statistical fit to the data. The resultant effects on the estimated LD_{50} were minimal. If, however, one extrapolates the three fitted functions to the low-dose region, considerable differences are seen. For example, at a particular low dose level, the estimated response rates are: probit = 2×10^{-13}, logit = 3×10^{-6}, Prentice's generalization = 7×10^{-4}. From these values we see how an arbitrary

model choice can produce highly variable extrapolated results.
Furthermore, the first two functions, which Bliss used, were later
improved upon by Prentice. The resulting estimated low-dose
response was not even between the two extremely different estimates
obtained from the original two models. Clearly, the model choice
must necessarily depend upon the biological mechanisms and not the
adequacy of the statistical fit. For additional details of this
example see Prentice (1) and Hoel (2).

The models used by statisticians for extrapolation purposes
have varying degrees of biological basis associated with them.
For example, I would classify the models used in the above flour
beetle example as tolerance distributions or threshold models
because it is assumed that each individual has a threshold of
response and these individual thresholds have a given distribution
in the population. Of course, we usually do not know what a reasonable distribution of thresholds in the population is and thus
are not really in a position to select a tolerance distribution
in a non-arbitrary manner.

In the next order of complexity we find the multi-hit and multi-
stage models. These models attempt to describe an affect as being
the result of a series of hits or mutations on a cell. The multi-
stage model for cancer developed by Armitage and Doll has received
the most attention (see 3). This model is consistent with most
epidemiology and bioassay data in spite of its simplistic description of the carcinogenesis process. The most important mathematical
features of this particular model are that the effects of duration
of exposure and the exposure rate are factored into a product of
two mathematical functions. In addition, the age-specific incidence
rate is most often an integer three to six power of age or duration
of exposure. This is precisely what is found for various cancer
types and sites in human data.

Recently, Whittemore (4) and Day and Brown (5) have looked
at the interesting question of which stage in the multi-stage
model is affected by a carcinogen. If an early stage is affected,
then the process changes very little upon removal of the carcinogen.
If, on the other hand, a late stage is affected, then withdrawal
of the exposure has immediate effects, as one would suspect. These
issues are of great importance when risk estimates are based upon
less than lifetime exposure data or when estimates for less than
lifetime exposure are based upon lifetime exposures. The numerical
effects of extrapolating over time can be considerable because
a multi-stage model implies a risk estimate with a power of three
to six on the duration of exposure.

Another mechanistic approach is to use the multi-hit model as
developed by Van Ryzin for the Food Safety Council (6). He estimates the number of hits directly from the data and applied the

model to toxic endpoints other than cancer. The multi-hit and multi-stage models were applied to a collection of nine different carcinogenesis dose-response data sets (NTA, aflatoxin B , DMN, vinyl chloride, BCME, saccharin, ethylenethiourea, dieldrin and DDT). The estimated number of hits ranged from 28 with NTA, to 1/2 for vinyl chloride. These extremes produce quite different extrapolations. In fact, the 1/2 hit for vinyl chloride is the first suggestion that linearity is not conservative and would underestimate the risk.

The comparison of the multi-hit and multi-stage models for the nine carcinogens in the Food Safety Council list of compounds yielded quite different low-dose estimates. For example, the "virtually safe dose" for NTA was three orders of magnitude higher using the multi-hit model than the multi-stage model, while for vinyl chloride the multi-hit model gave a dose six orders of magnitude lower than did the multi-stage model. All of this points to the critical importance of the model choice.

The most scientifically credible modelling approach involves the use of pharmacokinetics. Possibly the best example is vinyl chloride, where simple Michaelis-Menten kinetics for the activation of the compound completely describes the observed plateauing of the dose-response function at about a 20% tumor incidence. Once the administered dose is transformed by the simple kinetics, the tumor incidence becomes linear with dose (for details see 7,8). This transformation of dose removes the question of a less than linear response which was suggested by the multi-hit model. More importantly, it also indicates the problems one may encounter by naively applying an arbitrary mathematical function to biological data for extrapolation purposes.

After the administered dose is transformed into an effective dose through pharmacokinetics, the choice of the extrapolation function is still required. The vinyl chloride example suggested that linearity was consistent with the data. Other functions such as the probit which Gehring used in his original work could also be applied. As one might expect, the model choice still greatly impacts the risk estimate. For example, estimated dose rates for the probit and multi-stage differed by three orders of magnitude at a risk response level of 10^{-7} (see 8).

The accuracy of the pharmacokinetic measurements is an important aspect of the risk estimation process. Using Gehring's probit extrapolation after adjusting the dose can change the risk estimate by two orders of magnitude with less than a single order of magnitude change in the estimated Michaelis constant. The dependency on the Michaelis constant was observed to be much less when the multi-stage model was used in place of the probit function. Also numerically, the estimated dose associated with a risk of 10^{-7}

increased (using a probit function) from 0.5 (μg/liter) to 9700
(μg/liter) with the incorporation of the kinetics constant into
the model. The mulit-stage model's estimated dose decreased from
160 to 6 with the inclusion of the kinetics. This example clearly
illustrates the sensitivity possessed by the kinetic parameters
and the resultant numerical impact that the inclusion of kinetics
can make. Hopefully, the response for activated initiators will be
a simple function of effective dose, such as linear, after repair
has been considered. Although any real justification is lacking,
this approach has greater intuitive appeal than the more complicated
mathematical functions.

The incorporation of spontaneous or background rate into the
model is a statistical area which has received little attention.
The most likely reason is that the numerical implication of the
manner in which the background rate is treated has not been fully
developed. Two basic approaches to this problem have been used.
First, one may assume that the background is a mechanistically
independent process, in which case Abbott's correction is applied.
The second approach is to assume that the administered dose is
simply additive to the background. This latter approach has been
shown to generally produce a linear dose-reponse relationship at
low dose levels (9). Thus, linear versus non-linear responses at
low doses can create great numerical differences in risk estimates
and once again we stress that the choice of method is important.
Unfortunately, neither total independence nor total additivity is
probably true. However, if a small portion of the background is
additive, then the numerical effect is to consider the entire background as essentially additive. This observation is important
since it generally presses for linearity at low-dose levels (see
2,9).

Finally, let us consider the most difficult issue, that of
species extrapolation. It is assumed qualitatively that the rodent
is a good predictor of similar effects in man with regard to most
forms of toxicity. For carcinogenesis, only arsenic and possibly
benzene are human, but not rodent carcinogens. However, the list
of known human carcinogens is fairly short, consisting of about
18 chemicals and processes. The quantitative information is even
more sparse. Ideally, we would like to be able to predict the
potency of an agent in humans from the potency of that same compound
in rodents. This prediction has been done successfully for six
compounds where sufficient human data exists (10,11). Based
upon these data and associated pharmacological opinions, we may
extrapolate directly on either an equivalent daily dose per body
weight, or per surface area. Much additional work is needed in
this area, and should most definetely include comparative pharmacological considerations. Potentially, errors much greater than
those associated with high dose to low dose estimation are possible.
Some type of estimated precision associated with species scale-up

is needed to avoid false confidence in the quality and scientific soundness of risk estimation.

REFERENCES

1. R. L. Prentice, A generalization of the probit and logit methods for dose response curves, Biometrics 32:761-768 (1976).
2. D. G. Hoel, Incorporation of background in dose-response models, Fed. Proc. 39:73-75 (1980).
3. R. Peto, Epidemiology, multi-stage models, and short-term muta-genicity tests, in: "Origins of Human Cancer", . Hiatt, J. D. Watson, and J. A. Winsten, eds., Cold Spring Harbor Laboratory, Cold Spring Harbor, N.Y., 1403-1428 (1977).
4. A. S. Whittemore, The age distribution of human cancers for carcinogenic exposures of varying intensity, Am. J. Epidemiol. 106:418-432 (1977).
5. N. E. Day and C. C. Brown, Multistage models and primary prevention of cancer, J. Natl. Cancer Inst. 64:977-989 (1980).
6. Food Safety Council, Proposed system for food safety assessment. Food Cosmet. Toxicol. 16 (Supplement 2):109-136 (1978).
7. P. J. Gehring and G. E. Blau, Mechanisms of carcinogenesis: Dose response, J. Environ. Pathol. Toxicol. 1:163-179 (1977).
8. M. W. Anderson, D. G. Hoel, N. L. Kaplan, A general scheme for the incorporation of pharmacokinetics in low-dose risk estimation for chemical carcinogenesis: example - vinyl chloride, Toxicol. Appl. Pharmacol. 55:154-161 (1980).
9. K. S. Crump, D. G. Hoel, C. H. Langley and R. Peto, Fundamental carcinogenic processes and their impli-cations for low dose risk assessment, Cancer Res. 36: 2973-2979 (1976).
10. M. Meselson, In: National Academy of Sciences, National Research Council Report, Pest Control: An assessment of present and alternative technologies, Vol. 1, "Contemporary Pest Control Practices and Prospects: The Report of the Executive Committee", Washington, D. C. (1975).
11. D. G. Hoel, Low-dose and species-to-species extrapolation for chemically induced carcinogenesis, in: "Banbury Report 1: Assessing Chemical Mutagens: The Risk to Humans, U. K. McElheny and S. Abrahamson, eds., Cold Spring Harbor Laboratory, Cold Spring Harbor, N.Y., 135-145 (1979).

DISCUSSION

Q. Bond (BNL): You made the assumption that cancer was single-cell-in-origin. While I'm not disagreeing with you, I'd like to know just how firm you feel that assumption is.

A. Hoel (NIEHS): I must admit I lean on my more biologically-
oriented colleagues for that assumption and given a choice I would
make that type of assumption. The real point is that some assumptions
have to be made, as weak as they may be, and then a model prepared,
making sure that these assumptions are explicitely known. This is
a preferable approach to one where mathematical function are used
that are of nice statistical convenience but without any biological
relevance.

MOLECULAR DOSIMETRY AS A BRIDGE BETWEEN MAMMALIAN

AND NON-MAMMALIAN TEST SYSTEMS

William R. Lee

Louisiana State University
Baton Rouge, LA 70803 (U.S.A.)

There are two major differences between the estimation of
human health effects of a mutagen and of a carcinogen. First, with
a mutagen, the target tissue is known; it is only the dose to the
germ cells that contributes to the mutagenic risk to future generations. In contrast, any somatic tissue may be the relevant target
for a carcinogen. The knowledge that the germ cells are the target
tissue for mutagenesis is the major advantage that the geneticist
has in estimating mutagenic risk.

Second, in contrast to the epidemiological data from carcinogens, we do not have suitable epidemiological data to define the
health risk from mutagens. Even for the best known mutagen, ionizing
radiation, the large exposure of a human population in Hiroshima
and Nagasaki has not yielded a significant genetic effect that
will enable us to estimate the human health risk of ionizing radiation (1). The reason for this inability is the lack of sensitivity
of the epidemiological approach when the time between exposure and
and measurement of the effect is separated by at least one or more
generations and when the potential effects on health are as broad
as the health effects of spontaneously arising mutations. In fact,
any disease syndrome could be affected by the heredity of the patient
and therefore be subject to the detrimental effects.of deleterious
mutations.

In the absence of epidemiological data the estimation of health
risks due to mutagenicity must be based on extrapolation of laboratory data to man. Our confidence in extrapolating from laboratory
experimental test systems to man can be estimated by how well we can

527

extrapolate from one laboratory species to another. Within a number
of currently used test systems a relation is normally established
between the level of exposure to the mutagen and the end product--
mutation frequency--with no estimate of the amount of mutagen
reaching the genetic target. Comparisons among species based on
these data would be confounded by differences in the metabolism of
the mutagen coupled with differences in the genetic response to the
mutagen once it reaches the germ line. Since the genetic mechanism
has been conserved in evolution whereas physiological adaptations
vary substantially, extrapolation among species would be more
reliable if the genetic effects were identified free from
physiological differences among species. In other words, the
exposure-to-germ line dose relation should be separated from
the germ line dose to mutation frequency relation. We should
establish the physiological relation of exposure to germ line dose
in one set of experiments with mammals and define the relation of
germ line dose to mutation frequency in a different set of experiments using a well defined genetic test system. This requires a
clear distinction between exposure of the whole organism to an
agent and the dose that actually reaches the target tissue--the
germ cells.

To achieve an estimate of dose more relevant than simple
exposure of the organism, the extent of reaction of the mutagen
with a molecule in the germinal tissue can be measured. This
molecule, which has been referred to as the selected target molecule (STM) (2), need not be the genetically significant target (GST),
but ideally should react with the mutagen to form a stable selected
reaction product (SRP) within the nucleus of the germ cells. The
accumulated selected reaction product is a measure of the integral
over time of the reaction rate and should be proportional to the
formation of products of reaction at the genetically significant
target. The theory of the selected target molecule approach to
dosimetry was presented in a symposium in 1975 (3-5).

DNA is the obvious choice for the selected target molecule
with adducts to DNA being the selected reaction products. Not
only is DNA likely to contain the genetically significant target,
but also it provides a vehicle for measuring adduct formation in
the nucleus of the germ cell (6). Many sites within the DNA
molecule have been shown to be subject to alkylation. In early
work it was found that agents that alkylate predominantly by the
SN-2 mechanism alkylate principally ring nitrogens, especially the
N-7 of guanine (7-9). More recently it has been shown (10-14)
that it is more likely the less frequent alkylation of the ring
oxygen, possibly the O-6 of guanine, that is responsible for mutations induced by transitions. Agents that alkylate principally by
SN-1 mechanism alkylate more randomly within the DNA molecule and
produce a higher proportion of alkylated ring oxygens and phosphate
groups.

Fortunately, it is not necessary to identify the genetically significant target in order to use adduct formation of DNA as a dosimeter. It has been shown from theoretical considerations (3) that at the low concentrations required to retain cellular viability there should be a constant relationship among the array of alkylated sites within DNA except for steric effects. Furthermore, the distribution among alkylation sites within DNA for any one alkylating agent has been shown to remain constant among tissue from different species (15). Within a single germ cell stage steric effects should be similar. Therefore, for any one mutagen within a single germ cell stage total DNA alkylation should have a constant relationship to each of the alkylated sites within DNA and should serve as the most reliable selected reaction product for dosimetry (3,4,6). It is anticipated that some turnover of the selected reaction product will occur even in cells not undergoing repair processes, as for example, mature spermatozoa. Loss of labeled ethyl groups from stored sperm following alkylation of DNA with ethyl methanesulfonate was found to equal the in vitro hydrolysis rate (16). Therefore, the unstable reaction sites must be identified, and the rate at which they are lost must be used as a correction factor for the loss of total DNA alkylation between the time of formation and that of measuring the level of the SRP.

Adduct formation with a macromolecule in the nucleus of the germ cell is the vehicle by which we determine the dose to the target of interest for estimating the human health effects following exposure to mutagens. This method of molecular dosimetry has been developed for alkylating agents (17-20) and applied in our laboratory to ethyl methanesulfonate (EMS) (6,20), methyl methanesulfonate (MMS) and 1,2-dibromoethane (ethylene dibromide, EDB). The methodology of the molecular dosimetry that has been developed and applied to these alkylating agents can be extended to any agent that either directly or indirectly forms an adduct with DNA, or with slight modification of the techniques to any agent that directly or indirectly forms an adduct with nuclear protein.

Detection of low concentrations of adduct formation is for the present dependent upon labeling the potential adduct with a high specific activity radionuclide. Commercial companies are now offering custom synthesis of labeled compounds in which tritium is placed in a specific stable position by chemical synthesis at specific activities in excess of a Ci/mmol. Tritium labeled compounds produced earlier using the cheaper exchange methods are unsuitable for molecular dosimetry because of both low specific activity and ambiguity in the site of labeling. By combining specifically labeled mutagens at high specific activities with currently available sensitive tritium detection methods, dosage-response curves have been constructed using small quantities of Drosophila melanogaster germ cells (20).

Methods for determining the molecular dose as alkylations per nucleotide (AN) in the germ cells of the mouse have been developed (19). Because of the large amount of tissue available, and by using mutagens labeled at high specific activity, it is possible to measure the level of adduct formation in the germ cells of a mouse following exposure to levels of ethylene dibromide that are similar to or even below the maximum levels to which man is exposed environmentally (21). Therefore, the movement of the mutagen from the environment to the target tissue can be studied at levels relevant to human exposure, and the problems of extrapolating down from high exposure levels are avoided.

Injecting between 0.15 and 5.2 mg/kg of EDB interperitoneally, Sega (21) found that:

$$1 \text{ mg/kg}_{\text{i.p. exposure}} = 1 \times 10^{-7} \text{ AN testicular DNA}$$

There was a linear response between the airborne exposure given to the mice and the resulting testicular DNA alkylation. From inhalation data Sega (21) found that:

$$1 \text{ mg/kg} = 5.9 \text{ ppm-hr}$$

As an example we will assume a human exposure to 100 ppm-hr. An exposure of a mouse to 100 ppm-hr would be expected to yield 1.7 $\times$ 10^{-6} AN in testicular DNA. According to usual toxicology procedures, a factor of 10 is assumed in extrapolating from small mammals to man; therefore, if we apply this correction factor, the estimated human dose to testicular DNA would be 1.7 $\times$ 10^{-5} AN.

In estimating the genetic effects of ionizing radiation, data from large scale experiments with the mouse were used (22,23): However, it does not seem feasible to repeat these large scale experiments for each of the numerous chemicals that have now been shown to be mutagenic. It is hoped that representative classes of chemicals, classified according to the adduct transferred to DNA, will be thoroughly tested in both the mouse and D. melanogaster. It is concordance between these two well-developed test systems for germ cell mutagenesis that will provide confidence in extrapolating to man. Since it will be some time before a large scale comparative study with chemical mutagens can be completed, it is appropriate to review the comparisons with ionizing radiation between Drosophila and the mouse.

To make genetic comparisons among species, we need, in addition to common units of dosimetry, e.g. adduct formation or absorption of radiation, a common unit of mutational response. The unit of mutational response previously used was the locus; mutation frequency per locus was compared among different species, and from the number of loci in man the risk to man was estimated (24).

A claim has been made for the extrapolation among species based on the locus, but with a correction factor for the amount of DNA per nucleus (25). This use of the per locus mutation frequency was severely criticized (26) by showing that there was substantial variance in mutation frequency among loci within species that obscured the relationship among species. By accepting the large variance among loci within species, it is safer to develop risk estimate models based on large blocks of loci rather than individual ones in the Drosophila sex-linked recessive lethal test. There are 600-800 loci which can be studied simultaneously (27).

An alternate to the locus as a common unit for extrapolation of genetic risk to man is the spontaneous mutation frequency. An early empirical observation in mutagenesis was that Metazoa have similar spontaneous mutation frequencies per generation even though their generation times vary considerably. This observation has been strengthened with more recent data, indicating that the spontaneous mutation frequency is approximately 10^{-6} for humans and other animals (28), a value similar to the spontaneous per locus sex-linked recessive lethal frequency in Drosophila (29). By expressing mutagenicity in terms of multiples of the spontaneous mutation frequency, differences among laboratories in the rigidity of the mutation detection criteria are cancelled. Data from Drosophila are consistent with the assumption of proportionality between spontaneous and induced rates of mutation (30). Using the pure ratio of induced to spontaneous mutation frequency, the BEIR (22) and later UNSCEAR (23) reports developed the doubling dose or relative risk method to estimate the health risk to man.

For purposes of comparative mutagenesis it is appropriate to ask how well the results from different test systems compare in estimating the doubling dose of ionizing radiation. We now have a record of over a quarter of a century of regulatory activity based on estimates of the health risk due to the mutagenesis of ionizing radiation. The most recent estimate of the doubling dose based on data from the mouse systems is from 50 - 250 rems (22). If one uses the most extensive low level dosage response curve for oogonia in Drosophila (31) the doubling dose is computed to be 500 rems. If one compares doubling dose estimates derived from mouse and Drosophila data using current methods, the two are within a factor of 2 to 10 of each other.

While differences in estimates of doubling dose for ionizing radiation exist among test systems and presumably there are differences among species, these estimates fall within an order of magnitude of each other. This degree of concordance suggests that any one of these test systems could estimate the results of the other test systems within an order or magnitude and presumably could also estimate the risk to man within one order of magnitude of error.

We have used the <u>Drosophila</u> sex-linked recessive lethal test
to determine the relation between dose, AN, and mutation frequency
with the provision that the results be multiplied by a factor of
ten to cover the error term in extrapolating from <u>Drosophila</u> to man.

Exposure to environmental mutagens normally occurs over a long
period of time. The problem of computing the dose and mutational
response where the dose is given over a period of years can be
viewed as the problem of giving a very low dose during one cell
cycle. Spermatogonial stem cells in man undergo cell division
breaking up the protracted time of years into a series of cell cycles.
Within each cell cycle, if there is a constant probability that a
molecule will induce a mutation, and if there is not selection for
or against mutations once induced, then the total accumulated
frequency of mutations will be proportional to the dose (using the
definition of dose as the integral over time). Therefore, the
linear component to the dose response curve can legitimately be
extrapolated to much higher accumulated lifetime doses to determine
the mutation frequency per generation. It is only this linear component that can be used with lifetime dose to estimate the mutation
frequency per generation in man.

If a particular cell stage is uniquely sensitive to a mutagen,
as is the case for post-meiotic germ cell stages to many chemical
alkylating agents, then the dose acquired by this sensitive stage
must be used to estimate the additional risk per generation and
should be added to the generation total.

Previous work with EDB has shown that the mutation frequency
for a given exposure was low in mature sperm and in late spermatids
but rose to a higher frequency in mid-spermatid, spermatocytes,
and gonial cells (32). From the previously published work it
appears that the frequency observed in the early to mid-spermatid
stage is similar to that of earlier spermatocyte and gonial stages;
therefore, foe EDB any germ cell and late spermatid stage other
than mature sperm may be used for risk estimation. There are technical advantages for studying the spermatid stage rather than the
earlier gonial stage. Mutant clusters occur when an early gonial
cell undergoes replication. Clustering does not change the overall
estimate of risk nor the probability of detecting a mutation, but
it does substantially increase the variance. Therefore, the spermatid stage has been selected for determining a dose-response curve
for EDB and the frequencies found in the spermatid stage could be
applied to the stem cells in the testes. Females have not been studied with EDB. With other mutagens, including both ionizing radiation
and chemicals, females have been shown to be less sensitive per unit
of exposure than have males. If females are less sensitive per dose
to EDB than males, the consequence would be a reduction in the
estimate of mutation frequency in the population by a factor no
greater than two.

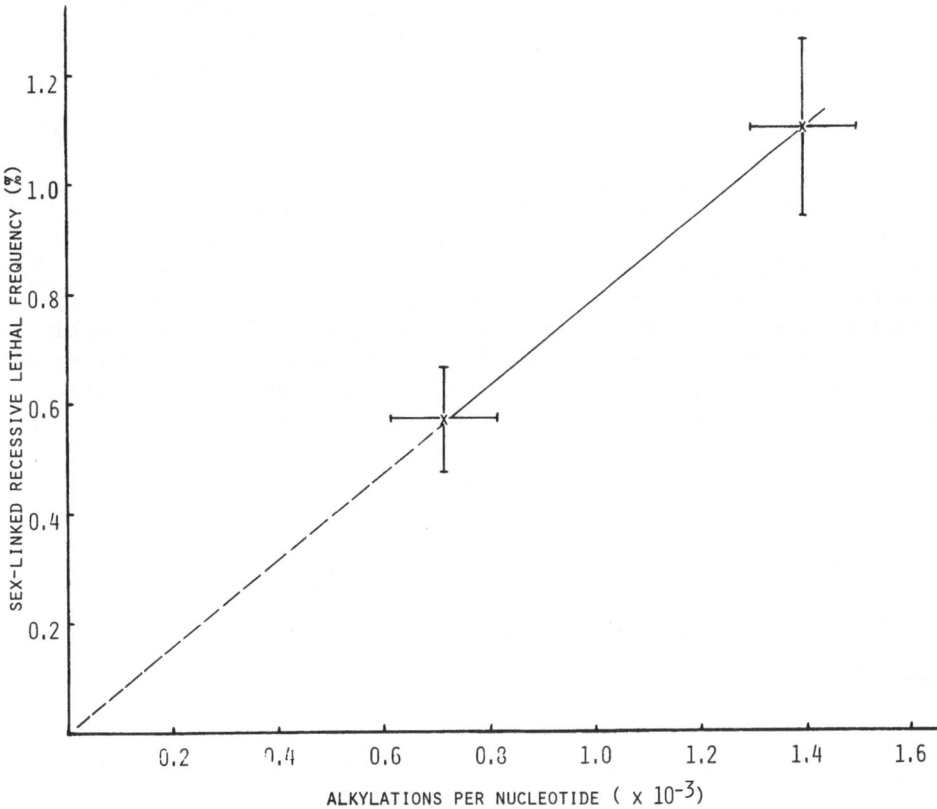

Figure 1. Relation of dose (alkylations per nucleotide) to
genetic response (sex-linked recessive lethal frequency) determined
in Drosophila germ cells that were spermatids at the time of exposure to EDB. Dose error equals twice the mean square error in
analysis of variance of six estimates at the two lowest exposure
levels. Response error equals Poisson standard deviation of number
of lethals observed at each level.

     A dose-response curve has been prepared for EDB using methods
we have previously published (20) with the modification that cells
transferred from the treated male to females during the second
two day mating were used for both dosimetry and genetic analysis;
therefore, the treated cells were in the spermatid stage (Fig 1).
The dose is expressed as alkylations per nucleotide, AN, and the
mutation response as percent of sex-linked recessive lethals. The
dose response curve for EDB, as in the case of our previously published work for EMS treatment of spermatozoa (20) is linear and
extrapolates to the origin. It should be noted that the lowest
dose level induced mutations at about five times the spontaneous

mutation frequency of the stock used and that it is within the
capability of the system to measure the molecular dose to germ cells
and the mutation frequency at a dose that would only double the
spontaneous mutation frequency. Because of the apparent linearity,
the doubling dose for EDB is estimated to be 1.5 x 10 AN, and the
doubling dose to man is estimated to be one-tenth of this value.
Therefore, the human exposure of 100 ppm-hr of EDB estimated to give
a dose of 1.7 x 10 AN in testicular DNA is nearly equivalent to
the estimated doubling dose for man of 1.5 x 10 . Using Table 50
in UNSCEAR (23) and using the caveat of proportionality between
human health effects for spontaneous and induced mutations, it is
estimated that for a human exposure of 100 ppm-hr EDB there would
be an additional 2.5 cases in the first generation and 14.5 oases at
equilibrium of major diseases of genetic origin per thousand live
born. Chromosomal diseases are not considered in this calculation,
for EDB does not induce chromosome aberrations at low exposure
levels. This method includes two adjustments of a factor of 10 each
(for a total factor of 100) in the direction of safety. We feel
this adjustment in the direction of safety is necessary and adequate
to protect the public. It is a matter of economics whether a closer
estimate of risk can justify the considerable increase in cost.

REFERENCES

1. J. V. Neel, H. Kato, and W. J. Schull, Mortality in the children of atomic bomb survivors and controls, Genetics 76:311-
326 (1974).

2. W. R. Lee, Comparison of the mutagenic effects of chemicals and
ionizing radiation using Drosophila melanogaster test systems, in: "Radiation Research, Biomedical, Chemical, and
Physical Perspectives – Proceedings of the 5th International
Congress of Radiation Research", O. F. Nygaard, H. I. Alder,
and W. K. Sinclair, eds., 976-983 (1975).

3. C. S. Aaron, Molecular dosimetry of chemical mutagens. Selection of appropriate target molecules for determining molecular dose to the germ line, Mutat. Res. 38:303-310 (1976).

4. W. R. Lee, Molecular dosimetry of chemical mutagens:
Determination of molecular dose to the germ line, Mutat.
Res. 38:311-316 (1976).

5. G. A. Sega, Molecular dosimetry of chemical mutagens. Measurement of molecular dose and DNA repair in mammalian germ
cells, Mutat. Res. 38:317-326 (1976).

6. W. R. Lee, Dosimetry of chemical mutagens in eukaryote germ
cells, in: "Chemical Mutagens", Vol. 5, A. Hollaender and
F. J. de Serres, eds., 177-202 (1978).

7. P. Brooks and P. D. Lawley, The reaction of mono- and di-
functional alkylating agents with nucleic acids, Biochem.
J. 80:496-503 (1961).

8. P. D. Lawley, Effects of some chemical mutagens and carcinogens on nucleic acids, Prog. Nucleic Acid Res. Mol. Biol. 5:89-131 (1966).

9. E. C. Miller and J. A. Miller, Mechanisms of chemical carcinogenesis: Nature of proximate carcinogens and interactions with macromolecules, Pharmacol. Rev. 18:805-838 (1966).

10. A. Loveless, Possible relevance of 0-6 alkylation of deoxyguanosine to the mutagenicity and carcinogenicity of nitrosamines and nitrosamides, Nature 223:206-207 (1969).

11. P. D. Lawley and C. J. Thatcher, Methylation of DNA in cultured mammalian cells by MNNG: The influence of cellular thiol concentrations on the extent of methylation and 6-oxygen atom and guanine as a site of methylation, Biochem. J. 116: 693-707 (1970).

12. L. Sun and B. Singer, The specificity of different classes of ethylating agents toward various sites of HeLa cell DNA in vitro and in vivo, Biochemistry 14:1795-1802 (1975).

13. B. Singer, All oxygens in nucleic acids react with carcinogenic ethylating agents, Nature 264:333-339 (1976).

14. P. D. Lawley, Some chemical aspects of dose-response relationships in alkylation mutagenesis, Mutat. Res. 23:283-295 (1974).

15. B. Singer, W. Bodeel, J. Cleaver, G. Thomas, M. Rajewsky, and W. Thon, Oxygens in DNA are main targets for ethylnitrosourea in normal and xeroderma pigmentosum fibroblasts and fetal rat brain cells, Nature 276:85-88 (1978).

16. W. R. Lee, P. M. S. Skinner, and F. C. Janca, Hydrolysis of ethylated bases in stored Drosophila sperm after EMS treatment, Environmental Mutagen Society 10th Annual Meeting, New Orleans, Abstract, (1979).

17. W. R. Lee and G. A. Sega, Dosage determination of chemical mutagens of the germ line of Drosophila melanogaster, Newsletter of the Environmental Mutagen Society 3:37 (1970).

18. G. A. Sega, P. A. Gee, and W. R. Lee, Dosimetry of the chemical mutagen ethyl methanesulfonate in spermatozoan DNA from Drosophila melanogaster, Mutat. Res. 16:203-213 (1972).

19. G. A. Sega, R. B. Cumming, and M. F. Walton, Dosimetry studies on the ethylation of mouse sperm DNA after in vivo exposure to ^{3}H-ethyl methanesulfonate, Mutat. Res. 24:317-333 (1974).

20. C. S. Aaron and W. R. Lee, Molecular dosimetry of the mutagen ethyl methanesulfonate in Drosophila melanogaster spermatozoa: Linear relation of DNA alkylation per sperm cell (dose) to sex-linked recessive lethals, Mutat. Res. 49:27-44 (1978).

21. G. A. Sega, On the effects of ethylene dibromide (EDB) in mammalian germ cells, Contractors report from Biology Division of Oak Ridge National Laboratories to EPA (1980).

22. BEIR II Report, Report of the Advisory Committee on the Biological Effects of Ionizing Radiation, National Academy of Sciences, National Research Council, Washington, D.C. (1980).

23. UNSCEAR, Sources and Effects of Ionizing Radiation (1977).

24. BEAR, Report of the Advisory Committee on the Biological Effects
     of Atomic Radiation, National Academy of Sciences, National
     Research Council, Washington, D.C. (1956).

25. S. Abrahamson, M. A. Bender, A. D. Conger, and S. Wolff,
     Uniformity of radiation induced mutation rates among differ-
     ent species, Nature 245:460-462 (1973).

26. A. Schalet and K. Sankaranarayanan, Evaluation and re-evalua-
     tion of genetic radiation hazards in man. I. Inter-specific
     comparison of estimates of mutation rates, Mutat. Res. 35:
     341-370 (1976).

27. S. Abrahamson, F. E. Würgler, C. DeJongh, and H. U. Meyer, How
     many loci on the X-chromosome of Drosophila melanogaster
     can mutate to recessive lethals?, Environ. Mutagen. 2:
     447-453 (1980).

28. L. L. Cavalli-Sforza and W. F. Bodmer, "The Genetics of Human
     Populations", W. H. Freeman, San Francisco, CA (1971).

29. W. R. Lee, S. Abrahamson, R. Valencia, E. S. von Halle,
     F. E. Würgler, and S. Zimmering, The sex-linked recessive
     lethal test for mutagenesis in Drosophila melanogaster,
     U.S. EPA Gen-Tox Report, submitted for publication (1981).

30. P. T. Shukla, K. Sankaranarayanan and F. H. Sobels, Is there
     a proportionality between the spontaneous and the x-ray
     induction rates of mutations?, Mutat. Res. 61: 229-248 (1979).

31. S. Abrahamson, Mutation process at low or high radiation doses,
     "Biological and Environmental Effects of Low-Level Radiation",
     Vol. I, IAEA Publication STI/PUB/409:1-8 (1976).

32. P. G. Kale and J. W. Baum, Sensitivity of Drosophila
     melanogaster to low concentrations of gaseous mutagens.
     II. Chronic exposures, Mutat. Res. 68:59-69 (1979).

DISCUSSION

Q. Baum (BNL): The doubling dose for ethylene dibromide sounds
like a very low number in relationship to its TLV value of, I
think, 20 ppm. Certainly, workers have been exposed to concentrations much higher than 100 ppm-hr in many, many cases. In light
of the correlation between carcinogenicity and mutagenicity and
even though we cannot detect mutations in the offspring of these
workers, we should have observed many tumors, we haven't. Do
you have any comments?

A. Lee (Lousiana): I would question the negative data on the
grounds that most of the epidemiological studies are based on
agricultural workers and I think those populations would be
difficult to follow accurately. If one limits the doubling dose for
genetic risk to the major genetic diseases, as was done for ionizing
radiation in the BEIR and UNSCAR Reports, then the risk for cancer
is higher than the risk for genetic effects. However, if one includes

all genetic effects, then the risk for cancer may be much less than the heritable risk.

Q. Malling (NIEHS): I would like to suggest that newer techniques, such as antibodies against specific adducts in DNA, would give one more resolving power in your studies. One would then be able to include human studies.

A. Lee (Lousiana): Yes, broadening the studies to get direct measurements of alkylation in man by the use of antibodies against adducts of DNA would be most useful. These techniques would also allow us to do some studies at very low dose levels.

IS RADIATION AN APPROPRIATE MODEL FOR CHEMICAL

MUTAGENESIS AND CARCINOGENESIS?

V. P. Bond

Associate Director for Life Sciences and Safety
Brookhaven National Laboratory
Upton, NY   11973   (U.S.A.)

INTRODUCTION

The principal effects of interest with "high-level radiation"
exposure (high doses and dose rates) are early "all-or-none" effects
such as illness and death. These effects are due to the severe
depletion of stem cell populations in the organs resulting from the
high average absorbed organ doses (1). Absorbed organ dose-response
curves for these all-or-none <u>organ effects</u> show the classic S-shape
characteristic of such relationships well known in pharmacology,
toxicology and medicine. The organ dose is the amount of energy per
gram of tissue (100 ergs/gram), averaged usually over the entire
organ.

For <u>single-cell effects,</u> however, e.g. chromosome abnormalties,
somatic cell mutation or transformation, cell death, and heritable
genetic alterations, a very different organ dose response function
is seen (2-4). Thus, if the average <u>organ</u> dose (or the average dose
to the medium for cells in culture) is plotted versus the fraction
of <u>cells</u> (not organs) responding, or versus the probability of the
all-or-none single-cell effect, the curve clearly shows a linear
function initially with a quadratic component at higher doses and
dose rates, e.g. curve A, Fig 1). The overall curve follows the
relationship:

$$I = \alpha D + \beta D^2$$

where I is the fraction of cells responding, D is the organ dose,
and alpha and beta are constants. This relationship for organ dose-
single cell effects cannot be forced to fit the classical S-shape
response curve for early multi-cell organ effects. This so-called

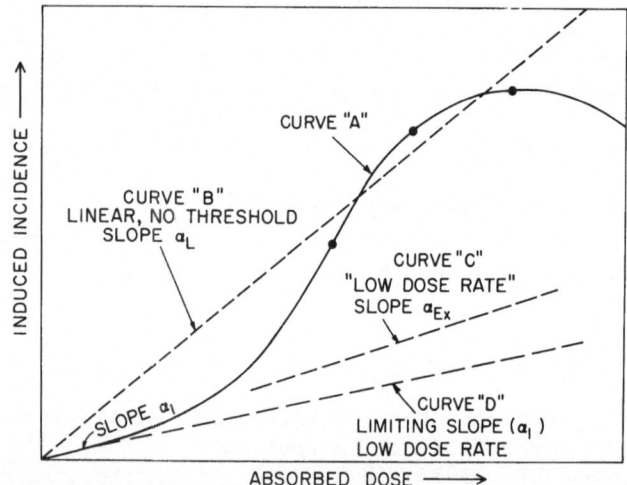

Figure 1. Schematic curves of incidence of a cellular all-or-none
effect vs. absorbed dose. The curved solid line for high absorbed
doses and high dose rates (curve A) is the "true" curve. The linear,
no threshold dashed line (curve B) was fitted to the four indicated
"experimental" points and the origin. Slope α_1 indicates the essen-
tially-linear portion of curve A at low doses. The dashed curve
C, marked "low dose rate," slope α_{Ex} represents experimental
high-dose data obtained at low-dose rates. This experimental low-
dose-rate curve may, in principle, at very low-dose rates, approach
or become indistinguishable from the extension of the solid curve
of slope α_1, the dashed curve D labeled "limiting slope for low
dose rate".

"linear quadratic" function has given rise to the "linear hypothesis".
The linear relationship referred to usually is not the "αD" term shown
as slope α_1 in Figure 1, but rather a chord, shown as curve B, drawn
from zero dose and zero excess incidence, to some arbitrary point on
the curvilinear part of the overall function. The fact that the initial part of the curve is linear for either the radiobiological curve
A or hypothetical curve B gave rise to the well-known expression, "any
amount of radiation, no matter how small, is potentially harmful".

The purpose of this paper is to show why the quadratic, or "linear
quadratic" relationship holds for organ dose-single cell radiation
effects and to explore the extension of this relationship to chemical
exposures in general. It will also be shown that, although the "$\alpha D +
\beta D^2$ relationship" may be unexpected for normal pharmacological-medical
dose-response relationships, a linear, no-threshold curve of this kind
is expected for all stochastic-type (accidental or risk) situations

with health consequences, e.g. all common accidents (5), including
exposure to "low-level radiation" (LLR). Examples in radiation biology will be given, and reasons will be proposed as to why some
chemicals may act in the same way. Critical experiments will be
suggested to determine if the organ dose-cell response relationships
for chemical mutagens and carcinogens follow the classical pharma-
cology-type or the stochastic-type curves. Earlier thoughts on the
subject have been published (2-4, 6-8).

The Stochastic or Risk Approach

Radiation effects on cells derive from chance or stochastic
collisions between a charged radiation particle, and the gross
sensitive volume (GSV) of a cell, i.e. the volume in which a particle must hit in order to have a non-zero chance of interacting with
the much smaller molecular target (Ref. 8). Any given organ dose
yields a number of particle-cell GSV collisions, called "events"
(9,10) or individual cell doses. These range in size from near zero
to a maximum determined by the quality of the specific radiation (see
Fig 2). The entire range of possible cell dose sizes cannot be
studied without employing a range of very low- to quite high-LET
radiations. With LLR, each event equates to a cell dose.

Although cell doses[1] are not measurable directly in the living
cell, the spectrum of cell doses per organ dose can be assessed by
constructing a cell phantom in the form of a spherical proportional
counter. The counter contains a tissue equivalent gas, the mass
of which is equal to the first-presumed and then measured mass of
the cell GSV. When exposed in a radiation field of any quality,
a wide spectrum of "events" or cell doses, such as that shown in
Figure 1 is obtained (10). Each point within the overall area represents a single cell dose of a given size. Also shown in the plot
is an S-shape curve representing the probability of an all-or-none
single cell (not organ) effect on the cell as a function of cell dose
(not organ dose) size. It is this cell dose-single cell effect curve,
only recently described (6-8), that is the analogue for radiation of
the usual threshold-type pharmacological organ dose-organ response
curve. The dose in this curve clearly is that to the individual cell,
and not an average to the organ or cell medium. The derivation of
the new curve is indirect and somewhat complicated, and is described
in detail elsewhere (6-8).

[1]Cell dose will be used here for ease and clarity, even though cell
event or hit may be preferable. Dose is usually reserved for situations in which the expected benefit derives from the effect of the
measured amount of the agent given. Mutagenic and carcinogenic effects are incidental to an agent's normal use.

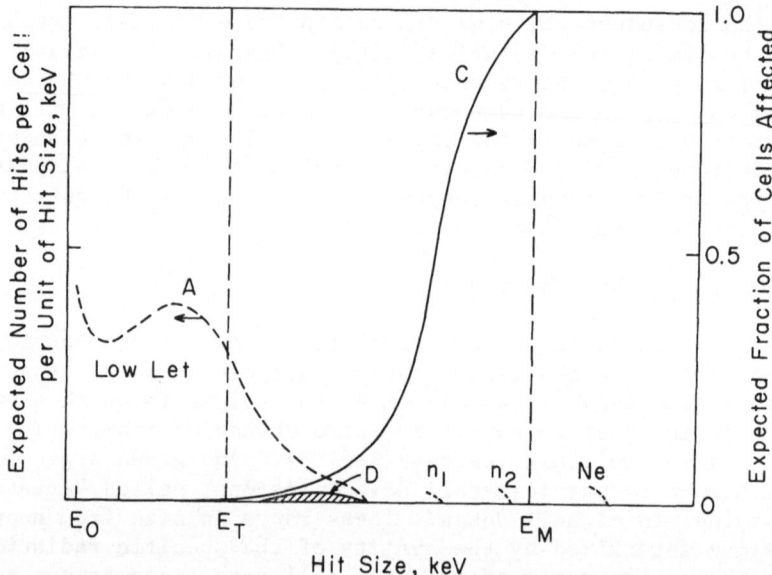

Figure 2. Schematic plot of number of events or hits per (simulated) cell vs. event or hit size in a tissue equivalent microdosimetric tissue equivalent proportional counter simulating the gross sensitive volume of a cell. The plot is for a low-LET radiation. E_T refers to the threshold for an all-or-none effect from a single hit. E_M is the hit size at which the probability of an all-or-none effect per hit equals 1.0. n_1, n_2, and Ne refer to the ends of the spectra for high-LET radiations. For curve C, the description is to the right of the figure. Curve C is for a heretofore-undescribed function showing the probability of a cellular all-or-none effect vs. the size of the event, or cell dose.

        The reason for the linear term in Equation 1 can now be seen by considering the event spectrum (cell hit, or cell dose spectrum) shown in Figure 2. The entire cell dose size spectrum is obtained with a single value of organ-absorbed dose. Thus, there is always a wide spectrum of very small to large cell doses for any given increment of average organ dose. Hence, no matter how small the average organ dose may be, and therefore the average cell dose, there is still a probability that the small organ dose may contain a large dose to one or a few cells. The large cell dose will contain an inherent probability of being above a threshold and of causing an all-or-none effect in the affected cell. Thus, there is some probability of the all-or-none cell effect no matter how small the average ab-

sorbed <u>organ</u> dose. The organ dose-single cell effect curve at low
doses for single-cell effects accordingly will be linear and without a threshold.

The total time over which cell doses are delivered is important
(2-4). With ionizing radiation, all individual cell doses are delivered "instantaneously". All effects in the low dose, "linear
kinetics" region are due to a single cell dose or are "single hit"
events. However, at higher organ doses and at high organ "dose
rates" (total time of dose delivery) two or more ineffective cell
doses can interact in a single cell and cause an effect by a multi-
hit mechanism. It is this multi-hit or cumulative mechanism, essentially absent with LRR, that gives rise to the "squared term", or
the quadratic part of the overall "linear-quadratic" curve. At high
"dose rates" it is the sum of the linear and the quadratic terms that
constitutes the total curve, while at very low dose rates only the
linear term is observed even though the organ dose may be moderately
high (see below).

Relevant Radiobiology

From the above explanations, one would expect average organ dose-
cell response relationships in radiobiology to follow the formula
given in Equation 1 and shown in Figure 1. The organ dose in this
case means the average dose to the cells or to the medium containing
the cells. The most definitive experimental results bearing on this
relationship were obtained from the <u>Tradescantia</u>, or spiderwort,
plant cell system (11-14). The phenotypic change of blue or blue-
violet to pink color of cells in the stamen hairs of plants heterozygous for wild-type provide a measure of genetic damage produced
by exposure to ionizing radiations and chemical mutagens. The number of cells changing from blue to pink color is an index of genetic
effect. In test crosses with a pink-flowered parental clone and in
crosses with its sibling F_1 plants, the results are consistent with
the hypothesis that pink pigmentation is determined by a pair of
alleles at a single locus and that blue is dominant to pink (14).
A stamen hair propogates from the terminal one or two hairs; hence
the number of mutable cells/hair is approximately two.

Results for x-irradiation experiments are shown in Figure 3. The
incidence of pink mutant events scored in the individual cell stamen
hairs is plotted against <u>organ</u> dose. A log-log plot is used to make
clear the extent and nature of the data at very low values of organ
dose, i.e. below 10 rads. The data indicate clearly the proportionality of organ dose and single cell incidence at low doses, and the
lack of a threshold for incidence of effect. The frequency of events
is extremely low. The data, up to about 100 or more rads, can be
represented well by the function shown in Equation 1. The flattening
of the curve due to "cell killing", and obviously important at higher
organ doses, is not considered here. The slope of the linear term is

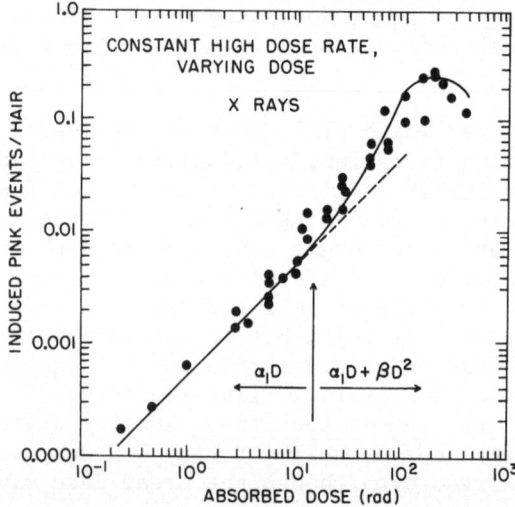

Figure 3. X-ray dose-response curve for induced pink mutations in
Tradescantia on a log log plot to show detail in the low-dose range.
The solid circles indicate experimental points at high-dose rate.
Note that the low dose portion of the solid curve and its dashed-
line extrapolation have a slope of unity corresponding to a linear,
no threshold dose-effect relationship. The increased slope at
higher doses indicates that the response involves a higher exponent
of dose (data of Sparrow et al. (11); Nauman et al. (12); Underbrink
et al. (13)).

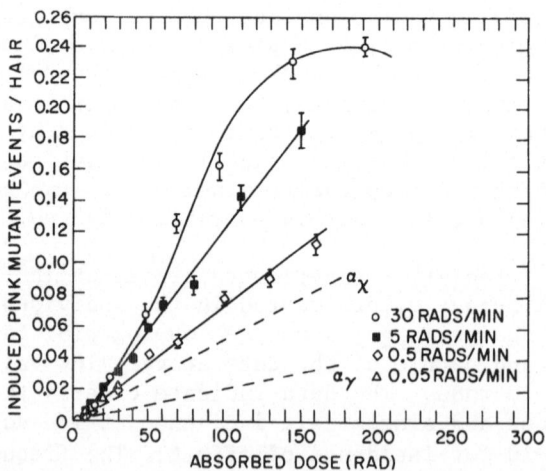

Figure 4. Dose-response curves for pink mutant events/hair after
x-irradiation at 0.05 and 0.5 rad min^{-1} (combined in one line)
and at 5 and 30 rad min^{-1}. The dashed lines represent the alpha
terms for x rays and gamma rays (from Nauman et al. (12)).

1.0 while the slope of the curvilinear portion is about 1.5. This
latter slope would presumably reach 2.0 if not precluded by normal
biological complications at higher doses, e.g. cell death.

The effect of organ dose rate is seen in Figure 4, in which is
plotted essentially the same data shown in Figure 3 using arithmetic
coordinates (upper curve). The two central curves with data points
represent organ dose rates lower than used for the uppermost curve.
The lower curve marked "X" represents the extension for x-rays of
the "αD" part of the low-dose curve in Figure 3, corresponding to
the same slope, $α_1$ and $α'_1$ shown in Figure 1. The lowermost curve
marked "γ" is obtained experimentally if hard γ-rays instead of
x-rays are used to determine the lower part of the curve in Figures
3 and 4. The influence of average organ dose rate (or exposure time)
is detailed in Figure 5 where an organ dose of about 80 rads was
delivered at progressively lower dose rates. The incidence per rad
at 80 rads decreases progressively as the dose rate is lowered
(exposure time lengthened), and the slope (incidence per rad at 80
rads) is seen to approach asymptotically the (gamma) incidence per
rad at low doses, as seen in Figures 3 and 4. Thus, the lower limit
of the incidence per rad (the slope) at very low doses is the same
as the lower limit of the incidence per rad (the slope) using high
doses delivered at low organ dose rates.

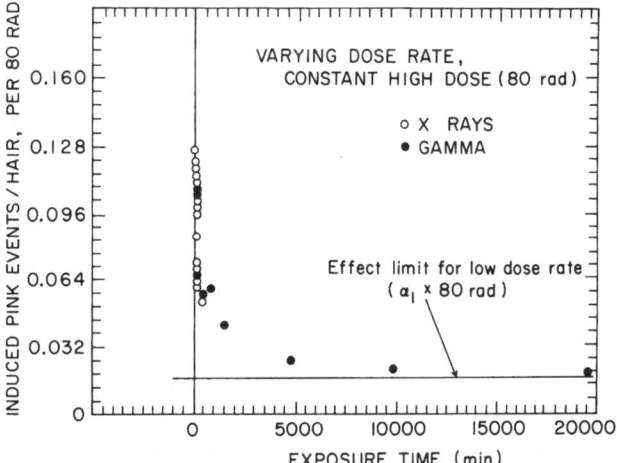

Figure 5. Effect of dose rate on the effectiveness of a single
large gamma-ray dose of about 80 rads for the induction of pink
mutations in _Tradescantia_. The horizontal line represents the
expected limiting low-dose-rate value for 80 rads. Note that the
effect at 80 rads decreases appreciably as the exposure time is
increased and approaches asymptotically the limiting "αD" value for
gamma radiation.

The linear and quadratic components of incidence (Fig 3 and 4) thus are separable either by lowering the organ dose or the organ dose rate (2,4). These two components are shown separately in Figure 6. The "sub-effect" damage of the quadratic component is repaired completely at low dose rates before it can contribute to a visible lesion. The linear component is without threshold and shows a constant increase in incidence per rad.

Figure 7 shows Tradescantia results for a typical high-LET (linear energy transfer) radiation--fast neutrons. The high-LET curve is linear at low doses. The relative biological effectiveness (RBE) of the neutrons compared to x-rays, analogous to the relative potency ratio, is large. The RBE in general increases as a function of LET.

Stochastic Approach For Chemicals?

From the above discussions and data, one can see why there is a linear or "αD" term with low-level radiation when average organ dose and not cell dose is used. The linear component is due to the fact that, for a given average organ dose and particularly at low doses, the distribution of the agent among the cells of the organ is not homogenous. Consequently, different cells receive widely different-size cell doses for the same average organ dose or for

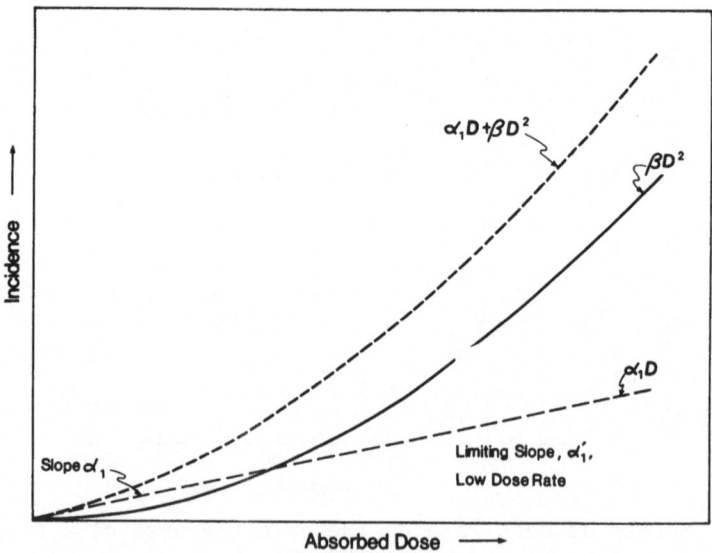

Figure 6. The linear-quadratic dose-response curve for Tradescantia, with the linear and squared components plotted separately (high-dose region of curve dominated by cell killing is ignored for purposes of this plot).

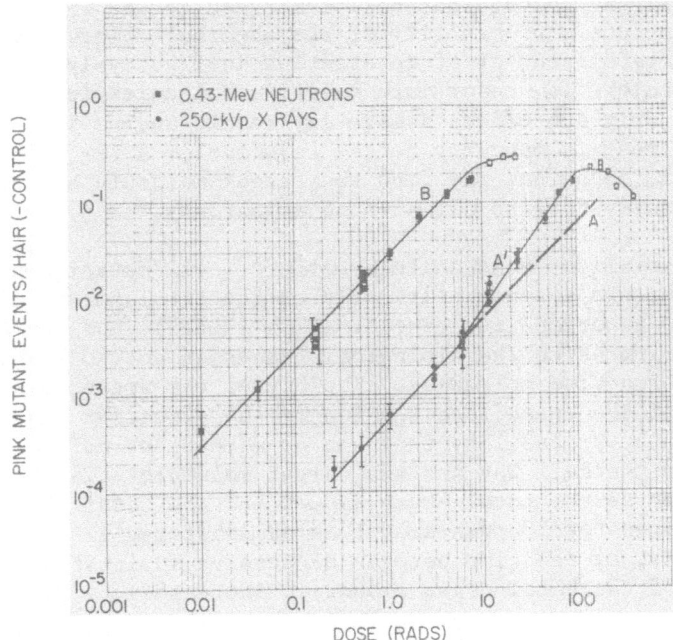

Figure 7. Plots of pink mutant events per hair (minus control) against dose for 250-kVp x-rays and 0.43 MeV neutrons. Note that the neutron curve is linear at low doses, but is to the left of the x-ray curve. This indicates that the neutrons are more effective than x-rays, per unit average organ dose, or absorbed dose.

the average cell dose. The processes are entirely by chance, i.e. stochastic, and all single-hit cell doses are delivered instantaneously. Thus, whether a given cell will show an all-or-none effect is dependent entirely on the size of the single-hit cell dose, i.e. the total number of electrons that were deposited within the cell GSV. The larger the number of electrons, the greater the probability that one or more of the electrons will "find" the molecular cell target and interact to produce the all-or-none effect.

Although some chemicals might interact in this same or in a similar fashion, cell doses are not delivered instantaneously, and there is not the obvious reason for the discrete cell amounts and therefor the inhomogeneity of cell dose sizes seen with radiation. The probability of a cell dose and a resultant all-or-none effect from a chemical, however, may well depend largely on the probability of one or more molecules getting through a number of barriers (including cell membranes) in order to interact with the appropriate molecular target(s) located in the GSV. Only a limited number of reactive molecules will find their way to the appropriate target cell[2], and or those, only a

fraction will get through the cell membranes to the GSV. Additional
obstacles include a probability of the chemical molecule being excreted or metabolized, or of interacting ineffectively with a non-
target molecule. If one or more reactive molecules interact with
the target, only sub-effect damage may result and be immediately
repaired. Thus, it may only be the rare one or a few molecules from
a given organ dose that may find the target molecule and produce the
damage necessary for the all-or-none effect.

Whether an all-or-none cellular effect (cell mutation or transformation) occurs depends critically on the total amount of the chemical given to an organ, its concentration ("LET"), and the time period
involved during which the biological processes act to inactivate or
excrete the reactive molecules. The higher the organ amount and concentration of the agent, and the longer the effective exposure period,
the greater the probability that one or more agent molecules will find
their way to the cell GSV and the target molecule. This probability
will decrease as the total organ amount and the total time period over
which the given total organ amount is administered is reduced (or
alternatively, as the time between successive administrations of
increments of the total organ amount is increased).

As the total organ dose is decreased to small values, or as the
dose rate for higher doses is decreased, two possible results may
occur. At the lower doses or lower dose rates, the probability may
be zero that any single reactive molecule or that a small number of
reactive molecules will interact with the target. In this case, there
will be at least a "pseudo-threshold" at low doses and, if the dose
rate is sufficiently small the slope of the dose-response curve will
become zero even at higher doses. The other alternative is that there
may be a "single-hit" component, i.e. there may be some irreducibly
small chance that a molecule of the agent may get through to the target molecule, no matter how small the single organ dose given in a
short period of time. Also, with at least some chemicals, there could
be some chemical or physical aggregation such that a few cells could
receive a relatively high dose (asbestos particles may represent an
extreme case). It is conceivable that there may be a "linear component" or a single-hit component with some probability of effect for
either very low organ doses or for higher total doses given at very
low dose rates.

The question of an absolute threshold or lack thereof is in fact

---

[2]Single cell autoradiographic evidence from tritium labeled compounds
suggests a wide spectrum of amounts of compound per cell with relatively low single administrations of compound to the cell medium.
Additional work is needed to characterize such spectra.

largely irrelevant for radiation and for chemical mutagens and carcinogens, as well as for other stochastic agents (5, 14). Unlike
most drugs, which are administered for effects beneficial to the
patient, the potential mutagenic or carcinogenic effects of an agent
are untoward side-effects. Thus, every effort is made to minimize
the risk of such adverse effects. The use of a stochastic or risk
agent (5, 14) almost always affords the benefit desired only if the
individual single-hit doses are above the threshold for serious harm.
For example, vehicle use would be absolutely safe if the maximum
speed were perhaps 3 MPH, so that the "dose" of kinetic injury on
accidental collisions could not exceed the threshold for serious injury or death. The vehicles would also be effectively useless. Thus,
the presence of an absolute threshold is largely academic.

With the practical use of chemicals at the minimum doses necessary
to achieve the benefit, the important question then is whether there
is a non-zero probability of such untoward effects as mutagenesis or
carcinogenesis? The critical test is to use that minimum amount as
a single, rapidly-administered organ dose, or in the "single hit"
mode. With the low probabilities frequently expected, this test may
be negative. Testing a low dose adequately, however, may be expensive if animals are used. A more useful and sensitive approach is
to give a single "minimally-useful" dose in the single hit mode just
described, and then repeat this administration at protracted intervals until a (much) larger total dose has been given. The time period between doses is critical and must be based on pharmacodynamic
considerations as well as on pragmatic considerations of obtaining
the desired benefit. If, with protracted administration of the small
increments, there is a "linear term" that is not zero, then mutagenic
or carcinogenic activity will have been established. An alternative
but somewhat less satisfactory approach based on our experience with
radiation is to administer the agent continuously but at a very low
dose rate. "Extrapolation" from results at high doses and dose rates
clearly would be useless in terms of demonstrating the presence or
absence of a linear term.

DISCUSSION

Although the above interpretations for chemical carcinogenesis
and mutagenesis are of course hypothetical, there is evidence (15-
21) that Equation 1 may hold for some chemicals. For example, the
evidence is good for dibromoethane in the Tradescantia system that
there may be a linear. as well a multi-hit component. Large
amounts of data are required in a relatively "sensitive" system to
detect the linear term at low doses, and/or higher doses at low dose
rates.

If the "$\alpha D + \beta D^2$" relationship does hold for some chemicals, then
the critical test for the presence of an "αD term" should then be

made for all potential mutagens and carcinogens shown to be active
at high doses and dose rates. In particular, a minimally-useful
single-hit organ-dose amount should be used in a fractionated manner to obtain a large total dose and thus a reasonable chance of
detecting the rare events involved. This approach would avoid the
fallacy of assuming that if an agent is active at high doses and
dose rates, then it must be active (presumably with the same effectiveness per unit dose) at very low doses. Furthermore it would
also be determined if the chemical shows a threshold. If, on the
other hand, the agent shows the linear term at normal usage levels,
then a non-zero risk of mutagenesis or carcinogenesis will have
been demonstrated. However, it is likely that the risk, i.e. the
slope of the dose-response curve with fractionation, at low doses
and/or dose rates will be lower than that predicted with the
"linear hypothesis".

With respect to the relevance of Tradescantia or any other single
cell system as a predictor for mutagenesis and carcinogenesis in
animals and man, it is assumed that even though actual mutational
rates can not be expected to apply, the cardinal principles of genetics largely transcend species and the particular environment in
which the cell is located. Relevance to carcinogenesis depends on
similarities between the genetic and transformational events occurring in a single cell. The evidence is extensive that many, but not
necessarily all, types of cancer are monoclonal in origin (22-26).
Hence, the initiating event for many if not most radiogenic cancers
may well be the transformation of a single cell (26). It is true
that large doses delivered at high dose rates may well perturb severely hormonal, immunological and other factors that may play a
role in the promotion of expression of a cancer but not in its cellular induction. Such perturbation would be inconsequential with
the use of LLR or of the low levels of chemicals discussed here.
Thus with LLR, the curve shapes and other relationships developed
for Tradescantia would be expected to apply in principle to animal
and human mutagenesis and carcinogenesis.

REFERENCES

1. V. P. Bond, T. M. Fliedner and J. O. Archambeau, "Mammalian
    Radiation lethality, A Disturbance in Cellular Kinetics",
    Academic Press, New York, N.Y. (1965).
2. V. P. Bond, Quantitative risk in radiation protection standards,
    Radiation and Environmental Biophysics 17:1-28 (1979).
3. V. P. Bond, A basis for estimating the risks of low-level
    radiation, Medical Physics Monograph Vol 5, "Biological
    Risks of Medical Irradiations", G. D. Fullerton, ed. (1980).
4. NCRP Report No. 64, Influence of dose and its distribution in
    time on dose-response relationships for low-LET radiations,
    April (1980).

5. Accident Facts. The National Safety Council, Chicago, I.L.
    (1980).
6. V. P. Bond, The meaning and role of risk in radiation and other
    safety practices, Proceedings, Symposium on "Radiation and
    Its Effects", Los Alamos, N.M., November (1980).
7a. V. P. Bond and M. N. Varma, The threshold-microdosimetric
    approach in radiation, Annual Meeting of the Radiation
    Research Society, Minneapolis, M.N., Abstract (1981).
7b. M. N. Varma and V. P. Bond, Microdose-response curve from
    exposure-incidence curves, Annual Meeting of the Radiation
    Research Society, Minneapolis, M.N., Abstract (1981).
8. V. P. Bond, The conceptual basis for evaluating risk from
    low-level radiation exposure, Proceeding 1981 NCRP
    Annual Meeting, to be published (1981).
9. H. H. Rossi, Energy distribution in the absorption of
    radiation, Adv. in Biol. and Med. Physics, Vol. II,
    Academic Press, New York, N.Y., 27-85 (1967).
10. H. H. Rossi, Radiation dosimetry, Vol. I, 2nd Edition, F. H.
    Attix, W. C. Roesch, eds., Academic Press, New York, N.Y.,
    p. 43 (1968).
11. A. H. Sparrow, A. G. Underbrink and H. H. Rossi, Mutations
    induced in Tradescantia by small doses of x-rays and
    neutrons: Analysis of dose response curves, Science
    176:916-921 (1972).
12. C. H. Nauman, A. G. Underbrink and A. H. Sparrow, Influence
    of radiation dose rate on somatic mutation induction in
    Tradescantia stamen hairs, Radiat. Res. 62:79-96 (1975).
13. A. Underbrink, A. Kellerer, R. Mills and A. Sparrow, Compari-
    son of x-ray and gamma ray dose response curves for pink
    somatic mutations in Tradescantia clones 02, Rad. Environ.
    Biophys. 13:295-303 (1976).
14. M. Emmerling-Thompson and M. M. Nawrocky, Genetic basis
    for using Tradescantia clone 4430 as an environmental
    monitor of mutagens. J. Heredity 71:261-265 (1980).
15. D. R. Foust, B. M. Bowan, R. G. Snyder, Study of human
    impact tolerance using investigations and simulations of
    free-falls", 21st STAPP, Car Crash Conference, New Orleans,
    L.A., Society of Automotive Engineers, Warrendale, P.A.,
    (1977). (See also proceedings of other STAPP conferences).
16. A. G. Underbrink, L. A. Schairer and A. H. Sparrow, Tradescan-
    tia stamen hairs: A radiobiological test system applicable
    to Chemical Mutagenesis, in: "Chemical Mutagens: Principles
    and Methods for Their Detection", Vol. 3., A. Hollaender,
    ed., Plenum Press, New York, N.Y., 171-207 (1973).
17. A. H. Sparrow, L. A. Schairer and R. Villalobos-Pietrini,
    Comparison of somatic mutation rates induced in Tradescantia
    by chemical and physical mutagens. Mutat. Res. 26:265-276
    (1974).
18. C. H. Nauman, A. H. Sparrow and L. A. Schairer, Comparative

effects of ionizing radiation and two gaseous chemical
mutagens on somatic mutation induction on one mutable and
two non-mutable clones of Tradescantia. Mutat. Res.
38:53-70 (1976).

19. International Agency for Research on Cancer, Ethylene
    Dibromide in: "IARC Monographs in the Evaluation of the
    Carcinogenic Risk of Chemicals to Man", V. 15, IARC,
    Lyon, 195-209 (1977).

20. C. H. Nauman, A. H. Sparrow, A. G. Underbrink and L. A. Schairer,
    Low-dose mutation response relationships in Tradescantia;
    Principles and comparison to mutagenesis following low-dose
    gaseous chemical mutagen exposure, Radiological Protection,
    First European Symposium on Rad-Equivalence. R. Chanet,
    ed., Commission of the European Communities, Luxemburg,
    EVR 5725e, 13-23 (1977).

21. C. H. Nauman, P. J. Klotz and L. A. Schairer, Uptake of
    tritiated 1,2 dibromoethane by Tradescantia floral
    tissues: Relation to induced mutation frequency in
    stamen hair cells, Environ. and Exp. Botany, 19:201-215
    (1979).

22. T. H. Ma, A. H. Sparrow, L. A. Schairer and A. F. Nauman,
    Effect of 1,2-dibromoethane (DBE) on meiotic chromosomes
    of Tradescantia, Mutat. Res. 58:251-258 (1978).

23. P. Fialkow, S. Gartler and A. Yoshida, Clonal origin of chronic
    myelocytic leukemia in man, Proc. Natl. Acad. Sci. USA
    58:1468-1471 (1967).

24. P. J. Fialkow, R. Jacobson and T. Papayannopoulou, Chronic
    myelocytic leukemia: Clonal origin in a stem cell common
    to the granulocyte, erythrocyte, platelet and moncity/
    macrophage. Am. J. of Med. 63:125-130 (1977).

25. G. Wiggans, R. Jacobson, P. Fialkow, P. Wooley, J. H. Macdonald
    and P. Schein, Probable clonal origin of acute myeloblastic
    leukemia following radiation and chemotherapy of colon
    cancer, Blood 52:659-663 (1978).

26. P. C. Nowell, The clonal evaluation of tumor cell populations.
    Science 194:23-28 (1976).

27. United Nations Document, Dose-response relationships for
    radiation-induced cancer, Report to the General Assembly
    UNSCEAR 30th Session, Appendix I. Document A/A.C. 82/R
    383 (1981).

SESSION X: REGULATORY RISK ASSESSMENT

D. C. Borg, Moderator

Chairman, Medical Department
Brookhaven National Laboratory
Upton, NY 11973 (U.S.A.)

The goal of the regulatory process in controlling genotoxic agents is to reduce human health hazards to "acceptable" or "safe" limits. The regulatory process is based on the premises that exposure to specific agents is a hazard to human health and that these agents can be identified, their genotoxic potency quantified, and their exposure levels controlled. Regulation is a complex process, only part of which depends on the available scientific data. Other deciding factors include societal judgments regarding the importance or desirability of the agent, the perception of risk, what constitutes acceptable risk, the feasibility of measuring and controlling the exposure conditions, and the economic impact of controls on industry and society.

The previous sessions have dealt with state-of-the-art in vitro and in vivo techniques for assessing the genotoxicity of airborne agents, current information on the genotoxicity of selected airborne agents, and approaches for extrapolating these data to human health effects. This final session will examine the epidemiological basis for determing that airborne (and other) agents constitute a carcinogenic health hazard to humans (the presentation by S. C. Morris of Brookhaven National Laboratory) and offer two approaches for the regulation of genotoxic agents.

Environmental genotoxic agents may be hazardous to human health in two different ways--as inducers of cancer in somatic cells and as inducers of mutations in germ cells. The first of the two regulatory presentations, by Peter Voytek of the Environmental Protection Agency, will deal with the latter in discussing EPA's policy for regulating human exposure to mutagenic agents in order to prevent an increase

in heritable diseases in man. T. M. Farber of the Food and Drug
Administration will then present the FDA's approach for regulating
exposure to genotoxic agents based on the Sensitivity of Method procedure for assessing carcinogenic risk. Although not strictly applicable to airborne agents, this method is of interest as a regulatory approach for setting safe limits on genotoxic exposure.

IMPACT OF ENERGY AND INDUSTRIAL POLLUTION ON PUBLIC HEALTH

Samuel C. Morris, III

Biomedical and Environ. Assessment Division
Brookhaven National Laboratory
Upton, NY   11973 (U.S.A.)

Among all major killer disease, cancer alone has steadily increased during this century (Fig 1). Moreover, those cancers more
likely to be environmentally produced seem to be increasing the
fastest (Fig 2). Figure 3 shows results from surveys of cancer incidence rather than mortality. Although there may be comparability
problems since the survey populations were different, these data
provide the best available evidence of positive cancer trends. The
First National Cancer Survey was made in 1937-39, the Second in 1947
-48, and the Third in 1969-71. The National Cancer Institute's Surveillance, Epidemiology and End Results (SEER) Program began in 1973.
The SEER results shown are for 1973-76. Incidence trends from these
surveys were examined in detail by Devesa and Silverman (1,2) and by
Pollack and Horn (3).

What changes have there been in exposure? We may not be sure
of exactly what to look for, and measurements of exposure are not
readily available. We can get some idea from production trends in
the chemicals of concern. As an example, Figure 4 shows increases
in annual production of synthetic organic chemicals.

A connection is easily made between the concurrent increases
of cancer and of various environmental/industrial pollutants observed over several decades. However, the link is not as clear as it
may seem at first. Environmental effects can result from general,
occupational, or personal pollution, e.g. cigarette smoking. Changes in the male/female mortality ratio give some clue to the environmental causes of cancer. Figure 5 is a contour plot of these
ratios. In 1900, there was little difference between male and female mortality rates in any age group. In the 1930's and 1940's
male mortality began to outpace female mortality and, by the 1960's,

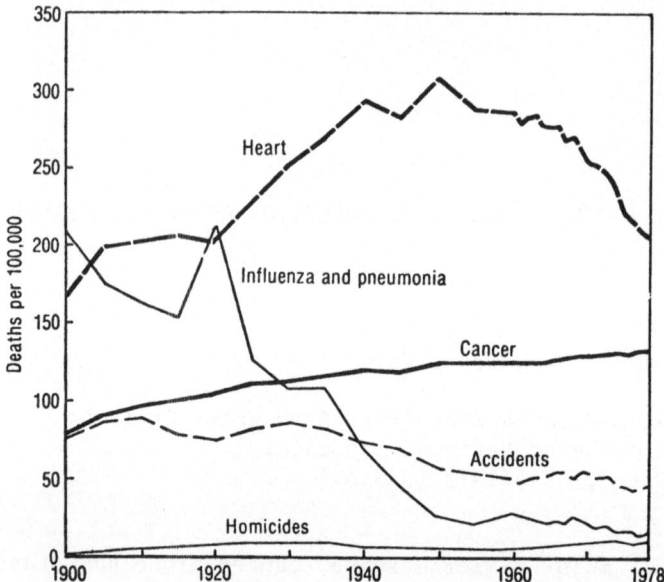

Figure 1.   Age-adjusted rates for major causes of death, 1900-1978.
From Council on Environmental Quality (39).

males had much higher mortality rates. The biggest differences occur
in the young adult ages, due primarily to accidents, while in middle
age, there is clearly an environmentally related peak. This peak is
due primarily to smoking, and possibly to occupational exposure. Also, this peak for men may be partly due to greater exposure to ambient
pollution. Recently, the rate difference between men and women seems
to be closing.

Higginson (4) has long professed that perhaps 80% of cancers are
environmentally related, where the 'environment' includes personal
environmental factors - smoking, diet, and behavior. While there is
clear evidence of cancer from occupational exposure and some evidence
of cancer in people living near point sources of pollution, there is
little evidence that low doses of industrial pollutants, to which the
general public are exposed, produce cancer. If associations do exist,
how can one find such associations with low level general exposures?
Among the available approaches are: international comparisons; time
trends; processes of elimination of other, more well-defined causes
of cancer, e.g. certain occupational exposures or smoking. One can
also use more classical epidemiological approaches--cross-sectional
analysis, case-control studies and time-space clustering.

Figure 6 shows an unusual international comparison, comparing
percentage of the labor force in manufacturing with cancer mortality

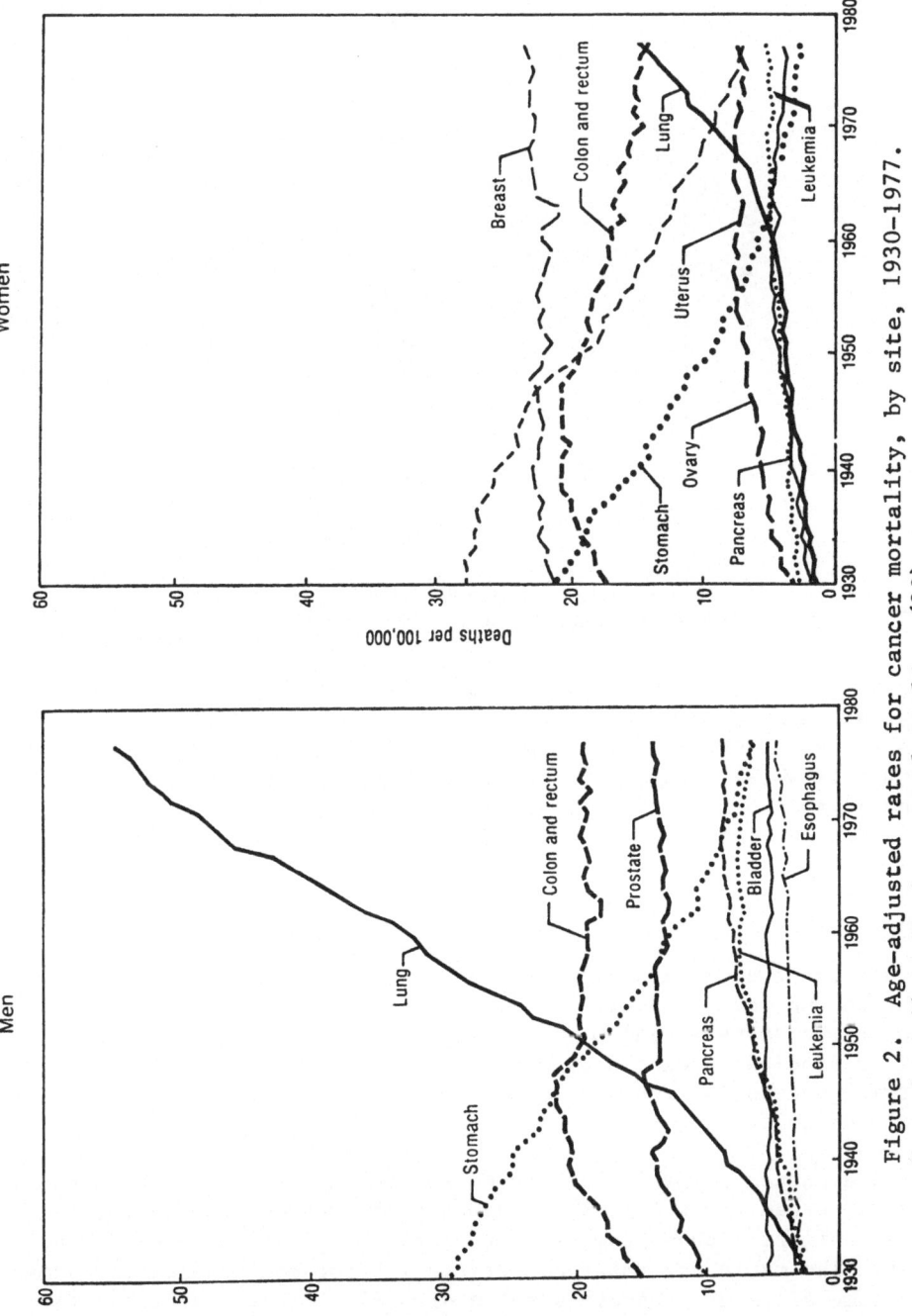

Figure 2. Age-adjusted rates for cancer mortality, by site, 1930-1977.
From Council on Environmental Quality (39).

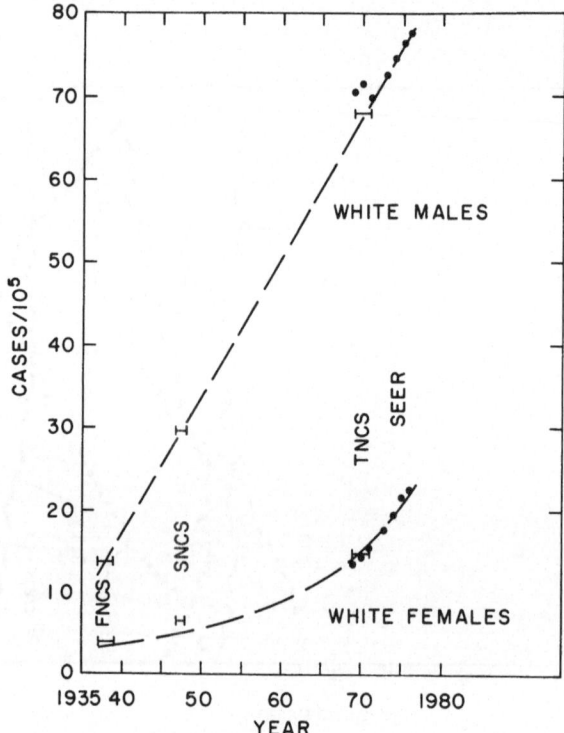

Figure 3.   Lung cancer incidence rates, U.S. Whites. Data for
First National Cancer Survey (FNCS), 1937-1939, Second National
Cancer Survey (SNCS), 1947-1948, and Third National Cancer Survey
(TNCS), 1969-1971, shown as lines, (from Devesa and Silverman(1))
standardized to U.S. 1950 population. Annual TNCS values and
Surveillance, Epidemiology and End Results (SEER) program data
from Pollack and Horn (3) standardized to U.S. 1970.

for different countries (5). A loose relationship is evident for
both men and women. Increased manufacturing seems associated with
greater increases in males, presumably due to occupational exposures.
Similar studies in examining cancer by type have shown a trend from
gastric cancers in less-developed countries to lung cancers in more-
developed countries.

    One phenomenon which has received much attention is the urban
factor. Urban areas typically have higher cancer rates than rural
areas. Figure 7 shows the urban to rural ratio for several cancer
sites. The ratio for total cancers is 1.6 for men and 1.4 for women,
with some sites at or above a ratio of 3 (6). Thus, the urban factor was readily linked to the effects of industrial pollution.
Goldsmith (6,7) has repeatedly indicated a number of inconsistancies

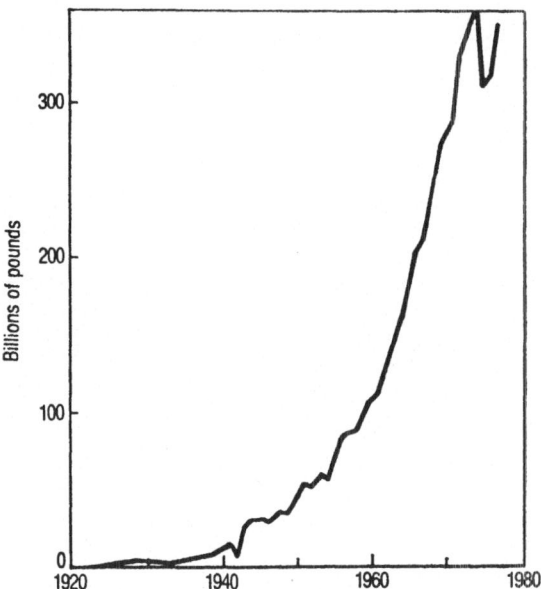

Figure 4.   Annual production of synthetic organic chemicals,
1920-1978.   Reprinted with permission of Science from Davis and
Magee (40), with acknowledgements to the Environmental Law Institute.

in the urban factor-industrial pollution hypothesis which, while not
causing outright rejection, certainly weakens the hypothesis considerably. It would be expected that: (a) The urban factor would be
greatest in the most polluted places, which is not the case; (b) Mi-
grants from rural to urban areas should show a cancer rate lower than
lifetime urban residents, which is also not the case; (c) There should
be a good correlation with pollution levels. This is difficult to
determine since pollution levels are well correlated with population
density, but correlations with pollution appear more tenuous than
those with population density; (d) Women should be affected as much
as men if the cause was ambient air pollution. This is generally not
the case, although women may be affected in proportion to their "spontaneous" rates (8). Of course, if one goes back to 1900 when men and
women apparently had similar "spontaneous" rates, this argument holds
little weight; and (e) A decrease in lung cancer should occur when
levels of known carcinogens, e.g. benzo(a)pyrene, decrease. This has
not occurred, although the long latent periods involved, uncertain exposures, and multiple agents--many possibly unidentified--weaken this
argument.

    Certainly, cigarette smoking is the most well known, and most
well documented, environmental cause of cancer. A particularly

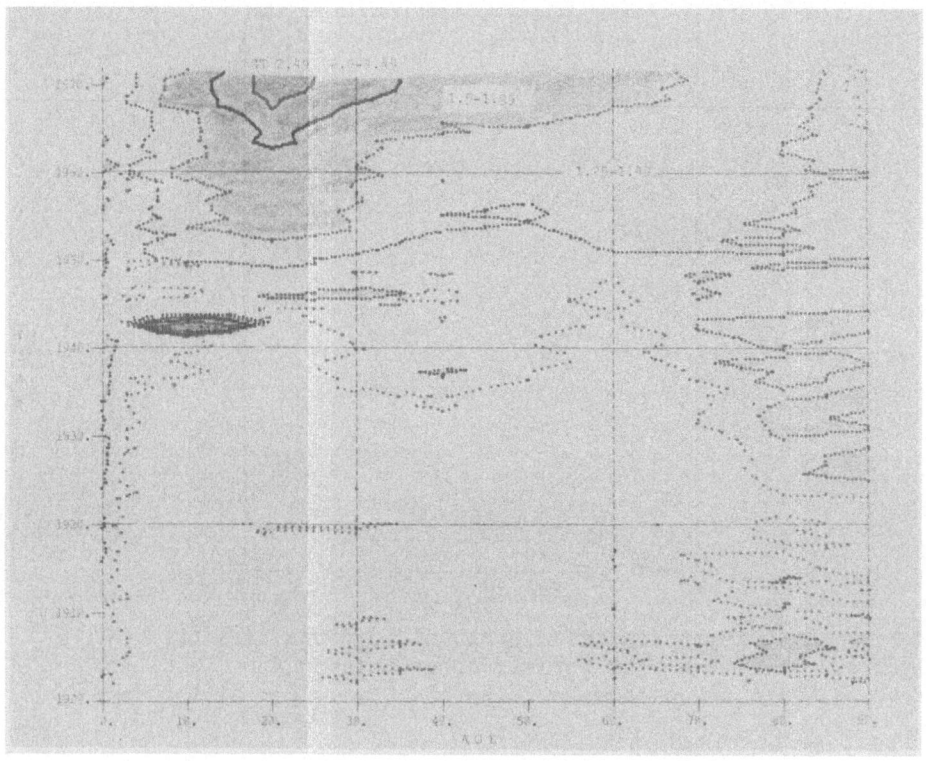

Figure 5.   Ratio between U.S. male and female mortality rate for
non-whites by age, 1900-1970.   Figure from S.R. Bozzo, Brookhaven
National Laboratory.

difficult problem is that the time trend for the development of modern industrial pollutants generally parallels the time trend for
smoking.   Moreover, most of the differences in lung cancer between
men and women are almost surely due to their differences in smoking
patterns.   Clearly a substantial portion of the total environmental
cancer is due to smoking.   Schneiderman (9) estimated 82% of lung
cancer in the Third National Cancer Survey (TNCS) and 74% of lung
cancer in the SEER data in men, and correspondingly 70% and 62% in
women, were due to cigarette smoking (Table 1).   The differences between the 1970 TNCS and the 1976 SEER estimates reflect differences
in the rate of smoking and differences in the cigarettes themselves.
Of course, many assumptions go into this table.   Nonetheless, some
interesting results are revealed.   The total annual changes in lung
cancer incidence between 1970-1976 was ∿1.7% per year.   If we sub
tract that fraction due to cigarettes, we find that the non-smoking
related lung cancer has increased between 6-11% per year (See Table
2).   It would appear from Schneiderman's analysis, that, although

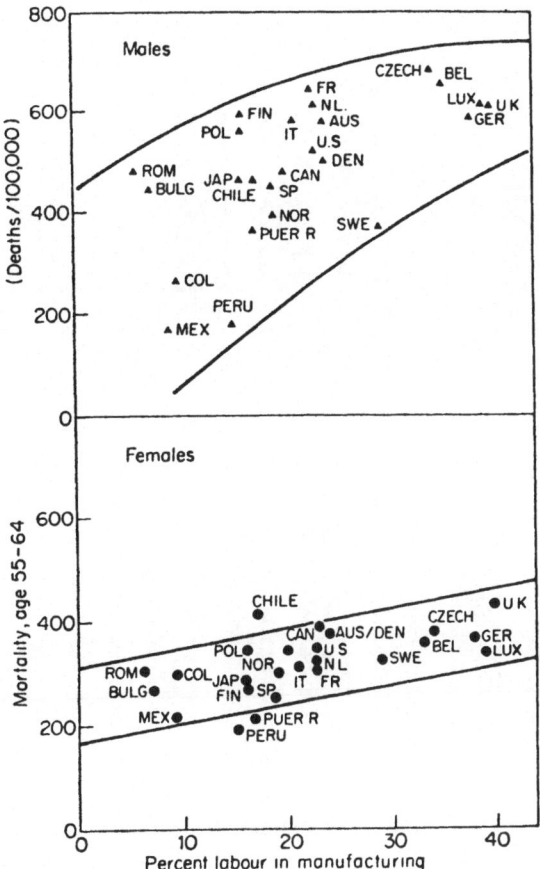

Figure 6. Relation between age-specific cancer mortality (1972–1973) and percentage labor in manufacturing (1940). From Harriss et al (5). Reprinted with permission of <u>Environment</u>.

cigarette smoking is by far the largest contributer to lung cancer, a fraction of lung cancers unrelated to smoking is increasing at a faster rate than the smoking-related fraction. Table 3 is a further analysis by Schneiderman which divides cancers into three classes: sites more likely to be environmentally related (respiratory, melanoma, bladder, kidney, liver, lymphoma and multiple myeloma), less likely (stomach, pancreas, intestine, and rectum) and unlikely (other sites). The annual rate of change (1970–1976) for those sites most likely to be environmentally related is by far the highest.

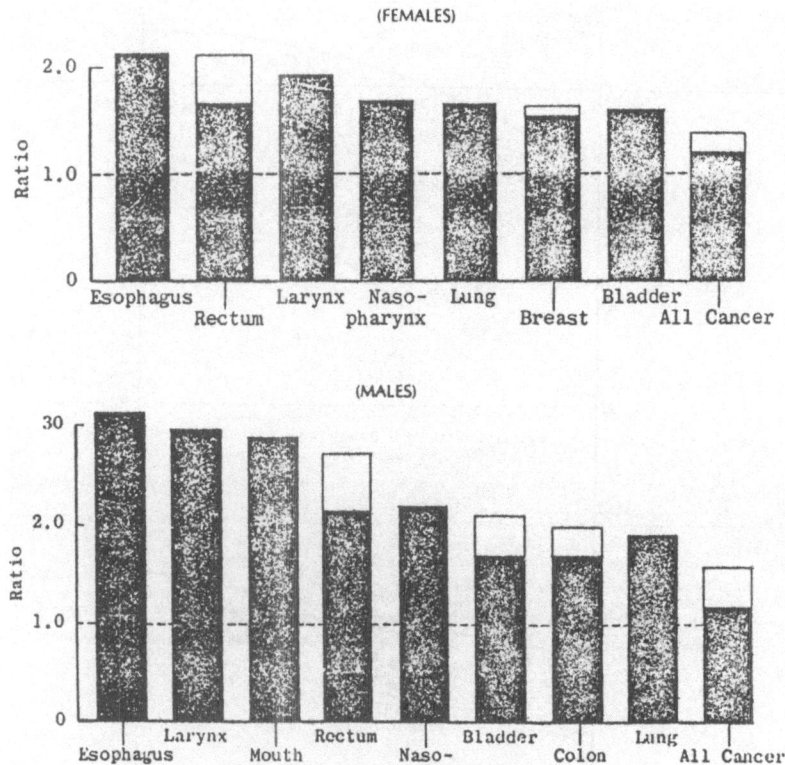

Figure 7.   Urban-rural county ratios of age-adjusted cancer mortality,
1950-1969.   From Goldsmith (7).   Reprinted with permission of the
J. Environ. Path. Toxicol.

Table 1.   Attributable Risk of Cancer due to Cigarette Smoking[a]

	Lung		Bladder		Pancreas	
	1969-1971	1976	1969-1971	1976	1969-1971	1976
Males	0.82	0.74	0.44	0.29	0.53	0.39
Females	0.70	0.62	0.25	0.15	0.35	0.25

[a]From M. A. Schneiderman (9).

Table 2.   Annual change in lung cancer rates[a]

New cases per 10^5 population		% change	% change per year
1969–71	1976		
Total 70.7	77.7	10.0	1.7
Non-smoking 12.4 Related with assumption of constant risk	17.0	36.8	6.1
Non-smoking 12.4 related with assumption of declining risk of smoking due to decreased tar and nicotine	20.5	65.3	10.9

[a]After M. A. Schneiderman (9).

        Another way to look at this is to directly examine the increase
in lung cancers among non-smokers. Enstrom (10) compiled data on
non-smokers; the numbers are not large and some of the comparisons
are open to question, but the conclusion from this source (Fig 8)
supports Schneiderman's analysis. There is a sharp increase in lung
cancer among male non-smokers and the same increase, although not as
well defined, for women.

        Turning briefly to occupational exposures and cancer, several
occupational cancers associated with specific jobs and/or to specific
agents to which the workers were exposed have been well documented.
The most well documented cases are usually those involving a rare
disease. For instance, it is easy to document that cases of angiosarcoma in vinyl chloride manufacture or of mesothelioma in asbestos
workers are occupationally related. The expected rate of such disease is nil, so that even 2 cases in one occupational exposure situation is strong evidence. Even in these cases, however, when detailed studies were made, it appears that the more common cancers,
e.g. lung cancer, are also linked to occupational exposure. Although
the evidence is weaker with the common cancers, the level of risk
may be higher than with the rare cancers. Thus, in over 5000 as-
bestos-insulation workers with more than 20 years exposure the
percentage increase in asbestosis over expected (nil) was much
greater than the percentage increase in lung cancer, but the total

Table 3.   Changes in cancer incidence: 1969–1971 through 1976

Cases per 10^5 age corrected to 1970[a]

White Males	1969–71	1976	Avg. Ann. Change, %
"More likely" sites[b]	55.0	69.5	+ 4.0
"Less likely" sites[c]	70.9	75.8	+ 1.1
Total "likely" sites	125.9	145.3	+ 2.4
Total others	216.6	228.7	+ 0.9
Total all sites	342.5	374.0	+ 1.5

[a]Smoking effects removed for lung, bladder, pancreas.
Changing smoking patterns considered.
Sunlight effect removed for melanoma.
"Others" includes all other sites plug smoking and sunlight-
attributed portions of lung, pancreas, and bladder, and melanoma.

[b]"More likely" includes respiratory, skin, bladder, kidney,
liver, lymphoma, and multiple myeloma.

[c]"Less likely" includes stomach, pancreas, intestine and rectum.

[d]Calculated by M. A. Schneiderman from TNCS and SEER data in
Toxic Chemicals and Public Protection (9).

impact from the excess lung cancer was twice that from asbestosis
(lung cancer--247 observed, 47 expected; asbestosis--94 observed,
none expected) (11). The increase in common cancers is much more
difficult to document. Without the clues provided by the rare
cancers an association may have been over-looked. The nature of
epidemiology is such that one usually needs several times the base
rate to document an effect. For example, lung cancer in coke oven
workers is well documented since the workers experience 3-7 times
the base mortality rate (12,13).

A report in 1978 by a group from the National Institute for
Occupational Safety and Health, the National Cancer Institute and
the National Institute of Environmental Health Sciences estimated
that 20% of cancer in the next few decades will be associated with
occupational exposures (14). They based their conclusions on the
effects of recent exposures not fully manifested in observed
cancers, estimates of industry-wide exposures based on studies of
high exposure cases, and inclusion of cancers produced jointly

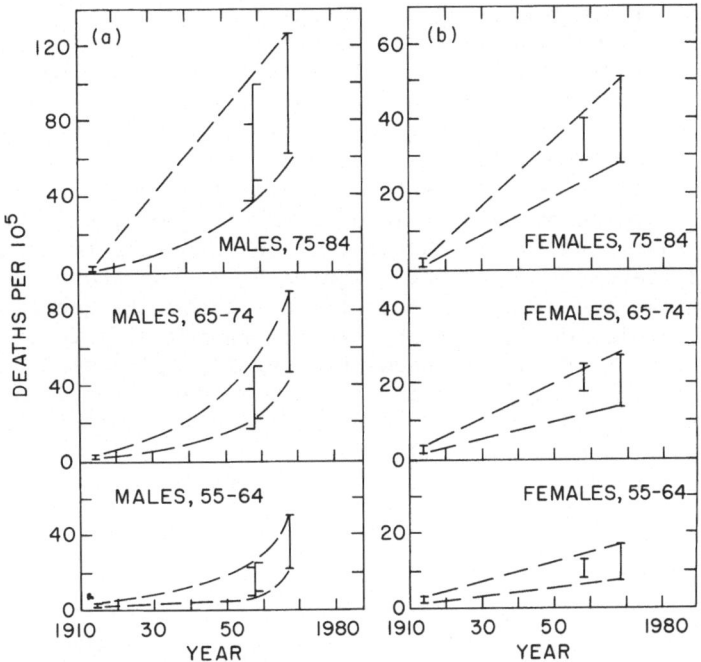

Figure 8.   Lung cancer mortality rates in white non-smokers; 95%
confidence intervals calculated by method of Haenszel et al. (41)
from data of Enstrom (10).

by occupational exposure and smoking. Since previous estimates
had placed occupational exposures in the 1-6% range, this report en-
gendered much controversy. A recent interagency committee report
has helped to place a proper perspective on the controversy (15).
However, the association of cancers with pollutant exposure in the
workplace is clear, regardless of the speculation on incidence rates.
Such exposures may act directly, or may increase the workers' risk
to cancer from other agents.

Strong arguements can be made to establish a basis for environmental cancer among the general public from industrial pollution.
There are several reports of workers bringing home--on their clothes
for example--toxic materials ultimately resulting in disease in their
families. There have been several studies of the families of workers
in the hydrocarbon industry. Two of the four studies found children
of such workers to be without an increased cancer risk. A third
study showed only a possibility of increased risk in one locale while
generally agreeing with those studies showing no increased risk (16-
19).

     A particularly broad-based study used to identify occupational
and public risks was the National Cancer Institute's Cancer Atlas,
U.S. Cancer Mortality by County: 1959-1960 (20). The Cancer Atlas
was presented as a tool for finding leads for further study (21-26).
It identifies counties with high cancer rates and may help to suggest hypotheses. The atlas also has been used in combination with
supporting evidence as a basis for demonstrating carcinogenic properties of various chemicals or industries. Fraumeni and his coworkers at the National Cancer Institute used the atlas to look at a
number of different industries. Figures 9-11 show these results for
paper, chemical and petroleum industries. These are generally corrected for population density or percent urbanization to account for
the urban factor. Figure 12 shows the results for arsenic. These
data are especially useful in an analysis we have been doing on
health effects of photovoltaic technologies using gallium arsenide.
Although Blot and Fraumeni (24) report a 15% increase in lung cancer
in counties with smelters, it is difficult to see how arsenic exposure alone could account for this (27).

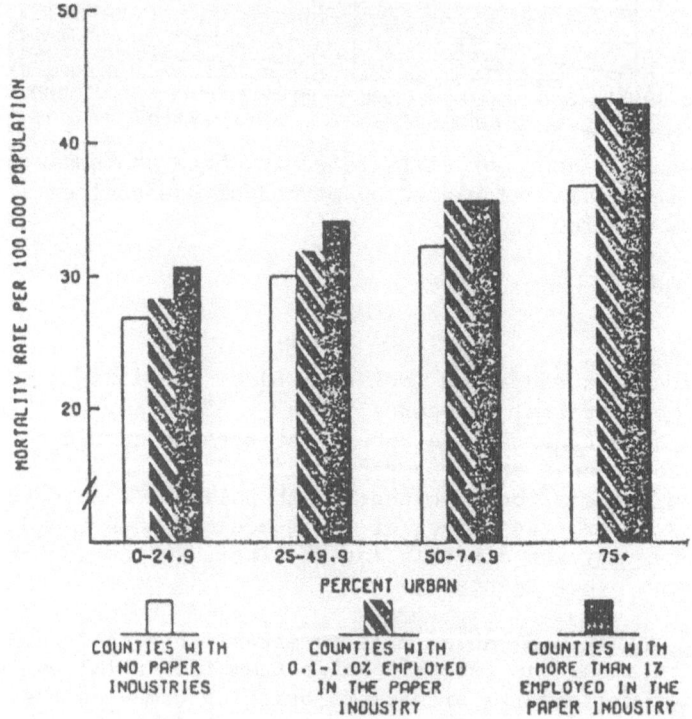

Figure 9. Average annual age-adjusted lung cancer mortality rates,
white males, 1950-1969, U.S. counties by percentage employed in
paper industry. Clear bar indicates counties with no paper industry;
striped 0.1-1.0% employed in paper industry; solid >1% employed in
paper industry. From Blot and Fraumeni (26). Reprinted with permission of the American Journal of Epidemiology.

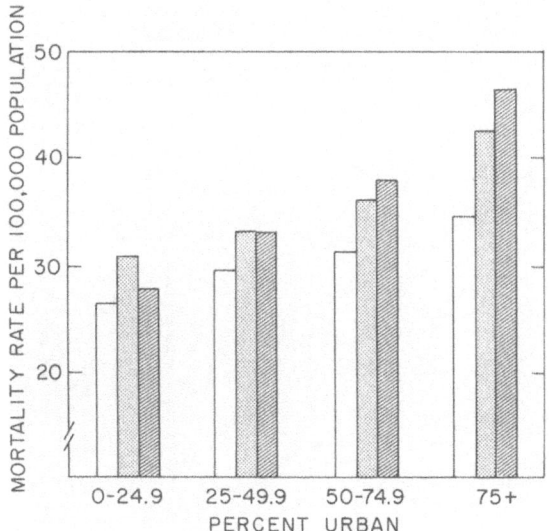

Figure 10. Average annual age-adjusted lung cancer mortality rates, white males, 1950-1969, U.S. counties by percentage employed in chemical industry. Clear bar indicates counties with no chemical industry; striped, 0.1-1.0% employed in chemical industry; solid, >1% employed in chemical industry. From Blot and Fraumeni (26). Reprinted with permission of the American Journal of Epidemiology.

     As one reaction to the Cancer Atlas, New Jersey quickly picked up the nickname "cancer alley". A recent report attempts to refute the basis for this nickname and makes some interesting points (28). The authors argue that because of its high population density, New Jersey, rather than being compared to other states, should be compared to urbanized areas like Nassau County or Westchester County, suburbs of New York City, or to the District of Columbia. However, New Jersey counties vary considerably in population density and urban character. The counties chosen for comparison have population densities 2 to 11 times that of New Jersey as a whole. Hopefully this should demonstrate that the so-called ecological approach to epidemiology, dealing with data on the basis of county vital statistics, is something with which one must be very careful. The need to be cautious does not mean that the approach should be abandoned. Our group has been active in developing, using and evaluating county level data (29-31). We believe it is cost effective and has some validity. An earlier study of the New Jersey data suggested that up to 4% of all cancers may have been associated with industrial pollution and indicated the potential adverse health effects of a now closed dye plant.

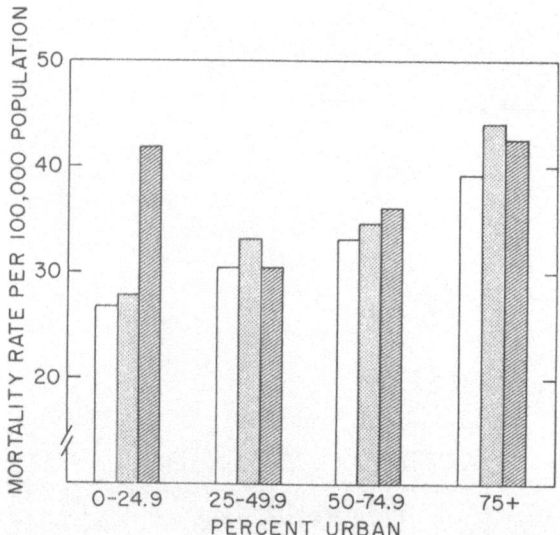

Figure 11. Average annual age-adjusted lung cancer mortality rates, white males, 1950-1969, U.S. counties by percentage employed in petroleum industry. Clear bar indicates counties with no petroleum industry; striped, 0.1-1.0% employed in petroleum industry; solid >1% employed in petroleum industry. From Blot and Fraumeni (26). Reprinted with permission of the American Journal of Epidemiology.

Definitive information can only come by examining cancer incidence in great detail. The National Cancer Institute has taken this approach in a follow-up to the Cancer Atlas. They have done in-depth studies in Georgia, a state which had exhibited high cancer rates in the atlas. Case-control studies established that the increased mortality in the atlas was at least partially due to occupational exposure in shipyards incurred by some 35,000 people during World War II (22). For policy, the distinction between an elevated cancer rate due to a general population exposure to pollutants in the air or in water or to occupational exposures which have not yet been linked to their true source is important.

The rare-type cancers are again useful as markers. There have been individual cases of mesothelioma among members of the general public with no known association with the asbestos industry (32). A case control study in a Louisiana parish, identified through the Cancer Atlas as having the highest lung cancer rates in the country, determined that the relative risk of lung cancer for people living within 1 km of an industrial plant (petroleum refinery, metal and lumber manufacturing and food cannery) was 1.9-3.2 times that of matched controls living greater distances from the industrial sites (33). In contrast, a study of Kaiser Foundation Health Plan members living in the San Francisco Bay Area showed no increased cancer in-

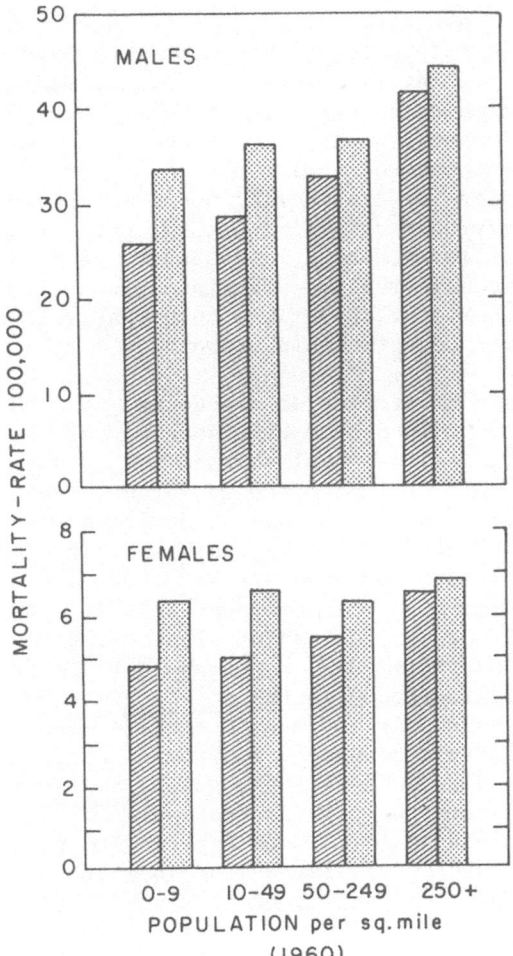

Figure 12. Average annual age-adjusted lung cancer mortality rates, white males and females, 1950-1969, U.S. counties by population density and presence of copper, lead or zinc smelting. Striped bar indicates counties with no copper, lead or zinc smelting, dotted bar counties with copper, lead or zinc smelters or refineries. From Blot and Fraumeni (24). Reprinted with permission of Lancet.

cidence among those living in areas characterized by a heavy concentration of petroleum and chemical industries compared to those members living in control areas (34).

There has been an interesting turnabout in a similar study done in Los Angeles. The first results clearly showed an increased rate of lung cancers in parts of the city where measured levels of benzo-(a)pyrene (BaP) were as much as 5 times that expected from automobile

emissions alone, presumably as a result of local petrochemical industries (35). An extrapolation was made from epidemiological results of British gas retort workers exposed to BaP with the conclusion that the increased cancer rate in these parts of Los Angeles
was compatible with what would be expected given these BaP levels
(36). A subsequent case-control study accounted for essentially all
of the excess cancers by considering occupational exposure (37).
Although this study does not eliminate the possibility that air
pollution caused some increased lung cancer, environmental exposure
can no longer be considered the principle cause of the excess cancers.
The effect, if it exists, seems less than what a linear extrapolation
from the gas retort workers would support.

In the assessment studies we are doing, we have taken dose-
response functions derived from high dose situations (frequently
occupational exposures) and applied them to low dose situations
where it is difficult or impossible to directly derive dose-response
functions. There are some problems in applying dose-response functions to situations different from those in which they were developed.
The specific pollutants involved are not quite the same. If one
applies a well-documented dose-response function for coke-oven workers to diesel exhaust or coal power plant emissions, using, for
example, cancers per ng/m^3 BaP or benzene soluble fraction of total
particulate, the differences in the relationship between the index
and the total exposure mix may make a considerable difference in the
response. The exposure situation and timing would also be much different. Uncertainties such as these should be expressed explicitly
in the results. Examples in which extrapolations can be validated--
or proved wrong--such as in the Los Angeles study, can be used as
guides in making such extrapolations or in establishing levels of
uncertainty. Such an extrapolation confines us to solid data in
man. A valuable use of non-human data might be in adjusting the
dose-response functions by comparing the in vitro test responses to
the mixtures from the known human situation, e.g. coke-oven workers,
to the mixture in the unknown one, e.g. public exposure to diesel
exhaust, wood smoke, or coal power plants. Direct quantitative extrapolation from animal studies involves even more uncertainties than
extrapolation from high dose human exposure situation to low dose
ones, but the ratio of effects to similar, but different, agents or
mixes may be useful.

In conclusion, there is little question that considerable cancer
is due to exposure to industrial and chemical pollutants, especially
among workers. That will probably continue to be the case. People
living in the immediate vicinity of industrial plants releasing pollutants are also at increased risk for cancer. The least documented
situation is whether there is an increased risk for cancers among
the general public due to widespread, low level chemical exposure
in air and water. Verification of this latter situation will remain

highly uncertain and of great or little importance depending on our
attitude. In most cases, we are talking about an extremely low risk
situation. The risk to the public from diesel exhaust is predicted
to be extremely low--an insignificant increase in the current risk
associated with urban air pollution (38). The introduction of a new
pollutant that is probably a carcinogen and to which millions will
be exposed is a difficult problem for policy makers to face. Added
to this problem is the uncertainty in the state of our current knowledge, and the possibility of greater than anticipated effects.

Although the risk to those exposed may be insignificantly low
on an individual basis, the aggregated risk to the millions of exposed individuals may be sufficient to justify considerable expen-
ditures on control measures or other technologies. In many situations we do not need to shut down something immediately to prevent
an epidemic of cancers, but society may want to balance costs and
benefits. The current high uncertainty in knowledge often means
that in such a balancing, science is outweighted by values and at-
titudes. Values will always be important, we must take care that
pseudo-arguments over science are not used to mask differences in
values. Hopefully, as we learn more, the role of science will increase and policy making will become more accurate.

ACKNOWLEDGEMENT

Work supported by Health and Environmental Risk Assessment
Program, Office of Health and Environmental Research, U.S. Department
of Energy. I thank S.R. Bozzo and L.D. Hamilton for their advise.

REFERENCES

1. S. S. Devesa and D. T. Silverman, Cancer incidence and mortal-
    ity trends in the United States, 1935-1974. J. Natl. Cancer
    Inst. 60:545-571 (1978).
2. S. S. Devesa and D. T. Silverman, Trends in incidence and mor-
    tality in the United States. J. Environ. Path. and Tox.
    3:127-155 (1980).
3. C. S. Pollack and J. W. Horm, Trends in cancer incidence and
    mortality in the United States, 1969-76. J. Natl. Cancer
    Inst. 64:1091-1103 (1980).
4. J. Higginson, Environment and Cancer. The Practitioner 198:
    621-630 (1967).
5. R. C. Harriss, C. Hoheuemser and R. W. Kates, The burden of
    technological hazards, in: "Enetgy Risk Management",
    G. T. Goodman and W. D. Rowe, eds., Academic Press, New
    York, 118-137 (1979).
6. J. R. Goldsmith, Effects of air pollution on human health,
    in: Air Pollution, 2nd ed., A. C. Sterm, ed., Academic

Press, NY, 1:547-615 (1968).

7. J. R. Goldsmith, The "urban factor" in cancer: smoking,
    industrial exposures, and air pollution explanations.
    J. Environ. Path. and Tox. 3:205-217 (1980).

8. S. R. Finch and S. C. Morris, Consistency of reported
    effects of air pollution on mortality, in: Advances in
    Environmental Science and Engineering, J. R. Pfafflin
    and E. N. Ziegler, eds., Gordon and Breach, London,
    106-117 (1979).

9. M. A. Schneiderman, As reported in Toxic Substances Strategy
    Committee, Toxic Chemicals and Public Protection, U.S.
    Government Printing Office, 161-165 (1980).

10. J. E. Enstrom, Rising lung cancer mortality among nonsmokers.
    J. Natl. Cancer Inst. 62:755-760 (1979).

11. W. J. Nicholson, Asbestos: the TLV approach. Annals NY Acad.
    Sci. 271:152-169 (1976).

12. J. W. Lloyd, Long-term mortality study of steelworkers, v.
    Respirating cancer in coke plant workers, J. Occup. Med.
    13:53-67 (1971).

13. C. K. Redmond, B. R. Strobino and R. H. Cypress, Cancer
    experience among coke by-product workers. Annals NY
    Acad. Sci. 271:102-115 (1976).

14. K. Bridbord, P. Decoufle, J. F. Fraumeni, Jr., D. G. Huel,
    R. N. Hoover, D. P. Rall, U. Saffiotti, M. A. Schneiderman,
    and A. C. Upton, Estimates of the fraction of cancer
    in the United States related to occupational factors.
    Institutes' testimony on proposed OSHA cancer policy. (1978).

15. Toxic Substances Strategy Committee, Toxic chemicals and Public
    Protection, U.S. Government Printing Office (1980).

16. S. L. Kwa and L. J. Fine, The association between parental
    occupation and childhood malignancy. J. Occup. Med. 22:
    792-794 (1980).

17. M. Zack, S. Cannon, D. Loyd, C. W. Heath, Jr., J. M.
    Falletta, B. Jones, J. Housworth and S. Crowley, Cancer
    in children of parents exposed to hydrocarbon-related
    industries and occupations. Am. J. of Epidemiology
    111:329-336 (1980).

18. J. Fabia and T. D. Thuy, Occupation of father at time of
    birth of children dying of malignant disease. Br. J.
    Prev. Soc. Med. 28:98-100 (1974).

19. T. Hakulinen, T. Salonen and L. Teppo, Cancer in the
    offspring of fathers in hydrocarbon-related occupations.
    Br. J. Prev. Soc. Med. 30:138-140 (1976).

20. T. J. Mason and F. W. Mckay, F. W., U.S. Cancer mortality by
    county: 1959-1969. DHEW Publication No. (NIH) 74-615,
    National Cancer Institute, Bethesda, MD (1974).

21. R. Hoover and J. F. Fraumeni, Jr., Cancer mortality in U.S.
    counties with chemical industries. Environ. Res. 9:196-207
    (1975).

22. W. J. Blot, J. F. Fraumeni, Jr., T. J. Mason and R. N., Hoover, Developing clues to environmental cancer: a stepwise approach with the use of cancer mortality data. <u>Environ. Health Perspectives</u> 32:53-58 (1979).

23. W. J. Blot, B. J. Stone, J. F. Fraumeni, Jr. and L. E. Morris, Cancer mortality in U.S. counties with shipyard industries during World War II. <u>Environ. Res.</u> 18:281-290 (1979).

24. W. J. Blot and J. F. Fraumeni, Jr., Arsenical air pollution and lung cancer. <u>The Lancet</u> ii:142-144 (1975).

25. W. J. Blot, L. A. Brinton, J. F. Fraumeni, Jr. and B. F. Stone, Cancer mortality in U.S. counties with petroleum industries. <u>Science</u> 198:51-53 (1977).

26. W. J. Blot, and J. R. Fraumeni, Jr., Geographic patterns of lung cancer: industrial correlations. <u>Am. J. Epidemiology</u> 103:539-550 (1976).

27. P. D. Moskowitz, L. D. Hamilton, S. C. Morris, K. M. Novak and M. D. Rowe, Photovoltaic Energy Technologies: Health and Environmental Effects Document BNL 51284, Brookhaven National Laboratory, Upton, NY (1980).

28. H. B. Demopoulos and E. G. Gutman, Cancer in New Jersey and other complex urban/industrial areas. <u>J. Environ. Path. and Tox.</u> 3:219-235 (1980).

29. S. C. Morris, K. M. Novak and C. E. Calef, Use of County Level Data in Health, Energy, Demographic, Environmental and Economic Analysis. Proc. of a computer-based conference. (BNL 51041) Brookhaven National Laboratory, Upton, NY (1979).

30. S. R. Bozzo, K. M. Novak, F. Galdos, K. Hakoopian and L. D. Hamilton, Mortality, migration, income and air pollution: a comparative study. <u>Social Science and Medicine</u> 13P:95-109 (1979).

31. F. R. Drysdale and C. E. Calef, The Energetics of the United States of America: An Atlas. BNL 50501-12, Brookhaven National Laboratory, Upton, NY (1977).

32. M. L. Newhouse and H. Thompson, Mesothelioma of the pleura and peritoneum following exposure to asbestos in the London area. <u>Br. J. Lud. Med.</u> 22:261-269 (1965).

33. C. L. Shear, D. B. Seale and M. S. Gottlieb, Evidence for space-time clustering of lung cancer deaths. <u>Arch. Environ. Health</u> 35:335-343 (1980).

34. C. D. Hearey, H. Ury, A. Siegelaub, M. K. P. Ho, H. Salomon and R. L. Cella, Lack of association between cancer incidence and residence near petrochemical industry in the San Francisco Bay area. <u>J. Natl. Cancer Inst.</u> 64: 1295-1299 (1980).

35. H. R. Menck, J. T. Casagrande and B. E. Henderson, Industrial air pollution: possible effect on lung cancer. <u>Science</u> 183:210-212 (1974).

36. M. C. Pike, R. J. Gordon, B. E. Henderson, H. R. Menck, and J. Soottoo, Air Pollution, <u>in</u>: "Persons at High

Risk of Cancer", J. F. Fraumeni, Jr., ed., Academic Press, New York, 225–240 (1975).

37. M. C. Pike, J. S. Jing, I. P. Rosario, B. E. Henderson and H. R. Menck, Occupation: "explanation" of an apparent air pollution related localized excess of lung cancer in Los Angeles County, in: "Energy and Health", N. E. Breslow and A. S. Whittemore, eds., SIAM, Philadelphia, 3–16 (1979).

38. R. G. Cuddihy, F. A. Seiler, W. C. Griffith, B. K. Scott and R. O. McClellan, Potential health and environmental effect of diesel light duty vehicles (LMF-8L), Inhalation Toxicology Research Institute, Lovelace Biomedical and Environmental Research Institute, Albuquerque, NM (1980).

39. Council on Environmental Quality, 11th Annual Report US Government Printing Office, Washington, DC (1980).

40. D. L. Davis, Cancer and industrial chemical production. Science 206:1356–1358 (1979).

DISCUSSION

Q. Hatch (Northrop): Let me ask just one question. You talked about some cancer incidences that have been rising and suggest putative explanations. One of your slides showed that the incidence of gastric cancers has declined or plateaued for the last 50 years. Would you care to speculate on why that one specific type of cancer rate might be going down and do you think it might be related to a possible decrease in mycotoxin contamination in food stocks?

A. Morris (BNL): I don't know. It's obviously very likely related to what we eat. Beyond that, I'm not sure I'm prepared to venture.

Q. Borg (BNL): Let me push you a little further on what you think about the urban factors. You showed us some bar graphs which indicated that John Goldsmith's explanation that the urban effect resulted from a collection of occupational exposures isn't the whole story. It appears that there is an urban factor above and beyond the occupational exposure. What urban factors are left?

A. Morris (BNL): Well, one that is talked about is differences in cigarette smoking between urban and rural areas.

Q. Borg (BNL): Does that bear out?

A. Morris (BNL): Surprisingly, there is not an overwhelming amount of data to support this possibility. We have looked at cigarette smoking, but in a somewhat different context. In comparing the overall effects of air pollution in different

cities, some people have attempted to correct for cigarette
smoking by using state data on cigarette consumption. The data
is more readily available on a state basis because the states
collect taxes on cigarettes. The difficulty with using state
data is that when you are comparing a city in the Midwest with a
city in the East, the state data from a rural state doesn't
necessarily go with the city that happens to be in the rural state.
To what degree this discordance occurs is just not clear. There's
all sorts of other potential urban factors, certain kinds of
viruses, for example, could be concentrated in an urban area where
people are more crowded.

Q. Borg (BNL): What about, at least in the last half of the
20th century, poverty and its attendent changes on dietary habits,
medical care and personal hygeine? Is poverty mostly an urban
affair in this half of the century?

A. Morris (BNL): I'm not sure that's the case.

Q. Borg (BNL): But poverty is associated with higher instances of
cancer, isn't it?

A. Morris (BNL): I'm not sure that poverty is necessarily concentrated in the cities. You could look at cancer incidence
differences between the inner cities and the suburban ring where
there is more of a contrast of poverty.

Q. Costa (BNL): First, to comment on the incidence of gastric
cancer. I once heard a lecture by John Weissberger of the American
Health Foundation. He speculated that the decrease in gastric cancer
was due to refrigeration. Refrigerators came into common use
about the time the decrease started.

    An early graph you presented showed that the incidence of
lung cancers is increasing. Is most of that increase non-smoking
related?

A. Morris (BNL): It would appear that the increase in lung cancer
not related to smoking is going up at a much faster rate, perhaps
five or ten times, then is the increase in the lung cancer
presumably related to smoking. On the other hand, the fraction
of lung cancers related to smoking constitutes the bulk of these
cancers, so that the total increase in cancer is much closer to
the increase in that fraction related to smoking.

Q. Goldstein (Cancer alley - Rutgers): Three quick questions:
First, on the important point you made about the rare tumors
caused by exposure to vinyl chloride and asbestos. You seemed to
imply that with vinyl chloride as well as with asbestos, you have

evidence that tumors other than hepatic angiosarcomas are occurring
in the exposed workers. That isn't my understanding of what NIOSH
has been finding. Second question, you said something about
arsenic, which you were going to get back to and you never did. I
think it's a very important point particularly if we go to more
coal use. Have you or any of the other investigators looking at
arsenic and lung cancer used skin cancer, which is also supposedly
related to arsenic exposure, as a way to confirm whether or not
the effects are truly there? And the third question is a general
one. You referred to Demopoulos' findings. What did you really
mean?

A. Morris (BNL): Let's start with the last question. I think
that there's a suspicion when you read Demopoulos' paper that he
is picking out numbers to suit his case. On the other hand, I
think that he makes some very valid points that the ecological
approach of epidemiology can be misleading.

Second, the analysis we've been doing with arsenic has been
strictly for lung cancers. The estimated increased incidence of
lung cancers are low enough that you will never find them. We
have estimated arsenic emissions for that part of the arsenic
production process which would be involved in a large photovoltaic
industry, arriving at minute fractions of cancer. For total
arsenic emission, we would expect something like four or five lung
cancers within a 50-mile radius of the production works, or a
couple of hundred lung cancers nationwide from all arsenic emissions.
Now, that's based on the application of a linear no-threshold dose
response function, which may overstate the case. We haven't done
a similar calculation for skin cancer. I don't know anyone who
has. Bernard Cohen from Pittsburgh made an interesting calculation
where he took the arsenic from coal production and followed it
throughout its lifetime, that is to say until it was eventually
deposited in deep-sea beds. On that basis, he estimated, assuming
that the population remains somewhat similar to what it is today,
that you would produce something like 70,000 cancers per plant-
year over the few hundred thousand years that the arsenic
remained in the biosphere.

With vinyl chloride, my understanding is that there is a lung
cancer increase which is comparable and perhaps bigger than the
increase in hepatic tumors.

Q. Goldstein (Rutgers): What about brain tumors?

A. Morris (BNL): The increase in brain tumors in not substantial
in size.

Q. Albertini (Vermont): I was recently at a symposium at Oak
Ridge where someone facetiously said that perhaps the most potent
carcinogen would be an effective cure of cardiovascular disease.

Since the death rate is recalcitrant at one, are these data corrected for the rather substantial increases in cardiovascular disease and other diseases in the last several years as causes of death? Might it be that other causes are dropping off?

A. Morris (BNL): Obviously, other causes of death are dropping off. Therefore, you're more likely overall to die of lung cancer because you're less apt to die of heart disease. These data are corrected for that.

Q. Hatch (Northrop): Didn't it appear, though, that your non-smoking population increase in lung cancers is in older age groups, as opposed to the age of risk for smokers?

A. Morris (BNL): That's right. I think that has to do a lot with the fact that it's much easier to see an increase where the population base is bigger and the base is simply a lot bigger in the older age groups. In order to detect an increase in lung cancer in 20 year-olds, the size of the population must be pretty big.

Q. Axelrod (Livermore): Correcting for age is not going to correct for a change in competitive risks. I suspect that you haven't corrected for competitive risks, but just for age.

A. Morris (BNL): That's right, there is an additional correction for the competitive risk of dying from heart disease.

POLICY AND PROCEDURES FOR USING MUTAGENICITY DATA IN ASSESSING

GENETIC RISK

Peter Voytek

U.S. Environmental Protection Agency
Reproductive Effects Assessment Group
Washington, DC   20460   (U.S.A.)

INTRODUCTION

Before a regulatory agency such as the U.S. Environmental
Proection Agency (EPA) can use mutagenicity data to assess genetic
risk, many societal and bureaucratic events must occur. First,
the Congress must recognize the need for environmental legislation
to protect the public and the environment against exposure to
harmful toxicants. This recognition is initiated primarily by
public episodes and awareness, such as the kepone and PCB incidences, which prompt the Federal Government to take appropriate
action through legislation.

The next step involves the actual drafting of the statute by
the Congress, with input coming from various sources, e.g. federal,
state, and local governments; the White House; environmentalist
groups; industry. Because of the diversity in professional back-
grounds and environmental interests, in addition to debates over
controversial issues involving the balancing of economic considerations against preservation of public health and the environment, many
years may pass before the statute is finally approved by both the
U.S. Senate and the House of Representatives. After this process
has occurred and if the statute is signed into law by the President
it goes to the appropriate agency in the Executive Branch which
has the authority to see that the mandate from Congress is
implemented. Within EPA there are seven statutes which gives the
Agency the authority to perform health risk assessments that can
be used in taking regulatory actions. They include: the Clean
Air Act, the Atomic Energy Act, the Safe Drinking Water Act, the

Federal Water Pollution Control Act, the Resource and Conservation
Recovery Act, the Federal Insecticide, Fungicide and Rodenticide
Act, and the Toxic Substances Control Act. Table 1 contains
examples of some verbage from a few of these acts. This type of
language which pertains to public health safety is written in
general terms and it becomes the Agency's responsibility to interpret
the wording in the statutes to forge workable test rules, guidelines,
and regulations. EPA's interpretation of these laws comes under
severe scrutiny and is often challenged in the courts by environmental groups who often feel the Agency's position is too weak in
protecting the public. On the other hand, representatives from
industry often feel the positions taken are too stringent and will
impair economic growth and stability.

EPA has interpreted the statutes mentioned above as giving the
Agency the authority to use mutagenicity data on environmental
pollutants to assess heritable genetic disease. Once this decision
was made, guidelines needed to be developed for mutagenicity risk.
An internal Agency Steering Committee was formed, which, under the
guidance of an expert in the field of genetic toxicology, was
charged with writing proposed guidelines for assessing mutagenic
risk. Since many program offices of EPA ultimately use assessments
for regulatory decisions, the members of the Steering Committee
were drawn from all parts of EPA to obtain their input and to gain
their support. The mutagenicity risk assessment guidelines went
through eight drafts within a year and then went out for external
review by nationally and internationally known experts in genetic
toxicology. Comments from these reviewers were used to revise the

Table 1. Wording in some EPA statutes as they may
          pertain to mutagenicity

1. Toxic Substances Control Act, (1976). "Testing Requirement
Rule-health and environmental effects for which standards for the
development of test data may be prescribed include carcinogenesis,
mutagenesis, teratogenesis...and any other effect which may present
an unreasonable risk of injury to health or the environment." p. 2007.

2. Resource Convervation and Recovery Act (1978). "regulating the
treatment, storage, transportation and disposal of hazardous wastes
which have adverse effects on health and the environment" p. 2798
"Identification and listing of Hazardous Waste Management...taking
into account toxicity...," p. 2806.

3. Federal Insecticide, Fungicide and Rodenticide Act (1978).
"unreasonable adverse effects on the environment means and unreasonable risk to man or the environment, taking into account the
economic, social and environmental costs and benefits of the use
of any pesticides." p. 6.

guidelines before the Administrator approved them.   After two and
a half years, the guidelines were finally published as'proposed
guidelines in the Federal Register on November 13, 1980.

Public comments received during a 90-day comment period will be
answered by the Agency's Reproductive Effects Assessment Group.
Then both comments and responses to comments will be reviewed by
the Agency's Science Advisory Board, which consists of senior
scientists outside of EPA who advise the Administrator on all
scientific matters which will impact on regulatory policy.   Therefore, from the time Congress recognizes the need for a government
action to protect the environment to the promulgation of the risk
assessment guidelines, several years may pass before the parts of
the law pertaining to health risks may be effectively put into
use.

HISTORY ON THE USE OF MUTAGENICITY DATA FOR HEALTH ASSESSMENT
PURPOSES

In 1976 EPA published interim Procedures and Guidelines for
Health Risk of Suspected Carcinogens (1).   These procedures and
guidelines proposed that short-term mutagenicity testing could be
used as supportive evidence that a chemical may be a potential
carcinogen to exposed humans.   Epidemiological evidence or positive
results from a whole mammal bioassay, however, was needed to
confirm a chemical's carcinogenic potential and only these data
could be used to estimate the cancer risk to humans.

In 1977 the Agency, based on mutagenicity data, issued a
Rebuttable Presumption Against Registration (RPAR) of the pesticide
ethylene oxide (2).   The RPAR process allows the Agency to assess
new data on a chemical already in use, and when the evidence indicates
the chemical may cause adverse health effects, the Agency can make
public these findings to alert the manufacturers and users.   This
procedure represents the first step taken by the Agency which can
lead to actions restricting the use of and exposure to the pesticide.
In 1978 the Agency proposed a battery of mutagenicity test protocols
which included tests for point mutations, chromosomal aberrations,
and primary DNA damage (3).   Negative results from this battery
of tests which cover most, if not all, genetic endpoints could be
used by the Agency to operationally define a chemical to be a
nonmutagen and with less confidence, to be a noncarcinogen.

PROPOSED AGENCY-WIDE MUTAGENICITY RISK ASSESSMENT GUIDELINES

In 1980, the Agency proposed Mutagenicity Risk Assessment
Guidelines (4).   These guidelines represented the first time a
regulatory agency proposed the use of mutagenicity data as a
toxicological endpoint to protect the public against exposure to

genotoxic environmental pollutants. The guidelines contained
sections on rationale for regulating chemical mutagens, concepts
relating to heritable mutagenic risk, rationale for using nonhuman
test systems and quantitative risk assessment methods. The purpose
of the guidelines was not only to describe how the Agency will use
mutagenicity data to assess heritable risk, but more importantly
to promote a better quality of genetic toxicity testing. This
effort was made to insure more reliable risk assessments on which
to base regulatory actions.

Thus far, mutagenicity risk assessments performed in EPA have
been of a qualitative nature. The weight-of-evidence-approach
involves examining all available mutagenic, toxicological, or
biochemical evidence indicating that the chemical in question or
its active form reaches germinal tissues. When the data are
limited to in vitro or nonmammalian test systems, only statements
as to the "intrinsic" mutagenicity of an agent can be made. If
additional toxicological data in whole mammals are available and
indicate the chemical can reach the germinal tissues and cause
testicular atrophy or abnormal sperm morphology, then a more confi-
dent statement can be made that the chemical may be a potential
human mutagen.

Unfortunately, most genetic toxicology studies on environmental
agents are of limited quantity, poor quality or were designed for
purposes other than assessing mutagenic activity. Until these
problems are overcome, the quality and usefulness of mutagenic
assessments will suffer.

At present, there are two approaches for estimating the risk
of heritable genetic disease resulting from environmental exposure
to genotoxic agents. The first approach involves whole mammal
bioassays, e.g. the mouse specific locus test (5) in which the
parental generation is exposed and visible mutations are sought
for in the offspring. These tests can demonstrate that an agent
has mutagenic activity, that it reachs the germinal tissues and
that mutations are transmitted to the next generation, i.e. are
heritable. Of course, the mammalian system is only a model for man
and may only approximate the various modifying parameters in humans
such as gonadal barrier, germ-cell stage specificity, pharmacokinetics, metabolism and repair mechanisms. However, if one assumes
that the rodent model is representative of events occuring in humans,
then extrapolations can be made to estimate the increased incidence
of heritable genetic disease at different exposure levels.

Because many of the whole mammal tests are in the developmental
stage, with the exception of the mouse specific locus test described
by Russell (5,6), the costs of evaluating a large number of chemicals
are high and there are not enough qualified laboratory facilities
to test more than a few chemical substances per year.

An alternative dosimetry approach (7) is being considered as another means to estimate mutagenic risks in humans. This second approach involves nonmammalian or in vitro test systems in which a correlation between induced mutation frequency and DNA alkylation is compared with data concerning exposure versus alkylation in germinal cell DNA of intact mammals. By using DNA alkylation as the common denominator, a relationship between mutation frequency and exposure can be obtained. These data can then be used to estimate the increased risks for heritable genetic diseases in humans by using the doubling dose technique (8).

Since it is not possible to conclude from the in vitro assays that mutations are actually occurring in germinal cells, the dosimetry approach is one step removed from the whole mammalian bioassay approach discussed above. Furthermore, it is only applicable to chemicals which cause gene mutations; clastogenic or non-disjunctional events are not ascertained. The technique does demonstrate that the chemical has mutagenic activity, reaches the target tissue, and interacts with the target molecule DNA. Therefore, human exposure to the chemical could cause mutations transmissible to future generations. An example of a quantitative mutagenic risk assessment using the dosimetry approach for the compound ethylene dibromide is presented later in this report.

In writing the current EPA mutagenicity guidelines, numerous research needs were identified that could provide better approaches to assess mutagenic risk. For example, many of the in vivo mammalian assays need to be validated with direct- and indirect-acting mutagens from different structural classes. A comparison of dose-response data among the various test systems can provide insight as to which assay is the most sensitive for particular classes of chemicals.

Many of the mammalian and microbial in vitro systems for point mutations are inexpensive and simple to perform. Theoretically, any of these tests could be used in the dosimetry risk assessment approach to obtain a correlation between induced mutation frequency versus alkylation to DNA. However, a comparison of the relationships among the various test systems with whole mammal bioassays is needed for verification. Adduct binding to DNA is measured radioactively by labelling the suspected mutagen and thus requires a purified compound. Since many environmental agents are complex mixtures with traces of unknown contaminants, the usual labelling techniques would not be applicable. Other methods of assessing interactions of environmental mixtures and DNA would then be needed to correlate induced mutation frequency in the short-term test system with DNA damage in germinal cells. Therefore, examination of other prospective DNA dosimeters is needed in order to use the dosimetry approach for estimating the increased heritable risks associated with environmental mixtures.

The monitoring of biological effects in humans can be used to provide evidence that environmental agents may have mutagenic potential. Monitoring tests which may be applicable involve examination of sister chromatid exchanges (SCE), micronuclei, sperm count, sperm motility and abnormal sperm morphology. Body fluids and waste products can also be tested for point mutational activity in experimental test systems for gene mutations. However, these procedures cannot be used for mutagenic risk assessments since they represent a toxicological effect which may only be associated with mutagenic events in germ cells. The selection and experimental design of such studies, as well as the interpretation of the results, should be carefully reviewed by scientists with experience in both risk assessment and in human biological effects monitoring. Although animal systems can be used to estimate mutagenic risk to humans, many uncertainties exist and assumptions must be made which can weaken the accuracy of the assessment. However, if an environmental agent has been shown to cause in whole mammal systems heritable mutations and to cause in man a toxicological response, e.g. SCE, which is indicative of DNA damage (the same toxicological response being observed in the test animals), then the estimation of mutagenic risk in humans from animal systems is strengthened and supported. These principles are summarized in Table 2.

COMPARISON OF MUTAGENIC RISK WITH CARCINOGENIC RISK USING ETHYLENE DIBROMIDE

In a recent EPA report (9), a quantitative mutagenic risk assessment for ethylene dibromide was performed using the dosimetry approach. Mutation frequency to alkylation of DNA was correlated in *Drosophila melanogaster*, while mice were used to correlate exposure levels (by inhalation) to DNA alkylation in male germ

Table 2.   Extrapolation of mutagenic risk from
           animal studies to humans

mammalian tests for mutagenicity	uncertainties / assumptions →	human risk assessment
+		
biological effects in humans and in the animal model	less uncertainties / less assumptions →	human risk assessment
biological effects monitoring in humans	unable to be / done at present →	human risk assessment

cells.   Table 3 outlines the procedure used to determine the
heritable genetic disease risk in humans associated with exposure
to ethylene dibromide.   Estimates of human inhalation exposure to
this pesticide at citrus fumigation stations varied from 0.4 to
151.8 mg/Kg/yr depending on the use and application technique (10).
For example, exposure of workers at one operation site was estimated
to be 9.4 mg/Kg/yr.   If people of reproductive age were exposed to
this level of ethylene dibromide for ten years, the total exposure
dose would be 94 mg/Kg/generation.   In mice, such an exposure would
lead to 1.20×10^{-5} alkylations/sperm nucleotide/generation (9).
In Drosophila melanogaster, this amount of alkylation would yield
0.8 doubling doses.   From the UNSCEAR report (8), it is estimated
that in humans, doubling the mutation frequency (the doubling dose)
would result in 2000 autosomal dominant and 500 irregular inherited
mutations (diseases) per million live births or, at equilibrium,
14,500 inherited mutations per million live births.   Therefore,
an exposure of 9.4 mg/Kg/yr of ethylene dibromide for ten years
was estimated to increase the incidence of human genetic disease
in the first generation by 0.20% and, at equilibrium, 1.16% (11).

Table 3.   Outline of the procedure used to estimate
heritable genetic disease risk in humans
for etheylene dibromide

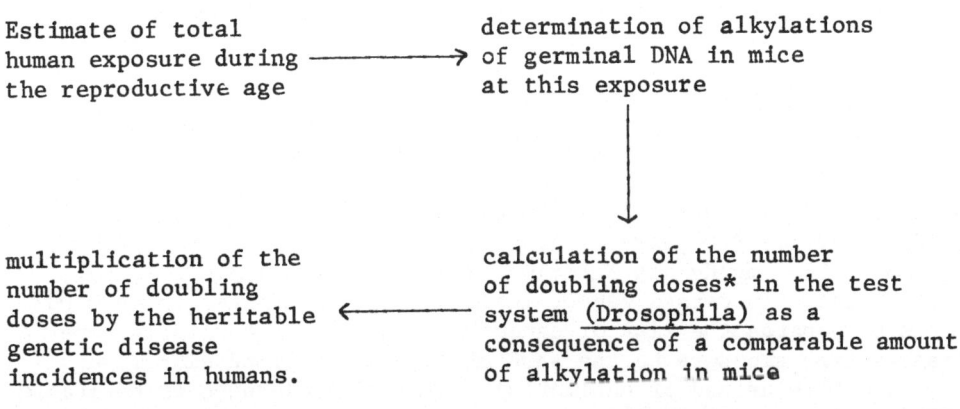

Estimate of total human exposure during the reproductive age ⟶ determination of alkylations of germinal DNA in mice at this exposure

multiplication of the number of doubling doses by the heritable ⟵ genetic disease incidences in humans.

calculation of the number of doubling doses* in the test system (Drosophila) as a consequence of a comparable amount of alkylation in mice

*A doubling dose is a dose that will induce a mutation frequency
equal to the spontaneous frequency.

Within the last decade much data have accumulated which
supports a strong correlation between chemical carcinogenicity
and mutagenicity (12). Indeed many scientist feel that all mutagens
are likely to be carcinogens and that the initiation step in chemical
carcinogenesis represents a somatic cell mutational event. In
this text, mutagens have been described as those agents which
can reach germ cell DNA and permanently alter gene expression. The
permanent change survives and can be expressed in subsequent generations. Since altering the mutation frequency is most likely to
increase the incidence of genetic disease in future generations
and, unlike cancer, not cease with the death of the affected
individual, regulatory agencies and the public are becoming more
concerned about long-term effects of environmental chemicals.

Thus far, cancer has been treated as the most sensitive
toxicological endpoint to be used to protect the public. This
concern is based primarily on the "no-threshold" concept of carcinogenicity; one molecular interaction between a chemical and
DNA can lead to a permanent alteration in genetic information,
i.e. a mutation, initiating cell transformation. This same "no
threshold" concept for chemicals causing gene mutations also
applies when assessing the risk to heritable genetic disease.
Instead of the DNA in somatic cells, the target molecule is DNA
in germinal tissue. From radiation studies in mice, it has been
shown that for an equivalent amount of ionizing radiation increased
incidences of cancer are of the same order of magnitude as the
increased incidences of heritable genetic disease (13). These
findings indicate that for ionizing radiation both toxicological
endpoints (cancer and heritable genetic disease) appear to have
the same sensitivity.

The EPA's Carcinogen Assessment Group has performed a risk
assessment on the carcinogenicity of ethylene dibromide from the
mouse cancer bioassay (15). By using the multistage model (14,15),
the following relationship was obtained.

$$P = 1 - e^{-(0.221)}$$

where D is human inhalation exposure to ethylene dibromide in mg/
Kg/day/lifetime (70 years). By using the lifetime daily exposure
of 3.68×10^{-3} mg/Kg/day/lifetime (9.4 mg/Kg/yr x 10/70 x 1/365),
the estimated increased cancer incidence was calculated to be 0.08%.
Table 4 compares the risk of cancer to the risk of heritable genetic
disease from exposure to the same amounts of ethylene dibromide.
By using this method of estimating increased incidences of heritable
genetic diseases at first generation and at equilibrium for ethylene
dibromide, the results suggest that incidences of harmful heritable
diseases may be higher than that of cancer from the same level of
exposure.

Table 4.   Comparison of heritable genetic risk as compared
to cancer risk to humans for ethylene dibromide

| Exposure | Cancer | Heritable Genetic Disease | |
		first generation	equilibrium
94 mg/Kg/generation		0.20%	1.16%
36.8 x 10^{-4} mg/Kg/d/lifetime	0.081%		

CONCLUSION

Considerable time elapses between the writing of an environmental
health statute and the implementation of that statute to protect the
environment and human health. This new discipline of "mutagenic
risk assessment" is in the developmental stages with respect to
scientific procedures and methods and is considered by many to lag
behind the legal recognition of environmental problems of heritable
genetic disease. Legally recognizing the potential harmful
consequences of exposure to environmental mutagens creates a need
for appropriate actions to be taken by regulatory agencies using
the best available scientific knowledge. The assumptions used to
estimate heritable disease risks should be openly stated and reviewed by experts in the scientific community before such information is used in making regulatory decisions. Furthermore, risk
assessments for environmental mutagens can provide greater protection
to future human health than if such agents were left uncontrolled
or untested. Equally as important, this process will stimulate
better scientific procedures and techniques to provide a more
precise and accurate estimates of risk in the future.

The comparative risk assessment on ethylene dibromide briefly
discussed here is intended to serve merely as an example of one
approach the Agency will consider in estimating the quantitative
heritable disease risk as a consequence of exposure to certain
environmental mutagens. Other considerations have been outlined in
the proposed Mutagenicity Risk Assessments Guidelines (4). This
area undoubtedly will stimulate many fruitful and controversial
scientific and regulatory considerations in the 1980's.

REFERENCES

1. U. S. Environmental Protection Agency, Interim procedures and
     guidelines for health risk and economic impact assessments of
     suspected carcinogens, Federal Register 51:21402-21405,
     May 25, 1976.

2. U. S. Environmental Protection Agency, Notice of RPAR for
    ethylene dibromide, Federal Register 42:63134-63160,
    December 14, 1977.

3. U.S. Environmental Protection Agency, Proposed guidelines for
    registering pesticides in the U. S. Hazard evaluation:
    Humans and domestic animals, Federal Register 43:37400-37403,
    August 22, 1978.

4. U. S. Environmental Protection Agency, Mutagenicity risk assess-
    ments proposed guidelines, Federal Register 45:74984-74988,
    November 13, 1980.

5. W. L. Russell, X-ray-induced mutations in mice, Cold Spring
    Harbor Symposium on Quantitative Biology 16:327-336 (1951).

6. L. B. Russell, P. B. Selby, E. Von Halle, W. Sheridan, and
    L. Valcovic, The mouse specific locus test with agents other
    than radiations: Interpretation of data and recommendations
    for future work, Mutat. Res. (in press) (1981).

7. pertinent references cited in 4.

8. United Nations Scientific Committee on the Effects of Atomic
    Radiation Sources and Effects of Ionizing Radiation, Report
    to the General Assembly, United Nations, New York, (1977).

9. W. R. Lee, Final report on EDB to the Toxicology Branch, Health
    Effects Division Office of Pesticides Program, U. S.
    Environmental Protection Agency (1980).

10. U. S. Environmental Protection Agency, Office of Pesticides Pro-
    grams, Ethylene Dibromide Position Document 2/3, (1980).

11. E. Jackson and P. Voytek, Comparison of estimated human muta-
    genicity versus carcinogenicity to ethylene dibromide (EDB),
    in preparation, (1981).

12. J. McCann and B. N. Ames, The Salmonella/microsome mutagenicity
    test: predictive value for animal carcinogenicity, in:
    "Origins of Human Cancer, Book C", H. H. Hiatt, J. D. Watson,
    and J. A. Winstin, eds., Cold Spring Harbor Conferences on
    Cell Proliferation, Cold Spring Harbor, NY, 1431-1450 (1977).

13. Biological Effects of Ionizing Radiation II Report, Report of
    the Advisory Committee, National Academy of Sciences, National
    Research Council, Washington, D. C. (1981).

14. P. Z. Daffer, K. S. Crump, and M. D. Masterman, Asymptotic theory
    for analyzing dose-response survival data with application to
    the low-dose extrapolation problem, Mathematical
    Biosciences 50:207-230 (1980).

15. U. S. Environmental Protection Agency, The cancer assessment
    group's updated quantitative risk assessment on ethylene
    dibromide, in preparation, (1981).

THE SENSITIVITY OF METHOD PROCEDURE AS A REGULATORY MECHANISM FOR APPROVAL OF CARCINOGENS

Theodore M. Farber

Food and Drug Administration
Washington, DC   20204 (U.S.A.)

Technology has focused our attention on many potential hazards of which we were unaware earlier in the century. Our awareness of possible hazards and their apparent causes has increased significantly. Our understanding of the biologic activity of chemicals, their mechanism(s) of action and their potential toxicologic activity has greatly increased. We no longer focus on acute, lethal effects but have become increasingly concerned about subtle health injuries which decrease the quality of our lives. Presently, the leading public concern about exposure to chemicals is the risk of cancer. Consequently, considerable effort in both the public and private sectors is devoted to the determination of what chemicals are or might be carcinogenic.

The efforts of the Federal Government to reduce the serious-ness of the cancer problem in our society is multifaceted. These efforts range from basic research on the causes of cancer, through efforts to improve the diagnosis and treatment of the disease, to the rehabilatitation of the cancer patient. The efforts of the regulatory agencies to limit the public's exposure to carcinogens centers around the identification of carcinogens, assessing the risk of cancer to humans, establishing regulatory priorities and undertaking regulatory activities. A regulatory agency's actions in these areas is determined by the specific mandates given these agencies by Congress.

The approach taken by the Food and Drug Administration (FDA) for the regulation of potential carcinogens as food and color additives can be basically summarized by Shakespeare who wrote, "The best safety lies in fear." Almost prophetically, Shakespeare

could have been describing food safety policy. Since 1958, our
fear of "poisons' in the food supply has a rigorous safety standard
which requires evidence that, with reasonable certainty, no harm
will be imparted by substances added to foods. In the case of
carcinogens, that fear has been tremendous. Therefore, no amount,
no matter how minute, of a carcinogen may be added to food. This
concept was included in the Delaney clause in the 1958 amendment
which represented, at that time, the ultimate application of a
concept of absolute zero risks associated with the food supply.
The Delaney Clause flatly prohibited the approval of any substance
found to be carcinogenic for use in the food supply of either man
or animals. Thus, in 1958, the FDA could not have approved the
use of a carcinogenic drug for animals if the compound resulted
in detectable residues in edible animal tissues.

However, at the time of the implementation of the Delaney
Clause, there were a great many applications already approved for
diethylstilbestrol (DES) use in animal feed. In fact, there were
enough applications to supply all the beef animals in the United
States, but the FDA could not approve an application for a new feed
containing DES. At that time the Department of Health, Education,
and Welfare, as well as the pharmaceutical industry and livestock
producers, believed that this application of the Delaney Clause
was excessively stringent and had a adverse effect on various
manufacturers of animal drugs and medicated feeds. Responding
to this consensus, Congress modified the 1958 anti-cancer Delaney
Clause as a part of the drug amendments of 1962. This amendment
permitted the use of carcinogenic compounds as animal feed additives
and veterinary drugs, provided that, "no residue of the additive
shall be found by methods approved by the secretary by regulation
in any edible portion of the animal after slaughter or in any food
such as milk or eggs yielded by or derived from the living animal."
This amendment of the Delaney Clause has been called by some "The
DES Proviso."

Regrettably, the legislative history regarding the definition
of "no residue" is unclear and no unanimity exists on the proper
interpretation of this phrase. Two intrepretations are certainly
possible. One interpretation is that absolutely no traces of a
carcinogen could be allowable in the edible tissue of food-
producing animals, which would seem to defy the raison d'etre for
the DES proviso. The other interpretation is that carcinogenic
agents may be approved if an approved assay revealed no residue.

What does "no residue" or "zero redidue" mean? In everyday
language the word "zero" is used synonymously with the words
"nothing" or "none" or the total absence of something. However, for
the scientist, this meaning is much more complex. For the scientist,
"zero" is an abstraction -- a concept for talking quantitatively
about existence. Zero is a quantitative phenomenon -- one that

requires appropriate measuring devices and reference points to demonstrate its value. From a scientific point of view, zero is a relative concept determined by the conditions under which it is measured. Even if an assay procedure is as sensitive as the current capability of analytical science, it cannot absolutely ascertain that a residue is completely absent from animal tissue. Thus, since 1962, the FDA has interpreted the revised Delaney Clause to mean that a carcinogenic compound could be approved for use in animals if examination of edible tissue by an FDA-approved assay showed no residues (i) at the limits of current analytical sensitivity, and, (ii) the residue was below a level which was considered to pose essentially no risk of cancer in humans.

Until recent years the FDA had defined "no residue" as a gravi-metric zero residue of 2 parts per billion. Two parts per billion was the lower level of assay sensitivity for DES, the most potent carcinogen known to that time to be used in food-producing animals. However, analytical chemists are now measuring parts per billion and sometimes per trillion and there is every expectation of ultimately measuring even smaller numbers of molecules. The zero of the last ten years has no meaning in today's scientific world. With each advance in analytical ability, zero has become smaller. Therefore, the FDA has had to operationally redefine this "no residue" level which has continually become a smaller target. A new regulatory approach regarding the approval of carcinogens is contained in the preamble to the Sensitivity of Method (SOM) document published in the Federal Register of March 20, 1979. The present proposed regulation more precisely define the operational definition of "no residue" in food products from treated animals. As proposed, the criteria in the SOM document apply only to drugs for food-producing animals, although it is certainly possible that in the future they could be adapted to all direct and indirect food additives for human foods as well as to many other agents present in our environment.

When sponsors initially seek approval for compounds they must provide the FDA with information for an initial hazard assessment or threshold assessment to determine if SOM applies. This infor-mation includes the proposed use of the compound, the chemical structure of the compound, toxicologic data on the agent and the results of in vitro mutagenicity studies on the chemical. Based on this data and information from the literature, the FDA will determine if the compound should undergo the data collection process required under SOM.

The SOM regulations specify several steps for gathering and evaluating the scientific information needed to assure that a compound deemed to be carcinogenic, or suspected of such after a threshold assessment, can be safely used in food-producing animals (Table 1).

Table 1. Six-step evaluation process

1. Target Species Metabolism

2. Test Species Metabolism & Comparison with Target Species

3. Toxicity Testing and Assignment of Safe Level of Exposure.

4. Selection of Target Tissues, Marker and Required Assay Sensitivity.

5. Regulatory Assay

6. Withdrawal Period

The first step consists of metabolic studies in the target species, i.e. the particular animal for which the proposed feed additive or drug is intended. These studies should determine which tissues may accumulate residues of the compound and the composition of this residue. The substance administered to target animals may not necessarily be the substance consumed by the persons who eat the edible products of target animals. Therefore, metabolites as well as the parent compound, should be evaluated in the threshold assessment. Additionally, the FDA may request that these metabolic studies also determine if administration of a particular compound results in elevated levels of endogenous compounds such as estrogens which might be considered potentially carcinogenic.

The sponsored compound itself must always be tested for carcinogenecity when it is determined, on the basis of the threshold assessment and the initial metabolism study, that a sponsored compound has the potential to contaminate edible tissues with a residue whose consumption may pose a human risk of carcinogenesis. Metabolic transformation or nonenzymatic degradation of a sponsored compound can lead to a number of tissue residues that cannot be obtained (either by isolation or synthesis) in sufficient amounts for carcinogenic testing. (These residues are referred to in this document as "intractable residues".) Testing the sponsored compound itself, therefore, provides an experimental means for acquiring data bearing on the carcinogenic potential of such residues.

Although the dominant criterion for selecting the test animal species or strains for chronic toxicity testing will be the degree to which a species or strain models man, applying a secondary criterion for selection can help address the problem of intractable residues. Specifically selection of the test animals can also be based on comparative metabolic approach (target animal versus test animal). Data from this approach, collected in step two, can be used to determine the extent to which a particular species or strain, due to the way these test animals metabolically convert the sponsored compound, will be exposed during testing to the same complement of residues to which man may be exposed in tissues derived from the target animals.

The parent compound and any metabolites found in the tissue residue which are deemed to be potentially carcinogenic by the FDA must then undergo chronic toxicity testing for carinogenicity in two species of laboratory test animals as mentioned under step 3. If these substances are determined to be carcinogenic, the results of chronic toxicity testing will be used for the statistical calculation of the operationally defined zero residue level. For each tested compound regulated as a carcinogen the appropriate data from the chronic dose-response studies shall be analyzed according to the linear extrapolation technique as described by Gross et al. (1) and Hoel et al. (2). The purpose of this analysis is to interpret the "no residue" requirement of the act by determining the lowest level of reliable measurement required for a regulatory assay to be approved for the monitoring of the total residue. The permissible level determined by the linear extrapolation mode for each test compound is to be expressed as a fraction of the total diet fed to the test animal calculated at the 99% confidence level for a maximum lifetime risk that is essentially zero but never expected to exceed one in one million. The lowest of all calculated acceptable levels for the parent drugs and/or its metabolites is to be designed as the required sensitivity of the method for tissue assay.

The linear extrapolation procedure was selected because it was the least likely method to underestimate risk and no complicated mathematical procedures or arbitrary selection of slope was required. However, other statistical methods may be considered if supported by an adequate scientific justification for their application. At this time, little, if any, research is being done to determine which extrapolation model is most appropriate. The arguments for each model rest almost entirely on evaluation of past experiments, none of which are particularly well-designed for the purposes for which they are now being used. Again, we are using a fundamentally metaphysical approach to the problems of this evaluation rather than an experimental one. Similarly, with respect to correlating the results of animal feeding studies with human consumption, a variety of scaling factors has also been advanced. The results vary dramatically, depending on which factor is used. Since we do not understand many of the metabolic differences between man and experimental animals, it becomes difficult to determine the best way of comparing results obtained in animal models with that of man. This is particularly true when we are considering substances of low potency and low dietary levels.

Pooling of data from various chronic tests using different sexes or strains is permitted if it can be demonstrated that the protocols are similar in design. If statistically significant biological differences in tumorigenic responses are seen between sexes or among species or strains, only subsets of data representing statistically and biologically compatible bioassay data may be combined. All tumors, benign and/or malignant shall be considered. The number of animals at risk may be adjusted for competing risks unrelated to compound-induced carcinogenesis only when the data clearly supports such an adjustment.

The last three steps found in Table 1 are required for the purpose of monitoring the edible tissue of food-producing animals for violative levels of residues.

In the fourth step, a metabolic study of the compound in the target animals must be initiated to identify both the tissue and the residue (called the target tissue and the marker residue) that can serve as an indicator to determine whether the operationally defined zero residue level has been met in all tissues.

The next step requires the development of a regulatory assay which is capable of measuring the marker residue in the target tissue at a level that assures that no residue is present above the defined zero level in all tissues.

Finally, a premarketing withdrawal period must be established that permits safe use of the compound in food-producing animals.

If at some point in this stepwise data collection and evaluation process it can be determined that the compound in question presents no risk of carcinogenesis in humans, the compound is then exempt from SOM and subject to only the general safety provisions for drugs used in food-producing animals.

The application of an SOM approach, particularly the application of a acceptable societal risk to the calculation of a virtually safe use of carcinogens and other toxic agents as described in Step 3, may be the wave of the future in regards to the regulation of not just drugs for food-producing animals but many other toxic agents perceived to have sifniciant benefits to mankind.

REFERENCES

1. M. A. Gross, Evaluation of Safety of Food Additives. Biometrics
        26:181 (1970).
2. D. G. Hoel, Estimation of Risks of Ineversible Delayed Toxicity.
        Journal of Toxicology and Environmental Health 1:133-151
        (1975).

THE HUMAN GENETIC RISK OF AIRBORNE GENOTOXICS: AN APPROACH

BASED ON ELECTROPHORETIC TECHNIQUES APPLIED TO MICE

F. M. Johnson and Susan E. Lewis*

Laboratory of Biochemical Genetics
National Institute of Environmental Health Sciences
Research Triangle Park, NC 27709 (U.S.A.)

INTRODUCTION

Among the most serious long-term adverse health effects potentially resulting from human exposure to airborne genotoxic agents is heritable damage. Environmentally-caused mutations transmitted through the gametes might not only persist for multiple human generations but, depending on the circumstances and the nature of the genetic alteration, may accumulate over time to be expressed as a gradual degradation in quality of life.

Laboratory exposures of mice to mutagens have potential for predicting the consequences of human environmental exposure. The detection of induced, electrophoretically-expressed mutations in the mouse is of particular relevance to human genetic risk assessment because mutants at numerous structual loci are identifiable. When spontaneous and induced mutants in known proteins can be compared with non-mutant types for structural and functional alterations, a basis is formed to explore the effects of mutations on fitness. In addition, the utilization of homologous loci in the mouse and man allows mutation rates and their effects to be assessed in different organisms on comparable genes.

There are a number of variables which have importance to the interpretation of data obtained from the mouse. Different strains may vary in spontaneous mutation rates while sensitivity to induced damage may vary among loci. Thus, the extrapolation of mouse data

*Research Triangle Institute, Research Triangle Park, NC 27709

to human risk estimates will not be without complication, and will require the eventual examination of genetically diverse strains of mice.

We have applied the electrophoretic mutation concept in three separate experiments (1,2,3) and have discussed the determination of mutation frequencies in some detail (4). These experiments have provided an opportunity to evaluate the sensitivity of the approach and to examine some of the potential problem areas which may pertain to future applications in mutagen testing and human genetic risk assessment.

MATERIALS AND METHODS

Two strains of mice, C57BL/6J and DBA/2J, obtained from the Jackson Laboratory, are utilized for electrophoretic analysis. The approach is similar to one described earlier (5,6,7) except that more assays are used in the current system. According to the experimental design (Fig 1), male mice from one strain are exposed to a mutagen and then mated to females of the other strain. An F_1

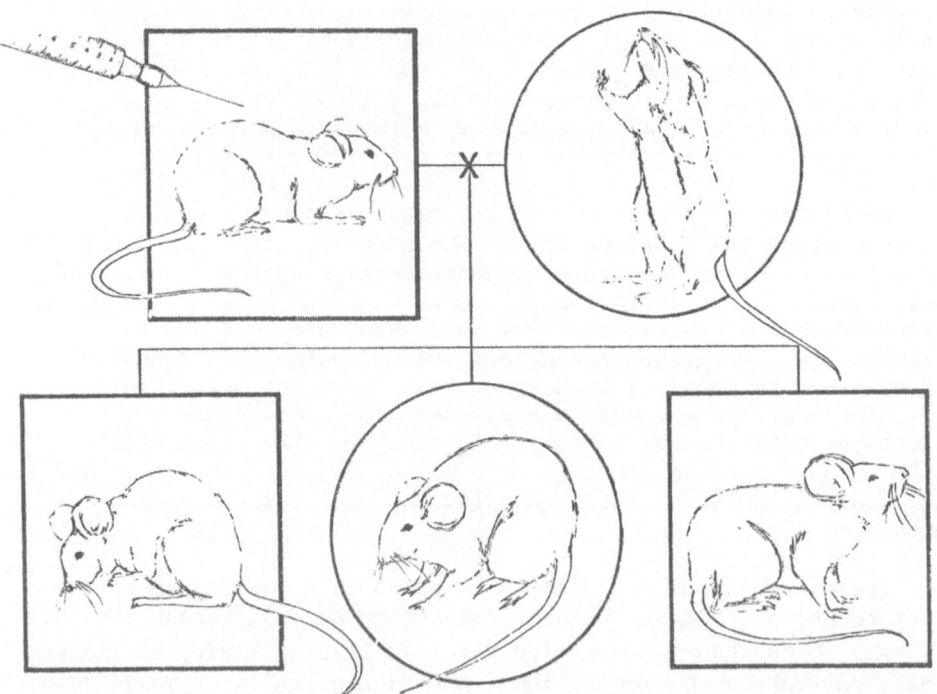

Figure 1. Diagram of experimental design; treatment and breeding of animals.

generation of progeny is obtained and all parents and offspring are
analyzed by a set of electrophoretic procedures, performed on red
cell lysates and kidney homogenates (1) (Fig 2). Other tissues
such as liver and muscle could also be analyzed. Since mortality
from surgery is very low, it is possible to mate suspected mutant-
bearing animals for the purpose of genetic analysis.

Gene products are visualized by histochemical staining after
electrophoretic separation in starch gels. Two groups of loci
have been studied. One group, which contains Mod-1 (coding for
malic enzyme), ES-1 and ES-3 (for two esterases), Idh-1 (isocitrate
dehydrogenase), Hbb (hemoglobin β), Pep-3 (a peptidase), Gpd-1
(hexose phosphate dehydrogenase), Gpi-1 (glucose phosphate
isomerase), Pgm-1 and Pgm-3 (two phosphoglucomutases), is particularly useful because the parental origin (maternal or paternal)
of any new mutant is indicated by an altered electrophoretic
pattern (Fig 3). Two types of mutation are revealed: those that
cause electrophoretic mobility to be altered and those resulting in
inactive or missing gene products (null alleles). Figure 3
diagrams the electrophoretic patterns determined by the locus
Pep-3 as an example. The enzyme PEP-3 from C57B1/6 mice is slower
in mobility than that from DBA/2 mice and both parental forms can
be seen in kidney samples from normal F_1 individuals. One mutant
found recently showed an F_1 pattern in which the DBA/2 parental
band was missing. Both parents were phenotypically (electrophoretically) normal and the variant was unique among several dozen

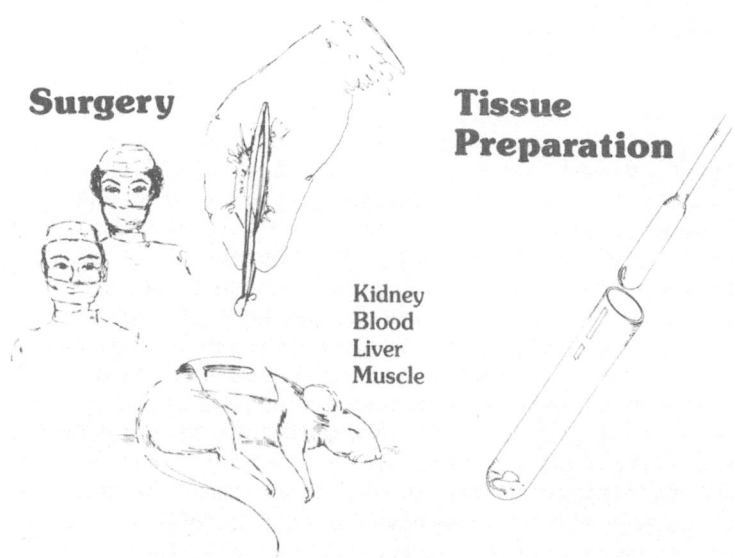

Surgery   **Tissue Preparation**

Kidney
Blood
Liver
Muscle

Figure 2.   Experimental design; surgery and tissue preparation.

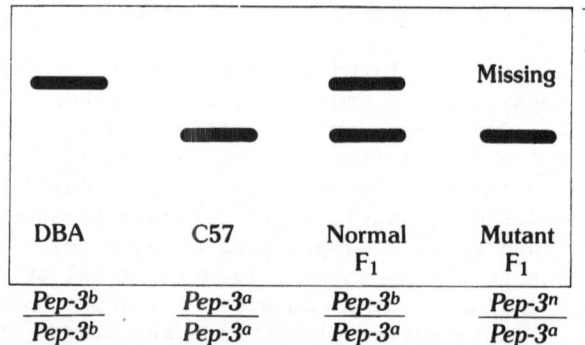

Figure 3.   Appearance of a null mutation which arose in the DBA/2 strain at the Pep-3 locus.

progeny produced by the DBA/2 male parent. Because it was the DBA/2 parental type that was missing in the F_1 sample, a mutation originating in the DBA/2 parent is indicated. In a like manner, mutants with altered electrophoretic mobility are traced to the parent of origin by noting which of the two parental forms have altered positions on the gels.

The electrophoretic system also includes a group of loci that do not show electrophoretic differences between the C57B1/6 and DBA/2 strains. These loci encode for 10 enzymes: two esterases, two lactate dehydrogenases, three peptidases, a phosphoglucomutase, a 6-phosphogluconate dehydrogenase and glucose-6-phosphate dehydrogenase. Another protein, hemoglobin-α, is also included in this group. The lack of electrophoretic differences for this group of proteins between the two parental strains make these loci somewhat less useful than those in the first group because the parental origin of any new mutant cannot be determined by an examination of the gel patterns. In addition, null mutants probably will not be readily distinguishable since their identification would be based on detecting a reduction in band intensity instead of an absence. Nevertheless, examination of the 11 additional proteins do add power to mutation detection by increasing the number of loci studied. Figure 4 illustrates how a mutant at one of the invariant loci, Pep-1, appears in the electrophoretic system.

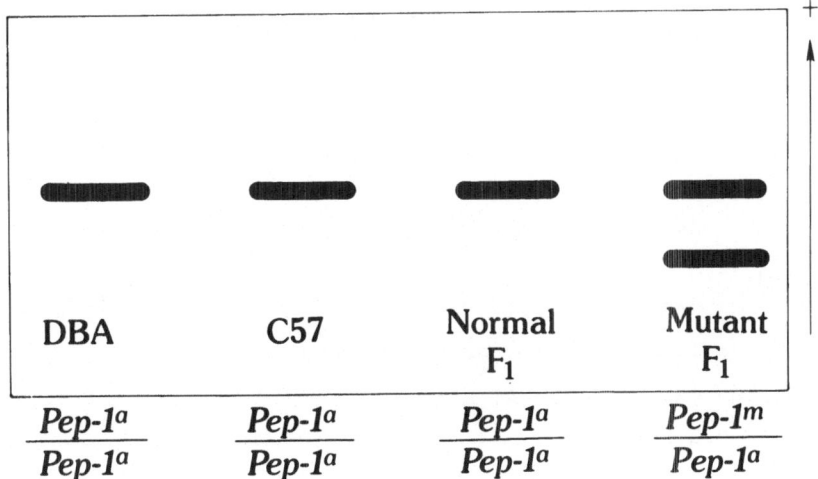

Figure 4. Example of an electrophoretic mutation at a locus (Pep-1) that is the same in C57Bl/6 and DBA/2 mouse strains.

RESULTS AND DISCUSSION

The largest body of induced, electrophoretic mutation data (3) has been generated with the use of a 250 mg/Kg b.w. exposure to ethylnitrosourea (ENU). Table 1 shows a summary of the recent work (3) to which our latest unpublished results have been added. The number of loci tested is calculated as 21 times the number of F_1 progeny analyzed from treated parents. However, the mutability of some of the invariant loci as well as the capability of the electrophoretic system to resolve differences at these loci may be considerably lower than for the loci which differ between the strains.

The mutation frequencies in Table 1 show ENU-treated males of both strains to have mutation frequencies elevated considerably over background. The average induced rate for DBA/2 mice is 2.5 x 10^{-4} while for C57Bl/6 mice, it is 5.1 x 10^{-4}. In comparison the spontaneous rate based on the same loci is probably less than 1 x 10^{-5} (1-4 and unpublished results).

Another set of frequency estimates (Table 2) can be calculated using only those loci at which there is genetic variation detectable between the C57Bl/6 and the DBA/2 strains of mice. This reduces the number of loci from 21 to 10 and results in calculated induced mutant frequencies of 3.6 x 10^{-4} for DBA/2 and 7.6 x 10^{-4} for C57Bl/6. Interestingly, the unweighted average, 5.6 x 10^{-4}, is almost exactly the same as found by Russell et al. (8) studying

Table 1.   Summary of mutant frequencies

Mutant Origin	No. F1 Animals	No. Loci Tested	No. Mutants	Mutation Frequencies
ENU-Treated DBA/2	1317	27,657	7	2.5×10^{-4}
ENU-Treated C57B1/6	657	13,797	7	5.1×10^{-4}
Total Treated	1974	41,454	14	3.4×10^{-4}
Spontaneous	--*	308,000	0	$<0.97 \times 10^{-5}$

*Spontaneous data are based on F_1 offspring of mutagen-treated and control group parents. Since all crosses having a mutagen-treated parent also involve an untreated parent, it is possible to separate mutants of spontaneous origin in the untreated parents from induced mutants in the treated parents when considering loci with allelic differences between the strains.

Table 2.   ENU-induced mutation rates based on loci that differ between strains

Mutant Origin	No. F1 Animals	No. Loci Tested	No. Mutants	Mutation Frequencies
DBA/2	1317	13,170	5	3.6×10^{-4}
C57B1/6	657	6,570	5	7.6×10^{-4}
DBA/2 & C57B1/6 (unweighted avg. of above)				5.6×10^{-4}
Visible Systems*	7584	53,088	28	5.3×10^{-4}

*Russell et al. (8)

the induction of mutations with the same dose of ENU in the visible marker testing systems. The 7 loci in that system also depend on genetic differences between the parents.

. The data in Tables 1 and 2, although not conclusive, suggest that there may be a substantial difference in mutation rates between the two strains. Such differences are of importance in risk considerations. However, in this particular case there is a possible confounding factor. Russell and Hunsicker (9) have suggested that the age at treatment could affect the induction of mutations with ENU and the C57B1/6 males were slightly younger than the DBA/2 males (8½-10½ weeks) vs 13 week old, respectively at the time of ENU injection.

Table 3 presents the distribution of ENU-induced and spontaneous mutants by strain and locus. Two loci account for half of the induced mutations, 3 at Pep-3 in C57B1/6 mice and 4 at Mod-1 in DBA/2 mice. All the mutations induced by ENU in male C57B1/6 mice were at different loci from those mutations induced in male DBA/2 mice, suggesting the possibility of strain-specific differences in locus

Table 3. Distribution of ENU-induced and spontaneous mutants

Loci	Induced C57B1/6	Induced DBA/2	Induced Combined	Spontaneous C57B1/6
Mod-1	0	4 (1,3)**	4	1 (0,1)
Es-1	1 (0,1)	0	1	0
Idh-1	3 (1,0)	0	1	1 (1,0)
Pep-3	3 (0,3)	0	3	0
Pgm-1	0	1 (0,1)	1	1 (0,1)
Pep-1	2 (1,1)	0	2	0
Hba	0	1 (1,0)	1	0
Pgm-2	0	1 (1,0)	1	0
Total	7	7	14	3

*Johnson et al., (4)
**(mobility mutants, null mutants)

sensitivity. However, the data are limited, and they show no
significant different from a Poisson distribution for the mutants
of either strain.

Using only the data for the 10 variant loci, a comparison of
the number of mutants affecting either electrophoretic or the null
classes shows that null mutations were detected about three times
as often as ones involving changes in electrophoretic mobility.
Table 4 lists the available data from the ENU experiments and
includes earlier results from the work with procarbazine (1,2).
The spontaneous mutants are not new mutants, but mutants found
to be pre-existing in the C57B1/6 parental animals (1). The 3 to

Table 4. Mobility and null mutants in induced
and spontaneous classes

Loci	Induced Mutants		Spontaneous Mutants	
	Mobility	Null	Mobility	Null
with electrophoretic variation between strains:				
Mod-1	1	3	0	1
Es-1	0	1	---	---
Idh-1	1	0	1	0
Hbb	1	0	---	---
Pep-3	0	4	---	---
Pgm-1	0	1	0	1
Totals	3	9	1	2
without electrophoretic variation between strains:				
Pgm-2	1	0	---	---
Pep-1	1	1	---	---
Hba	1	0	---	---
6-Pgd	1	0		
Totals	4	1		

1 null-mobility relationship applies to loci with electrophoretic differences between the strains (where nulls and mobility mutants presumably are detected with equal facility) but not to loci that have the same electrophoretic mobility in both strains (where mobility mutants are much easier to observe than null mutants). The data suggest, therefore, that nulls arise more frequently than mobility mutants. However, there is no obvious reason why a randomly occurring amino acid substitution (the most frequent type of mutation induced by ENU in structural genes (10,11)) should lead more often to loss of function rather than to a shift in electrophoretic mobility. Indeed, since the active site comprises only a fraction of the total amino acid sequence for most enzymes, nulls might be expected to arise infrequently relative to mobility mutants.

The null category is not necessarily discrete from the mobility class since it is possible that an enzyme could be both electrophoretically altered and reduced in activity. However, on the basis of activity, a mutant altered in both ways would be scored as a null mutation. To date, only one of our nulls appears to have altered electrophoretic mobility. One presumptive Mod-1 mutant, to be described in detail elsewhere, shows a trace of activity of intermediate mobility when in heterozygous combination with the normal "slow" form of the enzyme. This observation suggests the existence of functionally active heteropolymers of mutant and non-mutant enzyme, but inactive, electrophoretically fast mutant homopolymers. Thus, in reality our mutants are divided into a slightly more complicated way than is first apparent. One class is comprised of mutants with reduced activity which may or may not also be electrophoretically altered, while the second class contains only electrophoretic mutants which have retained relatively normal activity. Certainly the manner of mutant assignment will be reflected in the distribution of mutants between classes. However, had we added the one apparent null with altered mobility to the mobility class instead of the null class, it would make practically no change in the overall picture (Table 4). It will be interesting to reclassify all the mutants by various criteria as they are further characterized.

Another possible complication relevant to the determination of null mutants is the occurrence in one strain of a mobility mutant that is indistinguishable from the normal mobility type of the other strain. For example, with one strain homozygous for the allele coding for a fast electromorph (ff) and the other slow (ss), a mutation from f to s would result in ss F progeny phenotypically similar to the sn genotype expected from an f to null mutation. Only when the mutant was backcrossed to the ff genotype would the distinction be evident. Thus, the ss mutant-bearing F_1, backcrossed to an ff, would produce all fs progeny, but an sn F_1 genotype would be expected to result in 1/2 fs and 1/2 fn progeny. All putative mutants are routinely subjected to genetic analysis,

and no mutants of this type have yet been found among the loci
that differ between the strains. However, at Pgm-2, a locus that
is the same in the two tester strains, a mutant coding for an
electrophoretic type typical of another strain (SM/J) has been
identified (12).

Another possible explanation that could be offered to account
for the prevalence of nulls is that they result from alterations
in the sequence of nonstructural DNA, i.e. mutational changes in
leader and intervening sequences and in other possible regulatory
elements. With a large amount of regulating, nonstructural DNA,
the number of apparent null mutations originating from changes in
nonstructural DNA could equal or exceed that arising in the structural genes proper.

The molecular action of the causative substance involved in
the production of mutations will naturally influence the distribution of mutations among null and mobility classes. Agents causing
chromosome breakage, deletions and frameshifts, for example, would
be expected to produce more null mutations. ENU, however, primarily
causes base-pair substitutions (10,11). There is independent
evidence to suggest that spontaneous null mutations may arise more
frequently than electrophoretic mobility mutations. Spontaneous
mutants in Drosophila were found to arise at a rate of 1.28×10^{-6}
for mobility types but at 3.86×10^{-6} for nulls (13).

CONCLUSION

As many inborn errors of metabolism are caused by enzyme
deficiencies (14), severe heritable damage in humans could result
from the induction of null mutations in germ cells by environmental
mutagens. Mutagens act by a variety of mechanisms and the severity
of their effects depends upon the mechanism of action in addition
to the duration and extent of exposure. That null mutations
apparently comprise a large fraction of the total damage would
suggest that nulls should be subjected to further characterization,
i.e., examined for whether or not a gene product is synthesized
and what physiological consequences result. Should the indicators
be correct that electrophoretic analysis results in the identification of non-structural mutants, characterization may also prove
to be of value for gaining additional fundamental understanding of
molecular mechanism involved in gene regulation in mammals.

Detection of mutants by electrophoresis clearly leads to
the efficient identification of nulls, but more importantly the
technique also provides the means to characterize their nature
as well as the properties of mutations that cause alterations in
electrophoretic mobility. Since known gene products are analyzed
one can determine if and how the products are structurally altered.

Other features of the electrophoretic system which make it
valuable for purposes of risk estimation include sensitivity,
economy and versatility. With many of the loci being tested
in mice having known homologs in humans, laboratory results can
be reasonably extrapolated to humans. Many loci can be analyzed
from the same tissue samples, and there is the potential to
incorporate a considerable number of additional assays. Both
parental animals contribute mutation rate data and concurrent
spontaneous mutation rate data is continuously produced from
crosses having a mutagen-treated parent as well as from indenendent
control matings.

ACKNOWLEDGEMENT

        This work was supported by the U.S. Environmental Protection
Agency under contract 68-02-3626.

REFERENCES

1. F. M. Johnson, G. T. Roberts, R. K. Sharma, F. Chasalow,
    R. Zweidinger, A. Morgan, R. W. Hendren, and S. E. Lewis,
    The detection of mutants in mice by electrophoresis: Results
    of a model induction experiment with procarbazine, Genetics
    97:113-124 (1981).
2. F. M. Johnson and S. E. Lewis, Mouse spermatogonia exposed to a
    high multiply fractionated dose of a cancer chemotherapeutic
    drug: Mutation analysis by electrophoresis, Mutat. Res. 81:
    197-202 (1981).
3. F. M. Johnson and S. E. Lewis, Electrophoretically-detected
    germinal mutations induced in the mouse by ethylnitrosourea,
    Proc. Natl. Acad. Sci. USA 78:3138-3141 (1981).
4. F. M. Johnson and S. E. Lewis, Mutation rate determinations
    based on electrophoretic analysis of laboratory mice, Mutat.
    Res. 82:125-135 (1981).
5. L. R. Valcovic and H. V. Malling, An approach to measuring
    germinal mutations, Environ. Health Perspectives 6:201-205
    (1973).
6. H. V. Malling and L. R. Valcovic, A biochemical specific locus
    mutation system in mice, Arch. Toxicol. 38:45-51 (1977).
7. E. R. Soares, TEM-induced gene mutations at enzyme loci in the
    mouse, Environ. Mutagenesis 1:19-25 (1979).
8. W. L. Russell, E. M. Kelley, P. R. Hunsicker, J. W. Bangham,
    S. C. Maddux, and E. L. Phipps, Specific-locus test shows
    ethylnitrosourea to be the most potent mutagen in the mouse,
    Proc. Natl. Acad. Sci. USA 76:5818-5819 (1979).
9. W. L. Russell and P. R. Hunsicker, Possible effect of age at
    injection on mutation induction by ethylnitrosourea in the
    mouse, Genetics 94:s90-s91, Abstract (1980).

10. E. Vogel and A. T. Natarajan, The relation between reaction
     kinetics and mutagenic action of mono-functional alkylating
     agents in higher eukaryotic systems, II, Total and partial
     sex chromosome loss in Drosophila, Mutat. Res. 62:101-123
     (1979).
11. V. F. Simon, H. S. Rosenkranz, E. Zeiger, and L. A. Poirier,
     Mutagenic activity of chemical carcinogens and related
     compounds in the intraperitoneal host-mediated assay, J.
     Natl. Cancer Inst. 62:911-918 (1979).
12. V. M. Chapman, F. H. Ruddle, and T. H. Roderick, Linkage of
     isozyme loci in the mouse: Phosphoglucomutase-2 (Pgm-2),
     Mitochondrial NADP malate dehydrogenase (Mod-2), and
     Dipeptidase-1 (Dip-1), Biochem. Genet. 5:101-110 (1971).
13. R. A. Voelker, H. E. Schaffer, and T. Mukai, Spontaneous
     allozyme mutations in Drosophila melanogaster: Rate of
     occurrence and nature of mutants, Genetics 94:961-968 (1980).
14. J. V. Neel, H. Mohrenweiser, C. Satoh, and H. B. Hamilton,
     A consideration of two biochemical approaches to monitoring
     human populations for a change in germ cell mutation rates,
     in: "Genetic Damage in Man Caused by Environmental Agents",
     K. Berg, ed., Academic Press, N.Y., 29-47 (1979).

PHOTON-EMITTING MICROORGANISMS AS TEST OBJECTS FOR DETECTING

GENOTOXIC AGENTS.

Stanley Scher and Richard A. Wecher

Sonoma State University
School of Environmental Studies and
Department of Biology
Rohnert Park, CA   94928   (U.S.A.)

INTRODUCTION

Short-term tests with microorganisms provide a technically
simple, rapid and inexpensive approach to detecting environmental
agents that damage DNA (1). In the most widely used microbial
bioassay, specially constructed histidine auxotrophs of Salmonella
typhimurium are incubated with the chemical to be tested.
Mammalian tissue homogenates are included to metabolize promutagens
to their active form. The number of revertant colonies that develop
on petri plates after an incubation period of two days serves as a
measure of mutagenicity (2).

Similar bioassay procedures have been developed for a variety
of comparable endpoints using S. typhimurium and other microorganisms. These bioassays include the use of auxotrophic strains
of Escherichia coli (3) and Bacillus subtilis (4), forward mutations
to azaguanine resistance in S. typhimurium (5) and differential
growth inhibition in repair deficient and repair competent cells
of E. coli (6). Several assays for genetic events in eukaryotic
microganisms have also been developed (1).

Virtually all short-term microbial tests described to date
share a common feature: a dependence upon an incubation period
of approximately 48 hours either for growth and development of
colonies as evidence of a mutational event in microorganisms, or
for plaque production as an indicator of lysogenic induction of
lambda phage (7). In principle, it should be possible to determine
directly the quantity of a newly synthesized gene product as an

607

endpoint for mutagenicity bioassays. We report here on some
properties of Photobacterium phosphoreum mutants that suggest their
utility for detecting chemical agents that damage DNA.

Ulitzur et al. (8) have recently described a dark variant of
Photobacterium leiognathi that "reverts" to a luminous state when
treated with frameshift or base-substitution mutagens. However,
these authors failed to isolate luminous "revertants" from cultures
treated with intercalating agents such as acridine dyes. Cells of
the dark variant grown on a medium containing acriflavin were
capable of emitting light, but none of the progeny retained this
ability when transferred to a medium free of the mutagen. The
conditional dependence of photon emission from P. leiognathi cells
upon the presence of the mutagen in the medium provides evidence
for a transcriptional control of the luminescent system resulting
in changes in the phenotype, but not in the genotype.

In the present work, we have examined frameshift and base-substitution agents using P. phosphoreum strain PPL^{-1}. In this strain,
exposure to compounds from each of these classes of mutagenic agents
induced genetically stable progeny with near wild type photon
emission rates.

MATERIALS AND METHODS

The wild type strain of P. phosphoreum was isolated from
fishery offal at Bodega Bay, California. Identification was based
on light emission with a maximum near 500 nm, optimum growth
temperature and the fermentation of both glucose and maltose to
acid and gas. The mutant strain used in the experiments reported
here displayed biochemical identity with the wild type; however,
the rate of photon emission was reduced by several orders of
magnitude when compared with wild type controls. The growth
medium contained half strength "Instant Ocean", 5.0 g tryptone,
5.0 g yeast extract and 3.0 mL glycerol per Liter. Two percent agar
was added for solid media.

Suspension cultures were grown in unshaken flasks at 23-25°C.
to a density of 10^6-10^7 cells/mL. For direct measurement of
photon production, aliquots were aseptically transferred to
sterile borosilicate glass scintillation vials containing test
compounds. Absolute ethanol served as a solvent for chemicals
either poorly soluble or unstable in aqueous media. Final volume
per scintillation vial was 2.0 mL.

Photon emission was monitored with a liquid scintillation
counter (Beckman LS-150) using ambient temperature photomultiplier
tubes. The temperature on the conveyor ranged from 23-25°C.
Light emission from the population of cells in the scintillation

vials was integrated over 0.5 to 1.0 minute periods and recorded as counts per minute (CPM). Counting was performed in the $^{14}C + ^3H$ mode.

For plate incorporation assays, cells were grown as above and appropriate dilutions were spread on the agar media containing the test compound. Plates were incubated at 23-25°C for at least 18 hours before scoring. Visual scoring required dark adaptation. Bright clones were scored as those brighter than controls.

Plates were photographed in total darkness with a 35 mm Nikon camera using a 26 mm Nikon lens mounted 70-80 cm above the plane of the petri plates. This permitted up to 24 plates to be compared in the same frame. With an f-stop of 2.6, exposure times of 12-24 hours were used. Kodak Tri-x film was developed in Ethol UFG to an ASA of 1600 by doubling the recommended development time for ASA 400. The resulting negatives were printed at several increasing exposure times and prints chosen in which control plates appeared dark. Remaining colonies were scored as bright.

Tryptone and yeast extract were obtained from Difco, glycerol ACS reagent grade from Baker, hydroxylamine hydrochloride from Mallinckrodt; other chemicals used were from Sigma.

RESULTS

Properties of the photon emission system

The light-generating reaction in Photobacterium and other luminescent procaryotes (9,10) involves a luciferase-catalyzed oxidation of reduced flavin mononucleotide ($FMNH_2$) by molecular oxygen with the concomitant oxidation of a long chain aldehyde. Mutants in which the synthesis of one or more of these components is blocked are non-luminous (11).

Luciferase is relatively non-specific in its aldehyde requirement over the range of 8-16 carbon chain length. To determine if the PPL^{-1} strain is aldehyde limiting, 0.1% dodecanal was added to cells and incubated at 23-25°C. In Figure 1, photon emission from aldehyde treated cells of strain PPL^{-1}, untreated controls and wild type cells are compared. No detectable increase in light emission was obtained upon addition of dodecanal to PPL^{-1} cells.

Photon emission from cells exposed to frameshift agents

Proflavin and ethidum bromide produce frameshift mutations by intercalation. The effect of these compounds on photon emission from PPL^{-1} cells is presented in Figures 2 and 3. In addition, a series of dose-response curves for ethidium bromide

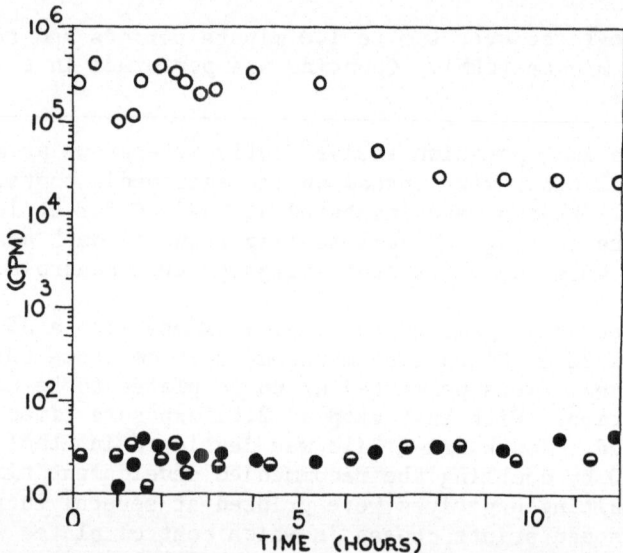

Figure 1: Photon emission from <u>Photobacterium phosphoreum</u> wild type and strain PPL^{-1}. Open circles, wild type; closed circles, PPL^{-1}; half-filled circles represent cultures to which 0.1 (v/v) dodecanal was added.

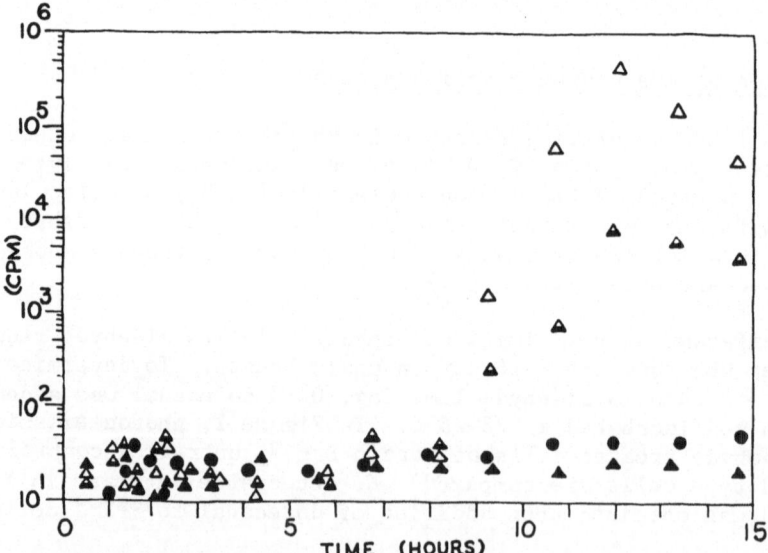

Figure 2: Kinetics of photon emission from <u>P. phosphoreum</u> strain PPL^{-1} following exposure to proflavin hemisulfate. Dose in µg/mL: solid triangles, 0.05; half-filled triangles, 0.5; open triangles, 5.0; solid circles, control.

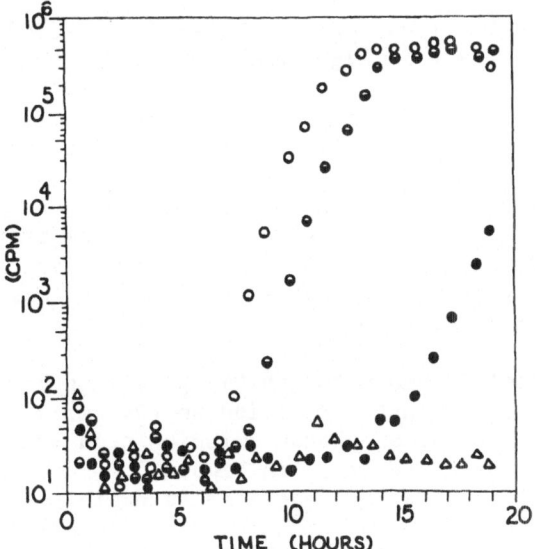

Figure 3: Kinetics of photon emission from P. phosphoreum strain
PPL[-1] following exposure to ethidium bromide. Dose in μg/mL:
solid circles, 5.0; half-filled circles, 10.0; open circles, 20.0;
open triangles, control.

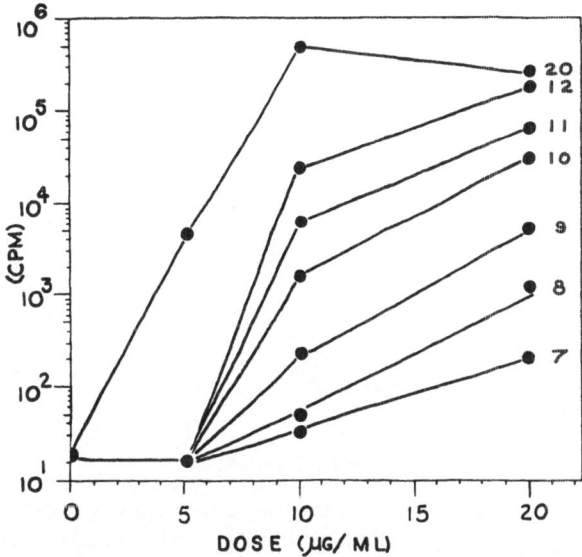

Figure 4: Dose response curves for ethidium bromide interpolated
from Figure 3 at times 7 to 20 hours.

is shown in Figure 4. These cells exhibited an extended lag
phase before an increase in light emission was detected.

The extended lag period suggests that protein synthesis may
be required for expression of the light-generating system in these
cells. An alternative, but less likely hypothesis is that proflavin
and ethidium bromide increase photon emission by acting on pre-
existing enzymes. According to this model, no new protein synthesis
would be required, and inhibitors of protein synthesis would be
ineffective in preventing the induction of photon emission by proflavin or ethidium bromide.

Chloramphenicol is known to inhibit protein synthesis by
blocking the peptidyl transfer reaction on 70S ribosomes. The
effect of chloramphenicol on photon emission from PPL^{-1} cells
exposed to proflavin is presented in Figure 5. In the absence
of chloramphenicol, proflavin treated cells exhibit a marked
increase in the rate of photon emission. In the presence of chloramphenicol, light emission from PPL^{-1} cells treated with proflavin
is prevented. These results argue against the hypothesis that
proflavin acts on pre-existing enzymes and suggests that chloramphenicol inhibits synthesis of enzymes necessary for expression
of the light-generating system. In addition, chloramphenicol may

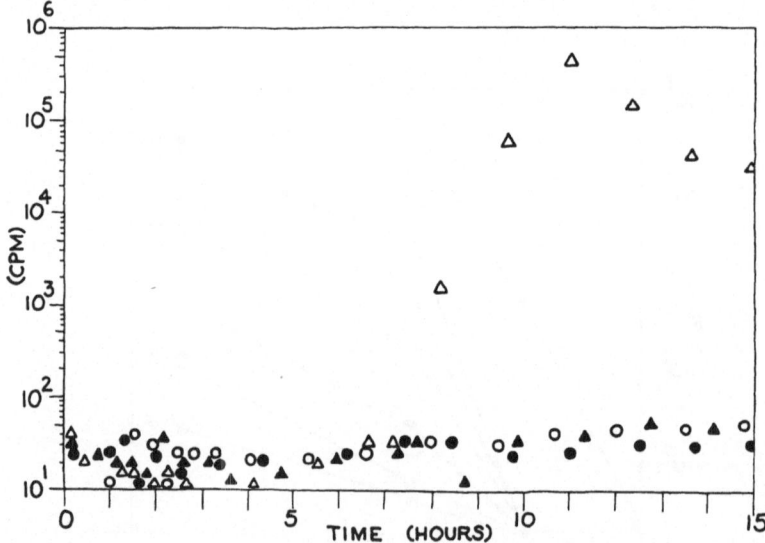

Figure 5: Effect of chloramphenicol on the kinetics of photon
emission from P. phosphoreum strain PPL^{-1}. Triangles represent
cultures treated with proflavin hemisulfate, 5.0 µg/mL; circles
represent controls; open symbols represent cultures receiving no
further additions; closed solid symbols received 5.0 ug/mL chloramphenicol.

act to inhibit the synthesis of inducible enzymes required for
metabolic activation of mutagens such as proflavin.

Plate incorporation assays

     Both base-substitution and frameshift agents have been
reported to increase the rate of luminescence from P. leiognathi
8SD18 cells (8). To determine the frequency of bright clones in
cells exposed to each of these classes of mutagens, dose-response
data for caffeine, hydroxylamine hydrochloride and ethidium bromide
were obtained by spreading PPL^{-1} cells at approximately 150 cells/0.1
mL on plates containing each of the chemicals at a dose range from
1 to 100 µg/mL. After incubation for at least 18 hours at 23-25 C.
colonies were scored visually or by photographing plates using light
emission from the bacterial colonies. In Figure 6, the percentage
of bright clones to survivors is plotted against the concentration
of the chemical. The resulting dose-response curves provide evidence
that both frameshift and base-substitution agents induce bright clones
in a large fraction of the cell population.

DISCUSSION

     We have reported here some observations on P. phosphoreum
strain PPL^{-1}. Additional information on a closely related species,
P. leiognathi strain 8SD18 is available for comparison (8).
Although the data are still fragmentary, we can begin to examine
the methodology and early results as they relate to the utility
of photobacteria as test objects for bioassay of genotoxic agents.

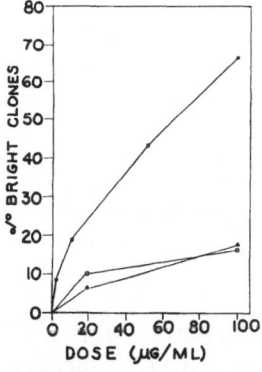

Figure 6: Percent of bright clones of P. phophoreum strain PPL^{-1}
among total survivors plated on media supplemented with ethidium
bromide, circles; hydroxylamine hydrochloride, squares; caffeine,
triangles.

Scintillation measurements

When wild type and PPL[-1] cells are plated and the resulting
colonies photographed, a weak light emission is observed from PPL[-1]
colonies. A more quantitative estimate of the light-generating
ability of these cells can be obtained from scintillation measurements. The data presented in Figure 1 indicate that the light
emission from PPL[-1] cells varies between 10^{-3} and 10^{-4} times the
intensity of the wild type. Thus PPL[-1] cells do not lack the genes
for synthesis of the light emission system; rather, the control of
gene expression for this system is set at an extremely low level.

In Figures 2 and 3 we have presented data for PPL[-1] cells
exposed to proflavin and ethidium bromide. The results show that
after an initial delay period, cells treated with these frameshift
agents display photon emission rates approaching that of wild type
cells. Differences in kinetics of photon emission for these
compounds suggest that scintillation measurements may provide information about early events in the expression of newly synthesized
gene products not easily obtainable by present techniques of bioassay. Scintillation measurements may also provide insights into
mechanisms of mutagenesis and closely related processes such as
metabolic activation, detoxication and repair that enhance or
diminish the rate of spontaneous and induced mutations. In addition,
measurements of photon emission kinetics may also facilitate the
analysis of co-mutagens and antimutagens.

Endogenous metabolic activation in photobacteria

When compared with S. typhimurium strains as test objects
(12), neither P. phosphoreum PPL[-1] nor P. leiognathi 8SD18 cells
require the addition of an exogenous source of metabolic activating
enzymes (such as the S-9 fraction of rat liver microsomes) for
conversion of proflavin or ethidium bromide to an active form.
Hastings and others (13,14) have noted that bacterial luciferase
may be viewed as a mixed function oxidase. This suggests that
photobacteria may possess one or more metabolic activating enzymes
not present in Salmonella. Ulitzur et al. (8) have reported that
20-methylcholanthrene and other frameshift agents increased luminescence for 8SD18 cells in the absence of S-9; however, addition of
S-9 enhanced the sensitivity of these cells to frameshift agents.

Plate incorporation assays

When grown on an agar medium supplemented with caffeine,
hydroxlamine hydrochloride or ethidium bromide, PPL[-1] cells yield
bright clones that emit photons at near wild-type rates. The
frequency of bright clones on such plates varies from 15 to 60%
of the surviving population. These yields appear to be orders of
magnitude greater than that expected from chromosomal mutations.

In this assay, bright clones are easily detected against a
background of non-luminous colonies. Accordingly, there is no
need to employ a selective medium. If the mutagen does not significantly reduce cell viability, then a large fraction of the treated
cells will form colonies; i.e. the plating efficiency may approach
100%. Thus, all of the original cells plated and all of their
progeny will be exposed to the mutagen and therefore subject to
mutagenesis. We have observed that colonies formed in this assay
may vary from each other in light emission. Such a heterogenous
population would be consistent with this interpretation.

Hydroxylamine and caffeine

Ulitzur et al. (8) and Ulitzur and Weiser (unpublished data)
have reported that hydroxylamine HCl and caffeine increase photon
emission rates from 8SD18 cells. Hydroxylamine enhances the rate
of deamination of cytosine to uracil over the spontaneous level.
It is therefore a base-modifying agent that induces base-substitution mutations. Caffeine acts synergistically with other mutagens
and can also interfere with excision repair processes (15).

When PPL^{-1} cells are exposed to these compounds in plate
incorporation assays, genetically stable clones are formed with
photon emission rates approaching wild type cells. In contrast,
these agents are inactive in the Salmonella/microsome test. Taken
together, these results suggest that Salmonella may possess detoxication enzymes, absent in photobacteria for demethylating caffeine
and/or transforming hydroxylamine to a non-mutagenic product.

Species differences

Both 8SD18 cells and PPL^{-1} cells exhibit low levels of light
emission in the absence of mutagens, and show marked increases in
the rate of photon emission when exposed to frameshift mutagens
and other gentoxic agents. However, there are several important
ways in which these strains differ from each other:

(i) Effect of C_{12} aldehyde: Addition of dodecanal to P.
leiognathi strain 8SD18 strongly stimulates cells to luminescence
at near normal rates (Ulitzur and Weiser, unpublished data). In
constrast, addition of this long chain aldehyde to P. phosphoreum
cells has no detectable effect on photon emission.

(ii) Stability of progeny of mutagen treated cells: In P.
leiognathi, photon emission is strictly dependent upon the continued
presence of the mutagenic agent. When cells of the 8SD18 strain
are treated with acriflavin and transferred to acriflavin-free
solid media, none of the colonies were capable of photon emission
(8). Cells of P. phosphoreum PPL^{-1} exposed to ethidium bromide
at 20 μg/plate, then grown in unsupplemented media and diluted to

approximately 10^{-4} molecules of ethidium bromide per cell yielded progeny that were capable of photon-emission when plated on ethidium bromide free agar.

Evidence for at least two different modes of action is beginning to emerge from studies of chemical agents that induce light emission in these photobacteria:

(i) In P. leiognathi strain 8SD18, compounds such as dodecanal restore normal levels of luminescence by relieving aldehyde-limitation. In this strain, compounds such as acriflavin may act to initiate transcription of one or more genes coding for long-chain aldehyde synthesis or other components of the light-generating system.

(ii) In P. phosphoreum strain PPL^{-1}, compounds such as ethidium bromide or hydroxylamine act to produce cells whose progeny continue to emit photons at near normal rates in the absence of the mutagen. In this strain, the genetic stability of the photon emission system in progeny of cells exposed to genotoxic agents provides evidence for a hertiable event.

Expression time

After cells are exposed to mutagenic agents, they require time for DNA replication and fixation of the genetic lesion. We have proposed that the requirement for colony development or plaque formation to amplify the phenotypic character of a mutated cell may be unnecessary if a newly synthesized product of a mutated gene is easily detectable. The experiments reported here focus attention upon light as an easily quantified gene product that can be measured non-destructively as an expression of mutagenicity. By substituting a physical method of amplifying the light emission properties of mutated cells for a biological amplification, it may be possible to shorten the expression time from days to hours or minutes.

SUMMARY

A new class of Photobacterium phosphoreum mutants that emit low levels of light have been isolated. When treated with known genotoxic agents, light emission increases by several orders of magnitude. Cells treated with genotoxins produce stable progeny. Dose-response curves provide evidence that both frameshift and base-substitution agents induce bright clones in a large fraction of the treated cells. An exogenous source of metabolic activating enzymes is not required for conversion of proflavin or ethidium bromide to an active form. Measurement of newly synthesized gene product(s) in these mutants may serve as an indicator of mutagenicity.

REFERENCES

1. M. Hollstein, J. McCann, F. A. Angelosanto, and W. W. Nichols, Short-term tests for carcinogens and mutagens, Mutat. Res. 65:133-226 (1979).
2. B. N. Ames, J. McCann, and E. Yamasaki, Methods for detecting carcinogens and mutagens with the Salmonella/mammalian-microsome mutagenicity test, Mutat. Res. 31:347-364 (1975).
3. M. H. L. Green and J. W. Muriel, Mutagen testing with TRP$^+$ reversion in Escherichia coli, Mutat. Res. 38:3-32 (1976).
4. H. Tanooka, Development and applications of Bacillus subtilis test systems for mutagens involving DNA-repair deficiency and suppressible auxotrophic mutations, Mutat. Res. 42:19-32, (1977).
5. T. R. Skopek, H. L. Liber, J. J. Krolewski, and W. G. Thilly, Quantitative forward mutation assay in Salmonella typhimurium using 8-azaguanine resistance as a genetic marker, Proc. Natl. Acad. Sci. (U.S.A.) 75:410-414 (1978).
6. E. E. Slater, M. D. Anderson, and R. S. Rosenkranz, Rapid detection of mutagens and carcinogens, Cancer Res. 31:970-973 (1971).
7. P. Moreau, A. Bailone, and R. Devoret, Prophage induction in Escherichia coli K 12 envA uvrB: A highly sensitive test for potential carcinogens, Proc. Natl. Acad. Sci. (U.S.A.) 73:3700-3704 (1976).
8. S. Ulitzur, I. Weiser, and S. Yannai, A new sensitive and simple bioluminescence test for mutagenic compounds, Mutat. Res. 74:113-124 (1980).
9. P. Baumann, L. Baumann, S. S. Bang, and M. J. Woolkalis, Reevaluation of the taxonomy of Vibrio, Beneckea and Photo-bacterium: abolition of the genus Beneckea, Current Micro-biol. 4:117-132 (1980).
10. G. M. Thomas and G. O. Poinar, Jr., A new entomopathogenic, nematophilic, bioluminescent bacterium, Xenorhabdus luminescens sp. nov. (Enterobacteriaceae:Eubacteriales) associated with the terrestrial nematode Heterorhabdites bacteriophora, Int. J. Syst. Bacteriol. 29:352-360 (1979).
11. J. W. Hastings and K. H. Nealson, Bacterial bioluminescence, Ann. Rev. Microbiol. 31:549-595 (1977).
12. J. McCann, E. Choi, E. Yamasaki, and B. N. Ames, Detection of carcinogens as mutagens in the Salmonella/microsome test: assay of 300 chemicals, Proc. Natl. Acad. Sci. (U.S.A.) 72:5135-5139 (1975).
13. J. W. Hastings and C. Balny, The oxygenated bacterial luciferase-flavin intermediate, J. Biol. Chem. 250:7288-7293 (1975).
14. J. W. Hastings, Bacterial bioluminescence light emission in mixed function oxidation of reduced flavin and fatty alde-hyde, CRC Critical Revs. in Biochem. 5:163-184 (1978).
15. J. Timson, Caffeine, Mutat. Res. 47:1-52 (1977).

DETECTION OF GENOTOXIC AIRBORNE CHEMICALS IN RAT LIVER CULTURE SYSTEMS

S. Ved Brat, C. Tong and G. M. Williams

Naylor Dana Institute for Disease Prevention
American Health Foundation
1 Dana Road
Valhalla, New York   10594 (U.S.A.)

INTRODUCTION

Airborne environmental agents are usually comprised of complex mixtures of chemicals of variable constitutions (1-4). The detection of potentially hazardous substances in these mixtures requires rapid and economical test systems. Most current testing involves the Ames Salmonella/microsome mutagenesis test (5) in which metabolism of the chemical is performed by an exogenous subcellular fraction from enzyme-induced rat liver, and the chemical is then assayed for its ability to induce mutations in various strains of Salmonella. Although this test is of proven value for the detection of mutagens (6), the need for a battery of tests to identify carcinogens is becoming recognized (7-11).

Liver cell systems are important components in an in vitro test battery because they provide intact cell biotransformation of xenobiotics and a variety of reliable and relevent endpoints for genotoxicity (11,12). Among the tests developed in our laboratory are: (a) the hepatocyte primary culture (HPC)/DNA repair test (13,14); (b) mutagenesis at the hypoxanthine-guanine phosphoribosyl transferase (HGPRT) locus in long term cultures of adult rat liver epithelial cells (ARL) (15,16); (c) hepatocyte-mediated mutagenesis of ARL cells (17); and (d) sister chromatid exchanges (SCE) in ARL cells (18). In addition, transformation assays are under development (19,20).

The present report deals with the application of rat liver culture systems to detect the genotoxicity of some complex environmental samples and a number of environmentally important

619

polycylic aromatic hydrocarbons. The ability of the HPC/DNA
repair test, the ARL/HGPRT mutagenesis assay, and the ARL/SCE assay
to detect these materials is presented.

MATERIALS AND METHODS

    Complex environmental mixtures were obtained from EPA as seven
coded samples of unknown composition. They were products of diesel
emission (1188, 1148), roofing tar pot emission (1256, 1266),
cigarette smoke condensate (1331, 1320) and coke oven emission
(1725). All four samples were dichloromethane-extracted. Benzo(a)-
pyrene (BaP) and benz(a)anthracene (BeA) were obtained from Sigma,
St. Louis. Benzo(e)pyrene (BeP) was procured from the NCI chemical
repository and pyrene from Pfaltz and Bauer. 7,12-Dimethylbenz(a)-
anthracene (DMBA) was purchased from Eastman Kodak and anthracene
from Eastman Organics, Rochester, NY.

    Procedures have been described in detail for preparation of
hepatocyte primary cultures (21) and the DNA repair test (11,14)
The adult rat liver epithelial cell/HGPRT mutagenesis assay has
been previously described (18,22). The sister chromatid exchange
assay in ARL cells (23) is a modification of the procedures as
described by Perry and Wolff (24) and Ved Brat et al. (25).

RESULTS

Genotoxicity of complex mixtures:

    Coded samples of environmental mixtures provided by the EPA
were assayed for DNA repair synthesis in hepatocyte primary cultures
or mutagenesis at the HGPRT locus in a rat liver cell line ARL18
(Table 1-3). BaP and DMBA were used as positive controls. All
four samples from one lot produced a dose-dependent positive response in the DNA repair test (Table 1). The coke oven emission sample produced the strongest consistent response (105.9 grains/nucleus)
while the cigarette smoke condensate had the weakest positive response (13.8 grains/nucleus). When three of these mixtures were
supplied under different code numbers and tested for mutagenicity as
well as HPC/DNA repair, all mixtures were again scored as positive
in the HPC/DNA repair test (Table 2), but only cigarette smoke condensate gave a positive mutagenic response (Table 3). Thus, the
HPC/DNA repair test appears to possess a greater sensitivity to
these types of mixtures than the ARL/HGPRT mutagenesis assay. This
difference could be due either to the fact that the DNA repair
assay measures damage over the whole genome as compared with the
single locus involved in the mutagenesis assay or to greater
metabolic activation by the hapetocytes.

Table 1. DNA repair synthesis activity in hepatocyte primary cultures after exposure to unknown samples in the HPC/DNA repair assay

Code #	Sample	Expt 1 Dose mg/mL	Expt 1 Grain/nucleus Mean ± S.D.	Expt 2 Dose mg/mL	Expt 2 Grains/nucleus Mean ± S.D.	Expt 3 Dose mg/mL	Expt 3 Grains/nucleus Mean ± S.D.
1188	Diesel emission	10^{-1}	Toxic	10^{-1}	28.1 ± 11.1	10^{-1}	39.8 ± 32.1
		10^{-2}	6.9 ± 7.2	5×10^{-2}	6.8 ± 2.7	5×10^{-2}	5.1 ± 4.4
		10^{-3}	0.1 ± 0.1				
1256	Roofing tar pot emission	10^{0}	Toxic	10^{-1}	Toxic	10^{-1}	Toxic
		10^{-1}	37.7	5×10^{-2}	20.5 ± 8.6	5×10^{-2}	22.0 ± 3.0
		10^{-2}	1.2 ± 1.1	10^{-2}	15.6 ± 7.6	10^{-2}	12.4
1331	Cigarette smoke condensate	10^{0}	Toxic	5×10^{-1}	Toxic	5×10^{-1}	Toxic
		10^{-1}	5.7 2.2	10^{-1}	11.9 ± 7.7	10^{-1}	13.8 ± 2.5
		10^{-2}	0	5×10^{-2}	4.2 ± 3.1	5×10^{-2}	7.1 ± 1.5
1735	Coke oven emission	10^{0}	Toxic	10^{0}	Toxic	10^{0}	Toxic
		10^{-1}	105.9 ± 46.0	10^{-1}	65.2 ± 22.3	5×10^{-1}	50.2 ± 25.0
		10^{-2}	53.7			10^{-1}	45.3 ± 14.8
	Benzo(a)pyrene		NR**	10^{0}	Toxic		NR**
				10^{-1}	24.9 ± 9.0		
	1% DMSO control		0		0.7 ± 0.1		0.1 ± 0.1

*Mean and standard deviation of triplicate coverslips exposed 1.5 hrs. after inoculation of hepato-cyates to test sample plus tritiated thymidine for 18 hrs.
*Not run.

Table 2. DNA repair synthesis activity in rat hepatocyte
          primary cultures after exposure to blind coded
          environmental mixtures for 1.5 hr plus tritiated
          thymidine for 18 hrs.

Code #	Sample	Dose (mg/mL)	Grains/nuclues Mean ± S.D.		
1148	Diesel emission	10^{0}		Toxic	
		10^{-1}	42.1	±	22.9
		10^{-2}	18.4	±	16.4
1266	Roof tar pot	10^{-1}		Toxic	
	emission	10^{-2}	38.4		
		10^{-3}	10.2	±	2.3
1320	Cigarette smoke	10^{0}		Toxic	
	condensate	10^{-1}	42.4	±	5.1
		10^{-2}	5.3	±	5.6
	Benzo(a)pyrene	10^{-1}	26.1	±	2.6
	0.1% DMSO Control		0.1	±	0.1

Genotoxicity of Polycyclic Aromatic Hydrocarbons:

A number of polycyclic aromatic hydrocarbons, both carcinogenic
and noncarcinogenic, were examined in three liver-derived assays
developed in this laboratory.

DNA repair synthesis: BaP, DMBA, and their analogues BeP,
pyrene, BaA and anthracene were examined for genotoxicity in the
HPC/DNA repair test using rat hepatocytes (Table 4). A dose-
dependent increase in DNA repair synthesis was observed in hepatocytes after exposure to both BaP and DMBA. The noncarcinogenic
analogues BeP and anthracene were consistently negative. BaA, which
is a weak initiating agent for mouse skin (26), was also positive
in this assay although negative results were reported previously
(14).

Mutagenesis: All six compounds were studied (Table 5) for
mutagenesis at the HGPRT locus in liver epithelial cells belonging
to cell line ARL 18 which is capable of metabolic activation
of a number of polycyclic aromatic hydrocarbons (18,27). Both
BaP and DMBA were highly mutagenic. Exposure of ARL 18 to 10^{-5}M
BaP was repeated in six separate experiments and to 10^{-5}M DMBA in
three separate experiments. Both compounds produced a consistent

Table 3. Induction of TG^r mutants by unknown samples in the ARL/HGPRT mutagenesis assay

Sample Code	Expt 1			Expt 2		
	Dose mg/mL	CFE% in^a control medium	TG^r mutant^b 10^{6}CFC Mean±S.D.	Dose mg/mL	CFE% in^a control medium	TG^r mutant^b 10^{6}CFC Mean±S.D.
1188 Diesel emission	10^0	Toxic		10^0	Toxic	
	10^{-1}	33.0	24 ± 13	5×10^{-1}	Toxic	
	10^{-2}	27.4	27 ± 11	2.5×10^{-1}	Toxic	
	10^{-3}	22.4	29 ± 22	10^{-1}	33.6	4 ± 10
1256 Roofing tar pot emission	10^{-1}	Toxic		10^{-1}	Toxic	
	10^{-2}	37.4	5 ± 6	5×10^{-1}	30.0	9 ± 7
	10^{-3}	45.7	12 ± 11	2.5×10^{-1}	37.8	27 ± 11
	10^{-4}	38.3	37 ± 14	10^{-2}	37.8	23 ± 16
1331 Cigarette smoke condensate	10^0	Toxic		10^0	Toxic	
	10^{-1}	34.0	233 ± 79	5×10^{-1}	Toxic	
	10^{-2}	40.4	8 ± 12	10^{-1}	40.6	140 ± 21
	10^{-3}	39.2	71 ± 30	5×10^{-2}	36.6	20 ± 5
	10^{-4}	32.0	0			
7,12 DMBA	2.5×10^{-3}	21.7	597 ± 58		NR^c	
	2.5×10^{-4}	34.8	513 ± 58			
	2.5×10^{-5}	33.6	129 ± 39			
Benzo(a)pyrene		NR		2.5×10^{-3}	41.0	300 ± 28
1% DMSO control		47.7	3 ± 4		45.1	16 ± 5

Table 3. Legend

[a]CFE-colony forming efficiency, defined as: number of colonies
 per flask/500 cells seeded per flask x 100.
[b]Mean and standard deviation of TG^r mutant colonies per 10^6
 colony forming cell, following a 12–14 day selection interval,
 defined as: number of mutant colonies per flask x $10^6/2.5x10^5$
 cells seeded per flask x CFE. Cells were plated for selection
 at a density of 10^4 cells/cm. A total of $1.5x10^6$ cells were
 plated.
[c]NR, not run.

Table 4. DNA repair activity in primary hepatocyte cultures
          after exposure to polycyclic aromatic hydrocarbons

Chemical	Dose (M)	Grains/nucleus* Mean ± S.D.		
Benzo(a)pyrene	10^{-3}	65.6	±	17.8
	$5x10^{-4}$	47.7	±	3.7
	10^{-4}	45.1	±	3.7
Benzo(e)pyrene	10^{-3}	1.1	±	0.4
	$5x10^{-4}$	0.5	±	0.5
	10^{-4}	1.2	±	0.9
Pyrene	$5x10^{-4}$	Toxic		
	10^{-4}	1.4	±	0.5
	$5x10^{-5}$	0.6	±	0.5
	10^{-5}	0.5	±	0.5
7,12-Dimethylbenz(a)anthracene	10^{-3}	Toxic		
	$5x10^{-4}$	38.9	±	10.1
	10^{-4}	38.9	±	4.9
	$5x10^{-5}$	22.0	±	2.1
Benz(a)anthracene	10^{-3}	Toxic		
	$5x10^{-4}$	17.2	±	6.0
	10^{-4}	14.8	±	2.6
	$5x10^{-5}$	0.6	±	
Anthracene	10^{-3}	1.9	±	1.4
	$5x10^{-4}$	1.5	±	0.6
	10^{-4}	0.3	±	0.2
0.1% DMSO Control		0.1	±	0.1

*Mean and standard deviation of triplicate coverslips exposed to
 the test agent plus ^{3}H-TdR for 20 hours.

Table 5. 6-Thioguanine resistant mutants induced by polycyclic aromatic hydrocarbons in adult rat epithelial cells, ARL 18

Exposure	Dose M	CFE* %	TG^r Mutants Per Flask Mean ± S.D.		Per 10^6 CFC* Mean ± S.D.	
Benzo(a)pyrene	10^{-4}	21.0	32.0 ± 4.7		610 ±	90
	10^{-5}	28.2	23.5 ± 2.0		335 ±	28
Benzo(e)pyrene	10^{-4}	23.3	0.5 ± 0.8		9 ±	14
	10^{-5}	26.6	0		0	
Pyrene	10^{-4}	25.8	0.3 ± 0.5		5 ±	8
	10^{-5}	31.6	0.2 ± 0.4		2 ±	5
7,12-Dimethyl-benzanthracene	10^{-4}	21.7	32.4 ± 4.5		597 ±	58
	10^{-5}	34.8	44.7 ± 5.0		513 ±	58
	10^{-6}	33.6	10.8 ± 3.3		129 ±	39
Benz(a)anthracene	10^{-4}	24.4	0.7 ± 0.8		11 ±	13
	10^{-5}	22.1	0.3 ± 0.5		6 ±	9
0.1% DMSO Control		30.9	0.3 ± 0.8		4 ±	11
		35.0	0.5 ± 0.6		6 ±	6

*   C.F.E. colony forming efficiency
** C.F.C. colony forming cells

and significantly positive response. This weak carcinogenic analogue BaA was not mutagenic like BeP and pyrene which are non-mutagenic as well as noncarcinogenic.

Sister chromatid exchange: All six polycyclic aromatic hydrocarbons were studied for their ability to induce SCE in ARL 18 (Table 6). The target cells exhibited a low, consistent level of baseline SCEs. In eight separate analyses, the baseline SCE frequency was 14.5 per cell with a standard deviation of 1.5, a level comparable to that observed in rat liver tumor cells (28) and in a newborn rat liver cell line (29). Both BaP and DMBA induced a significant increase in SCE with a dose-related response. The noncarcinogenic analogues BeP, pyrene and anthracene were negative. Although the SCE frequency did not show a dose-response, the weak carcinogen benz(a)anthracene was positive. This finding suggests that the weak carcinogens which are negative in other liver cell assays could be screened by SCE.

Table 6. SCE's induced by Benzo(a)pyrene,
7,12-Dimethylbenz(a)anthracene
and their analogues in ARL 18

Chemicals	Dose (M)	SCE/cell		
		Mean	±	S.D.
Benzo(a)pyrene	10^{-4}	103.30	±	22.75
	10^{-5}	59.75	±	16.96
	10^{-6}	16.15	±	3.83
DMSO		11.15	±	3.81
Benzo(e)pyrene	10^{-3}	15.75	±	6.03
Pyrene	10^{-3}	13.16	±	5.00
DMSO		14.15	±	5.64
7,12-Dimethyl-benz(a)anthracene	10^{-4}	56.20	±	13.50
	10^{-5}	40.40	±	17.60
	10^{-6}	29.53	±	8.60
DMSO		14.90	±	5.06
			±	
Benz(a)anthracene	10^{-3}	26.20	±	6.96
	10^{-4}	29.15	±	6.93
	10^{-5}	21.20	±	9.59
Anthracene	10^{-3}	15.25	±	5.05
	10^{-4}	16.75	±	4.02
DMSO		15.75	±	5.18

DISCUSSION

The present experiments have demonstrated that dichloromethane-
extracts of diesel emission, roofing tar pot emission, coke oven
emission and cigarette smoke condensates are genotoxic in the
HPC/DNA repair test. The cigarette smoke condensate was also
positive in the ARL/HGPRT mutagenesis assay unlike the diesel
emission and roofing tar pot emission samples. The polycyclic
aromatic hydrocarbons, BaP and DMBA were positive in the HPC/DNA
repair test, the ARL/HGPRT mutagenesis and the ARL/SCE assay.
As expected, their noncarcinogenic analogues, BeP, pyrene and

Table 7. Comparison of present results obtained in different assays employing rat liver culture systems with in vivo carcinogenicity and Salmonella mutagenicity of polycyclic aromatic hydrocarbons

	Rat Liver Culture Systems				Salmonella/Microsome Mutagenesis
	HPC/DNA Repair	ARL/HGPRT Mutagenesis	ARL/SCE	Carcinogenicity	
7,12-Dimethylbenzanthracene	+	+	+	+	+[a]
Benz(a)anthracene	+	-	+	+	+[a]
Anthracene	-	-	-	-	-[b]
Benzo(a)pyrene	+	+	+	+	+[a,b]
Benzo(e)pyrene	-	-	-	-	+[a,b]
Pyrene	-	-	-	-	-

[a]Hollstein et al. (30).
[b]LaVoie et al. (31) and personal communication.

anthracene were negative in all assays. Interestingly, the weak carcinogen BeA was positive in the HPC/DNA repair test and the ARL/SCE assay but not in the ARL/HGPRT mutagenesis assay. Thus, the results are consistent with the known carcinogenicity and bacterial mutagenicity of polycyclic aromatic hydrocarbons (Table 7).

The positive response of the weak carcinogens and a negative response of the non-carcinogens in the ARL/SCE assay and confirmation of these results in the HPC/DNA Repair test needs emphasis due to the fact that ARL/SCE is a relatively new assay. The positive response in this system is somewhat in contrast to the variability in the results obtained in Salmonella mutagenesis. Both benzo(e)-pyrene and benz(a)anthracene were found to be negative in Salmonella typhimurium TA1537 and TA1538 strains but were converted to frame-shift mutagens when metabolized by a rat liver homogenate.

Thus, the application of a multiple assay system consisting of the HPC/DNA repair test, ARL/HGPRT mutagenesis and the ARL/SCE assay in rat liver cells is a useful approach for the detection of airborne genotoxins. These cells are especially useful since they possess an intact cell capability to biotransform many xenobiotics. The known carcinogens/mutagens give a linear and parallel dose response in all three assays. Although a mechanistic relationship between these three molecular events is not fully understood, they comprise a battery of highly sensitive indicators of genotoxicity which correlates well with mutagenicity and carcinogenicity.

ACKNOWLEDGEMENTS

This research was supported by Contracts NO1-CP-55705 from the National Cancer Institute and 68-02-2483 from the U.S. Environmental Protection Agency. The authors thank Mrs. Nancy McNeilly and Mrs. L. Stempel for preparation of the manuscript.

REFERENCES

1. L. Fishbein, Atmospheric mutagens, in: "Chemical Mutagens", Vol. 4, A. Hollaender, ed., Plenum Press, N.Y., 219-319 (1976).
2. J. R. Huisingh, R. Bradow, R. Jungers, L. Claxton, R. Lweidinger, and S. Tejada, Application of bioassays to the characterization of the diesel particle emissions, in: "Application of Short-term Assays in the Fractionaliner and Analysis of Complex Environmental Mixtures", EPA Publ. 600/9-78-027, 381-418 (1978).
3. T. J. Hughes, E. Pellizzari, L. Little, C. Sparacino, and A. Kolber, Ambient air pollutants: Collection, chemical

characterization and mutagenicity testing, Mutat. Res. 76: 51-83 (1980).

4. C. E. Crisp, and G. L. Fisher, Mutagenicity of airborne particles, Mutat. Res. 76:143-164 (1980).

5. B. N. Ames, W. E. Durston, E. Yamasaki, and F. D. Lee, Carcinogens are mutagens: a simple test system combining liver homogenates for activation and bacteria for detection, Proc. Natl. Acad. Sci. USA 70:2281-2285 (1973).

6. J. E. McCann, E. Choi, E. Yamasaki, and B. N. Ames, Detection of carcinogens as mutagens in the Salmonella/microsome test. Assay of 300 chemicals, Proc. Natl. Acad. Sci. USA 72:5135-5139 (1975).

7. F. H. Sobels, Some problems associated with the testing for environmental mutagens and perspective for studies in comparative mutagenesis, Mutat. Res. 46:245-260 (1977).

8. B. A. Bridges, Short-term tests and human health - the role of DNA repair, in: "Environmental Carcinogenesis", P. E. Emelot and E. Kriek, eds., Elsevier/North Holland, Amsterdam, 319-327 (1979).

9. F. J. de Serres, Problems associated with the application of short-term tests for mutagenicity in mass screening programs, Environ. Mutagenesis 1:203-208 (1979).

10. V. A. Ray, Application of microbial and mammalian cells to the assessment of mutagenicity, Pharmacology Rev. 30:537-546 (1979).

11. G. M. Williams, The detection of chemical mutagens/carcinogens by DNA repair and mutagenesis in liver culture, in: "Chemical Mutagens" Vol. 6, A. Hollaender, ed., Plenum Press, N.Y., 61-79 (1980).

12. G. M. Williams, The use of liver epithelial cultures for the study of chemical carcinogenesis, Am. J. Pathology 85:739-754 (1976).

13. G. M. Williams, Carcinogen-induced DNA repair in primary rat liver cell cultures: A possible screen for chemical carcinogens, Cancer Lett. 1:231-236 (1976).

14. G. M. Williams, Detection of chemical carcinogens by unscheduled DNA synthesis in rat primary cell cultures, Cancer Res. 37: 1845-1851 (1977).

15. G. M. Williams, C. Tong, and J. J. Berman, Characterization of analog resistance and purine metabolism of adult rat liver epithelial cell 8-azaguanine-resistant mutants, Mutat. Res. 49:103-115 (1978).

16. C. Tong and G. M. Williams, Induction of purine analog-resistant mutants in adult rat liver epithelial lines by metabolic activation-dependent and -independent carcinogens, Mutat. Res. 58:339-352 (1978).

17. R. H. C. San and G. M. Williams, Rat hepatocyte primary cell culture mediated mutagenesis of adult rat liver epithelial

cells by procarcinogens, Proc. Soc. Exptl. Biol. Med.
156:534-538 (1977).

18. C. Tong, M. F. Laspia, S. Telang, and G. M. Williams, The use of
    adult rat liver cultures in the detection of the genotoxicity
    of various polycyclic aromatic hydrocarbons, Environ. Mut.
    (in press) (1981).

19. R. H. C. San, T. Shimada, C. J. Maslansky, D. M. Krieser,
    M. F. Laspia, J. M. Rice, and G. M. Williams, Growth charac-
    teristics and enzyme activities in a survey of transformation
    markers in adult rat liver epithelial-like cell cultures,
    Cancer Res. 39:4441-4448 (1979).

20. R. H. C. San, M. F. Laspia, A. I. Soiefer, C. J. Maslansky,
    J. M. Rice, and G. M. Williams, A survey of growth in soft
    agar and cell surface properties as markers for transforma-
    tion in adult rat liver epithelial-like cell cultures,
    Cancer Res. 39:1026-1034 (1979).

21. G. M. Williams, E. Bermudez, and D. Scaramuzzino, Rat hepatocyte
    primary cell cultures, III. Improved dissociation and attach-
    ment techniques and enhancement of survival by culture medium,
    In Vitro 13:809-817 (1977).

22. C. Tong and G. M. Williams, Definition of conditions for the
    detection of genotoxic chemicals in the adult rat liver
    epithelial cell/hypoxanthine-guanine phosphoribosyl transfer-
    ase (ARL/HGPRT) mutagenesis assay, Mutat. Res. 4:1-9 (1980).

23. C. Tong, S. Ved Brat, and G. M. Williams, Sister chromatid ex-
    change induction by polycyclic aromatic hydrocarbons in an
    intact cell system using adult rat liver epithelial cells,
    Mutat. Res. (in press) (1981).

24. P. E. Perry and S. Wolff, New giemsa method for differential
    staining of sister chromatids, Nature 251:156-158 (1974).

25. S. Ved Brat, R. S. Verma, and H. Dosik, Anthramycin-induced
    sister chromatid exchange and caffeine potentiation in
    chromosomes of indian muntjac, Mutat. Res. 63:325-334 (1979).

26. IARC Monographs on the evaluation of carcinogenic risk of the
    chemical to man, Certain polycyclic aromatic hydrocarbons
    Heterocyclic compounds (IARC, Lyon), Vol. 3, 45-68 (1973).

27. C. Tong and G. M. Williams, Variation in response in the
    adult rat liver epithelial cell/hypoxanthine-guanine
    phosphoribosyl transferase (ARL/HGPRT) mutation assay,
    Environ. Mutat. 1:148 (1979).

28. R. Dean, G. Bynum, D. Kram, and E. L. Schneider, Sister
    chromatid exchange induction by carcinogens in HTC cells.
    An in vivo system which does not require addition of
    activating factors, Mutat. Res. 74:477-483 (1980).

29. A. L. Meyer and B. J. Dean, Induction of sister chromatid
    exchanges in a rat-liver cell line with chemical carcino-
    gens, Mutat. Res. 91:47-50 (1981).

30. M. Hollstein, J. McCann, and N. Angelosanto, Short-term
    tests for carcinogens and mutagens, Mutat. Res. 65:133-226
    (1979).

31. E. LaVoie, V. Bedenko, N. Hirota, S. S. Hecht, and D. Hoffmann, A comparison of the mutagenicity, tumor-initiating activity and complete carcinogenicity of poly-nuclear aromatic hydrocarbons, in: "Polynuclear Aromatic Hydrocarbons", P. W. Jones, and P. Leber, eds., Ann Arbor Science Publishers, Ann Arbor, MI, 705-721 (1979).

THE INDUCTION OF SISTER CHROMATID EXCHANGES IN HUMAN LYMPHOCYTES

AND BACTERIAL MUTAGENESIS BY ORGANIC EXTRACTS OF URBAN AIRBORNE

PARTICLES

J. M. Lockard, C. J. Viau, C. Lee-Stephens, J. C.
Caldwell, J. P. Wojciechowski, H. G. Enoch* and
P. S. Sabharwal

School of Biological Sciences
University of Kentucky
Lexington, Kentucky 40506 (U.S.A.)

ABSTRACT

Although numerous other bioassays are available, the use of
in vitro test systems to detect the genotoxic effects of organic
airborne particles has been almost exclusively limited to the
Salmonella/mammalian microsome (Ames) assay. In addition to using
the Ames assay, the present study tested the organic fraction of
urban particulate matter in the sensitive sister chromatid exchange (SCE) assay, using cultured human peripheral lymphocytes
and hamster lung fibroblast (V79) cells. To provide metabolic
activation for detection of promutagens, either liver homogenate
from Aroclor-induced rats (S9) or feeder layers of irradiated
Syrian hamster fetal cells were added to the V79 system; the S9
fraction was included in the bacterial assay.

Total suspended particles were collected from ambient air
in Lexington, Kentucky, on precleaned glass fiber filters using
hi-vol samplers located approximately 10 m above ground level.
Organic soluble components were extracted by ultrasonication at
45 C in benzene and acetone, evaporated to dryness for weight
determination, redissolved in DMSO and stored under N_2 at -20°C.

All of four extracts tested induced linear dose-related
increases in the number of SCEs in human lymphocytes, causing a

*Center for Energy Research, Kentucky Department of Energy

doubling or more in the frequency of SCE above control levels with treatments as low as 40 to 60 μg/ml. In contrast, the V79 cells were mostly insensitive to the effects of the extracts, regardless of the absence or presence of either metabolizing system. All of four samples tested in the Ames assay induced dose-related increases in the number of revertants in strains TA 98 and TA 100. With only one exception, the addition of S9 had little or no effect on the mutation frequency observed.

The SCE assay in human peripheral lymphocytes appears to be a sensitive means of detecting the genotoxic effects of organic air pollutants. Furthermore, it is relatively rapid and inexpensive and requires smaller quantities of test chemicals than are needed for the Ames assay. Since it uses freshly isolated human cells, this assay is highly pertinent to the problems of prediciing potential risks of air pollutants to man.

AN EPIDEMIOLOGICAL STUDY OF CANCER

DEATHS IN PATHFORK, KENTUCKY

Samuel B. Gregorio

McDowell Cancer Network, School of Medicine
University of Kentucky
Lexington, KY   40536   (U.S.A.)

INTRODUCTION

A preliminary epidemiological study was conducted in response
to a letter which was initially sent to the Cancer Hopeline of the
McDowell Cancer Network on August 18, 1980 by Mrs. America T. Whittle,
postmaster of Pathfork, Harlan County in Eastern Kentucky. She and
her family have 51 years of service in that post office. Furthermore,
she and her husband once owned a grocery store in the community. Her
letter noted a high incidence of cancer deaths in a 1/4 square mile
area in Pathofork.

METHODS

A list of suspected cancer deaths was generated through letters
and interviews following the initial letter from Mrs. Whittle. Based
on this list, a study of death certificates was made from the files
of the Office of Health Data in Frankfort (DHR). A brief occular
environmental survey of the area was conducted and additional background information on Harlan County was compiled from other sources.

RESULTS

The study period was defined to cover the time between 1961 to
1980. There were 29 cancer deaths listed for this period. Of these,
20 were confirmed by death certificates. Five of the 9 unconfirmed
deaths were believed to have occurred in other states. The death

certificates of 10 other residents showed that they died of causes
other than cancer. Fifteen persons were said to have been treated
for cancer over the past 20 years; however, this remains to be confirmed by a study of their hospital records. No genetic relationships among the cancer deaths was found.

A. Data from death certificates

All of those who died of cancer were white. More than half
(55%) died between the ages of 61 and 70. The youngest was 28 years
old, while the oldest was 96 years old. There were almost as many
females as there were males who died of cancer (Table 1). Only one
male was unmarried and only one male had a wife who had career
training.

Approximately 25% of these cases had cancer of the lungs. This
group included miners, homemakers, and a funeral director. The death
certificates of three others indicated that they had generalized
cancer or metastasis. The rest had either cancer of the adrenal
gland, bladder, small bowels, colon, stomach, kidney, or urethra.
Almost half (40%) were homemakers or housewives, while 25% were
miners (Table 2).

B. Harlan County people

Harlan County covers an area of 469 square miles in Eastern
Kentucky. Only 3% of this land is devoted to farming. It has a
population of 41,400 people (1978). Approximately 21% of the people
are in wholesale or retail business, 19% are in professional mana-
gerial positions, 19% are in sales and clerical positions, and 11%
are in educational services. Seventeen percent of the population
are craftsmen or foremen, while the rest are in either manufacturing,
general services, construction, or government services.

The median level of education is 8.4 years. Only 4.3% of the
population has had 4 or more years of college education, while 24%
had 4 years of high school education. The median family income is
$4,682 while the per capita income is $1,593. There are 64 mineral
industries and 60 farms in the county (1972).

The average rate of cancer deaths in the county over the past
20 years is equal to that of the state, while that of Pathfork exceeds
both the county and the state. The increment of rate of cancer deaths
in the county is greater than that of the state, with Harlan increasing by 77 deaths per 100,000 population and the state by 47.
However, the cancer mortality rate in Pathfork far exceeds those of
the county and the state during the first and fourth quarters of this
20 year period (Table 3).

Table 1. Age, sex and race of cancer deaths

Date	Name	Race	Sex	Occupation	Age	Cancer Site
4-08-62	S.B.	W	M	Miner	63	Stomach
10-15-62	F.W.	W	M		68	Bladder
10-31-64	O.H.	W	M	Car Assembly Line	61	Lympho Sarcoma/Lymphoid Tissue
9-11-64	E.N.	W	F	Merchant	48	Ovarian; Int. Hemorrhage
7-10-65	E.E.	W	F	Housewife	62	General Metastatic Tumor
9-28-68	M.S.C.	W	F	Housewife	28	Adenocarcinoma Metastasis
6-09-69	J.D.N.	W	M	Car Assembly Line	71	Right Adrenal Gland
6-14-69	W.H.W.	W	M	Teacher	75	Prostrate
6-16-71	R.L.	W	M	Coal Miner	66	Small Bowel
6-08-73	S.V.	W	F	Homemaker	68	Renal Cells
2-02-73	F.H.	W	F	Homemaker	58	Gen. Carcinomatosis
6-15-74	B.T.	W	M	Miner	69	Lungs
11-29-76	W.S.S.	W	M	Accountant	58	Colon
9-11-77	M.W.	W	F	Housewife	86	Adeno Cancer Vaters Ampula (Hepatic-Pancr.)
5-06-77	C.A.H.	W	M	Funeral Director	61	Bronchogenic Carcinoma
10-21-78	J.L.	W	M	Miner	96	Prostrate
1-26-79	M.R.	W	F	Homemaker	61	Bronchogenic Ca of Left Lung
7-09-79	B.H.T.	W	F	Housewife	81	Lungs
9-04-79	S.L.	W	F	Housewife	70	Pelvic (Urethal)
8-30-80	C.F.	W	M	Coal Miner	69	Lung

Table 2. Occupation and cancer site

CANCER SITE	Accountant	Car Assembly Line	Funeral Director	Homemaker	Merchant	Miner	Teacher	Other	Total	%
Adrenal Gl.		1							1	5
Bladder								1	1	5
Small Bowel						1			1	5
Colon	1								1	5
Liver (Va. Amp.)				1					1	5
Lungs			1	2		2			5	25
Lympoid		1							1	5
Ovary					1				1	5
Prostrate						1	1		2	10
Kidney				1					1	5
Stomach						1			1	5
Urethra				1					1	5
Generalized				3					3	15
TOTAL	1	2	1	8	1	5	1	1	20	
Percent	5	10	5	40	5	25	5	5		

Table 3. Cancer mortality rates in Pathfork

Year	Pathfork		Harlan		Kentucky	
	Observation	Rate*	Observation	Rate*	Observation	Rate*
1961			43	103.9	4,380	143.5
1962	2	398.4	63	136.4	4,495	146.0
1963			53	119.4	4,410	144.0
1964	2	392.2	59	130.2	4,484	147.3
1965	1	194.9	77	159.1	4,665	152.0
1966			71	158.8	4,740	150.6
1967			59	135.3	4,891	153.5
1968	1	192.3	68	159.3	5,036	157.6
1969	2	382.4	70	168.3	5,066	157.5
1970			63	170.3	5,345	165.8
1971	1	189.0	76	204.3	5,436	167.3
1972			83	213.9	5,544	169.2
1973	2	372.4	78	195.0	5,450	164.8
1974	1	185.2	70	175.9	5,920	178.1
1975			69	173.8	5,994	178.7
1976	1	182.5	63	155.9	5,916	174.6
1977	2	362.3	68	165.5	6,307	184.0
1978	1	179.9	87	210.0	6,443	186.3
1979	3	535.7	81	198.0	6,599	188.6
1980	1	177.3	74	180.9	6,720	190.6

*Rate computed per 100,000 population of the area

C. Environmental Observations

    Geographical setting: Pathfork is located in a hollow surround-
ed by hills approximately 20 miles southeast of Harlan County. The
study area is shaped like a triangle. The post office is about mid-
point in this base. A Kentucky Utilities power line runs along the
community. A railroad track runs between the highway and Pucket's
creek. A railroad loading area used to be located across from the
post office.

    Industries: Pathfork is a mining community. Three coal compa-
nies have been in operation in the community within the past 20 years.
These industries employed approximately 75% of the men. One of the
mines was located 3 miles east of the post office; a second mine was
about 2 miles southeast and a third mine was about 1½ miles west.
One of the mines provided a commissary, school, church, restaurant,
beauty parlor, filling station, and a doctor and nurse. At least one
member out of every household worked for the industries. Starting
about 30 years ago, silt ponds were made. Water used in washing the
coal was collected in these ponds, subsequently treated with lime,
and then emptied into the creek. The couple interviewed claim that
the ecology of the creek changed drastically after the onset of mining operations.

    Other industries: There was a lumber company that was in opera-
tion in the community until 1940. A tannery is still in operation
some 20 miles southeast of Pathfork. Some male residents worked in
automobile factories in Cincinnati and Detroit.

    Environmental chemical exposure: About 15-20 years ago, Ken-
tucky Utilities started spraying defoliants to control weeks along
the power lines. Residents used to make their own soap with lye and
meatskins or old animal grease. Some 20 years ago, free samples of
a detergent, "Super Suds", was distributed in the community. There
has been no extensive use of chemicals to control flies, mosquitos,
or rodents. The use of agricultural chemicals has been limited.

    Liquor and tobacco: Harlan is a dry county. Tobacco consump-
tion is said to be heavy, mostly through smoking and some through
chewing. Marijuana is said to have been growing wild in the locality.

    Housing: Some residents have brick homes while others have
wooden frame houses. Oil heating was introduced about 20 years ago.
People once depended on fireplaces and cookstoves for home heating.
Some still do. Wood and coal are used as fuel for home heating.

    Water source: The community does not have a central water sys-
tem. Residents have individually drilled wells. The water has a
characteristic sulfur-iron rust taste.

Diet: The community produces its own vegetables. Food is sold in locally owned stores. Meat sources are beef, pork, chicken, rabbit, squirrel, and raccoon. Meats used to be salted and dried. Most meats are presently from outside the community. Many residents used to have their own milk cows. The couple interviewed indicated that there were times that the milking cow had the "staggers." At such times the milk was unfit for consumption. They supposed that the "staggers" was due to the cow's exposure to marijuana. Salt-cured bacon, potatoes, corn meal, and pinto beans were very common components of the diet. Pinto beans were said to have been more popular than any other kind of bean.

Sewage disposal: Most houses have their own cesspools constructed with concrete blocks. The first sanitary toilet bowls were introduced into the community about 1940. Some residents run their sewage into the creek. Although no major landfilling has been done, there is a garbage dump approximately 1 mile south of the post office.

DISCUSSION

From the available data, it appears that no direct relationship could be established between specific occupation and type of cancer. However, the occurrence of cancer deaths among homemakers (housewives) is noteworthy. This and the location of the community in relation to the coal mines, the railroad loading area, and the power lines strongly suggest that environmental factors (such as exposure to by-products of the mines, fumes from the train engine at the loading station, the defoliant, and other substances transported through run-off and subsurface water) play important roles in the epidemiology of cancer in Pathfork. Other factors such as materials used in building construction, water source, sewage disposal system, and diet could play a contributory role. Of particular interest is the high consumption of pinto beans, corn meal, salt-cured bacon, potatoes and milk from cows exposed to marijuana. A probable synergistic effect may occur due to these multiple agents.

In consideration of the nature of the data collected, the small numbers of cases and the variety of cancer types in this study, it is difficult to draw definitive conclusions. At best, these pre-liminary findings strongly indicate the need for a comprehensive epidemiological investigation that would include a more extensive study of death certificates from Harlan county, morbidity studies, familial studies, assays of body fluids, and samples from the environment for possible mutagens or carcinogens. Environmental studies could include samples of run-off and subsurface water, air, defoliant(s), and a study of the materials used in the construction of buildings.

CONCLUSIONS

A high rate of cancer mortality was reported in Pathfork, Harlan County between 1961 and 1980. Approximately 25% of these deaths had cancer of the lungs and another 15% had generalized cancer or metastasis. The rest had either cancer of the adrenal gland, stomach, small bowels, prostate, bladder, liver, lymphoid tissues, ovary, kidney, or urethra. Almost half (40%) were homemakers or housewives, while 25% were miners. More than half (61%) died between the ages of 61 and 70. The youngest person to die of cancer was 28 years old while the oldest was 96 years old.

Preliminary environmental observations revealed that three coal mines were in operation in the past 20 years at distances of 1½ miles to 3 miles from the community. These industries employed about 75% of the men in the community. Also noted was spraying of power lines in areas along the north side of the community, a railroad along the south side of the community with a loading area close to its center, and a lumber industry and a tannery some 20 miles away.

Although no direct relationship could be established between a specific occupation and specific types of cancer deaths, there appears to be strong indications that environmental factors might have played an important role in the epidemiology of cancer in the community. These factors could well be composed of airborne mutagens or carcinogens from the industries, spraying of defoliants, and fumes from the train. The substances could also have been transmitted through surface and subsurface water.

REFERENCES

1. Health Information Systems Branch, Kentucky Vital Statistics and Planning Data. Volumes: 1960 to 1979. Department for Human Resources, Bureau for Health Services.

2. U.S. Department of Commerce, County and City Data Book, Bureau of Census, (1972).

LIST OF PARTICIPANTS

Robert G. Ackerley	Texas Eastern Transmission Corp., P. O. Box 2521, Houston, TX 77001
Richard J. Albertini	Dept. Medicine & Oncology, Given Medical Bldg., C 312, Univ. of Vermont, Burlington, VT 05401
James W. Allen	Genetic Toxicology Div., Health Effects Research Lab., U.S. EPA, Research Triangle Park, NC 27711
Elizabeth Anderson	Dir., Office of Health and Environ. Assessment, U.S. EPA, RD 689, 401 M St., Washington, D.C. 20460
Colin F. Arlett	MRC, Cell Mutation Unit, Sussex Univ. Falmer, Brighton, BN1 9QC, United Kingdom
David Atkinson	SRI International, 333 Ravenswood Ave., Menlo Park, CA 94025
Michael C. Axelroa	Electronic Engineering Dept., Lawrence Livermore Laboratory, Livermore, CA 94550
James Barower	Safety and Environmental Protection Div., Brookhaven National Laboratory, Upton, NY 11973
Eugene D. Barber	Health Safety and Human Factors Lab, Eastman Kodak, Bldg. 320, Kodak Park, Rochester, NY 14650
Z. G. Bell	Mgr., Env. Health & Toxicol., PPG Industries, Inc., One Gateway Center, Pittsburgh, PA 15222
Daniel R. Benz	Medical Dept., Brookhaven National Laboratory, Upton, NY 11973
Roger Ben-Dyke	Dir., Inhalation Toxicol., Bio/Dynamics, P. O. Box 43, Mettlers Rd., East Millstone, NJ 08873
David M. Bernstein	Medical Dept., Brookhaven National Laboratory, Upton, NY 11973
Victor P. Bond	Director's Office, Brookhaven National Laboratory, Upton, NY 11973

643

Craig J. Boreiko   Chemical Indust. Inst. of Toxicol.,
                               P. O. Box 12137, Research Triangle
                               Park, NC 27709
Donal C. Borg,   Medical Dept., Brookhaven National
                               Laboratory, Upton, NY 11973
S. Ved Brat   Cytogenetic Unit, Naylor Dana Inst. for
                               Disease Prevention, Amer. Health
                               Foundation, Valhalla, NY 10595
James A. Brown, Jr.   Ford Motor Co., Chemistry Research,
                               2000 Rotunda Dr. P. O. Box 2053,
                               RM E-3198, Dearborn, MI 48121
Kathleen M. Burke   U.S. EPA, Office of Research & Develop-
                               ment, RD 683, 401 M. St., SW,
                               Washington, D.C. 20460
Evelyn G. Burtis   Exxon Corporation - REHO, P. O. Box 235,
                               East Millstone, NJ 08873

Bruce A. Carnes   Argonne National Laboratory, 9700 S. Cass
                               Ave., BIM, Argonne, IL 60439
Arland L. Carsten   Medical Dept., Brookhaven National
                               Laboratory, Upton, NY 11973
Charles N. Cawley   Cornell Univ., 205 MVR Hall, DEA Dept.,
                               Ithaca, NY 14853
David J. Chen   Genetics, Los Alamos Scientific
                               Laboratory, Los Alamos, NM 87545
William L. Chen   Dow Chem. Canada Ltd., Modeland Centre,
                               Sarnia, Ontario, Canada N7T 7K7
Constanza Chiappe   Dept. Biol. Sciences, Universidad de
                               Los Andes, Calle 18A Carrera 1E,
                               Apartado Aereo 4976, Bogota, D.E.,
                               Columbia, South America
Larry Claxton   Genetic Toxicol. Div., U.S. EPA,
                               Research Triangle Park, NC 27711
Mary K. Conner   636 Parran Hall, Univ. of Pittsburgh,
                               Pittsburgh, PA 15261
Milton J. Constantin   Comparative Animal Research Laboratory,
                               1299 Bethel Valley Rd., Oak Ridge, TN
                               37830
Daniel L. Costa   Medical Dept., Brookhaven National
                               Laboratory, Upton, NY 11973
Eugene P. Cronkite   Medical Dept., Brookhaven National
                               Laboratory, Upton, NY 11973
Karen Chytalo   NY State Dept. Environ. Conservation,
                               Bldg. #40, SUNY/Stony Brook, Stony
                               Brook, NY 11794
John Ciancia   U.S. EPA, Region 2, S&A Div., Edison,
                               NJ 08837

Robert Davis	U.S. EPA, Region 2, S&A Div., Edison, NJ 08837
Robert T. Drew	Medical Dept., Brookhaven National Laboratory, Upton, NY 11973
Theodore M. Farber	Chief, Food Animal Additives Bureau, Food and Drug Admin., 200 C St., S.W., Washington, D.C. 20204
Jack Foehrenbach	NY State Dept. Environmental Conservation, Bldg. #40, SUNY/Stony Brook, Stony Brook, NY 11794
Michael G Gabridge	W. Alton Jones Cell Science Center, Old Bard Rd., Lake Placid, NY 12946
Forrest L. Gager, Jr.	Philip Morris, USA, P. O. Box 26583, Richmond, VA 23261
Sheila M. Galloway	Litton Bionetics, 5516 Nicholson Ave., Kensington, MD 20795
Bernard D. Goldstein	Environ. & Community Medicine Dept., CMDNJ, Rutgers Medical School, Piscataway, NJ 08854
Alvin J. Greenberg	State of California - OSHA, 525 Golden Gate Avenue, 3rd Floor, San Francisco, CA 94102
Samuel B. Gregorio	McDowell Cancer Network, College of Medicine, Univ. of Kentucky, Lexington, KY 40536
Robert R. Guerrero	Pasadena Foundation for Medical Research, 99 N. El Molino Ave., Pasadena, CA 91101
Sonja B. Haber	Medical Dept., Brookhaven National Laboratory, Upton, NY 11973
Naomi Harley	Dept. of Environmental Medicine, NYU Medical Center, 550 1st Ave., New York, NY 10016
George Hatch	Northrop Services, P. O. Box 12313, Research Triangle Park, NC 27709
Jude Height	Chemical Systems Laboratory, Toxicology Bureau, DR DAR-CLB TC, Aberdeen, MD 21010
Anthony A. Herrmann	Dir., Employee H&S Affairs, Johnson & Johnson, 501 George St., New Brunswick, NJ 08903
David Hoel	Chief, Biometry Branch, NIEHS, P. O. Box 12233, Research Triangle Park, NC 27709
Seymour Holtzman	U.S. EPA, ORD (RD-682), Energy & Air Division, 401 M St., SW, Washington, D.C. 20460
Peter Infante	Office of Carcinogen Identif. & Classif., Health Standards Prog., U.S. Dept. of Labor, RM N 3718, 200 Constitution Ave., NW, Washington, D.C. 20210

Richard D. Irons   Dept. Pathology, Chemical Industry Inst.
                                    of Toxicology, P. O. Box 12137,
                                    Research Triangle Park, NC 27709

Franklin M. Johnson   National Institute of Environmental
                                    Health Science, P. O. Box 12233, Research
                                    Triangle Park, NC 27709
J. P. Jones   Dir., Health Sciences, Johnson & Johnson,
                                    501 George St., New Brunswick, NJ 08903
Robert Jungers   Environmental Monitoring System Lab.,
                                    U.S. EPA, Research Triangle Park, NC
                                    27711

Purushottam G. Kale   Safety & Environ. Protection Div.,
                                    Brookhaven National Laboratory, Upton,
                                    NY 11973
Ranjini Kale   Medical Dept., Brookhaven National
                                    Laboratory, Upton, NY 11973
Gerald L. Kennedy, Jr.   E. I. Dupont Co., Haskell Laboratory,
                                    Elkton Rd., Newark, DE 19711
Robert Kindya   GCA/Technology Div., 213 Burlington Rd.,
                                    Bedford, MA 01730
Judith Klotz   F. C. Hart Assoc., 155 Washington St.,
                                    Newark, NJ 07102
David Krahn   E. I. DuPont & Co., Haskell Laboratory,
                                    1007 Market St., Wilmington, DE 19898
Amy Kronenberg   Medical Dept., Brookhaven National
                                    Laboratory, Upton, NY 11973
Raymond S. Kutzman   Medical Dept., Brookhaven National
                                    Laboratory, Upton, NY 11973

A. P. Leber   PPG Industries, Inc., One Gateway Ctr.,
                                    Pittsburgh, PA 15222
William R. Lee   Dept. of Zoology, Louisiana State Univ.,
                                    Baton Rouge, LA 70803
Joellen L. Lewtas   Genetic Toxicology Div., Health Effects
                                    Research Laboratory, U.S. EPA, Research
                                    Triangle Park, NC 27711
June M. Lockard   College of Arts and Sciences, Univ. of
                                    Kentucky, Lexington, KY 40506
Judith B. Louis   N.J. Dept. of Environ. Protection
                                    (OCTSR), 190 W. State St., Trenton, NJ
                                    08625
Kenneth S. Loveday   Bioassay Systems Corp., 225 Wildwood Ave.
                                    Woburn, MA 01801

Te-Hsiu Ma   Western Illinois University, West Adams
                                    St., Macomb, IL 61455
Heinrich V. Malling   Laboratory of Biochemical Genetics,
                                    NIEHS, Research Triangle Park, NC 27709

Cesare Maltoni	Institute of Oncology, Bologna, Italy
Sharon Matthias	Occupational Hygiene Bureau, 2nd Floor, 9820 106th St., Edmonton, Alberta, Canada T5K 2J6
Kevin T. McCooey	E. I. Dupont, Haskell Laboratory, Elkton Rd., Newark, DE 19711
E. C. McCoy	Dept. Microbiology, NY Medical College, Basic Sciences Bldg., Valhalla, NY 10594
Leslie J. McGeorge	N.J. Dept. of Environ. Protection (OCTSR), 190 W. State St., Trenton, NJ 08625
Robert Mermelstein	Joseph C. Wilson Center for Technology, Xerox Corp., Rochester, NY 14644
Samuel C. Morris, III	Dept. of Energy and Environ., Brookhaven National Laboratory, Upton, NY 11973
Paul O. Nees	Hooker Chemical Co., Box 728, Niagara Falls, NY 14302
Gary O. Nelson	Lawrence Livermore Laboratory, L-520 T-22, Livermore, CA 94550
Gordon W. Newell	Assoc. Exec. Dir., Board on Toxicol. & Environ. Health Hazards, National Academy of Sciences, 2101 Constitution Ave., Washington, D.C. 20418
C. H. Powell	Mgr., Health Services Adm. PPG Industries, Inc., One Gateway Ctr., Pittsburgh, PA 15222
Otto G. Raabe	Radiobiology Lab., University of California at Davis, Davis, CA 91101
Ilene Raisfeld	Dept. Pharmacological Sciences, SUNY/ Stony Brook, Stony Brook, NY 11794
R. R. Raje	A & M Schwartz College of Pharmacy, Long Island Univ., Brooklyn, NY 11201
Gary M. Rand	Hooker Chemical Co., Corporate Toxicology, 222 Rainbow Blvd. N., Niagara Falls, NY 14302
Ronald E. Rassmussen	Dept. of Community & Environ. Medicine, Univ. of California at Irvine, Irvine, CA 92664
C. A. Reilly, Jr.	Argonne National Laboratory, 9700 Cass Ave., Argonne, IL 60439
Robert G. Reynolds	Northrop Services, Inc., Environ. Sciences, P. O. Box 12313, Research Triangle Park, NC 27709
Donald E. Rounds	Pasadena Foundation for Medical Research, 99 North El Molino Ave., Pasadena, CA 91101

Karen A. Schaich   Medical Dept., Brookhaven National
                                 Laboratory, Upton, NY 11973
Lloyd A. Schairer   Biology Dept., Brookhaven National
                                 Laboratory, Upton, NY 11973
Tomiko Shimada   Head, In Vitro Systems Facility,
                                 American Health Foundation, Dana Rd.,
                                 Valhalla, NY 10595
Ron Shiotsuka   Medical Dept., Brookhaven National
                                 Laboratory, Upton, NY 11973
Vincent F. Simmon   Genex Corp., Suite 1090, 6110 Executive
                                 Blvd., Rockville, MD 20852
Andrew Sivak   V.P., Bio/Medical Sciences, A. D. Little
                                 Inc., 20 Acorn Park, Cambridge, MA
                                 02140
Robert Snyder   Dept. of Pharmacology, Medical College,
                                 Thomas Jefferson University,
                                 Philadelphia, PA 19107
Louis Stang   Medical Dept., Brookhaven National
                                 Laboratory, Upton, NY 11973
J. Patrick Stone   Medical Dept., Brookhaven National
                                 Laboratory, Upton, NY 11973
Gary F. Strniste   Genetics, Los Alamos Scientific
                                 Laboratory, LS-3, MS 886, Los Alamos,
                                 NM 87545
William J. Suling   Southern Research Institute, 2000 Ninth
                                 Ave., South Birmingham, AL 35255
Michael Surgan   N.Y. State Dept. of Law, 2 World Trade
                                 Ctr., New York, NY 10047

Michael F. Terraso   Texas Eastern Transmission Corp.,
                                 P. O. Box 2521, Houston, TX 77001
Raymond R. Tice   Medical Dept., Brookhaven National
                                 Laboratory, Upton, NY 11973
Andrew M. Tometsko   Litron Laboratories, Ltd., 1351 Mt.
                                 Hope Ave., Rochester, NY 14620

Thomas F. Vogt   Medical Dept., Brookhaven National
                                 Laboratory, Upton, NY 11973
Peter Voytek   Reproductive Effects Assess Group,
                                 U.S. EPA, RD-680, 401 M St., S.W.
                                 Washington, D.C. 20460

James Walberg   The Rockefeller University, 1230 York
                                 Ave., Box 2, New York, NY 10021
David Weinstein   Hoffmann-La Roche, Inc., Kingsland St.,
                                 Nutley, NJ 07110
Otto White   Safety and Environ. Protection Div.,
                                 Brookhaven National Laboratory, Upton,
                                 NY 11973

Judith M. Wieland   Medical Dept., Brookhaven National
                             Laboratory, Upton, NY 11973

Paul O. Zamora   Inhalation Toxicology Research Inst.
                             Lovelace Foundation, P. O. Box 8590
                             Albuquerque, NM   87115

Rae Zimmerman   Assoc. Prof. of Planning, Grad. School of
                             Public Administration, New York University,
                             4 Washington Sq.,N., New York, NY 10003